Energy Systems

Series Editor:

Panos M. Pardalos, University of Florida, USA

For further volumes:
http://www.springer.com/series/8368

Steffen Rebennack • Panos M. Pardalos
Mario V.F. Pereira • Niko A. Iliadis
Editors

Handbook of Power Systems I

Springer

Editors
Dr. Steffen Rebennack
Colorado School of Mines
Division of Economics and Business
Engineering Hall
816 15th Street
Golden, Colorado 80401
USA
steffen.rebennack@gmail.com

Prof. Panos M. Pardalos
University of Florida
Department of Industrial and Systems Engineering
303 Weil Hall, P.O. Box 116595
Gainesville FL 32611-6595
USA
pardalos@ufl.edu

Dr. Mario V. F. Pereira
Centro Empresarial
Rio Praia de Botafogo
228/1701-A-Botafogo
CEP: 22250-040
Rio de Janeiro, RJ
Brazil
mario@psr-inc.com

Dr. Niko A. Iliadis
EnerCoRD
Plastira Street 4
Nea Smyrni 171
21, Athens
Greece
niko.iliadis@enercord.com

ISBN: 978-3-642-26468-9 ISBN: 978-3-642-02493-1 (eBook)
DOI 10.1007/978-3-642-02493-1
Springer Heidelberg Dordrecht London New York

Cover illustration: Cover art is designed by Elias Tyligadas

Cover design: WMXDesign GmbH, Heidelberg

Printed on acid-free paper

Springer is part of Springer Science+Business Media (www.springer.com)

To our families.

Preface of Volume I

Power systems are undeniably considered as one of the most important infrastructures of a country. Their importance arises from a multitude of reasons of technical, social and economical natures. Technical, as the commodity involved requires continuous balancing and cannot be stored in an efficient way. Social, because power has become an essential commodity to the life of every person in the greatest part of our planet. Economical, as every industry relates not only its operations but also its financial viability in most cases with the availability and the prices of the power.

The reasons mentioned above have made power systems a subject of great interest for the scientific community. Moreover, given the nature and the specificities of the subject, sciences such as mathematics, engineering, economics, law and social sciences have joined forces to propose solutions.

In addition to the specificities and inherent difficulties of the power systems problems, this industry has gone through significant changes. We could refer to these changes from an engineering and economical perspective. In the last 40 years, important advances have been made in the efficiency and emissions of power generation, and in the transmission systems of it along with a series of domains that assist in the operation of these systems. Nevertheless, the engineering perspective changes had a small effect comparing to these that were made in the field of economics where an entire industry shifted from a long-standing monopoly to a competitive deregulated market.

The study of such complex systems can be realized through appropriate modelling and application of advance optimization algorithms that consider simultaneously the technical, economical, financial, legal and social characteristics of the power system considered. The term technical refers to the specificities of each asset that shall be modelled in order for the latter to be adequately represented for the purpose of the problem. Economical characteristics reflect the structure and operation of the market along with the price of power and the sources, conventional or renewable, used to be generated. Economical characteristics are strongly related with the financial objectives of each entity operating a power system, and consist in the adequate description and fulfillment of the financial targets and risk profile. Legal specificities consist in the laws and regulations that are used for the operation of the power system. Social characteristics are described through a series of parameters that have to be considered in the operation of the power system and reflect the issues related to the population within this system.

The authors of this handbook are from a mathematical and engineering background with an in-depth understanding of economics and financial engineering to apply their knowledge in what is know as modelling and optimization. The focus of this handbook is to propose a selection of articles that outline the modelling and optimization techniques in the field of power systems when applied to solve the large spectrum of problems that arise in the power system industry. The above mentioned spectrum of problems is divided in the following chapters according to its nature: Operation Planning, Expansion Planning, Transmission and Distribution Modelling, Forecasting, Energy Auctions and Markets, and Risk Management.

Operation planning is the process of operating the generation assets under the technical, economical, financial, social and legal criteria that are imposed within a certain area. Operation is divided according to the technical characteristics required and the operation of the markets in real time, short term and medium term. Within these categories the main differences in modelling vary in technical details, time step and time horizon. Nevertheless, in all three categories the objective is the optimal operation, by either minimizing costs or maximizing net profits, while considering the criteria referred above.

Expansion planning is the process of optimizing the evolution and development of a power system within a certain area. The objective is to minimize the costs or maximize the net profit for the sum of building and operation of assets within a system. According to the focus on the problem, an emphasis might be given in the generation or the transmission assets while taking into consideration technical, economical, financial, social and legal criteria. The time-step used can vary between 1 month and 1 quarter, and the time horizon can be up to 25 years.

Transmission modelling is the process of describing adequately the network of a power system to apply certain optimization algorithms. The objective is to define the optimal operation under technical, economical, financial, social and legal criteria. In the last 10 years and because of the increasing importance of natural gas in power generation, electricity and gas networks are modelled jointly.

Forecasting in energy is applied for electricity and fuel price, renewable energy sources availability and weather. Although complex models and algorithms have been developed, forecasting also uses historical measured data, which require important infrastructure. Hence, the measurement of the value of information also enters into the equation where an optimal decision has to be made between the extent of the forecasting and its impact to the optimization result.

The creation of the markets and the competitive environment in power systems have created the energy auctions. The commodity can be power, transmission capacity, balancing services, secondary reserve and other components of the system. The participation of the auction might be cooperative or non-cooperative, where players focus on the maximization of their results. Therefore, the market participant focus on improving their bidding strategies, forecast the behavior of their competitors and measure their influence on the market.

Risk management in the financial field has emerged in the power systems in the last two decades and plays actually an important role. In this field the entities that participate in the market while looking to maximize their net profits are heavily concerned with their exposure to financial risk. The latter is directly related to the operation of the assets and also with a variety of external factors. Hence, risk mangers model their portfolios and look to combine optimally the operation of their assets by using the financial instruments that are available in the market.

This handbook is divided into two volumes. The first volume covers the topics operation planning and expansion planning while the second volume focuses on transmission and distribution modeling, forecasting in energy, energy auctions and markets, as well as risk management.

We take this opportunity to thank all contributors and the anonymous referees for their valuable comments and suggestions, and the publisher for helping to produce this volume.

February 2010

Steffen Rebennack
Panos M. Pardalos
Mario V.F. Pereira
Niko A. Iliadis

Contents of Volume I

Part I Operation Planning

Constructive Dual DP for Reservoir Optimization 3
E. Grant Read and Magnus Hindsberger

Long- and Medium-term Operations Planning and Stochastic Modelling in Hydro-dominated Power Systems Based on Stochastic Dual Dynamic Programming 33
Anders Gjelsvik, Birger Mo, and Arne Haugstad

Dynamic Management of Hydropower-Irrigation Systems 57
A. Tilmant and Q. Goor

Latest Improvements of EDF Mid-term Power Generation Management 77
Guillaume Dereu and Vincent Grellier

Large Scale Integration of Wind Power Generation 95
Pedro S. Moura and Aníbal T. de Almeida

Optimization Models in the Natural Gas Industry 121
Qipeng P. Zheng, Steffen Rebennack, Niko A. Iliadis, and Panos M. Pardalos

Integrated Electricity–Gas Operations Planning in Long-term Hydroscheduling Based on Stochastic Models 149
B. Bezerra, L.A. Barroso, R. Kelman, B. Flach, M.L. Latorre, N. Campodonico, and M. Pereira

Recent Progress in Two-stage Mixed-integer Stochastic Programming with Applications to Power Production Planning 177
Werner Römisch and Stefan Vigerske

Dealing With Load and Generation Cost Uncertainties in Power System Operation Studies: A Fuzzy Approach 209
Bruno André Gomes and João Tomé Saraiva

OBDD-Based Load Shedding Algorithm for Power Systems 235
Qianchuan Zhao, Xiao Li, and Da-Zhong Zheng

Solution to Short-term Unit Commitment Problem 255
Md. Sayeed Salam

A Systems Approach for the Optimal Retrofitting of Utility Networks Under Demand and Market Uncertainties 293
O. Adarijo-Akindele, A. Yang, F. Cecelja, and A.C. Kokossis

Co-Optimization of Energy and Ancillary Service Markets 307
E. Grant Read

Part II Expansion Planning

Investment Decisions Under Uncertainty Using Stochastic Dynamic Programming: A Case Study of Wind Power 331
Klaus Vogstad and Trine Krogh Kristoffersen

The Integration of Social Concerns into Electricity Power Planning: A Combined Delphi and AHP Approach 343
P. Ferreira, M. Araújo, and M.E.J. O'Kelly

Transmission Network Expansion Planning Under Deliberate Outages 365
Natalia Alguacil, José M. Arroyo, and Miguel Carrión

Long-term and Expansion Planning for Electrical Networks Considering Uncertainties 391
T. Paulun and H.-J. Haubrich

Differential Evolution Solution to Transmission Expansion Planning Problem 409
Pavlos S. Georgilakis

Agent-based Global Energy Management Systems for the Process Industry 429
Y. Gao, Z. Shang, F. Cecelja, A. Yang, and A.C. Kokossis

Optimal Planning of Distributed Generation via Nonlinear Optimization and Genetic Algorithms 451
Ioana Pisică, Petru Postolache, and Marcus M. Edvall

Index 483

Contents of Volume II

Part I Transmission and Distribution Modeling

Recent Developments in Optimal Power Flow Modeling Techniques 3
Rabih A. Jabr

Algorithms for Finding Optimal Flows in Dynamic Networks 31
Maria Fonoberova

Signal Processing for Improving Power Quality 55
Long Zhou and Loi Lei Lai

Transmission Valuation Analysis based on Real Options with Price Spikes 101
Michael Rosenberg, Joseph D. Bryngelson, Michael Baron, and Alex D. Papalexopoulos

Part II Forecasting in Energy

Short-term Forecasting in Power Systems: A Guided Tour 129
Antonio Muñoz, Eugenio F. Sánchez-Úbeda, Alberto Cruz, and Juan Marín

State-of-the-Art of Electricity Price Forecasting in a Grid Environment 161
Guang Li, Jacques Lawarree, and Chen-Ching Liu

Modelling the Structure of Long-Term Electricity Forward Prices at Nord Pool 189
Martin Povh, Robert Golob, and Stein-Erik Fleten

Hybrid Bottom-Up/Top-Down Modeling of Prices in Deregulated Wholesale Power Markets 213
James Tipping and E. Grant Read

Part III Energy Auctions and Markets

Agent-based Modeling and Simulation of Competitive Wholesale Electricity Markets 241
Eric Guerci, Mohammad Ali Rastegar, and Silvano Cincotti

Futures Market Trading for Electricity Producers and Retailers 287
A.J. Conejo, R. García-Bertrand, M. Carrión, and S. Pineda

A Decision Support System for Generation Planning and Operation in Electricity Markets 315
Andres Ramos, Santiago Cerisola, and Jesus M. Latorre

A Partitioning Method that Generates Interpretable Prices for Integer Programming Problems 337
Mette Bjørndal and Kurt Jörnsten

An Optimization-Based Conjectured Response Approach to Medium-term Electricity Markets Simulation 351
Julián Barquín, Javier Reneses, Efraim Centeno, Pablo Dueñas, Félix Fernández, and Miguel Vázquez

Part IV Risk Management

A Multi-stage Stochastic Programming Model for Managing Risk-optimal Electricity Portfolios 383
Ronald Hochreiter and David Wozabal

Stochastic Optimization of Electricity Portfolios: Scenario Tree Modeling and Risk Management 405
Andreas Eichhorn, Holger Heitsch, and Werner Römisch

Taking Risk into Account in Electricity Portfolio Management 433
Laetitia Andrieu, Michel De Lara, and Babacar Seck

Aspects of Risk Assessment in Distribution System Asset Management: Case Studies 449
Simon Blake and Philip Taylor

Index 481

Contributors

O. Adarijo-Akindele Department of Chemical and Process Engineering, University of Surrey, Guildford GU2 7XH, UK

Natalia Alguacil ETSI Industriales, Universidad de Castilla – La Mancha, Campus Universitario, s/n, 13071 Ciudad Real, Spain, Natalia.Alguacil@uclm.es

Madalena Araújo Department of Production and Systems, University of Minho, Azurem, 4800-058 Guimarães, Portugal, maraujo@dps.uminho.pt

José M. Arroyo ETSI Industriales, Universidad de Castilla – La Mancha, Campus Universitario, s/n, 13071 Ciudad Real, Spain, JoseManuel.Arroyo@uclm.es

Luiz Augusto Barroso PSR, Rio de Janeiro, Brazil

Bernardo Bezerra PSR, Rio de Janeiro, Brazil

Nora Campodonico PSR, Rio de Janeiro, Brazil

Miguel Carrión Department of Electrical Engineering, EUITI, Universidad de Castilla – La Mancha, Edificio Sabatini, Campus Antigua Fábrica de Armas, 45071 Toledo, Spain, Miguel.Carrion@uclm.es

F. Cecelja Department of Chemical and Process Engineering, University of Surrey, Guildford GU2 7XH, UK, f.cecelja@surrey.ac.uk

Aníbal T. de Almeida Department of Electrical and Computer Engineering, University of Coimbra, Portugal, adealmeida@isr.uc.pt

Guillaume Dereu EDF R&D, OSIRIS, 1, avenue de Gaulle, 92140 Clamart, France, guillaume.dereu@edf.fr

Marcus M. Edvall Tomlab Optimization Inc., San Diego, CA, USA

Paula Ferreira Department of Production and Systems, University of Minho, Azurem, 4800-058 Guimarães, Portugal, paulaf@dps.uminho.pt

Bruno da Costa Flach PSR, Rio de Janeiro, Brazil

Y. Gao Department of Chemical and Process Engineering, University of Surrey, Guildford GU2 7XH, UK

Pavlos S. Georgilakis School of Electrical and Computer Engineering, National Technical University of Athens (NTUA), Athens, Greece, pgeorg@power.ece.ntua.gr

Anders Gjelsvik SINTEF Energy Research, 7465 Trondheim, Norway, Anders.Gjelsvik@sintef.no

Bruno André Gomes INESC Porto, Departamento de Engenharia Electrotécnica e Computadores, Faculdade de Engenharia da Universidade do Porto, Campus da FEUP, Rua Dr. Roberto Frias 378, 4200-465 Porto, Portugal, bgomes@inescporto.pt

Quentin Goor UCL, Place Croix-du-Sud, 2 bte 2, LLN, Belgium, quentin.goor@uclouvain.be

Vincent Grellier EDF R&D, OSIRIS, 1, avenue de Gaulle, 92140 Clamart, France, vincent.grellier@edf.fr

H.-J. Haubrich Institute of Power Systems and Power Economics (IAEW), RWTH Aachen University, Schinkelstraße 6, 52056 Aachen, Germany, ubrich@iaew.rwth-aachen.de

Arne Haugstad SINTEF Energy Research, 7465 Trondheim, Norway, Arne.Haugstad@sintef.no

Magnus Hindsberger Transpower New Zealand Ltd, P.O. Box 1021, Wellington, New Zealand, magnus.hindsberger@transpower.co.nz

Niko A. Iliadis EnerCoRD Energy Consulting, Research, Development, 2nd floor, Plastira street 4, Nea Smyrni 171 21, Attiki, HELLAS, Athens, Greece, niko.iliadis@enercord.com

Rafael Kelman PSR, Rio de Janeiro, Brazil

A.C. Kokossis School of Chemical Engineering, National Technical University of Athens, Zografou Campus, 9, Iroon Polytechniou Str., 15780 Athens, Greece, akokossis@chemeng.ntua.gr

Trine Krogh Kristoffersen Risøe National Laboratory of Sustainable Energy, Technical University of Denmark, Denmark, trine.kristoffersen@risoe.dk

Maria Lujan Latorre PSR, Rio de Janeiro, Brazil

Xiao Li Center for Intelligent and Networked Systems (CFINS), Department of Automation and TNList Lab, Tsinghua University, Beijing 100084, China

Birger Mo SINTEF Energy Research, 7465 Trondheim, Norway, Birger.Mo@sintef.no

Pedro S. Moura Department of Electrical and Computer Engineering, University of Coimbra, Portugal, pmoura@isr.uc.pt

M.E.J. O'Kelly Department of Industrial Engineering, National University Ireland, Ireland

Panos M. Pardalos Department of Industrial and Systems Engineering, Center for Applied Optimization, University of Florida, Gainesville, FL 32611, USA, pardalos@ufl.edu

T. Paulun Institute of Power Systems and Power Economics (IAEW), RWTH Aachen University, Schinkelstraße 6, 52056 Aachen, Germany, tp@iaew.rwth-aachen.de

Mario Veiga Ferraz Pereira Power Systems Research, Praia de Botafogo 228/1701-A, Rio de Janeiro, RJ CEP: 22250-040, Brazil, mario@psr-inc.com

Ioana Pisică Department of Electrical Power Engineering, University Politehnica of Bucharest, Romania, ioanapisica@gmail.com

Petru Postolache Department of Electrical Power Engineering, University Politehnica of Bucharest, Romania

E. Grant Read Energy Modelling Research Group, University of Canterbury, Private Bag 4800, Christchurch 8140, New Zealand, grant.read@canterbury.ac.nz

Steffen Rebennack Colorado School of Mines, Division of Economics and Business, Engineering Hall, 816 15th Street, Golden, Colorado 80401, USA, steffen.rebennack@gmail.com

Werner Römisch Humboldt University, 10099 Berlin, Germany, romisch@math.hu-berlin.de

Md. Sayeed Salam BRAC University, Dhaka, Bangladesh, sayeed@bracuniversity.ac.bd

João Tomé Saraiva INESC Porto, Departamento de Engenharia Electrotécnica e Computadores, Faculdade de Engenharia da Universidade do Porto, Campus da FEUP, Rua Dr. Roberto Frias 378, 4200-465 Porto, Portugal, jsaraiva@fe.up.pt

Z. Shang Department of Process and Systems Engineering, Cranfield University, Cranfield MK43 0AL, UK

Amaury Tilmant UNESCO-IHE, Westvest 7, Delft, the Netherlands, a.tilmant@unesco-ihe.org

and

Swiss Federal Institute of Technology, Institute of Environmental Engineering, Wolfgang-Pauli-Strasse 15, 8093 Zurich, Switzerland, tilmant@ifu.baug.ethz.ch

Stefan Vigerske Humboldt University, 10099 Berlin, Germany, vigerske@math.hu-berlin.de

Klaus Vogstad Agder Energi Produksjon and NTNU Dept of Industrial Economics, klausvogstad@gmail.com

A. Yang Department of Chemical and Process Engineering, University of Surrey, Guildford GU2 7XH, UK, a.yang@surrey.ac.uk

Qianchuan Zhao Center for Intelligent and Networked Systems, Department of Automation and TNList Lab, Tsinghua University, Beijing 100084, China, zhaoqc@tsinghua.edu.cn

Da-Zhong Zheng Center for Intelligent and Networked Systems, Department of Automation and TNList Lab, Tsinghua University, Beijing 100084, China, zhaoqc@tsinghua.edu.cn

Qipeng P. Zheng Department of Industrial and Systems Engineering, Center for Applied Optimization, University of Florida, Gainesville, FL 32611, USA, zqp@ufl.edu

Part I
Operation Planning

Constructive Dual DP for Reservoir Optimization

E. Grant Read and Magnus Hindsberger

Abstract Dynamic programming (DP) is a well established technique for optimization of reservoir management strategies in hydro generation systems, and elsewhere. Computational efficiency has always been a major issue, though, at least for multi-reservoir problems. Although the dual of the DP problem has received little attention in the literature, it yields insights that can be used to reduce computational requirements significantly. The stochastic dual DP algorithm (SDDP) is one well known optimization model that combines insights from DP and mathematical programming to deal with problems of much higher dimension that could be addressed by DP alone. Here, though, we describe an alternative "constructive" dual DP technique, which has proved to be both efficient and flexible when applied to both optimization and simulation for reservoir problems of modest dimension. The approach is illustrated by models from New Zealand and the Nordic region.

Keywords Cournot gaming · Dual dynamic programming · Duality · Dynamic programming · Hydro-electric power · Reservoir management · Risk aversion · SDDP · Stochastic optimization

1 Introduction

The reservoir management problem for a hydrothermal power system is to decide how much water should be released in each period so as to minimize the expected operational cost, including the fuel cost of thermal plants, and the cost of shortages, should they occur. Because all reservoirs have limited storage capacities and because the inflows are uncertain, optimization of reservoir management becomes quite complex. In economic terms, though, it can be shown that the basic principle of reservoir management is to adjust generation until the marginal value of releasing

E. Grant Read (✉)
Energy Modelling Research Group, University of Canterbury, Private Bag 4800, Christchurch 8140, New Zealand
e-mail: grant.read@canterbury.ac.nz

S. Rebennack et al. (eds.), *Handbook of Power Systems I*, Energy Systems, DOI 10.1007/978-3-642-02493-1_1, © Springer-Verlag Berlin Heidelberg 2010

water equals the expected marginal value of storing water. The marginal value of releasing water is equal to the fuel cost of the most expensive thermal station used to meet residual demand after hydro generation or, during a shortage, to the marginal value of meeting loads that would otherwise not be serviced. The expected marginal value of the storing water depends on the optimal utilization of that stock in future periods, though.

Thus the art and science of reservoir optimization lies mainly in devising models to determine that optimal future utilization. No one technique seems ideal for all applications, but this paper describes a constructive dual dynamic programming (CDDP) technique that has been successfully applied in New Zealand and Norway. This technique was first implemented in the context of a relatively complex real problem, and not perhaps explained as clearly, or implemented as efficiently, as it might have been. Here we draw more on later papers to explain the basic concepts, focusing on ways in which CDDP may be implemented efficiently. In the process, we briefly survey some New Zealand developments, several of which have not been well reported in the literature, but then conclude by illustrating the approach with results from a Nordic model.

2 Background

Leaving aside heuristic and simulation techniques, two basic lines of development have been followed in reservoir optimization modeling over many decades now, one based on linear programming (LP) and the other on dynamic programming (DP). Comprehensive surveys are provided by Yeh (1985) and Labadie (2004).

LP models can readily capture much of the complexity of large reservoir and power systems, and can be further generalized to model integer and non-linear aspects. But LP is essentially deterministic, and Read and Boshier (1989) demonstrate that deterministic models may perform quite badly, and produce dangerously mis-leading recommendations, for systems with a high proportion of hydro power. This is due to two problems. The first problem is that determining the expected marginal value of storing water requires us to consider the optimal future utilization of water under a great number of possible future inflow sequences. And the second problem is that determining the optimal future utilization of water for any future period of any possible inflow sequence is just as complex a problem as the original problem we are trying to solve, for the current period. Thus, in principle, it requires us to consider the optimal future utilization of water under a great number of possible future inflow sequences, from that time forward. But each hypothetical future optimization must be conditioned on the inflows that would have been observed, under that inflow scenario, up to that time. And we must apply a "non-anticipativity restriction" to ensure that the current optimization does not assume that unrealistic foresight could be applied when making each such decision in future. Applying LP to a range of inflow scenarios would solve the first problem, but not the second, and Read and Boshier show that this still implies a significant bias to the results.

DP has also been widely used in reservoir management, and the literature has been surveyed by Yakowitz (1982) and by Lamond and Boukhtouta (1996). Its major advantage is that it can readily handle the stochastic situation described earlier by providing a consistent model of past and future decision-making under uncertainty. But it can do this only under the condition that the (conditional) probability distribution of "possible future inflow sequences," from any point in time forward, can be characterized simply. This allows the set of possible future situations for which release problems need to be solved to be described by a compact "state space." It can then be shown that, given reasonable assumptions about the value of water at the end of the planning horizon, an optimal release policy can be produced, not only for the current situation but also for all possible future situations, by a process of "backwards recursion" through this state space, from the end of the planning horizon back to the present.

The problem with DP, though, is that the state space can easily become much too large to deal with. The state space is just a grid of reservoir levels, for a single reservoir with uncorrelated inflows, but becomes two-dimensional if we account for a lag-one Markov model of inflow correlation. That is, a future decision problem must be solved for each possible combination of reservoir level and inflow in the previous period. Each additional reservoir adds a dimension to the state space, as does each inflow correlation. While the problem to be solved for each stage and state remains relatively simple, the number of such problems builds up as a power of the number of dimensions leading to the so-called curse of dimensionality.[1]

Accordingly, the general direction of development, in recent decades, has been towards methods that combine the best features of LP and DP, within a realistic computational limit. Thus LP models have been generalized to become Stochastic Programming (SP) models, in which non-anticipativity constraints force the model to choose realistic solutions for future decision periods, at least for a limited range of scenarios. The most successful development has then been the specialized "stochastic dual DP" (SDDP) model for reservoir management problems, as described by Pereira (1989) and Pereira and Pinto (1991). As might be expected from the name, that model utilizes insights from DP, and may be described as building up a solution to the dual of the DP formulation. But it is still based on LP/SP techniques. This brings very significant advantages in terms of the flexibility and efficiency with which system complexity, including multiple reservoirs and transmission systems, can be represented. On the other hand, and LP/SP-based model cannot deal with situations requiring non-linear modeling of intra-period behavior.

---

[1] The build-up is not as rapid as is sometimes supposed, since moving to a higher dimensional representation of the same system reduces the number of grid points in each dimension. Thus, retaining the same grid spacing, an N point representation in one dimension becomes at worst an $(N+1)^2/4$ point representation in two dimensions, an $(N+1)^3/27$ point representation in three dimensions, etc, not N^2, N^3 etc., as seems often to be suggested. For example, the two reservoir SPECTRA model uses a 6×12 storage grid. A three reservoir representation might use a $6 \times 8 \times 4$ grid, using 192 points, rather than 72. But the build-up is still substantial, and computational effort also increases because the dimensionality of each sub-problem in the recursion increases.

By way of contrast, the technique described here sticks much more closely to the DP philosophy, dealing directly with the DP dual and utilizing insights from LP/SP in that framework. Following Travers and Kaye (1997), we will refer to this whole class of methods as (Stochastic) "constructive dual dynamic programming", (S)CDDP, to distinguish it from Pereira's SDDP model, noted above. SDDP focuses on producing a good solution for the first period, and forms only approximate decision rules for later periods to the extent that this promises to significantly improve the initial decision. This is an iterative process, building up a progressively better approximation to the marginal water value (MWV) surface, in the form of "cuts" in a Benders decomposition of the SP problem. This process is repeated until the stopping criterion is met, typically being related to the quality of the decision rule approximation for the first period.

CDDP methods, on the other hand, solve an essentially exact dual to the standard DP formulation, using SP-like optimality conditions to "construct" the optimal dual solution directly, thus implicitly defining a DP-like operating policy over the entire state-space and planning horizon. As a consequence, while SDDP is applicable to higher dimensional problems, CDDP does not ultimately escape from the basic limitations of DP, particularly the curse of dimensionality. And it may not be able to model what may seem like minor changes to the system configuration without significant re-programming, if at all.[2] But it does have several advantages, for the problems where it is applicable.

This approach was first implemented in the two reservoir RESOP optimization model, described by Read (1989), but developed around 1984 as a module embedded in the PRISM planning model described by Read et al. (1987). Culy et al. (1990) embedded it in the SPECTRA model, used to manage reservoir releases in New Zealand from that time, and still used for a variety of purposes. Other developments have been sporadic, and often not reported in the international literature.[3] Read and Yang (1999) added an extra dimension to deal with inflow correlation. Cosseboom and Read (1987) reported an efficient application of this approach to coal stockpiling. Read and George (1990) described the technique in an LP framework, applicable to a wider class of problems involving a single storage facility. The model reported there was deterministic, but an efficient stochastic version was developed by Read et al. (1994).[4] Simultaneously, and independently, Bannister and Kaye (1991) described essentially the same (deterministic) technique. Travers and Kaye (1997) later generalized that approach, at least conceptually, to allow for a quite general multidimensional representation of the dual "marginal value" surface.[5]

---

[2] At least for earlier CDDP implementations, like SPECTRA, which were very fast, but very problem-specific. As discussed later, more recent implementations have been more generic.

[3] Related techniques were presented by Lamond et al. [1995] and Drouin et al. [1996].

[4] Velasquez [2002] describes a similar "generalized" DDP technique, utilizing Benders cuts as in SDDP, but imposing a DP-based state-space structure as in CDDP.

[5] Moss and Segall [1982] actually developed a conceptually similar algorithm, but described it in control theory terms and applied it to a very different problem.

Many of these models employ "intra-period pre-computations" to determine optimal operation of the power system, for a given release, within each decision period. In many cases, these imply some special structure for the dual problem that can be exploited to produce particularly efficient solution algorithms. A second group of models follows a slightly different path, being described in terms of "adding demand curves," basically because the intra-period pre-computations involved do not imply any easily exploitable special structure. First, Scott and Read (1996) showed how the intra-period optimization module could be replaced by a gaming module, and this approach was followed by Batstone and Scott (1998), Stewart et al. (2004), and Read et al. (2006). Second, an extra dimension was added to deal with risk aversion by Kerr et al. (1998).[6] Finally, the results of both these strands of research were embodied into two models. In New Zealand, Craddock et al. (1998) describe a model that adopts the Scott and Read approach to gaming, plus the Kerr et al. approach to risk aversion, in the context of a two-reservoir model similar to RESOP.[7] ECON also uses the "demand curve adding" approach for its ECON BID model but does not model gaming or risk aversion. As discussed later, it employs a heuristic extension of the two reservoir methodology to cover additional reservoirs when modeling the electricity markets in the Nordic countries (Norway, Sweden, Finland and Denmark), plus some of continental Europe.[8]

3 A Deterministic Single-Reservoir CDDP Algorithm

Most readers will be familiar with the classic DP formulation for managing a single reservoir due originally to Little (1955). Although we are mainly interested in the stochastic version of the DP formulation (SDP), we will first discuss the deterministic version, letting EF^t be the expected inflow in period or "stage" t, of a planning horizon, $t = 1, \ldots T$. This addresses the problem of finding a release policy defining the optimal release, q^t, for each period t, for example, for each week of the year, as a function of the "state" of the system at each stage. The method works, provided the "state space" properly defines the system "memory", by summarizing all information that could be known by a decision-maker, at that stage, and which would impact on the optimal decision. Assuming (for now) no correlation in inflows, the relevant "system memory" for the reservoir optimization problem is just the beginning-of-period reservoir storage level, s^t. So this becomes the "state variable" for the DP. The value function, $v^t(s)$, is central to the SDP method. It defines the value to be derived from the system, if optimally managed, over the remainder

6 While this development is incorporated into the RAGE/DUBLIN model, we will not describe it here because it could equally have been implemented using traditional DP.

7 This model became the "risk averse gaming engine" (RAGE) module of the DUBLIN model, intended as a replacement for SPECTRA, and now operated by ORBIT Systems Ltd.

8 ECON, now part of the Pöyry Group, also used this approach for an earlier, and simpler, model called ECON SPOT.

of the planning horizon, starting from storage level s in period t. Let $b^t(q^t)$ be the value derived in period t from release q^t. If we know the initial storage, S^0, and $V^{T+1}(s^{T+1})$, the value function for water stored at the end of the final period, T, the SDP optimization problem is[9]

$$\underset{q^{t=1,\ldots T}}{MAX} \quad \sum_{t=1}^{\mathrm{T}} b^t(q^t(s^t)) + V^{T+1}(s^{T+1}) \tag{1}$$

subject to

$$s^0 = \mathrm{S}^0 \tag{2}$$

$$s^{t+1} = s^t + \mathrm{EF}^{\mathrm{t}} - q^t(s^t) \quad for\,all \quad t = 1, \ldots T \tag{3}$$

$$\underline{\mathrm{S}}^{\mathrm{t}} \le s^t \le \overline{\mathrm{S}}^{\mathrm{t}} \quad for\,all \quad t = 1, \ldots T+1 \tag{4}$$

$$\underline{\mathrm{Q}}^{\mathrm{t}} \le q^t \le \overline{\mathrm{Q}}^{\mathrm{t}} \quad for\,all \quad t = 1, \ldots T \tag{5}$$

SDP then works from T back to the first period, recursively solving, for a regular grid of storage levels, s^t, in each period t

$$v^t(s^t) = \underset{q^t(s^t)}{MAX} \quad [b^t(q^t) + v^{t+1}(s^{t+1})]$$
$$\text{Subject to : } (3)^t, (4)^{t+1}, \text{and}(5)^t \tag{6}$$

This backward recursion actually produces an optimal release "policy," defining optimal releases for all values of s^t, for all $t = 1, \ldots T$, not just for a single storage trajectory, starting from S^0. This is not necessarily the best way to solve a deterministic problem. But backward recursion will be useful later when we wish to define a MWV function, allowing us to perform simulations covering the full range of possible storage levels that might be visited, in a stochastic environment. Solution of (6) may be relatively easy, but it still requires some kind of search over the possible release levels, to determine which one is optimal, and that search process must be repeated for each storage level in the grid, for each $t = 1, \ldots T$. The fundamental observation underlying CDDP is that the dual version of this problem is easier to solve than the primal version, given earlier.

Although Ben Israel and Flam (1989), and Iwamoto (1977), suggest some basic "inverse" concepts, the dual of this problem has received little attention in the literature. A DP state space is normally characterized as a set of discrete states, and so the value function is not even continuous, let alone differentiable. In our case, though, the discretisation of the state space is just an artefact of the primal DP formulation. In reality the state space is continuous, and Read (1979) shows that, provided the end-of-horizon value function and intra-period, cost/benefit functions are convex

[9] This formulation, and the CDDP algorithm, can be generalised by applying discounting in (1) or allowing for evaporation/leakage of s in (3). But, for simplicity, we will not do this here.

and differentiable, and so are all the value functions determined by backwards recursion. So we can assume that v^t and b^t have monotone non-increasing derivatives, mwv^t and mrv^t, with respect to s^t and q^t, respectively, which may be referred to as the marginal value of storing water, and of releasing water, respectively. Thus, ignoring bounds, the optimal solution to (6) is characterized by

$$mwv^t(s^t) = mrv^t(q^t(s^t)) = mwv^{t+1}(s^{t+1}) = mwv^t(s^t + \mathrm{EF}^t - q^t(s^t)) \quad (7)$$

In words, optimality requires that the marginal value of storing water at the beginning of period t, $mwv^t(s^t)$, equals the marginal value of releasing water during period t, $mrv^t(q^t(s^t))$, which equals the marginal value of storing water at the end of period t, $mwv^{t+1}(s^{t+1})$. Release and storage bounds, (4) and (5), may be accounted for in various ways, but (7) remains valid if we think of *mwv* and *mrv* as being set to zero for release or storage above its upper bound and to a very high "shortage" cost for release or storage below its lower bound.[10] Note that (7) provides an alternative form of the backwards recursion in (6), in which *mwv* curves are constructed directly by the recursion. This simplifies the optimization in (6) because *mwv* does not have to be inferred by differentiating v.

This "marginalistic DP" concept was utilized in a stochastic reservoir management optimization by Stage and Larsson (1961), and later in the STAGE model described by Boshier et al. (1983), and used by Read and Boshier (1989).[11] Although not described that way, it may be thought of as a "dual" approach to DP, in that it focuses primarily on optimizing the marginal value of water, and only secondarily infers the optimal release. It produces a set of MWV curves, where the MWV may be interpreted as a "fuel cost" for hydro.[12]

The contours of this MWV surface can then be interpreted as "operating guidelines," each corresponding to the marginal generation cost of a thermal station. It represents a critical storage level, below which water should be conserved by using that thermal station, as much as possible, in preference to releasing water from the reservoir.[13] Specifically, working from right to left, the guidelines in Fig. 1 represent the storage levels at which thermal stations with progressively higher marginal generation costs are to be used, in preference to hydro, while the leftmost (or "bottom") guideline(s) may represent the storage level at which (various levels of) load rationing become the most economic option.

Marginalistic SDP still followed the primal DP approach, though, of searching for an optimal release level for each beginning-of-period storage level in the

[10] The ambiguity arising when release or storage is exactly at its bound will be dealt with later.

[11] We ignore the forward simulation by which those models attempted to capture some aspects of inflow correlation, but later describe a related method, used in the ECON BID model.

[12] In practice, reservoir storage is often expressed in terms of stored energy, assuming constant (eg average) conversion efficiency, which allows storages to be aggregated conveniently.

[13] Guidelines are sometimes referred to as determining which thermal stations should be "base-loaded." But that is not quite correct, because the proportion of time over which a thermal station can generate is still limited by the load duration curve, even if it is used in preference to hydro.

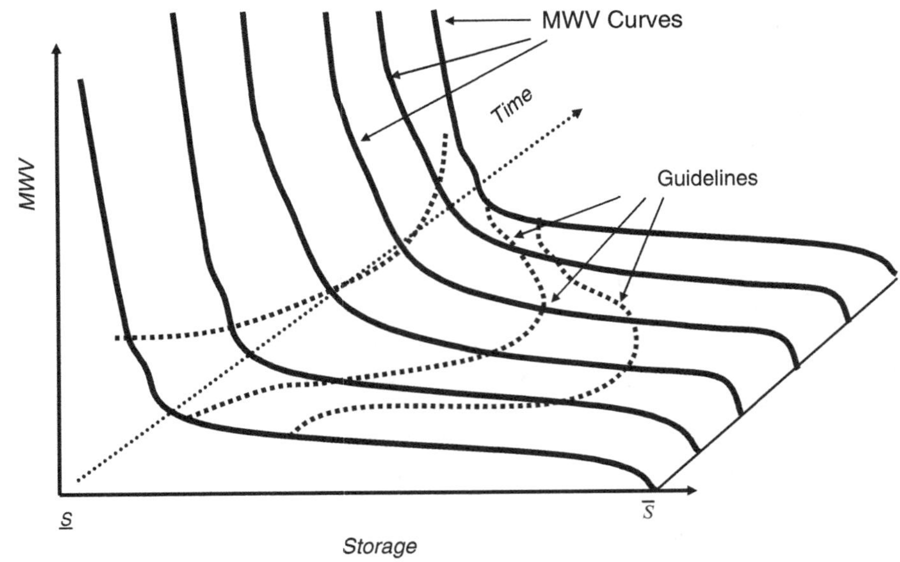

Fig. 1 Typical marginal water value (MWV) surface with guidelines

primal space, discretized in a conventional way. CDDP turns that around by directly constructing the MWV curve for each period from that in the next (chronological) period, thus implicitly finding the optimal beginning-of-period storage level corresponding to each release level. The required algorithm is based on the very simple observation that, when working backwards through the recursion, $mwv^{t+1}(s^{t+1})$ is already known at the time when q^t must be determined. Thus there is no need to search over the set of possible releases to determine whether a period t release would be optimal for a particular end-of-period storage level in period t, s^{t+1}. From (7) it may be seen that the set of such releases, $\mathcal{Q}^t(s^{t+1})$, is given by

$$\mathcal{Q}^t(s^{t+1}) = Q^t(mwv^{t+1}(s^{t+1})) = \{q^t : mrv^t(q^t) = mwv^{t+1}(s^{t+1})\} \quad (8)$$

Further, if s^{t+1} and q^t are known, it is easy to rearrange (3) to provide an expression for the corresponding s^t. Thus it is also easy to define $\mathcal{S}^t(s^{t+1})$, the range of beginning-of-period t storage levels from which it would be optimal to "aim at" end-of-period storage level s^{t+1} as the set from which s^{t+1} can be reached by adopting a release in the set $\mathcal{Q}^t(s^{t+1})$. Moreover, according to (7), the *mwv* for any beginning-of-period storage level in $\mathcal{S}^t(s^{t+1})$ must equal the *mrv* for the corresponding optimal release in period t, and also the *mwv* for end-of-period storage level s^{t+1} reached when that optimal release is implemented. Thus we have

$$\begin{aligned} &mwv^t(s^t) = mwv^{t+1}(s^{t+1}) && \textit{for all } s^t \in \mathcal{S}^t(s^{t+1}) \\ &\textit{Where}: \quad \mathcal{S}^t(s^{t+1}) = \{s^t = s^{t+1} - \mathrm{EF}^t + q^t : q^t \in \mathcal{Q}^t(s^{t+1})\} \\ & && \textit{for all } t = 1, \ldots \mathrm{T} \end{aligned} \quad (9)$$

This equation, which forms the core of all CDDP algorithms, creates a backward recursion in which the *mwv* curve for the beginning of each period is constructed directly from the *mwv* curve for the end of that period. This eliminates the need to search for optimal solutions to each DP sub-problem. At least in principle, it also eliminates the need to discretize the (primal) state space in any particular way. We call this the dual of the DP problem because, like the LP dual (and unlike marginalistic DP), it takes the key prices, *mrv* and *mwv*, as its primary variables and infers the key primary variables, q and s, from them. Further, if the intra-period optimization problem is an LP, as in the Read and George model discussed later, then the DP formulation implicitly defines a whole set of LP problems, starting from each possible initial storage level in each period. In that case, CDDP systematically produces the key dual variables for that whole set of LP problems.

In reality, the computational complexity of the algorithm depends on the efficiency with which the $\mathcal{Q}^t$ and $\mathcal{S}^t$ sets can be characterized and manipulated, and the variety of algorithms described later largely differ in this respect. And the CDDP algorithm has generally also been applied to stochastic problems. In general, then, the CDDP algorithm involves the following steps, each of which may be performed in a variety of ways, as discussed in the sections that follow:

1. Perform a set of intra-period optimizations to "pre-compute" $mrv^t(q^t)$, and identify the sets $\mathcal{Q}^t$ (mwv^{t+1}), for $t = 1, \ldots .T$, as in (8) above and Sect. 4 below.
2. Working backwards from an assumed monotone non-increasing end-of-horizon *mwv* curve, $MWV^{T+1}(s^{T+1})$, for each $t = T, \ldots .1$:
   (a) Apply (9) above to construct a beginning-of-period *mwv* curve for t from its end-of-period *mwv* curve, using either the "augmentation" or "demand curve adding" approach, as discussed in Sect. 5 below.
   (b) Perform an "uncertainty adjustment" to account for stochasticity, as discussed in Sect. 6 below.
3. If desired, simulate system performance, possibly using the optimized/pre-computed *mwv/mrv* functions as discussed in Sect. 7 below.

At the end of Sect. 2, we described how various CDDP implementations have adapted this general algorithm in different ways. The first group of models, following Read et al. (1987), dealt with situations in which the precomputed *mrv* functions had a natural structure, implying that the optimal release would either be constant or vary linearly over quite a wide range of storage levels. This implies a compact, but accurate, discretisation of the state space into regions delineated by "guidelines." Efficient algorithms then worked with each of those regions, rather than individual storage points, to form each new *mwv* surface in a process described as "guideline augmentation." The second group of models, following Scott and Read (1996), dealt with situations in which the precomputed *mrv* functions had no natural structure and reinterpreted the "augmentation" process in terms of adding demand curves for water. In both cases, models were developed that integrated the precomputation or uncertainty adjustment phases into the augmentation phase. Rather than describing each of these implementations, we discuss variations on each phase in turn, starting with simulation and precomputation because they are the easiest to understand. For simplicity we focus on the single reservoir case before moving on to discuss and illustrate multireservoir extensions.

4 Alternative Precomputations for Intra-period Optimization

One of the major advantages of DP is that it imposes no specific limitations on the structure of intra-period optimization problems or on the techniques used to solve them. CDDP is more limiting, but only inasmuch as the intra-period optimization must yield *mrv* as a monotone non-increasing function of release. Computationally, a single CDDP optimization run uses the results from each intra-period optimization only once, to construct the Q^t set for a particular s^{t+1}, and it may be efficient to perform this intra-period optimization in the context of the CDDP recursion, as in the Read and George model discussed later. But, while the structure of the *mrv* function produced by the intra-period optimization determines the type of CDDP algorithm that can be applied, the same CDDP optimization/simulation model could be implemented using precomputed Q^t sets produced by a variety of different intra-period optimization methodologies. So, conceptually, it is helpful to treat intra-period optimization as a distinct "precomputation" phase. This can also increase computational efficiency, because the computations can often be structured to produce all the required intra-period solutions in a very efficient manner, since each is just a small variation on the last. In fact this precomputation could be performed quite separately from, or even (with sufficient delay) in parallel to, the optimization. And the precomputed intra-period solution set could be re-use in later CDDP runs facing the same intra-period situations, or for simulation, as discussed in Sect. 7.

A simple intra-period optimization might just consist of using reservoir release, plus a "merit order" stack of thermal stations, to "fill" a load duration curve (LDC), perhaps adjusted for contributions from run-of-river or peaking hydro, wind, etc. If we think of *mwv* as defining the merit order position of reservoir release, an *mrv* curve can be computed efficiently by systematically setting *mwv* to the marginal cost of each successive thermal station, which thus defines *mrv*, and recording the corresponding change in release, to form a new step in the *mrv* curve. RESOP/SPECTRA actually uses a variation on that technique, using a well known "convolution" technique, due to Booth (1972) to model the way in which probabilistic breakdowns in the thermal system force more expensive thermal plant, and/or hydro, to cover significantly higher loads than might be predicted using a simple "de-rating" approach. Details may be found in Read et al. (1987), but this convolution process can be structured so as to be very efficient. The SPECTRA precomputations take only 0.14 s to produce a 100 MB file containing all the precomputed results necessary to run a CDDP optimization for 65 flow years, over a 468 week horizon, for a system with two hydro regions, 18 thermal marginal cost levels, and 15 load blocks, per week.[14]

Alternatively, though, exactly the same CDDP algorithm could have been applied using precomputations in which an intra-period LP is solved parametrically to determine the *mrv* curve for each period, as in Read and George (1990). The intra-period LP may not deal with unit breakdowns, but might be quite complex, meeting chronological load requirements subject to transmission system limits, etc. What matters, for this kind of step-based CDDP, is that parametric solution produces a stepped *mrv* function.

Scott and Read (1996) employ a very different precomputation, based on Cournot gaming, to represent the effect of imperfect competition in a deregulated market. Stewart et al. (2006) developed a similar methodology using offer curves precomputed using a different "supply function equilibrium" gaming model. This allows a complex, but important, real world issue to be studied in a way which would be very difficult, if not impossible, in an LP/SP-based model. Again, the details of the intra-period precomputation model do not matter, from a CDDP perspective, only that it produces a monotone non-increasing *mrv* curve. In these cases, though, the *mrv* curve does not have a distinct step structure and so, instead of the step-based "augmentation" algorithm introduced in RESOP, Scott and Read employ a "demand curve adding" process described in the next section. ECON's BID model uses this approach, too, even though it can employ either LP or NLP for intra-period optimization, and so will produce *mrv* curves with some kind of step structure.

5 Guideline Augmentation vs. Demand Curve Addition

Step 2 of the general CDDP algorithm given in Sect. 3 involves applying (9) above to construct a beginning-of-period *mwv* curve for t from its end-of-period *mwv* curve, using either the "augmentation" or "demand curve adding" approach. To understand the augmentation approach used in earlier CDDP models, consider the interpretation of (9) for the case of a single reservoir, where the intra-period benefit function can reasonably be approximated as piece-wise linear, thus implying a stepped *mrv* curve.[15] This is the case for any *mrv* curve produced by (parametric) solution of an intra-period LP, such as that of Read and George (1990), and also for the precomputed *mrv* curves used in RESOP. In that case, let $k = 1, \ldots K$ index the thermal stations in the order of decreasing marginal generation cost, m_k. Then level k in the mrv^t curve represents, ${q_k}^t$, the hydro release required if all thermal stations with marginal generation cost below c_k are used as much as possible to conserve hydro

[14] All SPECTRA computation times in this document are total elapsed times for a Pentium 4 HT at 2.6 GHz, with 512 MB RAM running Linux Red Hat 9.

[15] The approach described here could be generalized to deal with a piece-wise linear *mrv* curve, for example, implying a piece-wise quadratic approximation to the intra-period benefit function.

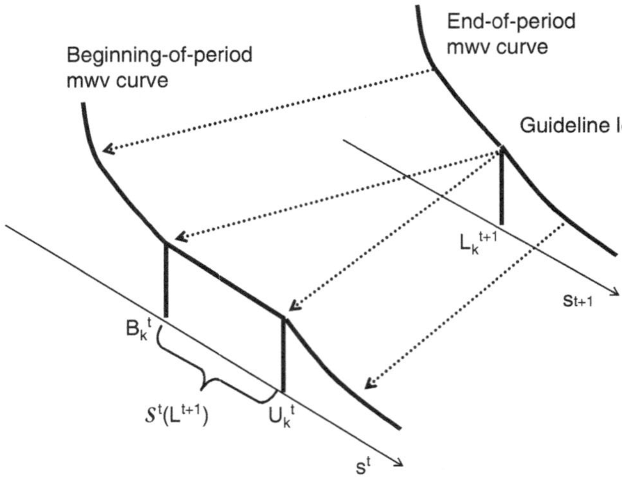

Fig. 2 Guideline augmentation

storage, while the remaining thermal stations are used only as much as necessary to cover the remainder of the LDC, after hydro release $q_k{}^t$. Let $b_k{}^t$ be the breadth of step k, representing the incremental contribution station k can make to conserving storage, by reducing release requirements.

In this case, the optimal release policy can be characterized in terms of guidelines of the type shown in Fig. 1.[16] Figure 2 shows the relationship between end-of-period and beginning-of-period *mwv* curves. Note the following:

(a) $\mathcal{S}^t(s^{t+1})$ will be a simple line interval, if there is a range of release levels for which *mrv* equals *mwv*(s^{t+1}). This will be the situation when end-of-period storage exactly equals L_k^{t+1}, the guideline level for station k, so that progressively greater release simply backs off more generation from that station, setting *mrv* to its marginal generation cost, c_k.

(b) $\mathcal{S}^t(s^{t+1})$ will be a single point, if there is a unique release level for which *mrv* equals *mwv*(s^{t+1}). This will be the situation for all end-of-period storage levels between guideline levels, with that optimal release level being q_{k+1}^{t}, between guideline levels L_k^{t+1} and L_{k+1}^{t+1}.

Also note that, even with maximum thermal support (i.e., $k = 1$), a release of $q_1{}^t$ is required if the LDC is to be covered.[17] But this will be offset by inflows of EF^t. So, to at least reach the minimum storage level for the beginning of period $t + 1$, we

[16] The situation is a little more complex where the underlying intra-period optimization is non-linear, because there may be a band of end-of-period storage levels over which the *mwv* matches the *mrv* derived from progressively backing off some thermals station, rather than a distinct "guideline" corresponding to a unique marginal cost. A generalized augmentation algorithm is possible, but "demand curve adding" may become more appropriate as discussed in Sect. 5.

[17] SPECTRA actually uses high-priced dummy thermal stations, representing load curtailment, so that the LDC is always notionally covered.

must have

$$MinStor^t = \underline{S}^t - \mathrm{EF}^t + q_1^t \tag{10}$$

If maximum thermal support were maintained for all beginning-of-period storage levels above *MinStor*, the beginning-of-period MWV for s^t would equal the end-of-period MWV for the point

$$s^{t+1} = s^t + \mathrm{EF}^t - q_1^t(s^t) \quad \text{for all} \quad t = 1, \ldots T \tag{3}$$

Rearranging this equation implies that the beginning-of-period *mwv* curve would be just the end-of-period *mwv* curve, shifted up or down so as to start from *MinStor*, rather than from the minimum physical storage level. Starting from a storage level below *MinStor*, shortage is expected, though, and the beginning-of-period *mwv* curve for that region must be formed by adding segments representing shortage costs. But, for storage levels above *MinStor*, less thermal support is optimal. Thus the CDDP "augmentation" algorithm proceeds by tracing up the beginning-of-period *mwv* curve, and performing the following operations:

- Inserting a "flat" section of breadth ${b_k}^t$ into that curve at the point where $mwv = mrv = m_k$. This is $\mathcal{S}^t(\mathrm{L}_k^{t+1})$ in the figure. The upper and lower limits of $\mathcal{S}^t(\mathrm{L}_k^{t+1})$, U_k^t, and B_k^t are called "augmented guidelines" and the process is called "augmentation." Starting from B_k^t, it will be optimal to "aim at" target level L_k^{t+1}, with maximum support from station k. Starting from U_k^t, it will be optimal to aim at target level L_k^{t+1}, with minimum support from station k (but maximum support from $k-1$).
- Then shifting the entire section of the end-of-period *mwv* curve lying above the guideline up, by b_k^t, to a higher beginning-of-period storage level. This forms a corresponding section of the beginning-of-period *mwv* curve, over which the optimal release is ${q_{k+1}}^t$ as above.

This process can be applied even if *MinStor* lies below the minimum physical level. This happens if, even when the reservoir starts period t empty, inflows are expected to be high enough that the LDC can be covered without maximum thermal support. In that case, part of the beginning-of-period *mwv* curve traced by the algorithm lies below the minimum physical storage level. Part may lie above the maximum physical storage level, too, if some thermal support is necessary to maintain the reservoir at its maximum level, even if it starts from that level. In both cases the *mwv* curve can be simply truncated to the physical storage limits. Alternatively, if expected inflows are high, the *mwv* curve tracing process may terminate at a storage level below the physical maximum. In that case, *mwv* can be set to zero for all higher beginning-of-period storage levels, reflecting the fact that spill is expected if storage lies in that zone.

This algorithm requires only minor arithmetic operations to insert steps into *mwv* curves, and this process is employed by RESOP, in a stochastic context. Read and George also apply it in a deterministic context, where the precomputations are produced by a parametric LP run. They note that, in that case, each *mwv* curve consists entirely of *mrv* steps added in when working back through future periods, with new *mrv* steps typically added to existing *mwv* steps. So the optimal policy tends to be dominated by a few steps, each covering a wide range of storage levels. This makes the deterministic problem very easy to solve, but implies a wide range of storage levels over which we are indifferent between loading some thermal station more heavily at an earlier date vs. later in the horizon, unless discounting or wastage makes delay preferable. But the stochastic situation is rather different as discussed in Sect. 6.

Read and Gorge focused on exploiting the step structure of their problem for maximum computational efficiency. But Scott and Read developed an algorithm that is potentially less efficient, but simpler and more general, because it does not assume any step structure in the *mrv* curve. We have talked, so far, about *mrv* and *mwv* curves as defining the marginal value of water released or stored for the future. But consider the inverses of the *mrv* and *mwv curve*, that is:

$$dcr^t(mrv) = \{q^t : mrv^t(q^t) = mrv\} \tag{11}$$

$$dcs^t(mwv) = \{s^t : mwv^t(s^t) = mvw\} \tag{12}$$

In economic terms, *dcr* may be described as the "demand curve for release," specifying the price that a power system manager would be prepared to pay for water released to meet current demand. Similarly, *dcs* is the "demand curve for storage," specifying the price that a reservoir manager would be prepared to pay for water stored to meet future demand.

Now, since water stored at the beginning of period t may be used to meet demand either in period t or in later periods, the demand curve for water stored at the beginning of period t, dcs^t (or equivalently, mwv^t), may be found by simply adding the demand curve for water in period t (mrv^t or dcr^t) to the demand curve for water stored at the end of period (dcs^{t+1} or mwv^t), as shown in Fig. 3.[18] So, ignoring uncertainty, the core CDDP algorithm becomes as follows:

1. Precompute the *dcr/mrv* curves over a (reasonably fine) grid, then
2. Working backwards from an assumed final DCS/MWV curve:
   - Add the intra-period *dcr* curve to the end-of-period *dcs* curve, using

$$dcs^t(mwv) = dcs^{t+1}(mwv) + dcr^t(mwv) \tag{13}$$

   - And then truncate to form a new *dcs* curve for the beginning of the period.

This algorithm can be readily applied to a variety of problems, without having to investigate the guideline/step structure applicable in each case or devise specialized algorithms to exploit it. Thus this approach is used by both the RAGE/DUBLIN

[18] Note that horizontal or vertical segments in the *dcr* curve insert "flats," or translated dcs^{t+1} segments, into dcs^t, just as for the augmentation process above. Otherwise, though, dcs^t is a composite, "stretched" in the storage dimension by the addition process.

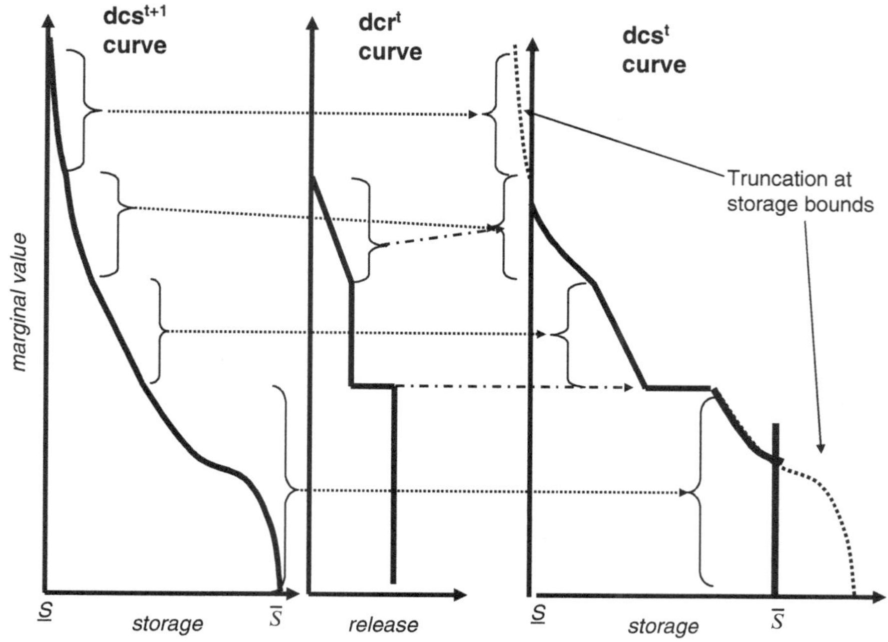

Fig. 3 Augmentation by demand curve addition

and ECON BID models. It uses a discrete set of grid of storage points, just as in standard DP. The grid must be fine enough, though, for *mwv* to maintain a reasonable approximation to the derivative of the underlying value function. Thus, it can be less efficient than the augmentation algorithm if the *dcr* actually does have a step structure. But, for each storage grid point, it can identify the optimal *mwv*, and hence the corresponding release decision, with a single addition. For the same grid, this seems clearly more efficient than applying standard DP, which must search to find the optimum.

6 Dealing with Uncertainty

Just as for regular DP, the stochastic version of the CDDP algorithm follows naturally from the deterministic version already described. To determine an optimal release for any period, including the first, we must determine how any water we leave in storage would best be utilized, under the whole range of situations in which we may find ourselves in the future. But we cannot know, or control, what state the system will actually be in, in future periods. Thus we need to find a release policy that determines the optimal release for any possible "state" we may find ourselves in. Initially, if we assume no correlation in inflows etc., the relevant "system memory" boils down to the reservoir storage level in the period, s^t. So this becomes the "state variable" for a basic DP formulation.

The stochastic version of the primal DP algorithm differs only from the deterministic version, in that the value function $v^{t+1}(s^{t+1})$ in (6) is replaced by its expected value. Similarly, here, we are concerned with matching *mrv* to the expected end-of-period *mwv*. There is an issue, though, with determining both of these. If we conservatively assume that a release decision, q^t, must be made at the beginning of t, before the inflows for that period are known, that decision must depend only on the initial storage state s^t. We can think of this as a decision to aim at an expected end-of-period storage level, es^t. But we cannot be sure of reaching that target or of achieving the expected benefits within the period. Alternatively, we could optimistically assume that inflows are known before decisions are made, in which case there would be a release decision, and benefit, for each possible inflow level.

In reality, release decisions may be reviewed more or less continuously, thus achieving results somewhere between these two extremes. So, as a compromise, RESOP actually assumes that the release decided at the beginning of the period is carried out, and provides the expected benefit, $b^t(q^t)$, unless storage reaches its upper or lower limits, in which case release is adjusted to keep storage feasible, as in (2).[19] But this means that, for some inflows m^{ht}, the marginal value of water, given that hydrology scenario h occurs, will turn out to be the *mwv* for the actual storage attained, when aiming at the end-of-period storage target, es^{t+1}.[20] For others, though, m^{ht} may turn out to be the *mrv* value corresponding to a release level adopted to keep storage within bounds. The beginning-of-period *mwv* must then equal the weighted average of these values. Formally, let F^{ht} be the deviation from the expected inflow, EF^t, under hydrology scenario h, occurring with probability P^h. Then

$$es^{t+1} = s^t + EF^t - q^t(s^t) \tag{3^e}$$

$$\begin{aligned} m^{ht} &= mrv^t(es^t + F^{ht} - \bar{S}^{t+1}) \quad && if \quad es^t + F^{ht} \geq \bar{S}^{t+1} \\ &= mrv^t(es^t + F^{ht} - \underline{S}^{t+1}) && if \quad es^t + F^{ht} \leq \underline{S}^{t+1} \\ &= mwv^t(es^t + F^{ht}) && otherwise \end{aligned} \tag{14}$$

$$emwv^{t+1}(es^{t+1}) = \sum_h m^{ht} P^{ht} \tag{15}$$

We then treat es^{t+1} just like a target storage level in the deterministic formulation, and so the backwards recursion is defined by

---

[19] Alternatively, we could assume a mid-week revision, but this would just produce the same formulation, with twice as many stages.

[20] Truncation at the storage bounds means that es^{t+1} is a target, not an expected value. In extreme cases, m^{ht} may have to be set to zero (for spill), or a very high value, if inflows and storage are both so low that shortage cannot be avoided. Both *emvw* and *mwv* are actually "expected" *mwv* curves, in the broader context. But the notation indicates that *emvw* is our expectation, at the beginning-of-period t and given the inflow distribution expected in period t, if we aim for es^{t+1}.

$$\mathcal{Q}^t(es^{t+1}) = \mathcal{Q}^t(emwv^{t+1}(es^{t+1})) = \{q^t : mrv^t(q^t) = emwv^{t+1}(es^{t+1})\} \quad (8)^e$$

$$\begin{aligned} &mwv^t(s^t) = emwv^{t+1}(es^{t+1}) \qquad \textit{for all } s^t \in \mathcal{S}^t(es^{t+1}) \\ &\textit{Where}: \quad \mathcal{S}^t(es^{t+1}) = \{s^t = es^{t+1} - \mathrm{EF}^t + q^t \;:\; q^t \in \mathcal{Q}^t(es^{t+1})\} \end{aligned} \quad (9)^e$$

RESOP implements this algorithm, with $emwv^{t+1}$ being calculated in a distinct "uncertainty adjustment" phase, in which mwv^{t+1} is determined by interpolation on the mwv^{t+1} surface, for each possible inflow variation (F^{ht}). This is rather clumsy, because it incurs the overhead of repeatedly translating between a guideline representation, which facilitates efficient augmentation, and a grid-based *mwv* representation, which facilitates efficient interpolation in the uncertainty adjustment. These translations also introduce a small error each time, and these errors accumulate as we work backwards through the planning horizon. But the process has been retained in RESOP, and hence in SPECTRA, because it is tolerably efficient and not too inaccurate if the *mwv* grid is not too coarse. Also, as discussed by Read and Boshier, interpolation on a *mwv* curve with the kind of convex shape shown in Fig. 1 will tend to raise *emwv*, thus introducing a bias toward conserving more water that may actually be considered desirable and/or realistic for simulations.[21]

The method does not fully exploit the potential of CDDP, though, and a superior approach was developed by Read et al. (1994) and, implicitly, used in subsequent models. Rather than running through a grid of es^t levels and applying (15) to each in turn, two alternative ways of constructing $emwv(es^t)$ may be suggested:

- First, Scott and Read (1996) express $mwv^{t+1}(s^{t+1})$ in its inverse form, as $dcs^{t+1}(mwv)$ then form *emvw* for the end of the previous period by simply adding multiple versions of the *mwv/dcs* curve, each shifted by F^{ht}, using

$$dcs^t(emwv) = dcr^t(emwv) + \sum_h (dcs^{t+1}(emwv) - \mathrm{F}^{ht})P^{ht} \quad (13)^e$$

- Second, Read et al. (1994) express $mwv(s^t)$ as a set of steps, and note that, if mwv^{t+1} changes by Δ_k at level s_k^{t+1}, then (3) implies that $emwv^t$ must change by $\Delta_k P^h$ at target storage level:

$$e\hat{s}_k^{ht}(s_k^{t+1}) = s_k^{t+1} - F^{ht} \quad \text{for } h = 1, \ldots\ldots H \quad (16)$$

Thus $emwv^t$ can be formed by running through the steps of the mwv^{t+1} curve, forming H new steps in the $emwv^t$ curve for each step of the mwv^{t+1} curve. Or it can be formed by simply merging H lists of $e\hat{s}_k^{ht}$ points into a single list, defining all the steps in the $emwv^t$ curve.

[21] In fact a 6 × 12 storage grid has been found sufficiently accurate for most purposes. The entire optimization process, including both augmentation and uncertainty adjustments, takes only about 0.18 s, for a two reservoir problem with 468 stages and 65 inflow levels at each stage. Other models, such as ECON BID, use a much finer grid, and so should be more "accurate."

The apparent accuracy of this second approach is actually spurious, because the hydrology distribution is not really discrete, and it will quickly produce a great many steps. But the list merging process can be made quite efficient, and Read et al. show that the number of steps can be limited by replacing many small *mwv* steps with a few larger *mwv* steps. This produces a piece-wise linear approximation to the underlying value function, conceptually similar to that in SDDP, although by a very different process. If the smallest step size is set to match that in the *dcs* representation, this representation requires significantly fewer steps, because many steps will be larger than the minimum size. But linear *mwv* segments may be used instead, thus producing what is effectively a quadratic approximation to the underlying value function. Using linear *mwv* segments means that flats often need to be inserted in the middle of a linear *mwv* curve segment. So, linear interpolation is required to find an exact solution to (7). But this is not burdensome, and allows a more accurate approximation to be maintained with a smaller number of segments in the *mwv* list.

With this modification, the stochastic version of the step-based algorithm becomes more like the RESOP algorithm, which also employed piece-wise linear *mwv* approximations between its augmented guidelines.[22] Both the step-based and demand curve adding algorithms can be further improved, though, by integrating the augmentation process into the "uncertainty adjustment." Rather than adding H dcs^{t+1} curves to form $edcs^t$, then add in the dcr^t curve to form dcs^t, we can simply add all $H+1$ curves in one integrated process. Or, rather than sort H lists of mwv^{t+1} steps to form $emwv^t$, then add in a list of steps from the mvr^t curve to form mwv^t, we can simply add all $H+1$ step lists in one integrated process.

Finally, all our discussion, to this point, has assumed that inflows are not correlated over time. If inflows are correlated, though, current inflows should impact on release decisions. Another dimension is often added to the state space of primal DP formulations, representing the previous period's inflows, which are assumed to impact on future inflows via a lag-one Markov process. In that case, P^{ht}, v^t, and q^t in (1) all become conditioned on F^{t-1}. So P^{ht} becomes $P^{ht|i}$, where i, the inflow observed in period $t-1$, is also drawn from the discrete set, $h = 1, \ldots H$. Serial correlation is not as important in the New Zealand system as in some others and was not accounted for in the original RESOP implementation.[23] But Read et al. (1994) generalize the step-based algorithm to account for correlation efficiently.

---

[22] Either way, the result is qualitatively different from the deterministic case. First, the uncertainty adjustment tends to break up any large flat areas in the *mwv* curve, making it much less likely that they will increase in size as new steps from the *mrv* curve are added. Second, as discussed by Boshier and Read, the effect of taking a weighted average over a convex *mwv* curve of the kind shown in Fig. 1 will be to raise *emwv*, as we move back through the recursion, so that new *mrv* steps are inserted at higher storage levels, implying a distinct preference for earlier thermal generation, as a precaution, to conserve storage.

[23] A heuristic was later added whereby the inflow distribution assumed for optimization purposes could be spread, so as to better reflect the cumulative impact of correlation on the aggregate flow distribution over time. An offset could also be applied during simulations, effectively treating the expected surplus/deficit implied by a Markov model as if it was already in the reservoir.

If, at the beginning of period t, it is believed that the probability of each F^{ht} occurring depends on the previous period's inflow, we will need to form a different end-of-period $emwv^{it}$ curve, and make a different release decision, $q^{it}(s^t, F^{it})$ for each inflow state, i. But all the end-of-period $emwv^{it}$ curves, for all prior inflow states i, will be created as probability weighted sums of the same H beginning-of-period mwv^{t+1} curves for $t + 1$, each shifted up or down by the corresponding current period inflow value, F^{ht}. All that differs is that the $emwv^t$ curve for each past inflow state, i, will be formed by placing a different weight, $P^{ht|i}$ (equal to zero if a particular transition is impossible), on shifted mwv^{t+1} curve h. Thus the steps of each beginning-of-period *emwv* curve lie in exactly the same places, and the list sorting algorithm only needs to be applied once to form all those steps. Extra effort is required only to apply differing probability weights when updating each $emwv^{it}$ and to insert *mrv* steps in differing places, as they become critical for each $emwv^{it}$. So, while dealing with correlation adds an extra "dimension" to the DP formulation, CDDP can cope with that extra dimension with a much lesser increase in computational effort than might be expected.

Read and Yang (1999) come to similar conclusions with respect to their CDDP model of a single reservoir problem with a continuous lag-one Markov inflow process (as opposed to the discrete Markov transition matrix assumed earlier). They show that the optimal release strategy can still be expressed in terms of a set of guidelines, which have a distinctive shape, and that augmented beginning-of-period guidelines for each period can be formed using a specific algebraic transformation of the end-of-period guidelines for that period. Combining this augmentation with an uncertainty adjustment produces a RESOP-like algorithm for the correlated case. Read et al. (2006) take a different approach, more suited to a situation where inflow correlation, or in this case price correlation, can be described by an over-arching tree structure, in which branches occur only in some decision periods and might (or might not) rejoin after several decision periods have elapsed. It is still possible to perform CDDP by working backwards through such a structure, with composite *emwv* curves being formed at each branch point as probability weighted combinations of the *emwv* curves derived by working backwards through each branch. But that approach is equally applicable to primal DP and will not be discussed further here.

7 Efficient Simulation Using CDDP Precomputations

Given the complexity of the reservoir optimization problem, simulations are very expensive to perform if we must re-solve the whole reservoir optimization problem at each step of the simulation. CDDP eliminates this requirement by optimizing *mwv*, and hence q, for all storage levels in all periods. But it can also eliminate the need to repeat intra-period optimization, because the mrv^t curves define release levels as a function of *mwv* via (4). So, any number of simulation runs, $j = 1, \ldots J$, can be performed by just repeating one simple step, all starting from $s^{j0} = S^0$,

$$s^{jt+1} = s^{jt} - q^t(s^{jt}) + \mathrm{EF}^t + \mathrm{F}^{jt} \tag{17}$$

The precomputation runs will also determine other system performance measures that may be of interest in the simulation, such as fuel usage, transmission loadings, etc. Provided those characteristics are stored during the precomputation phase, a probability distribution can also be formed for them during the simulation runs, or even after the simulation is completed. Precomputed *mrv* curves in the original RESOP model, or any LP-based model, have a stepped structure, implying an optimal policy in which the storage space is divided into regions, over which q^t is either constant, at a step value, or varies linearly between two such values. If $q^t(s^{jt})$ lies in step k, it may be found by interpolation, applying weights ${w_k}^{jt}$ and$(1 - {w_k}^{jt})$ to the upper and lower step solutions, with ${w_k}^{jt} = 1$ in many cases. Since the same step structure applies to any other characteristics determined by the precomputation, all we need to do, during the simulation, is to accumulate the weights. Then, at the end, we can reproduce probability distributions for any characteristic by simply applying the accumulated, probability weighted, step weights, *ew*. So if ${x_k}^t$ is, say, the thermal generation level precomputed for step k of the mrv^t curve, then the simulated probability of that generation level actually occurring is given by

$$ew_k^t = \sum_j P^j w_k^{jt} \tag{18}$$

Performing simulations in this way can obviously be very efficient. On the other hand, it may often be desirable to perform a more comprehensive simulation, accounting for factors that were ignored during the precomputation. In the case of SPECTRA, the comprehensive simulation, discussed in Sect. 8, proved to be fast enough. So the original simulation process, based on the approach discussed here, was de-commissioned, and computation times are no longer available.

8 Adding Reservoirs

All the CDDP models in commercial use optimize at least two reservoirs. Although the two-reservoir RESOP module in SPECTRA has seen no development effort since its initial implementation 25 years ago, it still solves very quickly. A computation time of the order of 0.03 s per optimization year, for the precomputation/optimization phase, suggests ample opportunity for generalization by adding more reservoirs. But, while CDDP offers significant efficiency improvements, which assist in reducing computational requirements when addressing higher dimensional problems, it still suffers from the "curse of dimensionality," like all DP-based models. It also faces specific conceptual and practical problems when generalized to higher dimensional problems.

Conceptual problems arise in generalizing methods relying on the specific structure of the *mwv* surfaces to achieve computational efficiency. In a one-dimensional state space, the structure of the *mrv* curve, and hence the *mwv* curve, can be characterized in terms of *mrv/mwv* being constant, or varying linearly, over each line interval, thus forming "steps" or "segments". In N dimensions, though, the *mwv* surface must be characterized by a set of N-dimensional polytopes. If the intra-period decision space has a simple structure, the *mwv* surface, representing the accumulation of successive *mrv* surfaces backwards over time, may also be quite simple. Thus Cosseboom and Read demonstrate an extremely simple CDDP algorithm for a coal stockpiling problem, which is two-dimensional, but only has one guideline, corresponding to a critical stock/destock decision. Travers and Kaye pursue a more general algorithm. In general, though, the polytopes need not have any particular shape, and it will be difficult to store or access an exact representation of this kind of surface in a systematic way or to perform searches or interpolations, for example, as in the CDDP methodology described previously.[24]

Moreover, although DP only needs to deal with one N-dimensional value surface, for an N-dimensional problem, CDDP must deal with N marginal value surfaces, each of N dimensions. For a two-reservoir problem, for example, the *mwv* for each reservoir varies as a function of the storage level in both reservoirs. This is a proper reflection of the real operational situation, but it means that identifying a complete set of operational guidelines now requires us to perform N-dimensional interpolations on N *mwv* surfaces, or on the intersection of such surfaces. This may be illustrated with reference to Fig. 4, which represents part of a guideline diagram of the sort produced by the two-dimensional RESOP model of the New Zealand power system. One axis represents storage in the North Island, while the other represents storage in the South Island.

Guidelines are shown for North Island thermal stations, and these may be determined by tracing out contours of the North Island *mwv* surface alone.[25] A pair of transfer guidelines is also shown, indicating the regions of the South/North Island storage diagram within which various inter-island transfer strategies will be optimal. Broadly, if storage levels are too far out of balance, it will be profitable to use hydro in one island, as much as possible, to conserve storage in the other. Between the transfer guidelines, storage levels are more or less in balance,[26] and the gain to be made by trying to balance them more closely is more than offset by transfer losses on the inter-island link. Such guidelines can only be identified, though, by comparing the two *mwv* surfaces, and plotting contours where the *mwv* in one island equals that in the other, adjusted for marginal losses.

These guidelines may then be augmented to form an upper/lower transmission guideline pair from each guideline, as shown. Operationally, these augmented guidelines define the set of beginning-of-period storage pairs from which it is optimal to adjust South or North Island release so as to aim at the corresponding end-of-period target guideline. The augmentation phase of RESOP works directly with

---

[24] One approach would be to use approximating cutting planes, as in SDDP.

[25] Similar guidelines could be produced for South Island thermal stations, if there were any.

[26] Thermal guidelines slope at around 45° here, because generation policy is driven only by total national storage. They level out when South Island storage is so extreme that transfer must be maximized, in one direction or the other, irrespective of incremental South Island storage.

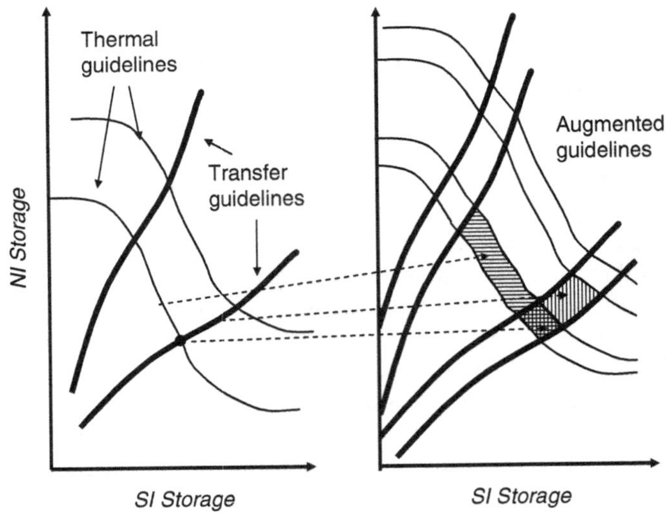

Fig. 4 Guideline augmentation for two reservoirs

these diagrams, but it does so by discretizing the state space in one storage dimension. Thus it effectively applies the one-dimensional algorithm already described, to a set of *mwv* curves, except that part of the *mwv* curve must be translated diagonally, reflecting a South/North island storage trade-off when the transfer guideline is encountered. While not exact, this process is robust and efficient. But generalization to deal with higher dimensional problems would require further conceptual development, which has simply not been attempted.

Thus SPECTRA remains a two-reservoir model more due to lack of conceptual development than due to concern about computation times. But one reason for this is that the game theory-based precomputations used in later developments produced less-structured *dcr* surfaces. So those models employed the less structured "demand curve adding" approach, which is more readily generalized to higher dimensions. When extending the "demand curve adding" approach to two or more reservoirs, we must add *dcr/dcs* "surfaces" rather than one-dimensional *dcr/dcs* "curves." Leaving aside the precomputations, this approach reduces the CDDP "optimization algorithm" to a generic "surface adding" problem, susceptible to analysis by experts in computational methods, without any specialized knowledge of optimization theory. Computational requirements will still build up, though, and will ultimately prove limiting. To date, heuristics have been used to extend models beyond two reservoirs, both in SPECTRA and the ECON BID model, as described later.

9 Current Models: RAGE/DUBLIN and ECON BID

Before the market reforms, a decade ago, SPECTRA was central to the operation of the New Zealand electricity sector, being extensively used for operational, planning, pricing, and policy purposes. Although it does not model some aspects of market behavior (e.g., "gaming"), it is still used for a wide variety of studies, mainly because it still runs very efficiently on modern hardware. Adding the times given above gives a total run time of 4.5 s for a 9 year planning horizon, coming up to around 5 s if the inflow dataset has to be converted to represent its energy content for changing reservoir configurations. But SPECTRA has been used for 25 years now, and has previously been described by Read (1989). It does not exploit the full generality of the CDDP approach, as described earlier. Two more recent CDDP-based model packages are worth describing in more detail as they cover some of the aspects not exploited by SPECTRA and have seen little previous mention in the academic literature.

As noted earlier, many of the developments described here were incorporated into the New Zealand RAGE/DUBLIN model, the basis of which is described by Craddock et al. (1999). This model allows for two regions, each containing an aggregate reservoir assumed to be controlled by a different participant. Transmission between these two regions is possible, subject to capacity constraints. The reservoir-based hydro system is complemented by a fringe of geothermal, wind, and non-storage hydro, plus a group of thermal stations.

Only one reservoir owner is allowed to exercise risk aversion,[27] but all participants of significant size, both hydro and thermal, are involved in a Cournot game, thus ensuring monotone *mrv* curves. Since gaming strategy is heavily influenced by contract levels and elasticities, data on both must be supplied. Such data is often private, and/or subjective, as is the degree to which it influences real decision-making. Based on a set of contract level assumptions, the model itself can calibrate the demand elasticities used to ensure that the resulting total load matches the one projected with traditional forecasting tools.

RAGE/DUBLIN defines end-of-period *dcs/mwv* surfaces over a regular grid of storage/wealth triplets but, like RESOP, recognizes that adding in the intra-period *dcr/mrv* surface effectively "shifts" the *dcs* in all three dimensions, by a lesser or greater amount, depending on the optimal release/profit triplet implied by the *dcs* and *dcr* for each storage/wealth triplet. Such shifts distort the regular grid and, as in RESOP, some effort has been required to re-form a regular grid in a robust and efficient manner. Heuristics are employed to model inflow correlation and to resolve situations where the Cournot game has no unique solution.[28] For simulation

[27] The New Zealand system is rather asymmetric, with the system in one island being purely hydro and dominated by one reservoir system containing a large proportion of national storage.

[28] During the development of RAGE, it was realized that stable, globally optimal, solutions could not be guaranteed when transmission links are involved, a view subsequently confirmed by Borenstein et al. [2000]. But this has not proved particularly troublesome in practice. And we note that pursuing such solutions does not necessarily improve the realism of market simulation since, if

purposes, heuristics are also used to apportion aggregate regional releases between constituent reservoirs, in such a way as to balance spill probabilities, etc.

The model has been used, for example, in analyses for the New Zealand Commerce Commission in merger and acquisitions cases. Optimization takes 3 min for a two reservoir model using weekly time steps, each with three load blocks, over an annual planning horizon, with 20 hydro sequences, over an 8×12 storage grid. Simulation takes 1 min using eight reservoirs. Calibration, when needed, takes approximately 30 min (All on a Pentium 4 computer with 3.2 GHz, 2 GB RAM).

Finally, we turn to the BID model developed by ECON around 2005/2006. This model has not previously been described in the academic literature, but documentation is provided by Bell et al. (2007). It was developed for use by transmission companies and regulators in the Nordic region and the North-Western part of continental Europe to assess the economic impacts of large, inter-regional transmission investments. While the Nordic region is largely hydro dominated (50% of the annual energy demand is covered by hydro electric generation in a normal inflow year, but for Norway alone, the percentage is 99%), the continental European system is largely thermal based. Hence, sufficiently detailed modeling of both hydro and thermal aspects is needed, particularly to assess the economic viability of new interconnectors between thermal-dominated and hydro-dominated parts of the modeled area.

The model has two layers. The second layer can simulate market outcomes down to an hourly time resolution, producing a more realistic price structure in thermal dominated areas. That model also includes an approximation of unit commitment issues such as start-up costs and lower efficiencies during part-load operation of thermal power plants and is based on the work by Weber (2005).

But we will focus on the first layer, which uses CDDP to optimize operating strategy for the hydro reservoirs. This layer uses the demand curve adding approach of Sect. 5 implemented for two reservoirs. However, some versions of the model cover around 12 countries, with several of those being split into smaller regions, and many inter-regional transmission bottlenecks. Since the economics of transmission investment depends on a good representation of price differences between regions, separate reservoirs for all main regions were thought desirable.

In a conventional two reservoir implementation, *dcr* would be calculated for various release combinations from each of the two aggregate reservoirs and *dcs* calculated similarly. But BID uses a different aggregation when optimizing *dcs* for each reservoir, treating that reservoir as "primary" and aggregating all others into one "complementary" reservoir. In a model with three actual reservoirs R_1, R_2, and R_3, the first primary reservoir would be R_1, complemented by the sum of R_2 and R_3. This two-reservoir model is solved as normal, but the resulting *dcr* and *dcs* are only used to guide releases from R_1. Then the *dcr* and *dcs* calculations are

they cannot be found by theoretical analysts, they can probably not be consistently found by market participants either.

repeated, with R_2 then R_3, being "primary." Thus, adding a new reservoir increases the computational burden linearly, not exponentially.

This worked well when only the Nordic market was modeled, because the storage levels of neighboring reservoirs had a large impact on each particular reservoir. However, when hydro systems in the Alps (France, Switzerland and Austria) were added, the approach had to be modified, because the storage level for a reservoir in the Alps would have a high influence on the value of water in other reservoirs in the Alps, but almost no impact on reservoirs in the Nordic countries. Thus a matrix of multipliers (which could be zero) was used to determine the weight placed on each reservoir when forming the aggregate complementary reservoir to match each primary.

A second feature of ECON BID is its modeling of inflow forecasting. Clearly, in any given period, hydro producers will have the ability to adjust inflow forecasts from a week to several months ahead, based on current conditions, such as ground water levels and snow pack levels. In the Nordic countries, for example, snow melt accounts for around half the annual inflow, and snow pack levels are measured continuously during winter to provide information to hydro producers. Modeling such forecasting behavior via a simple one-period autoregressive process would add additional dimensions to the *dcs* curve, and not be particularly accurate. Potentially "snow" reservoirs could have been added and linked to the hydro reservoirs in each region. This was done by Hindsberger (2005), but within a SDDP framework rather than CDDP.

Instead, ECON BID recalculates the *dcs* for each inflow series in a given simulation to reflect the accuracy with which the producers can forecast the future inflows and the impact this has on the *dcs*. For each inflow series to be simulated in a given model run, $h = 1, \ldots, H$, inflow forecasting is modeled within BID, as follows. Assume that we have an "expected" probability distribution of inflows Σ for the time horizon under consideration, obtained, for example, from historical inflow series. Then, for each h, and period $t = 1 \ldots T$ in the simulation time horizon:

1. A conditional inflow distribution is calculated from period t forward, given the expected inflow distribution Σ, the inflow series h used in this particular simulation, and the assumed deviation of the snow-pack level from normal at period t of series h. User specified weights determine the rate at which the conditional distribution is assumed to revert to the historical distribution, Σ.
2. *dcs* is then calculated, from T back to t, using this conditional inflow distribution, but only the *dcs* for period t is retained and this *dcs* is only used in simulating sequence h. Thus a model run will calculate $H^*T^2/2$ *dcs* curves, and retain HT curves, as opposed to calculating and retaining T *dcs* curves without inflow forecasting. In practice, the weight placed on each conditional distribution is 0% for most periods, and so most of the calculations to form each *dcs* can be stored and re-use, thereby reducing overhead markedly.

Finally, ECON BID was found to give too little diurnal price variation, compared with historical prices, particularly in hydro dominated regions, even after adjustments to model unit commitment costs and the inflexibility of nuclear plants. Within a given region, there may be hundreds of power plants from many different hydro schemes, each having different capacities, reservoirs, operational features, and relative storage levels in a given period. In general, the storage levels for the reservoirs in a given region will be distributed around an average relative storage level (i.e., how full the aggregate reservoir is in percentage terms). Thus, while the MWV structure may be essentially the same for each reservoir in a region, actual MWVs, in any period, may differ significantly from the "average" MWV from that region's *dcs*, because relative storage levels differ significantly in that period.

Because of the size of the modeled system, ECON BID is not able to model the technical aspects of each plant individually, but it does try to model the economic effect of the technical aspects. Rather than take just one MWV (and thus a single bid) for all the hydro in a region, based on the average relative reservoir level, BID samples several reservoir levels from a (user defined) storage level distribution, and uses the corresponding MWVs to construct a structured bid for the region in each period. This was found to give a better intra-day price profile, although it did require some calibration by the user to match historical price patterns, and may need recalibration as the system changes.

The BID model has been used in several published studies. Hindsberger (2007) uses the model to assess the value of demand response, while von Schemde (2008) investigates the impacts on prices from large-scale wind power developments in Sweden. Finally, Damsgaard et al. (2008) uses the model to examine whether abuse of market power had taken place in the Nordic power market.

Fig. 5 shows the MWV surface over the cause of a year for one reservoir assuming that the complementary reservoir is half full. It can be seen that the shape of the water value curve changes over time. For a particular week, varying the storage level of the complementary reservoir will shift the marginal value of water down, if the complementary storage is more than half full, or up, if it is less than half full.

Fig. 6 from Hindsberger (2007), shows the predicted wholesale power price for western Denmark. The seasonal price pattern shows the influence of hydro dominated systems in Norway and Sweden to the north, as determined by the MWV surface above. But this is overlaid by large diurnal price variations due to the largely thermal-based system of continental Europe, exacerbated, in this case, by high wind generation, which in hours with transmission constraints suppressed the regional price down to a level around Euro40/MWh, even in Winter.

This model is slower than the other two reported here. The example above had a total of 16 regions, but these were aggregated down to three for the hydro optimization phase. The storage grid is rather detailed though, with 14 points in the *dcr* and 40 steps in the *dcs*, for each storage[29] The code has subsequently been made faster, but the version used in this study took about an hour to calculate monthly MWV curves, for 1 year, using five load blocks per month, using an AMD Opteron 280 machine (2.4 GHz, 2 GB RAM) running Windows XP.

[29] Subsequently increased to 22, and 100, respectively.

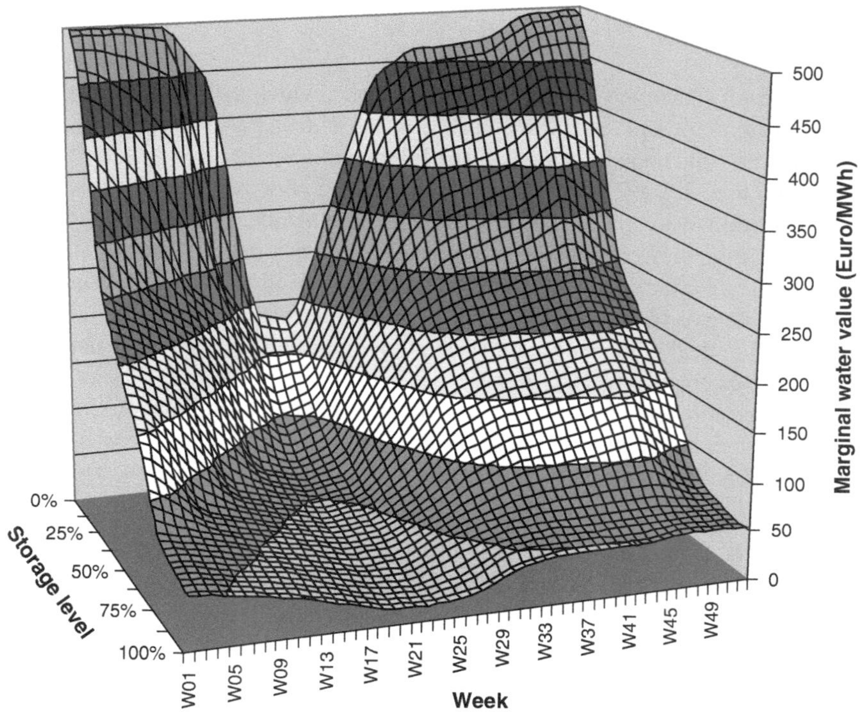

Fig. 5 MWV surface from ECON BID

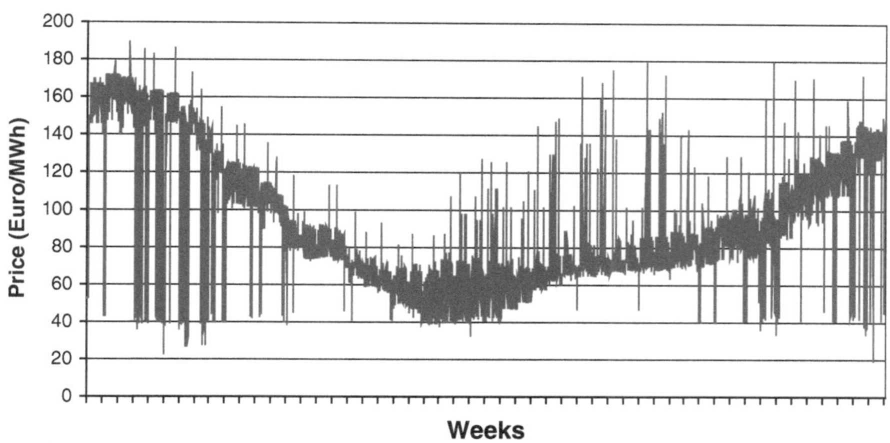

Fig. 6 Power price projection from ECON BID

10 Conclusion

We have briefly surveyed a range of CDDP models, which have seen limited previous exposure in the academic literature. Each has its own particular advantages and disadvantages, and none fully exploits the full potential of the approach. It should be recognized that, since CDDP attempts to produce a comprehensive representation of the multidimensional *mwv* surface, rather than a locally accurate approximation, it does not ultimately escape from the "curse of dimensionality." Thus an LP-based model such as Pereira's SDDP should be preferred for problems with a significant number of reservoirs. That approach also has the major advantage of being readily generalized to deal with changes in system configuration. On the other hand, computation times for SPECTRA suggest that optimization problems of significantly higher dimension could be tackled with quite "acceptable" computation times. And heuristics, such as the ones in DUBLIN and ECON BID, can also be used. In any case, for the problems where CDDP may be said, the following is applicable:

- It is conceptually simple and can be very efficient
- It produces a complete operating policy, covering the entire state space and planning horizon, rather than just for the scenarios considered in the optimization, thus making it easy to perform simulation studies
- It does not assume linearity and can be generalized to model non-linear risk aversion, for example
- It can efficiently accommodate considerable complexity in the intra-period release optimization model, provided this does not imply any increase in state-space dimensionality
- In particular, it can be readily generalized to allow the intra-period optimization to account for factors such as unit breakdowns, or gaming, which are difficult to model in an LP/SP framework.

Acknowledgements The authors thank Gavin Bell and Arndt von Schemde for data on ECON BID, Matthew Civil for data on SPECTRA, and Nick Winter for data on RAGE/DUBLIN.

References

Bannister CH, Kaye RJ (1991) A rapid method for optimization of linear systems with storage. Oper Res 39(2):220–232

Batstone S, Scott T (1998) Long term contracting in a deregulated electricity industry: simulation results from a hydro management model. ORSNZ Proceedings. pp. 147–156

Bell G, Hamarsland GD, Torgersen L (2007) ECON BID 1.1 Manual. ECON Report 2007–011. Available from www.econ.no

Ben-Israel A, Flam SD (1989) Input optimization for infinite horizon discounted programs. J Optim Theory Appl 61(3):347–357

Booth RR (1972) Optimal generation planning considering uncertainty. IEEE Trans Power Apparatus Syst PAS-91(1):70–77

Borenstein S, Bushnell J, Stoft S (2000) The competitive effects of transmission capacity in a deregulated electricity industry. RAND J Econ 31(2):294–325

Boshier JF, Manning GB, Read EG (1983) Scheduling releases from New Zealand's hydro reservoirs. Trans Inst Prof Eng New Zealand, 10 No. 2/EMCh, July, 33–41
Cosseboom PD, Read EG (1987) Dual dynamic programming for coal stockpiling. ORSNZ Proceedings. pp. 15–18
Craddock M, Shaw AD, Graydon B (1999) Risk-averse reservoir management in a de-regulated electricity market. ORSNZ Proceedings. pp. 157–166
Culy J, Willis V, Civil M (1990) Electricity modeling in ECNZ re-visited. ORSNZ Proceedings. pp. 9–14
Damsgaard N, Skrede S, Torgersen L (2008) Exercise of market power in the nordic power market. In: Damsgaard N (ed) Market power in the nordic power market. Swedish Competition Authority
Drouin N, Gautier A, Lamond BF, Lang P (1996) Piecewise affine approximations for the control of a one-reservoir hydroelectric system. Eur J Oper Res 89(1):63–69
Hindsberger M (2005) Modeling a hydrothermal system with hydro and snow reservoirs. J Energy Eng 131(2):98–117
Hindsberger M (2007) The value of demand response in deregulated electricity markets. IAEE Conference Proceedings, Wellington
Iwamoto S (1977) Inverse theorems in dynamic programming. J Math Anal Appl 58:13–134
Kerr AL, Read EG, Kaye RJ (1998) Reservoir management with risk aversion. ORSNZ Proceedings, pp. 167–176
Labadie JW (2004) Optimal operation of multireservoir systems: state-of-the-art review. J Water Resour Plann Manag 130(2):93–111
Lamond BF, Boukhtouta A (1996) Optimizing long-term hydro-power production using markov decision processes. Int Trans Oper Res 3(3–4):223–241
Lamond BF, Monroe SL, Sobel M.J. (1995) A reservoir hydroelectric system: exactly and approximately optimal policies. Eur J Oper Res 81(3):535–542
Little JDC (1955) The use of storage water in a hydro-electric system. Oper Res 3:187–197
Moss F, Segall A (1982) An optimal control approach to dynamic routing in networks. IEEE Trans Automat Contr 27(2):329–339
Pereira MVF (1989) Stochastic operation scheduling of large hydroelectric systems. Electric Power Energy Syst 11(3):161–169
Pereira MVF, Pinto LMG (1991) Multi-stage stochastic optimization applied to energy planning. Math Program 52:359–375
Read EG (1979) Optimal operation of power systems. PhD thesis, University of Canterbury
Read EG (1986) Managing New Zealand's oil stockpile. New Zealand Oper Res 14(1):29–50
Read EG (1989) A dual approach to stochastic dynamic programming for reservoir release scheduling. In: Esogbue AO (ed) Dynamic programming for optimal water resources system management. Prentice Hall, NY, pp. 361–372
Read EG, BoshierJF (1989) Biases in stochastic reservoir scheduling models. In Esogbue AO (ed) Dynamic programming for optimal water resources system management. Prentice Hall, New York, pp. 386–398
Read EG, Culy JG, Halliburton TS, Winter NL (1987) A simulation model for long-term planning of the New Zealand power system. In: Rand GK (ed) Operational research. North Holland, New York, pp. 493–507
Read EG, George JA (1990) Dual dynamic programming for linear production/inventory systems. J Comput Math 19(12):29–42
Read EG, George JA, McGregor AD (1994) Dual dynamic programming with lagged variables. ORSNZ Proceedings. pp. 148–153
Read EG, Stewart P, James R, Chattopadhyay D (2006) Offer construction for generators with intertemporal constraints via markovian dynamic programming and decision analysis. Presented to EPOC Winter Workshop, Auckland. Available from www.mang.canterbury.ac.nz/research/emrg/
Read EG, Yang M (1999) Constructive dual DP for reservoir management with correlation. Water Resour Res 35(7):2247–2257

Scott TJ, Read EG (1996) Modeling hydro reservoir operation in a deregulated electricity sector. Int Trans Oper Res 3(3–4):209–221

Stage S, Larsson Y (1961) Incremental cost of water power. AIEE Transactions (Power Apparatus and Systems), Winter General Meeting

Stewart PA, James RJW, Read EG (2004) Intertemporal considerations for supply offer development in deregulated electricity markets. Presented to the 6th European IAEE conference: Zurich. Available from www.mang.canterbury.ac.nz/research/emrg/

Travers DL, Kaye RJ (1998) Dynamic dispatch by constructive dynamic programming. IEEE Trans Power Syst 13(1):72–78

Velasquez JM (2002) GDDP: generalized dual dynamic programming theory. Ann Oper Res 117:21–31

von Schemde A (2008) Effects of large-scale wind capacities in Sweden. Report 2008–036, ECON Poyry, Oslo, Norway. Available from www.econ.no

Weber C (2005) Uncertainty in the electric power industry – methods and models for decision support. Int. Ser Oper Res Manag Sci, 77

Yakowitz S (1982) Dynamic programming applications in water resources. Water Resour Res 18(4):673–696

Yeh WW-G (1985) Reservoir management and operations models: a state-of-the-art review. Water Resour Res 21(13):1797–1818

Long- and Medium-term Operations Planning and Stochastic Modelling in Hydro-dominated Power Systems Based on Stochastic Dual Dynamic Programming

Anders Gjelsvik, Birger Mo, and Arne Haugstad

Abstract This chapter reviews how stochastic dual dynamic programming (SDDP) has been applied to hydropower scheduling in the Nordic countries. The SDDP method, developed in Brazil, makes it possible to optimize multi-reservoir hydro systems with a detailed representation. Two applications are described: (1) A model intended for the system of a single power company, with the power price as an exogenous stochastic variable. In this case the standard SDDP algorithm has been extended; it is combined with ordinary stochastic dynamic programming. (2) A global model for a large system (possibly many countries) where the power price is an internal (endogenous) variable. The main focus is on (1). The modelling of the stochastic variables is discussed. Setting up proper stochastic models for inflow and price is quite a challenge, especially in the case of (2) above. This is an area where further work would be useful. Long computing time may in some cases be a consideration. In particular, the local model has been used by utilities with good results.

Keywords Energy economics · Hydro scheduling · Stochastic programming

1 Introduction

Finding optimal operational strategies for a large hydrothermal power system with a large fraction of hydropower is a very demanding problem, both theoretically and computationally, since it is stochastic and usually large-scale. One major development in this area is the method of stochastic dual dynamic programming (SDDP) (Pereira 1989; Pereira and Pinto 1991). In this text we shall describe adaptions of this method in a Nordic context.

Norway has about 99% hydropower. In the Nordic countries, Denmark (with no hydropower), Sweden, Finland and Norway have a liberalized common power

A. Gjelsvik (✉)
SINTEF Energy Research, 7465 Trondheim, Norway
e-mail: Anders.Gjelsvik@sintef.no

S. Rebennack et al. (eds.), *Handbook of Power Systems I*, Energy Systems,
DOI 10.1007/978-3-642-02493-1_2,

market, in which hydropower constitutes about 50% of average generation. Because of the high fraction of hydro, market prices may depend very much on the hydrological situation, and taking inflow stochasticity into account is therefore essential. If we consider only a small entity that cannot influence the prices (a price taker), it also becomes necessary to model price stochasticity.

As is well known, it is common to separate the scheduling task in at least three steps: The long-term scheduling, with an horizon of 3–5 years or longer; the medium-term or seasonal scheduling looking 1–2 years ahead; and the short-term scheduling with a horizon of a few days to 1 week. The long- and medium-term scheduling problems are stochastic. The long-term scheduling sets end conditions for the medium-term scheduling, for example in terms of marginal water values, and the medium-term scheduling results are input to the short-term scheduling.

The long-term scheduling problem is frequently approached using some variant of the water value method (Stage and Larsson 1961; Lindqvist 1962), which is based on dynamic programming (see also the dual concept in Scott and Read (1996)). An overview of other models based on stochastic programming is given in Wallace and Fleten (2003). A variety of methods for hydrothermal scheduling (other than SDDP) are also reviewed in Labadie (2004).

In practice, the water value method can only be applied to systems with a very small number of reservoirs. It is therefore often applied to aggregated one-reservoir models of more complicated hydro systems and for simulation purposes supplemented by heuristics. Using SDDP techniques, however, allows stochastic optimization for multi-reservoir systems, which means that more realistic and detailed models can be dealt with.

As examples of application of SDDP and/or related techniques, we mention Tilmant and Kelman (2007), where inflow modelling is also discussed, and Iliadis et al. (2006) and Aouam and Yu (2008). In Philpott and Guan (2008), the convergence of the SDDP-algorithm is discussed, and a theoretical convergence proof is given. In addition to Pereira (1989) and Pereira and Pinto (1991), descriptions of the algorithm can be found in Tilmant and Kelman (2007) and de Oliveira et al. (2002).

In this chapter, we shall deal with two different scheduling models based on SDDP. One is a 'local' model, for a system confined in a geographical area that can be covered by a single power balance equation (without internal transmission bottlenecks), and typically owned by a single power company. It is usually assumed that the system is not large enough to influence market price, so that we have a 'price taker' case. This means that the market price must be dealt with as an exogenous stochastic variable. This model is mainly aimed at medium-term scheduling, but is also used for long-term scheduling.

The other model to be discussed here arises when the SDDP approach is applied to a 'global' system model much similar to the EMPS model (Botnen et al. 1992).

In Sect. 2, we describe elements of a mathematical model of a hydrothermal power system. In Sect. 3, a SDDP-based solution algorithm for the local model is dealt with. To handle the exogenously given price, a combination of SDDP and ordinary stochastic dynamic programming (SDP), originally described in Gjelsvik and Wallace (1996), is used. The method is also described in Gjelsvik et al. (1997), and in more detail in Gjelsvik et al. (1999), and a similar combination of SDDP and SDP was also used in Iliadis et al. (2006). This local scheduling algorithm is the main topic of this chapter. In Sect. 4, we discuss extensions to the basic model, such as an approximate way of handling head variations, and incorporation of risk control.

The global model is briefly outlined in Sect. 6. For this model inflow modelling becomes harder than that in the local case. Although the SDDP approach can deal with many reservoirs, it is not so easy to handle a many-dimensional multivariate inflow process. This is discussed in Sect. 7.

Section 8 deals with some computational issues, and Sect. 9 discusses and sums up some experiences with these models.

2 Basic Power System Model

2.1 Introduction

Most of the material in this section is general for hydropower system modelling; however, price modelling mainly applies to the local model.

The SDDP solution algorithm can be seen as a dynamic programming approach where future costs are represented by hyperplanes (often referred to as cuts), and it relies on linear programming. A fundamental requirement is that the problem must be linear or at least convex. In the model presentation that follows here, we therefore strive to obtain linear or piecewise linear relationships. Fortunately, most relations are close to linear.

It is necessary to use a finite time horizon at time T. For medium-term scheduling, T is usually up to 2–3 years ahead; for long-term scheduling one would use 3–5 years or more. The study period is divided in discrete time steps indexed t, with $t \in [1, \ldots, T]$, usually of length 1 week.

2.2 Power Station Model

Let Q be the release of water through a hydropower plant during a time interval, and let P be the corresponding electrical energy generated. It is assumed that

$$P = f(Q)\frac{h}{h_0} . \tag{1}$$

We take the function $f(Q)$ as piecewise linear, specific for each power station. For the SDDP algorithm, $f(Q)$ must be a concave function. h is the water head and h_0 a nominal reference head.

It is not possible to handle variable head directly in the SDDP algorithm (at least for a cascaded power system). Therefore, the head correction factor h/h_0 in

(1) must be applied with estimated values of h. In many Norwegian power stations this is a fair approximation, since the head often is quite large compared to the head variations. In principle, optimizing with variable head might lead to a non-convex problem, not suitable for SDDP. A study for a single power station is given in Bortolossi et al. (2002). For an example of a non-linear model for variable head, see Mariano et al. (2008).

In Gjelsvik and Haugstad (2005) a heuristic to deal with head variations was described, as used in connection with a hydropower system with cascaded reservoirs. In Sect. 4.1 we shall review this approach.

Generation in thermal power stations is modelled in a simplified manner, as a set of buying options, each with a fixed marginal cost.

2.3 Reservoir and Inflow

Let q_t be the vector of inflows in time step t and V_t the vector of reservoir volumes at the end of t. Further, let Q_t and s_t be vectors of reservoir releases and spills, respectively. With V_0 given, the water reservoir balances can be written as

$$V_t = V_{t-1} - H_1 Q_t - H_2 s_t + q_t, \tag{2}$$

with the condition that

$$\underline{V}_t \le V_t \le \overline{V}_t \tag{3}$$

for $t = 1, \ldots, T$. Here H_1 and H_2 are suitable incidence matrices that describe where releases and spills go, and $\underline{V}_t$ and $\overline{V}_t$ are (possibly time-dependent) limits.

Inflow sequences often show strong sequential correlation, so that q_t depends heavily on q_{t-1}. In stochastic dynamic programming, the stochastic term for the time interval t must depend only on the state at the beginning of the time interval t and not on earlier history. As is well known, this can be arranged by state space enlargement, whereby 'previous' inflows are included in the system state vector.

Historical inflow time series differ in length, but often up to 70 years or more are available. Variations in load etc. are treated as coupled to the inflow; one may speak of 'weather years'. If there are S observed weather years, we construct S parallel inflow scenarios by picking T weeks starting in the first year, then T weeks starting in the second year, and so on. Let q_t^i be the inflow of the ith scenario in week T.

Usually the inflow has strong seasonal variations. One reason for this is the accumulation of snow during winter, with the following spring melt. We try to eliminate the seasonal variations by normalizing computing normalized weekly inflows $\{z_t^i\}$ as

$$z_t^i = \frac{q_t^i - \overline{q_t}}{\sigma_t} \quad \text{for} \quad i = 1, \ldots, S, \quad t = 1, \ldots, T, \tag{4}$$

where $\overline{q_t}$ is the mean inflow in week t averaged over observed years and σ_t is the corresponding sample standard deviation.

A first-order auto-regressive model (AR1) is then used to represent the series $\{z_t^i\}$. Usually there are several series, so that z_t is a vector, whose components have been individually normalized as shown. The model is then

$$z_t = \phi z_{t-1} + \varepsilon_t \quad \text{for} \quad t = 1, \ldots, T. \tag{5}$$

Here ϕ is the transition matrix and the vector ε_t is the model error, or 'noise'. The elements of ϕ and $\text{cov}(\varepsilon_t)$ are estimated by a regression approach, minimizing the sum $\sum_t (z_t - \phi z_{t-1})^T (z_t - \phi z_{t-1})$. It is now assumed that the model error ε_t is independent of z_{t-1}.

The linear AR1 inflow model (5) is not always very good, but it is a compromise between accuracy and computational feasibility. In practice, it has been found that it is best to split the data into different seasons and to fit a separate model for each season.

From the fitting of (5) we have a (usually multivariate) sample distribution of the error term ε_t. For use in the optimization model, we must approximate this distribution by a discrete probability distribution with K discrete values $\varepsilon_t^1, \ldots, \varepsilon_t^K$ and corresponding probabilities ψ_k, where K is not too large.

$$\Pr(\varepsilon_t = \varepsilon_t^k) = \psi_k \quad \text{for} \quad k = 1, \ldots, K \quad \text{and} \quad t = 1, \ldots, T. \tag{6}$$

One way of doing this is to carry out a model reduction by applying principal component analysis (PCA) (Johnson and Wichern 1998) to the sample $\{\varepsilon_t\}$ for $t = 1, \ldots, T$. The principal components are transformed variables constructed so that they are independent, taken over the sample. Only the principal components that contribute most to the total variance are kept (typically 3). After the PCA has been carried out, the distribution of each principal component kept is approximated by a small number of discrete points. Finally the discrete points obtained are transformed back to the axes of the original normalized data and combined.

An example of the use of principal components analysis for inflow modelling, followed by discretization, is da Costa et al. (2006). In Jardim et al. (2001) clustering techniques are used for constructing representative discrete noise terms.

In addition to the above-mentioned method, we have also implemented an approach whereby we obtain the $\{\varepsilon_t^k\}$ by sampling from the collection of sample errors $\{\varepsilon_t\}$ directly.

Inflow modelling will be further dealt with in Sect. 7.

2.4 *Power Balance and Objective Function*

We deal here with the power balance for a 'local' system. It can be generalized by setting up several such balances and introducing transmission variables. Let u_t denote a vector of decision variables for time step t, containing releases Q_t, spills s_t, thermal generation w_t and transactions outside the spot market. Let c_t be the cost vector associated with u_t. We also write $x_t = [V_t^T, z_t^T]^T$ for (the continuous part of) the system model state vector at the end of time step t. Further we define, for $t = 1, \dots, T$,

y_t^+ – Sale to the spot market
y_t^- – Purchase from the spot market
y_t – Vector $y_t^T = [y_t^+, y_t^-]$
p_t – Spot price (weekly average)
δ_t – Transmission charge
d_t – Firm power demand

The power balance is then

$$A_t u_t - y_t^+ + y_t^- = d_t \quad \text{for} \quad t = 1, \dots, T. \tag{7}$$

The cost L_t for one realization in one time step becomes

$$L_t(u_t, y_t, p_t) = c_t^T u_t + (p_t + \delta_t) y_t^- - (p_t - \delta_t) y_t^+. \tag{8}$$

In the power balance (7), the hydro and thermal generations, as well as power transactions outside the spot market, are contained in $A_t u_t$; the matrix A_t contains the power plant conversion factors corresponding to the piecewise linear models in (1).

If a better time resolution in the market description is desired, the hours in a week are grouped into several 'load periods'. Hours with similar prices are lumped together in the same load period, for instance one for night and one for day. Instead of one power balance in (7), there is one for each load period, but the load periods do not follow each other sequentially.

In practice, the transmission cost δ_t is mostly neglected, but we have so far retained it in the model, to make the model more general. We also want to be able to deal with a limited market, and so we introduce limits $\underline{y}_t$ and $\overline{y}_t$ to the y_t vectors; see (19) below. Market limitations may stem from transmission limits, for instance. The model may also be run without a spot market, but with a local load or a local market. In some cases, price elasticity in the spot market may be included by splitting the market into steps with decreasing price for increase in sales. In most cases, though, the market is considered infinite, and so the limits are set at some sufficiently large value and will not be binding.

The firm power demand d_t represents obligations in the local area. However, the firm power demand is zero in the case where all generation is considered sold in the spot market, which is the most common modelling case in the Nordic market. If present, d_t may be considered to vary with the inflow according to the 'weather year'. This causes a difficulty with the SDDP algorithm that we are to use, so that partially we have to use averages; see a remark in Sect. 3.2.

If there is a firm power demand, its influence on the hydro schedules depends on the situation. If market limitations on y_t do not become binding, the hydro schedules will be unaffected by a change in d_t. Otherwise, an increase in d_t will in general lead to higher average storage levels in the reservoirs.

The value of the water remaining in the reservoirs at the horizon must be subtracted from the cost. Let this value be given by a function $\Phi(x_T)$.

Estimating $\Phi(x_T)$ is one of the challenges in the scheduling task. We use results from an aggregated long-term model of the system for this purpose, supplemented by heuristics for distribution between the reservoirs (Johannesen and Flatabø 1989). To a large extent, however, Φ is a function of total storage in the system. Ideally, the horizon T should be set as far away as possible to minimize the influence of errors in $\Phi(x_T)$, but here computing time must also be taken into account.

2.5 Price Modelling

We deal here with the local model, where the spot price is regarded as an exogenous stochastic variable. Price variations can be quite strong, as shown in Fig. 1. Price forecasts can be obtained in several ways. In the Nordic market, forecasts are often obtained by simulations with the so-called EMPS model (Botnen et al. 1992), which is a long-term model covering many areas and several years with the spot price as an internal variable. When using the EMPS model, each price scenario corresponds to a historical weather year.

Time series from such forecasts show that the spot price has a strong sequential correlation. As with the inflow, it is then necessary to include a price state in the system state description, since we intend to use dynamic programming. However, since our objective function (8) contains the product term $p_t y_t$, we cannot expect the future cost functions (see (3.2)) to be convex functions of reservoir and price variables; hence they cannot be represented by cuts, as is necessary when applying SDDP.

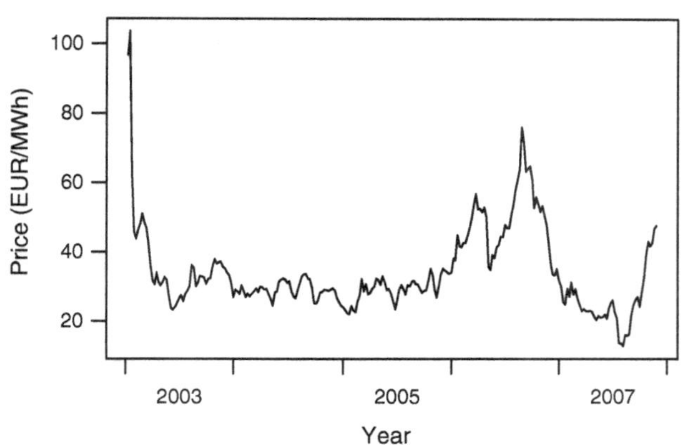

Fig. 1 Weekly average of Nord Pool spot price 2003–2007 (source Nord Pool)

Therefore, while reservoir and inflow states are treated as continuous variables, we combine SDDP with ordinary stochastic dynamic programming with discrete states with regard to p_t, see Sect. 3.1. Thus, p_t is represented by a set of M points $\zeta_t^1, \ldots, \zeta_t^M$.

To establish the probability distribution for price and inflow, we simplify by considering the stochastic processes for inflow and price as independent of each other and using the marginal probability distributions for each.

Broadly speaking, it is reasonable that more water leads to lower prices and the other way round. However, it would be a challenge to include this in the model.

One should also consider that we are dealing with a local part of the total system. To take a Norwegian example: the price will depend on the hydrological state of Norway, Sweden and Finland as a whole, but the system under consideration in the model may be covering only a couple of rivers, where the inflow may not fully follow the trend in the total system.

To describe the transitions between the established discrete price values from 1 week to the next, we use the following Markov model:

$$\Pr(p_t = \zeta_t^j | p_{t-1} = \zeta_{t-1}^i) = \rho_{ij}(t) \quad \text{for all} \quad i, j \in [1, M]. \tag{9}$$

Thus, $\rho_{ij}(t)$ is the probability that $p_t = \zeta_t^j$, given that p_{t-1} was ζ_{t-1}^i.

The numerical values of the transition probabilities are established from a set of price scenarios the following way (Mo et al. 2001b):

First, the price values within each week are grouped in M groups, and ζ_t^i is taken as the mean value of the $N_i(t)$ price values in the ith group in week t. This way, all ζ_t^i are established for all t in the data period. It is recorded which scenarios go into each group in each week, and an estimate $\tilde{\rho}_{ij}$ of ρ_{ij} is then taken as the fraction of the scenarios from the ith group at time $t-1$ that belong to group j at time t. The probabilities obtained this way do not involve the actual values p_t and may not give correct sample conditional means for the price at time t, however. Given that $p_{t-1} = \zeta_{t-1}^i$, $E\{p_t | p_{t-1} = \zeta_{t-1}^i\}$ should be equal to the average price in week t of the scenarios belonging to the ith group at time $t-1$. For this reason, improved values $\{\rho_{ij}\}$ are computed by minimizing a weighted sum of the squared deviations $\{\rho_{ij} - \tilde{\rho}_{ij}\}$ and the squared deviations in the conditionally expected price. One seeks ρ_{ij} so as to find

$$\min \left\{ \sum_{i=1}^{M} \left(\sum_{j=1} \rho_{ij}(t)\zeta_t^j - E\{p_t | p_{t-1} = \zeta_{t-1}^i\} \right)^2 + \omega \sum_{i=1}^{M} \sum_{j=1}^{M} (\rho_{ij}(t) - \tilde{\rho}_{ij}(t))^2 \right\}, \tag{10}$$

subject to the constraints

$$\sum_{j=1}^{M} \rho_{ij}(t) = 1 \quad \text{for all } i \tag{11}$$

$$\sum_{i=1}^{M} \rho_{ij}(t) N_i(t-1) = N_j(t) \quad \text{for all } j \tag{12}$$

$$0 \leq \rho_{ij} \leq 1 \quad \text{for all } i \text{ and } j. \tag{13}$$

Here ω is an appropriate weight factor; the second sum of squares in (10) is considered necessary to obtain a unique solution.

2.6 Overall Local Model

From the elements described, the local model can be summarized as follows. Find an operating strategy that gives u_t from x_t, such that

$$\min E \left\{ \sum_{t=1}^{T} L_t(u_t, y_t, p_t) - \Phi(x_T) \right\}, \tag{14}$$

subject to the constraints

$$x_t = F_t x_{t-1} + G_t u_t + \varepsilon_t \tag{15}$$

$$A_t u_t + B_t y_t = d_t \tag{16}$$

$$\underline{x}_t \leq x_t \leq \overline{x}_t \tag{17}$$

$$\underline{u}_t \leq u_t \leq \overline{u}_t \tag{18}$$

$$\underline{y}_t \leq y_t \leq \overline{y}_t \tag{19}$$

for $= 1, \ldots, T$ and x_0 and p_0 given, and with probability distributions given by (9) and (6). L_t is given by (8).

The expectation E is to be taken over both inflow and price. Equation (15) is the transition equation for the states except the price state, and contains (2) and (5). Equation (16) contains the power balances (7), and it may also be generalized to include other constraints that are not coupled in time. A_t, B_t and F_t, G_t are matrices of suitable dimensions. Reservoir limits, equipment ratings etc. are contained in (17)–(19).

3 Solution Method for the Local Model

3.1 Overview

As already mentioned, the solution method that we have chosen is a combination of stochastic dual dynamic programming and ordinary stochastic dynamic programming (SDP) (Dreyfus 1965). The ordinary SDP part is introduced to take care of the

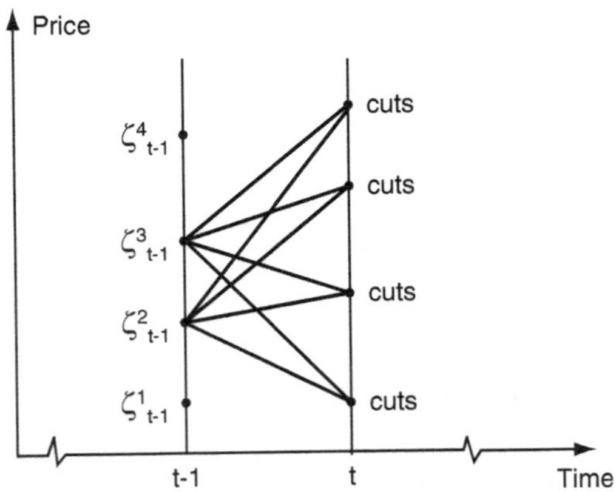

Fig. 2 View of the dynamic programming part of the combined approach, in the time-price plane

price process, which is modelled as described in Sect. 2.5. The reservoir and inflow states are treated as continuous variables and dealt with using hyperplanes, as in the ordinary SDDP algorithm.

A hybrid SDP/SDDP approach was also used in Iliadis et al. (2006).

The approach is visualized in Fig. 2, where the price dimension is shown schematically. In ordinary table-based SDP, there would be a number at each discrete state point, giving the expected future cost going from this state. Here, at each discrete price point there is now instead a set of cuts representing the expected future cost function as a function of the continuous state variables.

As mentioned in Sect. 2.5, the correlation between inflow and price 1 week ahead is neglected.

3.2 Solution by a Dynamic Programming Approach

The solution algorithm for the model established has been described in great detail in Gjelsvik et al. (1999). We outline it here.

As indicated in Fig. 2 we consider a time interval t, with the initial state given by x_{t-1} and p_{t-1}. There are K realizations of the inflow noise $\varepsilon_t = \varepsilon_t^k$, and for each of these M possible price values $p_t = \zeta_t^i$, $i = 1, \ldots, M$. We assume here that we learn ε_t and p_t immediately before the decisions for this time step are to be carried out. One justification for this is that with an interval of 1 week, it is possible to adjust to changing conditions, and a further reason is that usually at a time close to the actual operation more accurate forecasts are available than those assumed in our stochastic models. Let $\alpha_t(x_t|\zeta_t^j)$ be the expected future cost function at the end of time period t, at the system state $[x_t, \zeta_t^j]$. (The expected future cost function is the expected cost in going from the given state at the end of time interval t to an allowed final state using an optimal strategy). Applying the Bellman optimality principle (Dreyfus 1965), we obtain from (14) and (9) and (6) the recursive equation

$$\alpha_{t-1}(x_{t-1}|p_{t-1}=\zeta_{t-1}^i)=\sum_{j=1}^{M}\sum_{k=1}^{K}\rho_{ij}\psi_k \min\Big[L_t(u_t,y_t,\zeta_t^j)+\alpha_t(x_t|p_t=\zeta_t^j)\Big] \tag{20}$$

for all $t\in[1,T]$ and all $i\in[1,M]$. The constraints (15)–(19) must be satisfied for each transition. For each possible outcome $(\varepsilon_t^k,\zeta_t^j)$ separate decisions (u_t^{kj},y_t^{kj}) are made, and the final state obtained is x_t^{kj}.

In Pereira (1989), Pereira and Pinto (1991) and Gjelsvik et al. (1999), it is shown that, with a linear model, the expected future cost functions are piecewise linear functions of x and can be represented by hyperplanes in the x state space, which also means that these functions are convex.

We define $\alpha_{t-1}^{kj}(x_{t-1})=\min[L_t(u_t,y_t,\zeta_t^j)+\alpha_t(x_t|\zeta_t^j)]$ in (20). With the hyperplane representation, (20) then decomposes into single-transition sub-problems of the following form:

With x_{t-1}, $p_{t-1}=\zeta_{t-1}^i$, $p_t=\zeta_t^j$ and $\varepsilon_t=\varepsilon_t^k$ given, find

$$\alpha_{t-1}^{kj}(x_{t-1})=\min\Big[L_t(u_t,y_t,\zeta_t^j)+\alpha\Big], \tag{21}$$

with the constraints (15)–(19) and

$$\left.\begin{array}{c}\alpha+(\mu_t^{j1})^T x_t\geq\gamma_t^{j1}\\ \vdots\\ \alpha+(\mu_t^{jR})^T x_t\geq\gamma_t^{jR}\end{array}\right\}. \tag{22}$$

In (22) $\mu_t^{j1},\ldots,\mu_t^{jR}$ and $\gamma_t^{j1},\ldots,\gamma_t^{jR}$ define R hyperplanes (cuts) that represent the expected future cost function at the price point $p_t=\zeta_t^j$. For an exact representation, an extremely large number of cuts would usually be required. Therefore, an approximate representation is used, where one starts from zero or few hyperplanes and adds them iteratively to get an improved strategy, as in the ordinary SDDP algorithm.

It is assumed that the single-transition sub-problem described in (21)–(22) has a feasible solution; this is ensured by artificial variables. We note that the price of the previous week, p_{t-1}, does not enter the sub-problem. Therefore, in (20) it is not necessary to solve the sub-problem for all M^2 combinations of i and j on the backward run. One solves for the M different $p_t=\zeta_t^j$; then the initial state $p_t=\zeta_t^i$ enters through averaging with the transition probabilities ρ_{ij}.

One main iteration of the algorithm consists of a forward simulation and a backward recursion:

1. *Forward simulation.* The system is simulated from the initial state, with the given scenarios for price and inflow, using the strategy (hyperplanes) obtained so far. For each week t in inflow scenario s, the operation is found by solving (21) with x_{t-1} and p_{t-1} from the final state of the previous week, and ε_t and p_t from the observed values for this scenario. For values p_t that differ from the defined

points ζ_t, linear interpolation in the hyperplanes of the neighbouring points is used. The cost for each scenario is computed, and the average of these costs gives an upper bound for the operating cost. The reservoir trajectories $\{x_t^s\}$ for all t and s are stored.

2. *Backward recursion.* At the horizon $t = T$ the expected future cost function is given by $-\Phi(x_T)$. Consider a general time step, as indicated in Fig. 2, with the expected cost functions given at time t. For each discrete $p_t = \zeta_t^j, \quad j = 1, \ldots, M$ along the price axis, one solves the single transition sub-problems (21) with $x_{t-1} = x_{t-1}^s$ for the trajectories for all $s = 1, \ldots, S$ and all K inflow transitions in time step t. From this, an improved expected future cost function is constructed at each price point ζ_{t-1}^i at the end of week $t-1$ by adding new hyperplanes computed from the dual variables of (21) and (22) and the transition probabilities $\{\rho_{ij}\}$, as described in Gjelsvik et al. (1999). Proceeding step by step backwards from $t = T$ to $t = 1$, one obtains an updated strategy, and a lower bound for the cost. If converged, then stop, otherwise go to 1.

Convergence means that the simulated expected operating cost from step 1 comes 'close' to the lower bound from step 2. In practice, there is usually a gap, so that convergence is mainly judged by monitoring the maximal change in a reservoir trajectory from one main iteration to the next, prescribing a minimum and a maximum iteration number.

In the procedure outlined, we make a modification for the inflow, in that on the forward run we use the 'observed' inflow-price scenarios. This heuristic is intended to take care of any coupling between inflow and price when averaged over longer periods. It may, however, lead to gaps between upper and lower cost estimates, because the fitted inflow and price models used on the backward run (5) and (9) may not be fully consistent with the observed scenarios. Some numerical values are given in Sect. 9.

In the case with firm power $d_t, t = 1, \ldots, T$, that is modelled as inflow-dependent, we use averages of the firm power demand d_t in the backward run of the algorithm to avoid state dependencies. This may also contribute to a cost gap.

Apart from the 'outer' dynamic programming treatment of the spot price state, the approach is similar to that of the ordinary SDDP algorithm, and the same computational approach can be used for this part. To solve the single-transition sub-problem, a relaxation approach is used for the future cost hyperplanes and the reservoir balances, as in Røtting and Gjelsvik (1992). The LP problems actually solved are quite small, see Sect. 8. There is a limit to the number of hyperplanes allowed for each of the M price values; after reaching that, hyperplanes that are infrequently binding are overwritten. This is a crucial part of the algorithm.

Usually an initial set of reservoir trajectories is available at the start of the solution process, so that one can start with the backward recursion step of the algorithm. The inflow loop is put innermost in the calculations, since this only changes the right-hand side of the single-transition problem of minimizing (21) with constraints. Each problem with a new inflow is started from the solution of the previous one using the dual algorithm of linear programming. When the price p_t changes, both the cost row and the set of cuts changes; in this case, an all slack basis is used for start.

4 Extensions to the Local Model

In this section, we describe a few extensions to the medium-term local scheduling model as described in Sects. 2 and 3.

4.1 Head Variations in Medium-term Scheduling

As mentioned in Sect. 2.2, the SDDP algorithm cannot deal directly with variable head, as the problem then may become non-convex. In this section, we outline how variable head can be taken approximately into account in the local model, using a semi-heuristic approach. The method is described more closely in Gjelsvik and Haugstad (2005). It is based on an expansion around a nominal reservoir operating schedule. Release is considered fixed, and sensitivities of economic gain with respect to small changes in reservoir levels are calculated and added to the cost function.

Consider a hydropower system with n reservoirs. For $i = 1, \ldots, n$ we look at the ith reservoir, with a storage of V_t^i at the end of week t and a water surface elevation of h_t^i, referred to sea level, say. We assume that there is a power plant with output P_t^i immediately downstream of reservoir i. If there is a reservoir below this plant, let j be its number, and let k be the number of any upstream plant. In general, we assume that the outlet of a plant in a downstream reservoir is submerged; otherwise, the contribution to head sensitivity is zero.

As before we assume that generated power depends linearly on water head and is obtained from (1)

$$P_t^i = f^i(Q_t^i)\frac{h_t^i - h_t^j}{h_0^i}, \tag{23}$$

where h_0^i is the nominal head for plant i.

We now consider the situation where the volume V_t^i is changed by a small amount ΔV_t^i, without changing the release q_t^i (the change can be thought of as being brought about by a different operation at earlier stages). The influence of ΔV_t^i on generation in the downstream and upstream plants can be shown to be

$$\frac{\partial P_t^i}{\partial V_t^i} = \frac{\partial P_t^i}{\partial h_t^i}\frac{\partial h_t^i}{\partial V_t^i} = \frac{1}{A_t^i}\frac{P_t^i}{h_t^i - h_t^j} \quad \text{and} \quad \frac{\partial P_t^k}{\partial V_t^i} = -\frac{1}{A_t^i}\frac{P_t^k}{h_t^k - h_t^i}, \tag{24}$$

where $A_t^i = (\partial h_t^i / \partial V_t^i)^{-1}$ is the current surface area of reservoir i.

We take the prevailing market price of power, λ_t, as the marginal value of the generation change.

We now assume that we have available nominal reservoir operation schedule with nominal values of $\{P_t^i\}$, $\{\lambda_t\}$, $\{V_t^i\}$, $\{h_t^i\}$ and $\{A_t^i\}$ for all t and i. Using the above formulas, we may then approximately account for the cost change due to variable head by use of extra cost terms containing ΔV:

$$\sum_{t=1}^{T}\sum_{i=1}^{n}\tilde{c}_t^i \Delta V_t^i, \tag{25}$$

where the $\tilde{c}$- coefficients are to be determined. Using (24) we find

$$\tilde{c}_t^i = \frac{\lambda_t}{A_t^i}\left[-\frac{P_t^i}{h_t^i - h_t^j} + \frac{P_t^k}{h_t^k - h_t^i}\right] \quad \text{for all } i \text{ and } t \ . \tag{26}$$

A $\tilde{c}$-coefficient may be positive or negative. As expected an increase in reservoir storage decreases cost associated with the downstream plant (i), but increases cost in the upstream plant (k). As seen from (24), this also depends on the nominal generations P_t^i and P_t^k.

Use of the sensitivities derived above has been implemented in the model described in Sect. 3. In this model, the cost functions must be convex, and all state dependency must be contained in the hyperplanes. It is therefore not possible in general to have different model coefficients for various system states. Therefore, the mean values of the sensitivity coefficients above are used, where the mean is taken over the various inflow scenarios.

As indicated earlier, calculations are carried out in two steps. First, the scheduling program is run without head coefficients. From the releases and reservoir and price trajectories obtained from this run, a full set of head sensitivities $\{\tilde{c}_t^i\}$ is calculated for each inflow scenario. For each week, the sensitivities are averaged over the different inflow scenarios. The mean values of the sensitivities are then used to perturb the cost function according to (25) in a second run. The second run is then generally considered as giving the final schedule. Repeated recalculation of the head sensitivities based on rerunning the program with the last calculated sensitivities is possible, but the sensitivities could in principle oscillate from one calculation to the next. Tests indicate, however, that results may converge after a few repetitions of the recalculation procedure. Furthermore, recalculation does not seem necessary for a good result.

The above perturbation is a kind of heuristic. However, simulations with this correction have given reservoir trajectories that look more realistic than without, and the simulated economic result generally improves.

4.2 Risk Control

The local model has also been extended to allow for dynamic hedging using forward delivery contracts and a mechanism for risk control (Fleten 2000; Mo et al. 2001a).

The basic idea is to consider the accumulated profit. The study period is divided into suitable sub-periods, say quarters, with a profit target for each sub-period. There is a penalty for not meeting the profit target. This penalty function is specified by the user, and it can be shown that this is another form of a utility function that defines the user's risk aversion. Mathematically, the accumulated profit is introduced as a state in the state vector x in (15). The profit state is additive and linear, and so goes with the reservoir equations in the basic model, with a zero inflow.

Forward contracts can be bought or sold for every suitable period within the study period. The net amounts of forward contracts for each future week are also defined as state variables, and buying and selling of such contracts are included in the control vector u_t. This increases the computational burden, but the advantage is that the trading in forward contracts is handled dynamically. A closer description can be found in Mo et al. (2001a).

The study (Iliadis et al. 2006) gives some examples of risk management using an SDP/SDDP approach.

4.3 *Use of the Results from the Local Model*

Results from the local model are used in two ways. First, marginal values or release volumes for the first 1 or 2 weeks are used as input to the short-term scheduling and the daily spot market bidding process. Hyperplanes can be transferred directly to a mathematical model for short-term scheduling if desired (Fosso and Belsnes 2004).

Second, the output from the local model is used for various estimates, such as the hydro generation over the study period, for predicting reservoir levels, in risk analysis and for maintenance scheduling.

5 A Numerical Example

We give here a brief example of application of the local model. The example is taken from Gjelsvik and Haugstad (2005) and is used to illustrate the effect of the correction heuristics for variable head. The system is shown in Fig. 3. The mean annual generation is about 3,270 GWh. There are four plants with significant variations in head (5, 7, 8 and 10). Average market price is 15.4 EUR/MWh in this case. Run with and without head coefficients, the model shows an average increase in income when head coefficients included are of 0.13 EUR/MWh, compared to the schedule without head coefficients. Generation increases by about 1% on average, despite an increase in spilled water.

Inclusion of head coefficients change the strategy for operation of reservoirs considerably. For reservoirs 6 and 7 this is shown in Fig. 4. There is no plant immediately below reservoir 6, and when taking head into consideration, the model transfers more water to reservoir 7 to increase the head.

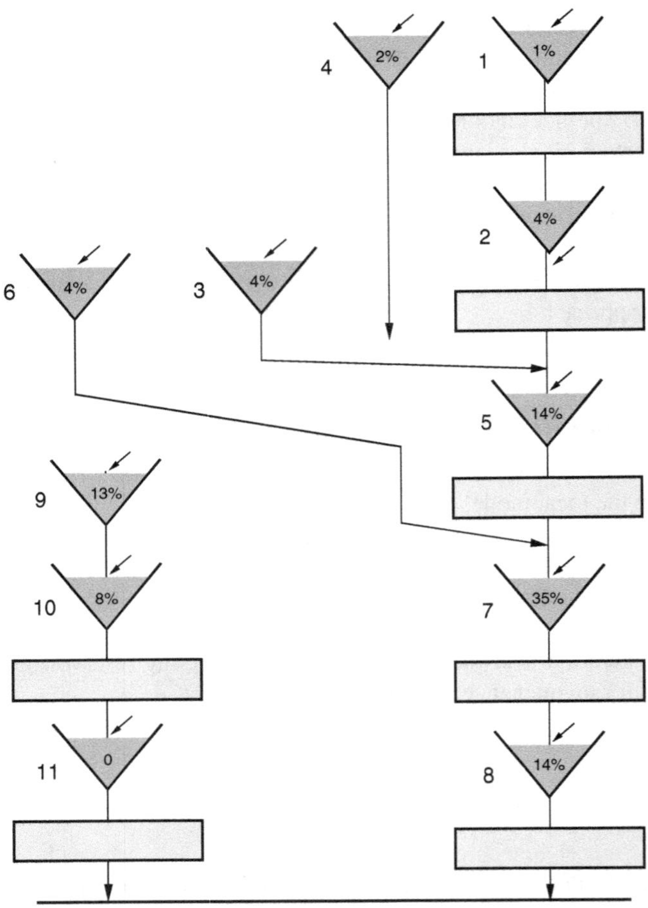

Fig. 3 Case system. *Cones* are reservoirs and *boxes* are power stations. Numbers in reservoir symbols show each reservoir's share of the total reservoir volume. *Boxes* indicate plants. Reservoirs 5, 7, 8 and 10 have a maximum head variation of 45, 70, 17 and 26 m, respectively. Nominal head for all these plants is in the range of 300–500 m

6 A Global Scheduling Model

The SDDP algorithm has also been implemented in an optimization model for a 'global' system, for instance corresponding to the Nord Pool market area and northern Europe. This model is similar to the EMPS system model Botnen et al. (1992), in that the system in each area is lumped together and represented as a single reservoir and a single power station during the optimization phase. An advantage compared to the standard EMPS model is that interconnections between areas are better described in the optimization phase. The model differs from the local model of Sect. 3, in that there are several busbars (one for each area) and that the market price is an internal variable. An external price mechanism has been implemented, though;

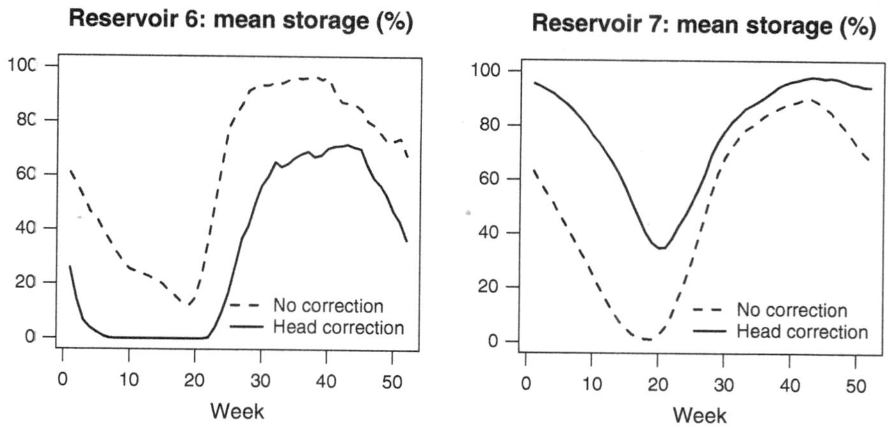

Fig. 4 Mean levels in reservoirs 6 and 7 with and without head correction

this may be useful for describing uncertainties in costs of alternative resources, such as oil or coal.

This model typically has on the order of 20 areas and 40 inflow series. Each area has a power balance similar to (7), with transmission modelled as a transport. The high dimensionality of the inflow process is a difficulty with this model. Since this model is intended to cover a much larger area than the model of Sect. 2, the inflows have more differing characteristics and are not so easily described by a few principal components. This will be further discussed in Sect. 7.

A few extensions have been implemented in the global model, such as internal markets for green certificates (Mo et al. 2005) and CO_2-quotas (Belsnes et al. 2003).

Results from the global model would typically be used for price forecasts and simulation studies from a given initial state. Convergence seems to be slower for this model, and the computing times can be several days on a single processor. Reasons for slower convergence may be that the reservoirs are quite large and less constrained than in the local model, and oscillations in power transfers from one iteration to the next. Also, it is harder to get a good inflow model, as will be discussed in the next section.

7 More on Stochastic Inflow Modelling

Setting up stochastic models for processes involved in stochastic hydropower scheduling usually requires some compromises between accuracy and tractability. In this section, we look at some difficult points of inflow modelling.

In modelling inflow for hydro scheduling, there are several requirements that one should try to meet: The model should give a set of discrete inflow values for each stage, with corresponding probabilities. The model should be as simple as possible to reduce computational burden, but it should also be sufficiently accurate and unbiased. For SDDP applications, the inflow model must preserve convexity.

In Sect. 2.3 we introduced a first order (vector) auto-regressive model (5) for the (normalized) inflow. This is supposed to take care of the sequential correlation. In a geographically dispersed system, it can be difficult to fit a single multivariate model, since different areas may have different inflow characteristics. Fitting an AR1 time series requires that the inflow process is a weakly stationary process, that is, with statistical properties that are independent of time. Even after removal of seasonally varying mean values this is often not the case. The problem can to some extent be avoided by splitting the data into several seasons and fitting the model to each season separately, ensuring proper handling where the seasons join.

If enough data are available, as is frequently the case, it would probably be best to carry out an individual regression analysis for each week.

A difficulty that is sometimes observed is that the residuals for a given week may depend on the initial state. This may happen because a time series is not 'stationary enough', but also when using linear regression directly on a single week. This means that the computations in (21) become biased, since the same distribution of ε_t is used for all initial states. A rather special case is shown in Fig. 5.

The data here come from a 16-area model of Denmark, Sweden, Finland and Norway, intended for use with the global model of Sect. 6. They are for week 24, at the end of the snow melting period. Sample residuals ε_{24} are plotted against normalized inflow values z_{23} at the beginning of the week. Although the distribution for Area 4 is fairly independent of z_{23}, the plot for Area 10 has a conical shape. In this case, it is not correct to use the same distribution of ε_t irrespective of z_{23}; here the model will give too large inflows for higher values of the previous week. One reason for the difficulty is probably that the chosen week 24 is around the average snow melting peak, and so it is reasonable that for the highest inflows there will be a reduction afterwards. It is not clear how this can be dealt with using a linear model.

In the case of several inflow series to the system, one may want to carry out a 'model reduction' to get a noise vector of lower dimensionality. In Sect. 2.3, it was outlined how this can be done by principal components. The success of this procedure varies. In practice, one can retain at most three or four principal components when the total number of noise vectors is to be kept at a reasonable level. (For example, Four principal components represented by 3 points each gives 81 different

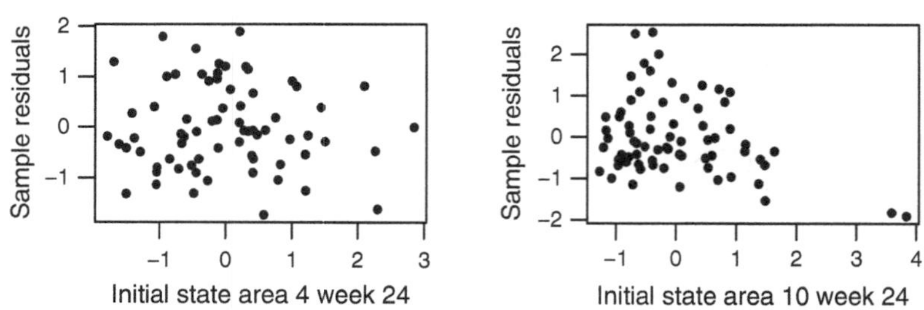

Fig. 5 Residual dependence on initial state. *Left*: Area 4. *Right*: Area 10

inflow cases.) A typical choice can be three principal components, represented by 5, 3 and 2 points, respectively, giving 30 combinations. It turns out that SDDP runs are not always very sensitive to the choice. For a geographically concentrated ('local') system, three principal components may work quite well, as in most cases there is a lot of cross-correlation between the inflows.

For a global model as outlined in Sect. 6, model reduction is more difficult. The inflow series shows less spatial correlation here. For the 16-area case mentioned earlier, it was found that seven principal components were required to cover 90% of the variance. Three principal components cover 75% of the variance. A problem may sometimes arise here, in the case of small areas where most of the variation in the inflow is contained in one or two principal components that are neglected. The inflow variation in such an area may almost disappear. A noise vector with too low variance may give strategies that are too optimistic.

As an alternative in this situation, we have experimented with a direct Monte Carlo approach: We construct the inflow noise vectors by sampling from the sample residuals ε_t available after the fitting of the VAR1-model. Here we avoid systematically cutting off some of the noise space by dropping principal components. On the other hand, it is not clear what a suitable sample size is. One must also check the sampled vectors, so that outliers in the sample of residuals ε_t are not included.

A further difficulty with a linear inflow model such as (5) is that it may generate negative inflows, particularly in weeks where the average inflow is low compared to the standard deviation. We cannot change this to a non-negative value for the scenario in case, since we need consistence to avoid non-convexity. In the SDDP algorithm, this is handled by penalty variables. However, this gives a (further) inconsistency between the forward and backward runs and may slow down convergence.

Negative inflows could be avoided by working with the logarithm of the inflow ($\log q$ instead of q). However, it can be shown that this transformation leads to non-convexity.

In connection with the algorithm description in Sect. 3.2, we mentioned that in the forward SDDP run, we use 'observed' inflow scenarios, which may not be fully consistent with the inflow model used on the backward recursion. This may lead to slower convergence. In a special case, similar to the model of Belsnes et al. (2003), an effect was directly visible in the results. It was with a special version of the global model of Sect. 6, where an internal quota market for CO_2 was modelled as a reservoir. The marginal price of the quota came out with a time variation that was not in line with standard economic theory, due to the inflow inconsistencies. Later simulations of this case showed that if 1,000 consistent inflow scenarios generated from the inflow model (5) were used, the incorrect price behaviour would disappear.

We have not carried out much other direct comparison to study the effect of 'observed' inflows vs. inflows generated by sampling in the stochastic inflow model. In scheduling with the local model, however, we believe that it is most realistic to use observed series of inflow and price, since that helps keep the correct coupling between inflow and price.

Summing up, it is not always easy to find a good inflow representation. Especially in the case of a global model with many inflows that are weakly correlated, the dimensionality of the noise space seems to be a problem. For the local model, the coupling between inflow and price should be further investigated.

8 Computational Issues

The SDDP computations are computer-intensive, and more so with the addition of the price state (Sects. 2 and 3). Depending on the model and the level of detail, the computer time on a Pentium IV computer or similar is in the range half an hour to more than a day. This is the case when weekly time steps are used everywhere. The computations are usually stopped after a given number of main iterations, typically 50–100. The number S of inflow scenarios is typically 50–70, and the number of discrete inflow values at each time step in the backward computations usually is at most 30. The number of hyperplanes stored for each future cost function usually is around 1,000. Solution of the linear programming sub-problems is carried out by general LP software, either a commercial solver or open software.

In a case with 28 reservoirs and 13 power stations, with four load periods, the model for a single transition (a single case of (15) through (19)) has around 460 variables and 33 to some 60 constraints (varying with the number of cuts entered into the active model.) (In this case, relaxation is not used for the reservoir balances but only for the cuts, since most reservoirs are rather small.) The number of simplex iterations may vary from very few (including zero and one), when starting from an advanced basis, up to a few hundreds when an initial basis is not available. One may think that the most usual commercial packages may not be optimal for these problems, since they are primarily intended and tuned for much larger problems. On the other hand, if one wants a finer time resolution within the week, possibly with separate power and reservoir balances for each sub-interval, then the number of rows and columns grows approximately linearly with the number of sub-intervals, and the model for a single transition becomes larger.

Computer time varies much with the size of the system, the length of the study period (T), the number of 'load periods', the number of inflow scenarios and price model levels (M), and the number of discrete inflow values at each stage (K). For the model size mentioned above, with 75 inflow scenarios, 7 price points and $K = 10$ discrete inflow values, the computing time is around 10 h on a single-core Pentium-class processor, using 50 iterations, and solving about 108 million small single-transition LP problems of the kind (21) with the associated constraints. The study period was two and a half year. Smaller and simpler systems can be solved in less than an hour.

Parallel processing has been implemented to reduce computing time. Multicore computers allow this to be easily applied by utilities. The reduction in computation time is almost proportional to the number of cores, but this has not been tested on large-scale parallel computers.

Another way of 'parallel processing' is also used by utilities. The idea is to split a large system with several watercourses into one (sub-)system for each watercourse and optimize each watercourse on a separate computer. This is a perfect decoupling for a price taker with no transmission limits to the market.

9 Discussion

The scheduling algorithm for a local system described in Sect. 3 has been implemented for medium and long-term hydro scheduling. The main advantage of such an algorithm (as with the ordinary SDDP algorithm) is that one can provide stochastic optimization with a detailed model of the hydropower system, so that one obtains reliable incremental water values for each reservoir. The algorithm has been used by power companies for some years and applied to systems with sizes ranging from 4 to over 50 reservoirs.

The global model of Sect. 6 has also been implemented, but is less used.

An obstacle to the use of both implementations is the computer time requirements. Some utilities have started using a parallel version of the local model.

The modelling of the stochastic processes inflow and price seems to be a major area where improvements are wanted. First there is the dimensionality problem. Especially with the global model of Sect. 6, there may be many almost independent inflows to deal with. This requires a relatively high number K of discrete inflow cases to be dealt with in the backward recursion; giving long computing times. An alternative to principal component analysis is sampling from the residuals of the inflow model. A linear regression for the inflow next week may sometimes be problematic, as shown in the example in Sect. 7.

For the fitting of the price model, usually only 50–75 scenarios of 'observed' series are available. This gives rough estimates of the transition probabilities. Some utilities have parametric price models that produce more than one price scenario for each inflow year. The price model can then be generated directly from the parametric model or estimated from a much larger number of price scenarios. As seen in Sect. 2.5, joint modelling of inflow and price remains a challenge.

In computations, we observe gaps between the upper and lower cost bounds. The upper bound, obtained in the forward pass, depends on the inflow scenarios, and is subject to sample variations, so that the 'gap' may even become negative. However, there is usually a positive gap. As mentioned in Sect. 3.2, the inflows used on the forward run are not fully consistent with the inflow model used on the backward run, since we use observed inflows and prices for the forward run. This, combined with the simplifications in price/inflow modelling, probably gives the main contribution to the gap. The size of the gap varies. For small models with a single inflow series, we have sample values in the range 1–2% of the cost, while in larger and more complicated systems with several inflow series, the gap can be in the range 10–15%.

A special problem is that of constructing the final value function $\Phi(x_T)$. If no good estimate for this function is available, the strategy obtained may not be optimal for the last year or so of the study period, and simulation results from this period may be misleading. In our implementation the final value function is based on aggregated water values from a standard SDP model that is distributed to individual reservoirs using heuristics. One should ensure that the study period is long enough.

10 Conclusion

This chapter reviews work on the application of SDDP-based algorithms for hydro scheduling, with some extensions, in the Nordic countries.

It seems that there is room for improvements, particularly in the stochastic models for inflow and price. This is especially the case when there are many independent inflows and in applications to risk management, green certificates or quota modelling. However, the present models work, and in particular the local model gives good results. Parallel processing helps shorten computing time.

Acknowledgements The authors thank International Centre for Hydropower, Trondheim, for permission to use material from Gjelsvik and Haugstad (2005).

References

Aouam T, Yu Z (2008) Multistage Stochastic hydrothermal scheduling. In: 2008 IEEE international conference on electro/information technology, Ames, IA, 18–20 May 2008. IEEE, NY, pp. 66–71

Belsnes MM, Haugstad A, Mo B, Markussen P (2003) Quota modeling in hydrothermal systems. In: 2003 IEEE Bologna PowerTech Proceedings, IEEE, NY

Bortolossi HJ, Pereira MV, Tomei C (2002) Optimal hydrothermal scheduling with variable production coefficient. Math Meth Oper Res 55(1):11–36

Botnen OJ, Johannesen A, Haugstad A, Kroken S, Frøystein O (1992) Modelling of hydropower scheduling in a national/international context. In: Broch E, Lysne DK (eds) Hydropower '92. A.A. Balkema, Rotterdam

da Costa JP, de Oliveira GC, Legey LFL (2006) Reduced scenario tree generation for mid-term hydrothermal operation planning. In: 2006 international conference on probabilistic methods applied to power systems, vols. 1, 2, Stockholm, Sweden, 11–15 Jun 2006. IEEE, NY, pp. 34–40

de Oliveira GC, Granville S, Pereira M (2002) Optimization in electrical power systems. In: Pardalos PM, Resende MGC (eds) Handbook of applied optimization. Oxford University Press, London, pp. 770–807

Dreyfus SE (1965) Dynamic programming and the calculus of variations. Academic Press, New York

Fleten SE (2000) Portfolio management emphasizing electricity market applications: a stochastic programming approach/Stein-Erik Fleten. PhD thesis, Norwegian University of Science and Technology, Faculty of Social Sciences and Technology Management

Fosso OB, Belsnes MM (2004) Short-term hydro scheduling in a liberalized power system. In: 2004 international conference on power systems technology, Powercon 2004, Singapore, IEEE

Gjelsvik A, Haugstad A (2005) Considering head variations in a linear model for optimal hydro scheduling. In: Proceedings, Hydropower '05: The backbone of sustainable energy supply, International centre for hydropower, Trondheim, Norway

Gjelsvik A, Wallace SW (1996) Methods for stochastic medium-term scheduling in hydrodominated power systems. Tech. Rep. A4438, Norwegian Electric Power Research Institute, Trondheim, Norway

Gjelsvik A, Belsnes MM, Håland M (1997) A case of hydro scheduling with a stochastic price model. In: Broch E, Lysne DK, Flatabø N, Helland-Hansen E (eds) Procedings of the 3rd international conference on hydropower, Trondheim/Norway/30 June–2 July 1997. A.A. Balkema, Rotterdam, pp. 211–218

Gjelsvik A, Belsnes MM, Haugstad A (1999) An algorithm for stochastic medium-term hydrothermal scheduling under spot price uncertainty. In: 13th power systems computation conference: Proceedings, vol. 2

Iliadis NA, Pereira MVF, Granville S, Finger M, Haldi PA, Barroso LA (2006) Benchmarking of hydroelectric stochastic risk management models using financial indicators. In: 2006 power engineering society general meeting, vols. 1–9, pp. 4449–4456. General Meeting of the Power-Engineering-Society, Montreal, Canada, 18–22 Jun 2006

Jardim D, Maceira M, Falcao D (2001) Stochastic streamflow model for hydroelectric systems using clustering techniques. In: Power tech proceedings, 2001 IEEE Porto, vol. 3, p 6. doi:10.1109/PTC.2001.964916

Johannesen A, Flatabø N (1989) Scheduling methods in operation planning of a hydro-dominated power production system. Int J Electr Power Energ Syst 11(3):189–199

Johnson RA, Wichern DW (1998) Applied multivariate statistical analysis. Prentice Hall, New Jersey

Labadie J (2004) Optimal operation of multireservoir systems: State-of-the-art review. J Water Resour Plann Manag-ASCE 130(2):93–111

Lindqvist J (1962) Operation of a hydrothermal electric system: a multistage decision process. AIEE Trans III (Power Apparatus and Systems) 81:1–7

Mariano SJPS, Catalao JPS, Mendes VMF, Ferreira LAFM (2008) Optimising power generation efficiency for head-sensitive cascaded reservoirs in a competitive electricity market. Int J Electr Power Energ Syst 30(2):125–133. doi:10.1016/j.ijepes.2007.06.017

Mo B, Gjelsvik A, Grundt A (2001a) Integrated risk management of hydro power scheduling and contract management. IEEE Trans Power Syst 16(2):216–221

Mo B, Gjelsvik A, Grundt A, Kåresen K (2001b) Hydropower operation in a liberalised market with focus on price modelling. In: 2001 Porto power tech proceedings, IEEE, NY

Mo B, Wolfgang O, Gjelsvik A, Bjørke S, Dyrstad K (2005) Simulations and optimization of markets for electricity and el-certificates. In: 15th power systems computation conference: Proceedings, PSCC

Pereira MVF (1989) Optimal stochastic operations scheduling of large hydroelectric systems. Electr Power Energ Syst 11(3):161–169

Pereira MVF, Pinto LMVG (1991) Multi-stage stochastic optimization applied to energy planning. Math Program 52:359–375

Philpott AB, Guan Z (2008) On the convergence of stochastic dual dynamic programming and related methods. Oper Res Lett 36(4):450–455. doi:10.1016/j.orl.2008.01.013

Røtting TA, Gjelsvik A (1992) Stochastic dual dynamic programming for seasonal scheduling in the Norwegian power system. IEEE Trans Power Syst 7(1):273–279

Scott TJ, Read EG (1996) Modelling hydro reservoir operation in a deregulated electricity market. Int Trans Oper Res 3(3–4):243–253. doi:10.1111/j.1475-3995.1996.tb00050.x

Stage S, Larsson Y (1961) Incremental cost of water power. AIEE Trans III (Power Apparatus and Systems) 80:361–365

Tilmant A, Kelman R (2007) A stochastic approach to analyze trade-offs and risks associated with large-scale water resources systems. Water Resour Res 43, w06425. doi:10.1029/2006WR005094

Wallace SW, Fleten SE (2003) Stochastic programming models in energy. In: Ruszczyński A, Shapiro A (eds) Handbooks in operations research, vol. 10. Elsevier, Amsterdam

Dynamic Management of Hydropower-Irrigation Systems

A. Tilmant and Q. Goor

Abstract This chapter compares the performance of static and dynamic management strategies for a water resources system characterized by important hydropower and agricultural sectors. In the dynamic approach, water for crop irrigation is no longer considered as a static asset but is rather allocated so as to maximize the overall benefits taking into account the latest hydrologic conditions and the productivities of other users throughout the basin. The complexity of the decision-making process, which requires the continuous evaluation of numerous trade-offs, calls for the use of integrated hydrologic-economic models. The two water resources allocation problems discussed in this paper are solved using stochastic dual dynamic programming formulations.

Keywords Hydropower · Irrigation · Water transfers · Mathematical programming · Dual Dynamic programming

1 Introduction

There is a growing demand worldwide for energy and water due to population growth, industrialization, urbanization, and raising living standards. In both sectors, the typical response has been to augment supply by constructing more power plants and transmission lines and more hydraulic infrastructures such as dams and pumping stations. But this strategy has reached a limit in many regions throughout the world. In the energy sector, for instance, raising concerns due to global warming and energy security have favored the development and implementation of

A. Tilmant (✉)
UNESCO-IHE, Westvest 7, Delft, the Netherlands
e-mail: a.tilmant@unesco-ihe.org
and
Swiss Federal Institute of Technology, Institute of Environmental Engineering, Wolfgang-Pauli-Strasse 15, 8093 Zurich, Switzerland
e-mail: tilmant@ifu.baug.ethz.ch

S. Rebennack et al. (eds.), *Handbook of Power Systems I*, Energy Systems,
DOI 10.1007/978-3-642-02493-1_3, © Springer-Verlag Berlin Heidelberg 2010

measures promoting efficiency and conservation. In the water sector, a similar trend is observed with the so-called closure of river basins, that is, when available water resources are fully committed. More attention is now given to strategies attempting at increasing the productivity of water through temporary reallocation, either within the same sector or across sectors. This third strategy lends itself to water market and market-like transactions where high value water users would financially compensate low value water users for the right to use their water. Lund and Isreal (1995) review types of water transfers such as permanent transfers, dry-year options, spot markets, water banks, etc.

Most of the studies on water reallocation reported in the literature are focusing on agriculture-to-urban water transfers whereby farmers are financially compensated by industries and/or municipalities for increasing the availability of water through temporary and/or permanent transfers. According to Molle et al. (2007), the rationale for agriculture-to-urban water transfer is essentially economic: the productivity of water in urban uses is generally much higher than in agriculture. Moreover, agriculture can also cope with a larger variation in supply. Examples of agriculture-to-urban water transfers can be found in Booker and Young (1994); Ward et al. (2006); Rosegrant et al. (2000).

To the best of our knowledge, there is no study that analyzes the economic rationale of water transfers from the agricultural to the hydropower sector. The vast body of literature on hydropower scheduling and multipurpose multireservoir operation usually assumes that irrigation water demands are constant quantities that must be met as long as there is enough water in the system (Labadie 2004; Yeh 1985). In the *constraint method*, one of the most common multiobjective methods, irrigation withdrawals, is considered as additional constraints to reflect the priority given to the agricultural sector and the (nearly) constant water demands. This method is used in Tilmant and Kelman (2007) to assess the hydrological risk in the multireservoir system of the Euphrates river basin. The SOCRATES model, developed by Pacific Gas and Electricity, to schedule its power plants also represents consumptive uses such as irrigation as constraints (Jacobs et al. 1995).

As mentioned early, this modeling approach is consistent with the traditional priority given to the agricultural sector by relevant government agencies in charge of water resources allocation. Although this administrative allocation mechanism is likely to remain at the heart of many countries' water policy, it is interesting to analyze the benefit/cost ratio of such a static management approach, especially in the context of high energy prices and river basin closure. As a matter of fact, as fuel prices keep rising and water resources are fully allocated, allocation mechanisms are likely to become more and more scrutinized by a public whose expectation for better performance is likely to increase. Assessing the extent and frequency of agricultural-to-hydropower water transfers should lead to more informed decision on the sharing of an increasingly scarce resource (water). The opportunity cost of those fixed entitlements is also a key indicator, which should be traded-off against equity considerations. Finally, agricultural-to-hydropower water transfers constitute a single policy instrument that can simultaneously improve the efficiency of both sectors (energy and water), and therefore promoting the development and analysis of integrated management approaches.

A dynamic management approach of a hydropower-irrigation system will increase the productivity of water by continuously adjusting allocation decisions (releases and withdrawals) based on the hydrologic status of the system and the productivities of other users in the basin. Such a dynamic management approach is commonly used in the hydropower sector. In a regulated electricity market, for instance, an independent system operator (ISO) produces a dispatch based on a least-cost criterion (also called "merit-order" operation): hydropower plants are dispatched so as to minimize the expected operating costs of the hydrothermal electrical system over a given planning period (e.g., 5 years). This exercise is regularly updated according to the status of the system, which includes the storage levels in the reservoirs and the relevant hydrologic information. In deregulated electricity markets, hydropower companies dynamically manage their assets, which now also include a portfolio of contracts, by generating energy, selling/purchasing energy on the spot market, and selling/purchasing contracts (Scott and Read 1996; Fleten 2000; Barrosso et al. 2002). Again, these decisions are regularly updated as hydrologic conditions, spot prices, and financial position change with time.

This paper compares the performance of a hydropower-irrigation system under a static and a dynamic management approach using two Stochastic Dual Dynamic Programming (SDDP) formulations. In the static approach, net economic returns from hydropower generation are maximized under fixed entitlements to the irrigation areas. These constraints are removed in the second formulation where allocation decisions now include irrigation withdrawals, which are chosen to maximize the net economic returns from both hydropower generation and irrigation. Allocation decisions and net economic indicators are estimated from simulation results and then compared. This paper is organized as follows. Sections 2 and 3 present the SDDP model corresponding to the first and second formulations, respectively. Section 4 describes the hydropower-irrigation system in the Euphrates river in Turkey and Syria. Simulation results are then analyzed in Sect. 5.

2 Stochastic Dual Dynamic Programming

The hydropower operation problem can be mathematically represented as a multi-stage, stochastic, nonlinear optimization problem. In stochastic dynamic programming (SDP), a technique well suited to solve such problems, release decisions r_t are made to maximize current benefits f_t plus the expected benefits from future operation, which are represented by the recursively calculated benefit-to-go function F_{t+1}:

$$F_t(s_t, h_t) = \underset{q_t|h_t}{E} [max\{f_t(s_t, q_t, r_t) + \underset{h_{t+1}|h_t, q_{t+1}}{E} F_{t+1}(s_{t+1}, h_{t+1})\}]. \quad (1)$$

As we can see in (1), the vector of state variables typically includes the storage level s_t at the beginning of time period t and a hydrologic variable h_t representing various hydrologic information and the probabilistic relationship with the inflows q_t. Tejada-Guibert et al. (1995) and Kim and Palmer (1997) discuss the various options for the hydrologic state variable h_t. To solve the hydropower operation problem with SDP, the state variables must be discretized so that an approximate solution of SDP can be developed by evaluating the functional equation at the grid points only. The drawback of this discrete approach comes from the fact that the computational effort W required to solve the SDP model increases exponentially with the number of reservoirs J, making it unsuitable to handle systems involving more than 3–4 reservoirs (Johnson et al. 1993).

SDDP, an extension of the traditional discrete stochastic dynamic programming (SDP), can handle a large state space, that is, a large number of reservoirs, by constructing a piecewise linear function to approximate the benefit-to-go function. This is done through sampling and decomposition; for each sampled value of the state variables, a hyperplane is constructed and provides an outer approximation of the benefit-to-go function. These hyperplanes are equivalent to Benders cuts in Benders decomposition (Kall and Wallace 1994). Intuitively, the computational effort should be reduced since the value of F_{t+1} can now be derived by extrapolation instead of interpolation as in SDP. In other words, a limited number of values for the state variables (points) are now sufficient to provide an approximation of F_{t+1} as illustrated in Fig. 1. The SDDP algorithm was first developed by Pereira (1989) and Pereira and Pinto (1991) to dispatch the power plants of the Brazilian vast hydropower system. This algorithm is now used in several hydro-dominated countries such as Norway and New Zealand (Scott and Read 1996; Mo et al. 2001; Kristiansen 2004). A somewhat similar, though different, technique is the so-called constructive dual DP (CDDP), which essentially solves the dual of the traditional primal DP (Yang and Read 1999; Scott and Read 1996). CDDP differs from the dual DP approach of Pereira and Pinto (1991), in that it defines the whole marginal value surface exactly, while SDDP constructs a locally accurate approximation.

To further save computation time, the SDDP algorithm first starts with a small number of points and then recursively constructs the cuts corresponding to these points starting at the last stage T and moving backward until the first stage t_0. The cuts generated during this backward optimization phase are then evaluated by checking whether the approximation they provide is statistically acceptable or not. This is done by simulating forward the system with (1) the cuts generated in the previous backward optimization phases and (2) several hydrologic sequences, which can be historical and/or synthetically generated. If the approximation is not acceptable, then a new iteration starts and a new backward optimization phase is implemented with a larger sample, which now also contains the points the last simulation went through.

In SDDP, the structure of the one-stage optimization problem (1) is particular: it must be a convex program, such as a linear program (LP), so that the Kuhn–Tucker conditions for optimality are necessary and sufficient. In particular, the gradient of the objective function is equal to the linear combination of the gradients of the binding constraints and their corresponding dual information. Moreover, the hydrologic state variable is typically a vector of p previous flows $q_{t-1}, q_{t-2}, \ldots, q_{t-p}$, which are then used to generate q_t using a build-in periodic autoregressive model PAR(p) with

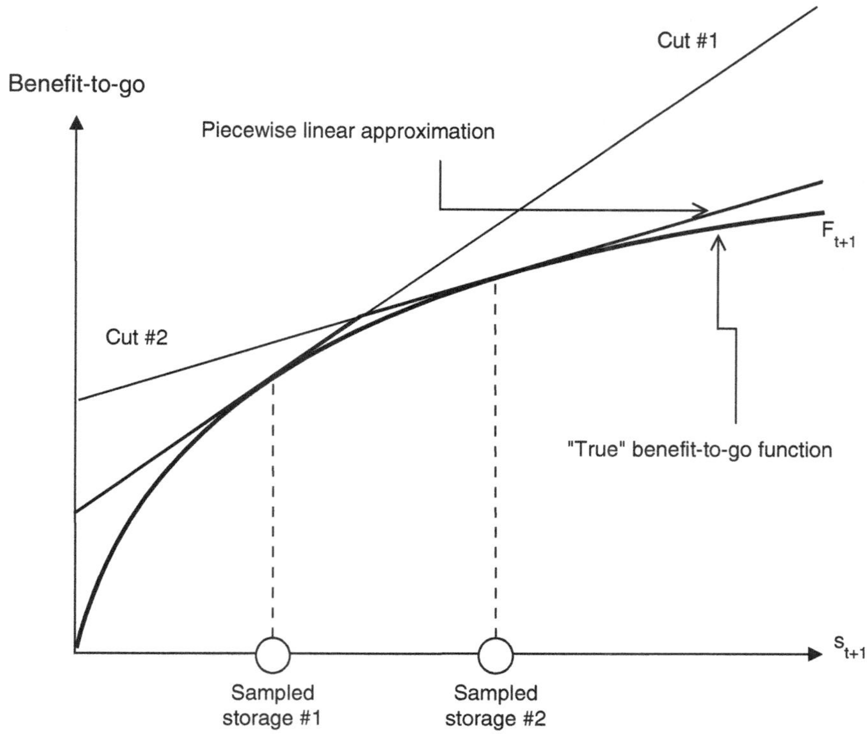

Fig. 1 Piecewise linear approximation of F_{t+1}

cross-correlated residuals. From these restrictions on the mathematical structure of the one-stage optimization problem, it is possible to evaluate the derivatives of the objective function F_{t+1} with respect to the state variables (s_{t+1}, q_t) from the derivatives of the binding constraints and the dual information, which are both available at the optimal solution at stage $t + 1$. Details concerning the analytical determination of the gradients of the functional equation F_{t+1} can be found in Tilmant and Kelman (2007).

When dealing with hydropower systems, the one-stage optimization problem is not convex because the production of hydroelectricity is a nonlinear function of the head (storage) and release variables. One way to remove this source of nonconvexity is to assume that the production of hydro-electricity is dominated by the release term and not by the head (or storage) term (Archibald et al. 1999; Wallace and Fleten 2003). This assumption is valid as long as the difference between the maximum and minimum heads is small compared to the maximum head. If the system is strongly nonlinear, alternative methods such as Neuro-DP can be employed to solve the multireservoir operation problem. In Neuro-DP, the one-stage optimization problem no longer needs to be convex and the benefit-to-go function is approximated by an artificial neural network, which must be trained prior to solving the optimization problem (Castelletti et al. 2006). Another promising option is to use the Q-Learning method of reinforcement learning to derive the optimal operating policies through a forward looking depth-first procedure that alleviates the curse of dimensionality found in discrete SDP (Lee and Labadie 2007). However, since both Neuro-DP and reinforcement learning were implemented on rather small systems (with 3 and 2 reservoirs, respectively), it is still unclear whether they would be able to deal with a larger system, such as the one used to illustrate the SDDP approach.

With this linear assumption, the one-stage optimization problem becomes a LP problem and the decomposition scheme can be implemented. For a system of J price-taking hydropower plants, a typical immediate benefit function (2) can be written as

$$f_t(s_t, r_t, q_t) = \tau_t \sum_j (\pi_t^h(j) - \theta(j))c^h(j)r_t(j) - \xi_t' x_t, \tag{2}$$

where $c^h(j)$ is the production coefficient associated with the jth hydropower plant (MW/m^{3}s^{-1}), $\pi_t^h(j)$ is the short-run marginal cost of the (remainder of) hydrothermal electrical system to which power plant j contributes (US\$/MWh), τ is the number of hours in period t, $\theta(j)$ is the variable O&M cost of power plant j (US\$/MWh), ξ is a vector of penalty coefficients (US\$/unit deficit/surplus), and x_t is a vector of deficits and/or surpluses that must be penalized (e.g., spillage losses, minimum flow requirements).

With the above immediate benefit function and using L cuts to approximate F_{t+1}, the one-stage optimization problem (1) becomes

$$F_t(s_t, q_{t-1}) = max\,\{f_t(s_t, q_t, r_t) + F_{t+1}\} \tag{3}$$

subject to

$$s_{t+1} - C_R(r_t + l_t) = s_t + q_t - e_t(s_t) - i_t, \tag{4}$$

where C_R is the connectivity matrix, i_t is a vector of irrigation water withdrawals, l_t is a vector of spills, and e_t is a vector of evaporation losses, which are assumed to vary linearly with the known initial storage levels. This equation assumes that there is no lagging and attenuation of reservoirs releases because the monthly time step is large enough to ignore travel times between reservoirs. The next constraints specify lower and upper bounds on storages and releases.

$$\underline{s}_{t+1} \le s_{t+1} \le \overline{s}_{t+1} \tag{5}$$

$$\underline{r}_t \le r_t \le \overline{r}_t \tag{6}$$

$$f_t(s_t, r_t, q_t) = \tau_t \sum_j (\pi_t^h(j) - \theta(j))c^h(j)r_t(j) - \xi_t' x_t \tag{7}$$

$$\begin{cases} F_{t+1} - \varphi_{t+1}^1 s_{t+1} \le \gamma_{t+1}^1 q_t + \beta_{t+1}^1 \\ \vdots \\ F_{t+1} - \varphi_{t+1}^L s_{t+1} \le \gamma_{t+1}^L q_t + \beta_{t+1}^L, \end{cases} \tag{8}$$

where φ_{t+1}^{l}, β_{t+1}^{l}, and γ_{t+1}^{l} are the parameters of the expected lth cut approximating F_{t+1}.

3 SDDP Model with Irrigation Benefits

The objective function of the SDDP model (3) does not include the economic benefits associated with irrigation; the previous model treats irrigation water demands as "requirements" or "needs." The inelasticity of irrigation water demands imply that irrigation withdrawals are simply subtracted from the mass balance equation, independently of the status of the system. The Lagrange multipliers λ_w associated with the constraints (4) therefore correspond to the marginal water values reflecting hydropower only, that is, the saved costs due to avoided thermal production. The next section describes an extension of the SDDP model that simultaneously maximizes the net benefits from both hydropower generation and irrigation, and therefore providing optimal allocation decisions, which now include reservoir releases and irrigation withdrawals. The proposed extension allows one to assess the extent and frequency of water transfers in a hydropower-irrigation system and to identify the source (e.g., irrigation demand site) and sink (e.g., downstream power station and/or irrigation demand site).

The net economic benefits from the agricultural sector are estimated from demand functions developed for each irrigation demand site. Gibbons (1986), Young (2005), Ward and Michelsen (2002) review methods for assessing farmer's demand for irrigation water. In Booker and Young (1994), marginal benefit functions for the Colorado River system are obtained from a linear programming model that seeks to maximize irrigator profit for various water supplies and salt discharges. Regression analyses are then used to fit second-order polynomials to the data (the net benefits and the corresponding water withdrawals). In Rosegrant et al. (2000), the profit from agricultural demand sites relies on the residual method and on a nonlinear empirical relationship between crop yield and water application. Quadratic benefit functions are fitted to the results of income-maximizing farm behavior models for various water supplies in Ward and Pulido-Velazquez (2007).

The net benefit from the agricultural sector, denoted $f_{t_B}^{i}$, is the sum of the benefits obtained at each irrigation demand site d as a function of the volume of water $y_{t_B}^{d}$ that has been delivered to that site during the irrigation season, that is, from stage t_A to t_B. Hence, we assume that irrigation benefits depend solely on the total volume of water allocated to the crops and not on the timing of those allocation decisions. However, to avoid unreasonable timing of irrigation supplies, additional constraints on y_t and on irrigation withdrawals i_t are added to the model. Note that these site-specific net benefit functions $f_{t_B}^{i,d}(y_{t_B}^{d})$ must be linear or at least approximated by piecewise linear functions $\hat{f}_{t_B}^{i,d}(y_{t_B}^{d})$ to be compatible with the Benders decomposition scheme in SDDP:

$$f^i_{t_B}(y_{t_B}) = \sum_{d=1}^{D} \hat{f}^{i,d}_{t_B}(y^d_{t_B}), \tag{9}$$

where D is the number of irrigation demand sites. For simplicity we assume one irrigation net benefit function per irrigation demand site, though the model can handle several crops and also includes a relationship between crop yield deficit and irrigation supplies. In that case, each crop is represented by a state variable y_t and is characterized by a specific planting date (t_A), harvest date (t_B), irrigation efficiency (ϵ), farm-gate price, maximum irrigated area, variable costs, and a yield reduction coefficient. In practice, however, the full model has been rarely implemented due to the difficulty in gathering detailed agronomic data, and the increase in computation time when a large number of crops must be considered.

The immediate benefit function $f_t(.)$ can now include up to three terms: (1) net benefits from hydropower generation, (2) penalties for violating operating constraints, and (3) revenues from agricultural products. Note that the third term can only be observed at the end of the growing season, when agricultural products are harvested and sold, that is, at stage t_B. During the rest of the year, the immediate benefit function $f_t(.)$ has only the first two terms: the revenues from the production of hydroelectricity plus the penalties for not meeting operating constraints:

$$f_t(s_t, q_t, r_t, y_t) = \begin{cases} \tau_t \sum_j (\pi^h_t - \theta^h_t) c^h(j) r_t(j) - \xi'_t x_t & \text{if} \quad t \neq t_B \\ \tau_t \sum_j (\pi^h_t - \theta^h_t) c^h(j) r_t(j) & \\ \quad + f^i_t(y_t) & \\ \quad - \xi'_t x_t & \text{if} \quad t = t_B \end{cases} \tag{10}$$

To incorporate the benefits from irrigated agriculture in the objective function of the SDDP algorithm, a new D-dimensional state variable y_t must be included in the state vector. As explained earlier, we assume that y^d_t is the volume of water diverted to the irrigation site d from the beginning of the irrigation season (stage t_A) until the current stage t. In a sense, y^d_t is a "dummy" reservoir of accumulated water for irrigation, which is being refilled during the irrigation season and then depleted at the end of that season (stage t_B) when crops are harvested and sold. In other words, y^d_t is the sum of the crop water requirements over the time interval $[t_A, t_B]$ at site d (Fig. 2).

An irrigation demand site also has its own topology in terms of return flows, that is, the possibility that losses may eventually drain back somewhere downstream, in a reservoir different from the source. Each irrigation demand site, that is, reservoir of accumulated water, is therefore characterized by a net benefit function, a topology, an efficiency, and a percentage of return flows. Denote $\alpha(d)$ as the percentage of irrigation losses for the irrigation demand site d that will drain back to the river system. Let C_R be the connectivity matrix of the reservoir system and C_I be the connectivity matrix of the irrigation system, that is, $C_I(j, d) = \alpha$ when reservoir j receives return flows from the irrigation site d and/or $C_I(j, d) = -1$ when water is diverted from reservoir j to the irrigation site d.

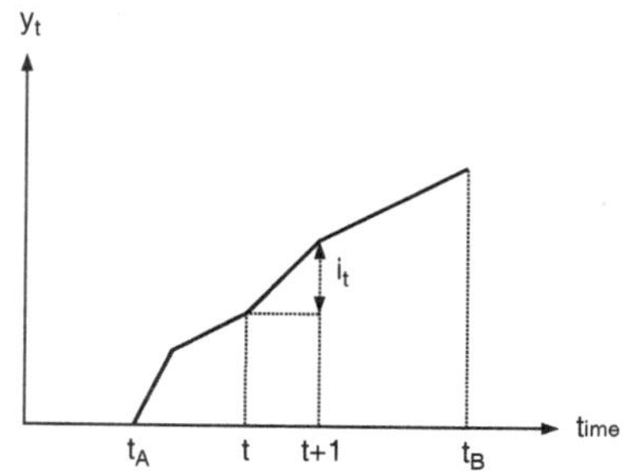

Fig. 2 Reservoirs of accumulated water for irrigation

With the new state variable y_t and the above definitions, the objective function of the one stage SDDP optimization problem becomes

$$F_t(s_t, q_{t-1}, y_t) = max\{f_t(s_t, q_t, r_t, y_t) + F_{t+1}\} \tag{11}$$

subject to the new water balance constraint, which takes into account the topology of irrigation return flows:

$$s_{t+1} - C_R(r_t + l_t) - C_I(i_t) = s_t + q_t - e_t(s_t). \tag{12}$$

Although the lagging of irrigation return flows could in principle be considered in the forward simulation phase of the SDDP algorithm, we chose to ignore it to keep the mathematical formulations of the backward optimization and forward simulation phases identical.

$$\underline{s}_{t+1} \leq s_{t+1} \leq \overline{s}_{t+1} \tag{13}$$
$$\underline{r}_t \leq r_t \leq \overline{r}_t \tag{14}$$

The next constraints provide lower and upper bounds on the monthly volumes of water diverted from the reservoirs to the irrigation areas.

$$\underline{i}_t \leq i_t \leq \overline{i}_t \tag{15}$$

Lower and upper bounds on the accumulated volume of water diverted to the irrigation areas must also be specified:

$$\underline{y}_{t+1} \leq y_{t+1} \leq \overline{y}_{t+1} \tag{16}$$

Equation (17) updates the water balance of the "dummy" reservoir y_t at stage t. We can see that y_t is the net accumulated volume of water since irrigation losses are subtracted (ϵ is the irrigation efficiency).

$$y_{t+1} - \epsilon i_t = y_t \tag{17}$$

Finally, the new Benders cuts approximating F_{t+1} have an additional variable, the vector y_{t+1}, and the associated vector of slopes is denoted η_{t+1}:

$$\begin{cases} F_{t+1} - \varphi_{t+1}^{1} s_{t+1} - \eta_{t+1}^{1} y_{t+1} \leq \gamma_{t+1}^{1} q_t + \beta_{t+1}^{1} \\ \vdots \\ F_{t+1} - \varphi_{t+1}^{L} s_{t+1} - \eta_{t+1}^{L} y_{t+1} \leq \gamma_{t+1}^{L} q_t + \beta_{t+1}^{L}. \end{cases} \tag{18}$$

To determine the values of φ_{t+1}^{l}, β_{t+1}^{l}, η_{t+1}^{l}, and γ_{t+1}^{l}, one must imagine that at stage $t+1$ the triplet $(s_{t+1}^{\circ}, q_t^{\circ}, y_t^{\circ})$ is sampled. From that triplet, K vectors of inflows q_{t+1}^{k} are generated by the following autoregressive model of order one:

$$\begin{aligned} q_{t+1} = \mu_{t+1} + \rho_{t+1,t} \frac{\sigma_{t+1}}{\sigma_t} \left(q_t^{\circ} - \mu_t\right) \\ + \xi_t \sigma_t \sqrt{1 - \rho_{t+1,t}^{2}} \end{aligned} \tag{19}$$

where ρ, μ, and σ are the estimated lag-one autocorrelation, mean, and standard deviation associated with the inflows to the reservoirs. Note that residuals ξ_t are cross-correlated. The water balance constraint can then be calculated

$$\begin{aligned} s_{t+2} - C_R(r_{t+1} + l_{t+1}) - C_I(i_{t+1}) \\ = s_{t+1}^{\circ} + q_{t+1}^{k} - e_{t+1}(s_{t+1}^{\circ}) \end{aligned} \tag{20}$$

and the corresponding dual information is $\lambda_{w,t+1}^{l,k}$, which is a $(J \times 1)$ vector, where J is the number of reservoirs. At that stage $t+1$, the L cuts which constitute the upper bounds to the true future benefits function F_{t+2} are also constraints

$$F_{t+2} - \varphi_{t+2}^{l} s_{t+2} \leq \gamma_{t+2}^{l} q_{t+1} + \beta_{t+2}^{l}, \quad l \in [1, \ldots, L] \tag{21}$$

and the dual information is the $(L \times 1)$ vector $\lambda_{c,t+1}^{l,k}$. The dual information of the optimization problem at stage $t+1$ can be used to derive the vector of slopes φ_{t+1}^{l} with respect to the storage state variable s_{t+1} to approximate the future benefit function F_{t+1} at stage t:

$$\frac{\partial F_{t+1}^{k}}{\partial s_{t+1}} = \lambda_{w,t+1}^{l,k}. \tag{22}$$

Taking the expectation over the K artificially generated flows, we get for the jth element of the slope vector φ_{t+1}^{l}

$$\varphi_{t+1}^{l}(j) = \frac{1}{K} \sum_{k=1}^{K} \lambda_{w,t+1}^{l,k}(j). \tag{23}$$

Similarly, the jth element of the vector of slopes γ_{t+1}^{l} with respect to the hydrologic variable q_{t+1} can be obtained from the dual information associated with the constraints (20) and (21), that is, from the vectors $\lambda_{w,t+1}^{l,k}$ and $\lambda_{c,t+1}^{l,k}$ respectively,

$$\begin{aligned}\frac{\partial F_{t+1}^{k}}{\partial q_t} &= \frac{\partial F_{t+1}^{k}}{\partial q_{t+1}}\frac{\partial q_{t+1}}{\partial q_t}\\ &= \left(\lambda_{w,t+1}^{l,k}(j) + \sum_{l=1}^{L}\lambda_{c,t+1}^{l,k}\gamma_{t+2}^{l}(j)\right) \times \left(\frac{\sigma_{t+1}(j)}{\sigma_t(j)}\rho_{t,t+1}(j)\right)\\ &= \gamma_{t+1}^{l,k}(j)\end{aligned} \tag{24}$$

The expected slope with respect to the inflows can then be determined by

$$\gamma_{t+1}^{l}(j) = \frac{1}{K}\sum_{k=1}^{K}\gamma_{t+1}^{l,k}(j). \tag{25}$$

As for the other gradients, η_{t+1}^{l} is also derived at stage $t+1$ from the Lagrange multipliers $\lambda_{y,t+1}$ associated with the constraints (17):

$$\frac{\partial F_{t+1}^{k}}{\partial y_{t+1}} = \lambda_{y,t+1}^{l,k}(j) \tag{26}$$

The expected slope of the lth cut with respect to y_{t+1} can be calculated

$$\eta_{t+1}^{l}(j) = \frac{1}{K}\sum_{k=1}^{K}\lambda_{y,t+1}^{l,k}(j). \tag{27}$$

Note that jth element of the vector of constant terms becomes

$$\begin{aligned}\beta_{t+1}^{l}(j) = \frac{1}{K}\sum_{k=1}^{K} F_{t+1}^{k} - \varphi_{t+1}^{l}(j)\, s_{t+1}^{\circ}(j)\\ - \gamma_{t+1}^{l}(j)q_t^{\circ}(j) - \eta_{t+1}^{l}(j)y_{t+1}^{\circ}(j),\end{aligned} \tag{28}$$

where y_{t+1}° is the vector of sampled accumulated water for irrigation, s_{t+1}° is the vector of sampled storage volumes at stage $t+1$, and q_t° is a vector of sampled flows at stage t.

As mentioned earlier, the SDDP algorithm is organized around two phases. A backward optimization phase generates the Benders cuts whose approximation power must then be evaluated. This evaluation is carried out at the end of a forward simulation phase. Because the set of Benders cuts (18) provides only an approximation to the multidimensional benefit-to-go functions, simulating the system will give a lower bound $\underline{Z}$ to the solution of the multistage decision making problem. Let M be the number of hydrologic sequences used in simulation, the expected lower bound on the optimal solution is given by

$$\mu_{\underline{Z}} = \frac{1}{M} \sum_{t=1}^{T} f_t^m(s_t, q_t^m, r_t, y_t) = \frac{Z^m}{M}, \tag{29}$$

where $f_t^m(.)$ is the immediate benefit at stage t for the hydrologic sequence $m \in [1, 2, \ldots, M]$. The standard deviation of the estimated lower bound can also be calculated

$$\sigma_{\underline{Z}} = \sqrt{\frac{1}{M-1} \sum_{m=1}^{M} \left(Z^m - \mu_{\underline{Z}}\right)^2}. \tag{30}$$

The 95% confidence interval around the estimated value of $\underline{Z}$ is given by

$$\left[\mu_{\underline{Z}} - 1.96\frac{\sigma_{\underline{Z}}}{\sqrt{M}}, \mu_{\underline{Z}} + 1.96\frac{\sigma_{\underline{Z}}}{\sqrt{M}}\right]. \tag{31}$$

On the other hand, at the end of the backward optimization phase, that is, at stage one, the function $F_1(s_1^\circ, q_0^\circ, y_1^\circ)$ overestimates the benefits of system operation over the planning period when the sampled values for the state variables are s_1°, q_0°, and y_1°:

$$\overline{Z} = F_1(s_1^\circ, q_0^\circ, y_1^\circ). \tag{32}$$

If $\overline{Z}$ is inside the confidence interval, then the approximation is statistically acceptable and the problem is solved. Otherwise, a new iteration is needed: a new backward recursion is implemented and the natural candidates for the sampled values of s_t and y_t are the storage volumes and accumulated volumes the previous simulations pass through. This backward phase is then followed by a new forward simulation, which will exploit the cuts that have been generated during the previous backward recursions.

The SDDP model described in this section is coded in MATLAB and relies on the solver CLP to solve the one stage optimization problems (11)–(18). The next two sections describe the implementation of the SDDP model (11)–(18) on the cascade of multipurpose reservoirs in the Euphrates river (Fig. 3).

4 SDDP Model of the Euphrates River in Turkey and Syria

The Euphrates flows from Turkey to Iraq where it merges with the Tigris River before discharging into the Persian Gulf. Irrigated agriculture has always been an important activity in this region; some of the first hydraulic societies have emerged in the fertile plains delineated by these two rivers. More recently, the headwaters have attracted the attention of water planners, and major irrigation and hydropower

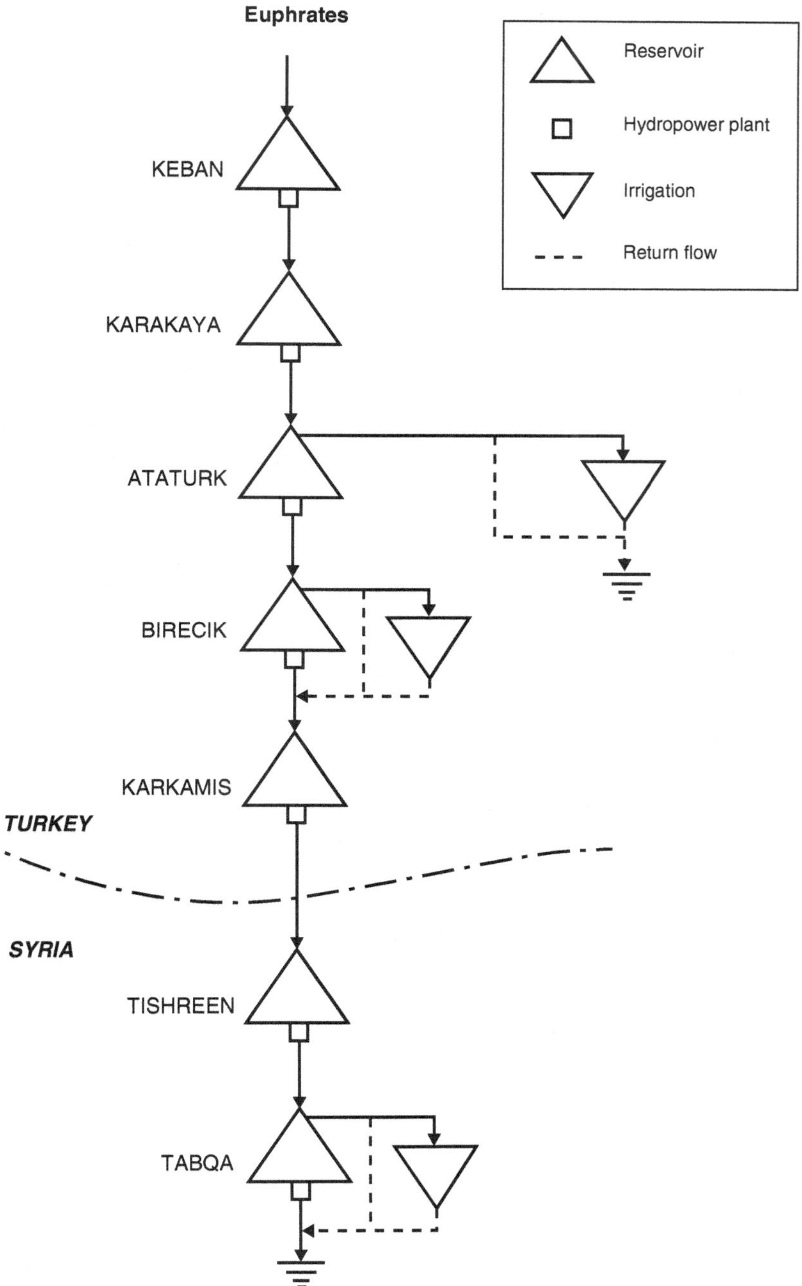

Fig. 3 Schematization of the hydropower-irrigation system

Table 1 Majors GAP dams considered in the SDDP model

Name	Capacity (MW)	Storage capacity (km^3)	Irrigation
Keban	1,240	31.0	
Karakaya	1,800	9.58	
Ataturk	2,400	48.7	Yes
Birecik	672	1.22	Yes
Karkamis	180	0.157	
Tishreen	630	1.88	
Tabqa	880	14.16	Yes

schemes have been developed over the last 30 years. In Turkey alone, the Great Anatolia Project (GAP) is a large-scale project that involves the construction of 22 reservoirs and 19 hydropower stations. In Syria, the Tabqa scheme diverts the Euphrates water to irrigate the left and right banks of the Euphrates while generating 2.3 TWh/a of energy. More details on these two rivers can be found in Beaumont (1996) and Kolars and Mitchell (1994).

The SDDP model of the Euphrates River in Turkey and Syria includes seven reservoirs and their hydropower plants and three irrigation districts supplied by Ataturk, Birecik, and Tabqa reservoirs. Table 1 lists the main characteristics of these infrastructures. See Tilmant et al. (2008) for a complete description of the system together with the data sources and main assumptions.

As explained in the introduction, two SDDP formulations will be developed and their allocation policies compared. The first formulation seeks to maximize the net benefits from hydropower generation by considering irrigation withdrawals as fixed quantities driven by crop water requirements. This first formulation has been described in detail in Sect. 2. The second formulation introduces more flexibility in the allocation mechanism by seeking to maximize both the net benefits from hydropower generation and crop irrigation. Details concerning this second formulation can be found in Sect. 3. In both cases, the planning period covers 60 months. The number of inflow branches (K) and simulation sequences (M) is identical and set to 20. After convergence, the SDDP model provides allocation policies, that is, releases, storage volumes, spills, and irrigation withdrawals as a function of the system status (s_t, q_{t-1}, y_t). Those policies are analyzed in the next section.

5 Analysis of Allocation Policies

Neoclassical economic theory advises allocating water to its most productive uses, thereby maximizing the productivity of the available water. In a system involving consumptive (irrigation) and non-consumptive (hydropower) users, a trade-off must be found at each stage (t) between diverting, releasing, and keeping the water in storage for future uses. The "temporal" trade-off, that is, the balance between immediate and future uses, is achieved when the future and immediate marginal water values are equal. Note that these marginal water values correspond to the Lagrange multipliers associated with the mass balance equations (4) and (12). The release and withdrawal decisions give rise to a "spatial" trade-off; at a particular reservoir j, the equilibrium between withdrawal and release is reached when the lateral productivity is identical to the sum of downstream productivities.

In the system depicted in Fig. 3, the upstream farmers face a coalition of downstream consumptive users and are therefore likely to see their entitlements curtailed, especially during dry years when marginal water values increase throughout the basin. This phenomenon is illustrated in Fig. 4, where we can see the statistical distributions of annual water transfers for the three irrigation districts (Ataturk, Birecik, and Tabqa). As expected, the Ataturk irrigation district suffers the most; on average, 900 hm^3/a of irrigation water is reallocated downstream. Note that 400 hm^3/a will be reallocated 75% of the time, indicating that expanding the irrigation area in the upstream part of the basin is not economically sound and would lead to significant opportunity costs in terms of benefits forgone for the energy sector. The examination of Fig. 4 also reveals that the most downstream irrigation district, Tabqa, is affected by water transfers only 25% of the time.

Those agricultural-to-hydropower water transfers increase the total hydropower generation of the system by 6.4% on average, but reduce the overall irrigated area throughout the basin by 22.2% (Figs. 5 and 6). While the overall energy generation is on average higher, the contribution of hydropower plants to this increase is unequally distributed. Ataturk, with the largest installed capacity, remains the largest

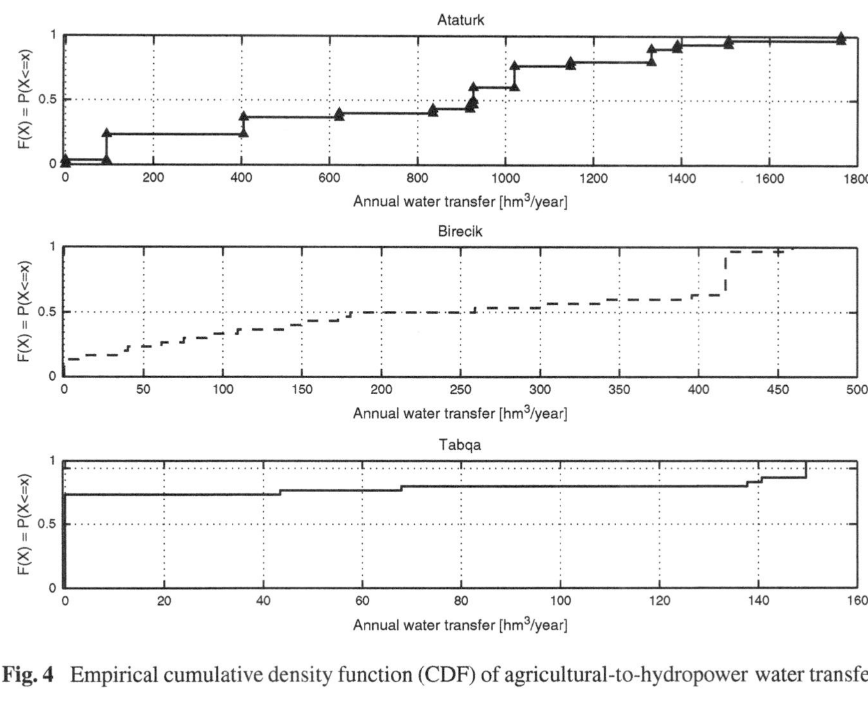

Fig. 4 Empirical cumulative density function (CDF) of agricultural-to-hydropower water transfers

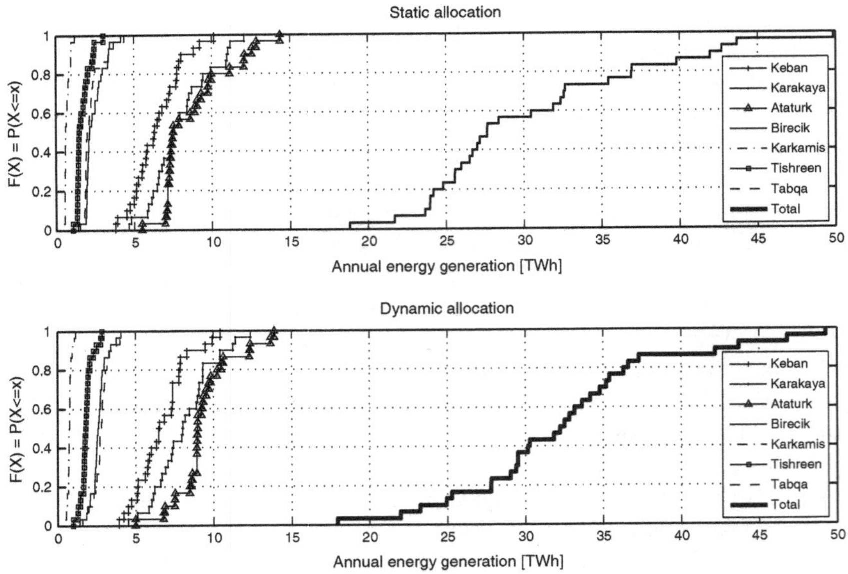

Fig. 5 Empirical cumulative density function (CDF) of annual hydropower generation for static and dynamic water allocation

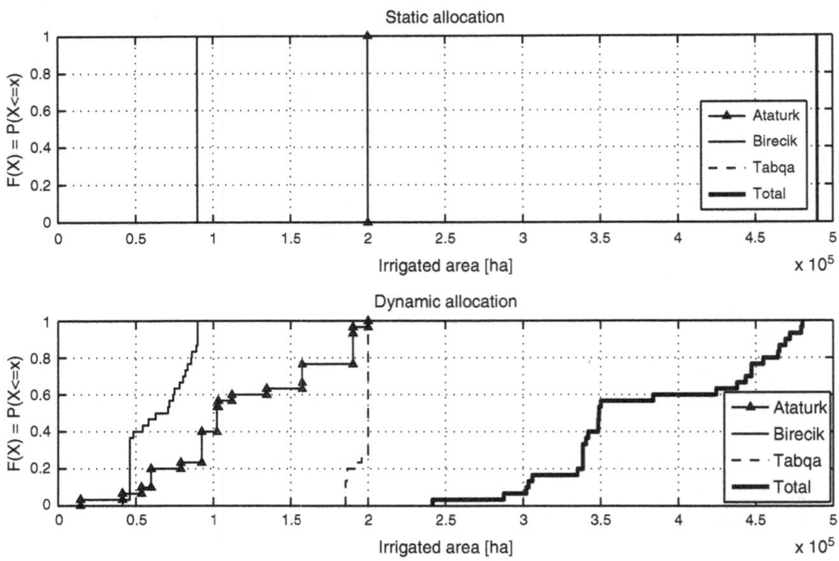

Fig. 6 Empirical cumulative density function (CDF) of irrigated land for static and dynamic water allocation

contributor of the system, with an increase of 6.7% on average (Fig. 5). Tabqa, the latest hydropower plant of the cascade, increases on average its hydropower generation by 23.1% as it benefits from the water transfers from the entire system. As a matter of fact, irrigation is, on average, reduced by about 40% and 27% at the Ataturk and Birecik irrigation demand sites, making water available for hydropower generation downstream.

The observed reallocation of water from the agricultural to hydropower sector increases the total benefits of the system operation by 6% on average. However, this overall increase masks a significant difference between both sectors; the average annual additional benefit for hydropower is around 93.4 million US$ while the agricultural sector loses 29.5 million US$. Such a dynamic allocation process is inherently inequitable unless a proper compensation mechanism is developed, whereby upstream farmers would be compensated for increasing the availability of water downstream. One option would be to share the additional benefits obtained by the hydropower sector among the different irrigation districts based on the volumes of water that have been reallocated. The individual contribution of each power plant to such financial compensation would be proportional to the productivity of that power plant and the additional volume of water released through the turbines.

6 Conclusion

As the competition for water is likely to increase in the near future due to socioeconomic development and population growth, water resources managers will face hard choices when allocating water between competing users. When crop irrigation is involved, water is usually allocated by a system of annual rights to use a fixed, static volume of water, which is typically less than what farmers would expect. Such a static management approach may have significant opportunity costs when large agricultural areas are located in the upper reaches of the river basin. A dynamic approach for water allocation is proposed where irrigation water demand is no longer considered as a static asset but is rather allocated dynamically taking into account the productivities of downstream users, which vary according to the status of the water resources system. The two water resources allocation problems, static and dynamic, are compared using stochastic dual dynamic programming (SDDP) formulations on a cascade of seven reservoirs on the Euphrates river in Turkey and Syria. Moving from static to dynamic water allocation leads to water transfers from the agricultural to the hydropower sector. Hydropower generation increases by 6.4% while irrigated areas are reduced by 22%. The annual net benefit from system operation obtained using the dynamic approach are on average 6% higher than those derived using the static formulation. The higher benefits from hydropower sector could be used to financially compensate the agricultural sector for the loss of revenues resulting from such water transfers.

References

Archibald TW, Buchanan CS, McKinnon KIM, Thomas LC (1999) Nested benders decomposition and dynamic programming for reservoir optimisation. J Oper Res Soc 50:468–479

Barrosso L, Fampa M, Kelman R, Pereira M, Lino P (2002) Market power issues in bid-based hydrothermal dispatch. Ann Oper Res 117:247–270

Beaumont P (1996) Agricultural and environmental changes in the upper euphrates catchment of turkey and syria and their political and economical implications. Appl Geogr 16:137–157

Booker J, Young R (1994) Modeling intrastate and interstate markets for colorado river water resources. J Environ Econ Manage 26:66–87

Castelletti C, de Rigo D, Rizzoli A, Soncini-Sessa R, Weber E (2006) Neuro-dynamic programming for designing water reservoir netwok management policies. Control Eng Pract 15:1031–1038

Fleten SE (2000) Portfolio management emphasizing electricity market applications - A stochastic programming approach. NTNU, Trondheim, Norway

Gibbons D (1986) The Economic Value of Water. Resources for the Future, Washington, DC, USA

Jacobs J, Freeman G, Grygier J, Morton D, Schultz G, Staschus K, Stedinger J (1995) Socrates : A system for scheduling hydroelectric generation under uncertainty. Ann Oper Res 59:99–133

Johnson S, Stedinger J, Shoemaker J, Li C, Tejada-Guibert A (1993) Numerical solution of continuous-state dynamic programs using linear and spline interpolation. Oper Res 41:484–500

Kall P, Wallace S (1994) Stochastic Programming. Wiley, NY, USA

Kim Y, Palmer R (1997) Value of seasonal flow forecasts in bayesian stochastic programming. J Water Resour Plann Manage 123:327–335

Kolars J, Mitchell W (1994) The Euphrates river and the southeast anatolia project. Southern Illinois University Press, Carbondale, USA

Kristiansen T (2004) Financial risk management in the hydropower industry using stochastic optimization. AMO-Adv Model Optim 6:17–24

Labadie JW (2004) Optimal operation of multireservoir systems: State-of-the-art review. J Water Resour Plann Manage 130:93–111

Lee J, Labadie J (2007) Stochastic optimization of multireservoir systems via reinforcement learning. Water Resour Res 43, doi:10.1029/2006WR005627

Lund J, Isreal M (1995) Water transfers in water resources systems. J Water Resour Plan Manage 121:193–204

Mo B, Gjelsvik A, Grundt A (2001) Integrated risk management of hydropower scheduling and contract management. IEEE Trans Power Syst 16:216–221

Molle F, Wester P, Hirsch P, Jensena J, Murray-Rust H, Paranjpye V, Pollard S, van der Zaag P (2007) Water for Food, Water for Life: A Comprehensive Assessment of Water Management in Agriculture, Earthscan and Colombo: International Water Management Institute, London, pp. 585–624

Pereira M (1989) Optimal stochastic operations of large hydroelectric systems. Electr Power Energy Syst 11:161–169

Pereira M, Pinto L (1991) Multi-stage stochastic optimization applied to energy planning. Math Programming 52:359–375

Rosegrant M, Ringler C, McKinney D, Cai X, Keller A, Donoso G (2000) Integrated economic-hydrologic water modeling at the basin scale: The maipo riverbasin. Agric Econ 24:33–46

Scott T, Read E (1996) Modelling hydro reservoir operation in a deregulated electricity market. Int Trans Oper Res 3:243–253

Tejada-Guibert A, Johnson S, Stedinger J (1995) The value of hydrologic information in stochastic dynamic programming models of a multireservoir system. Water Resour Res 31:2571–2579

Tilmant A, Kelman R (2007) A stochastic approach to analyze trade-offs and risks associated with large-scale water resources systems. Water Resour Res 43(W06425):doi:10.1029/2006WR005,094

Tilmant A, Pinte D, Goor Q (2008) Assessing marginal water values in multipurpose multireservoir systems via stochastic programming. Water Resour Res 44(W12431):doi:10.1029/2008WR007,024

Wallace S, Fleten S (2003) Stochastic programming models in energy. In: Ruszczynski A, Shapiro A (eds) Stochastic programming, Handbooks in operations research and management science, vol. 10. North-Holland

Ward F, Michelsen A (2002) The economic value of water in agriculture: concepts and policy applications. Water Policy 4:423–446

Ward F, Pulido-Velazquez M (2007) Efficieny, equity, and sustainability in a water quantity-quality optimization model in the rio grande basin. Ecol Econ 66, doi:10.1016/j.ecolecon.2007.08.018:23–37

Ward F, Booker J, Michelsen A (2006) Integrated economic, hydrologic, and institutional analysis of policy responses to mitigate drought impacts in rio grande. J Water Resour Plann Manage 132:488–501

Yang M, Read E (1999) A constructive dual dynamic programming for a reservoir model with correlation. Water Resour Res 35:2247–2257

Yeh W (1985) Reservoir management and operations models: a state-of-the-art review. Water Resour Res 21:1797–1818

Young R (2005) Determining the Economic Value of Water - Concepts and Methods. Resources of the Future, Washington, USA

Latest Improvements of EDF Mid-term Power Generation Management

Guillaume Dereu and Vincent Grellier

Abstract To optimize mid-term power generation management, Electricité de France (EDF) has developed a set of computer tools, which provide an order of magnitude of the supply and demand balance for the few following years. They also compute hydraulic reservoir management strategies for short-term unit commitment to correctly handle hydraulic reservoirs. This set of tools must be used every day with 500 scenarios and must give results in less than half an hour. Calculation durations are really critical. That is the reason why we had to study in depth the modeling and the optimization methods to reduce calculation durations as much as possible.

In this set of tools, a new simulator has been developed with numerous improvements. Those improvements consist in a more precise generation unit modeling and a better generation unit commitment optimization. Those allow, for instance, to take into account piecewise linear Bellman function to optimize hydraulic reservoir commitment and to use either exact methods of resolution or several heuristics to optimize generation unit commitment. The first kind of heuristics begins with a choice of thermal power stations that are in use at a specific time, then allocates a part of the total production to each power plant. The second type is based on a branch and bound algorithm to solve a mixed integer program (MIP). Moreover, this article compares the performances of several linear solvers on an industrial problem.

Keywords Mid-term · Modeling · Simulation · Unit commitment

1 Introduction

To optimize mid-term power generation management, Electricité de France (EDF) has developed a set of computer tools, which provide an order of magnitude of the supply and demand balance for the few following years. They also compute

V. Grellier (✉)
EDF R&D OSIRIS, 1, avenue de Gaulle 92140 Clamart, France
e-mail: vincent.grellier@edf.fr

S. Rebennack et al. (eds.), *Handbook of Power Systems I*, Energy Systems,
DOI 10.1007/978-3-642-02493-1_4,

hydraulic reservoir management strategies for short-term unit commitment to correctly handle hydraulic reservoirs. The currently used set of mid-term optimization tools has some limitations, hence the decision to develop a new set of tools. This set includes many functional improvements of its simulator module. Those consist in a more precise generation units modeling and a better generation unit commitment optimization.

This article and the methods presented herein are as follows:

- Allow to take into account piecewise linear Bellman function to optimize hydraulic reservoir commitment in the simulator
- Describe several heuristics to optimize the electrical unit commitment in the simulator
- Compare the performances of several linear solvers on an industrial problem.

This set of tools must be used every day with 500 scenarios and must give results in less than half an hour. Calculation duration are really critical. That is the reason why we had to study in depth the modeling and the optimization methods to reduce calculation duration as much as possible.

2 EDF Mid-term Power Generation Management

2.1 EDF Generation Units

EDF has the first generation fleet in Europe with a total of 98.8 GW of installed capacity. The varied range of EDF facilities mixes all forms of energy: nuclear, thermal, hydraulic, and other renewable energies. Most of the electricity generated by EDF in France is produced by nuclear power plants: 87% in 2005. The 58 nuclear power plants, located in 19 different sites, provide a total generating capacity of 63.1 GW. Thermal power made up to 5% of EDF's output in 2005. The thermal power stations (coal, fuel oil, and gas) supply additional power to meet peaks in demand, in particular during cold periods. The largest source of renewable energy are EDF hydroelectric power plants, which made up 8% of EDF output in 2005. Around 500 hydroelectric power plants, linked to 250 reservoirs, are to be found throughout France.

2.2 Mid-term Power Generation Management Purposes

Mid-term power generation management has two main purposes:

- To define an order of magnitude of the supply and demand balance for the few following years for many scenarios of uncertainties (between 500 and 50,000) by calculating indicators of volumes of coal, fuel oil and gas, costs, volumes of shortage, margins (available capacity minus demand), marginal costs (cost of making up one more MW). These indicators are calculated as the mean of all scenarios or as risk indicators (means of the 1 or 5% worst scenarios for instance). Those indicators make it possible to forecast the quantity of fuel oil, coal, and gas to be bought. Shortage volumes, margins, and marginal costs allow to buy forward or future contracts to limit risks of shortage.

- To calculate hydraulic reservoir management strategies for short-term unit commitment. Indeed, short-term unit commitment uses only 2 days of data and needs indicators not to use too much water. Mid-term power generation management computes Bellman functions for each hydraulic reservoir using dynamic programming. Those Bellman functions are used in short-term unit commitment instead of a cost of water.

2.3 Mid-term Power Generation Management Tools

EDF mid-term power generation management set of tools is composed of four main tools as illustrated in Fig. 1:

"Uncertainty generators" create scenarios of each type of uncertainty: demand, generation unit availability, hydraulic inflows, spot market prices, and quantities that can be bought or sold. These scenarios are combined together to allow for correlations between different uncertainties.

The "Global optimizer" calculates Bellman functions for hydraulic reservoirs using dynamic programming. Considering the great number of reservoirs (around 250), it is impossible to calculate Bellman functions for such a high dimension. To solve this problem, hydraulic reservoirs are aggregated in three independent

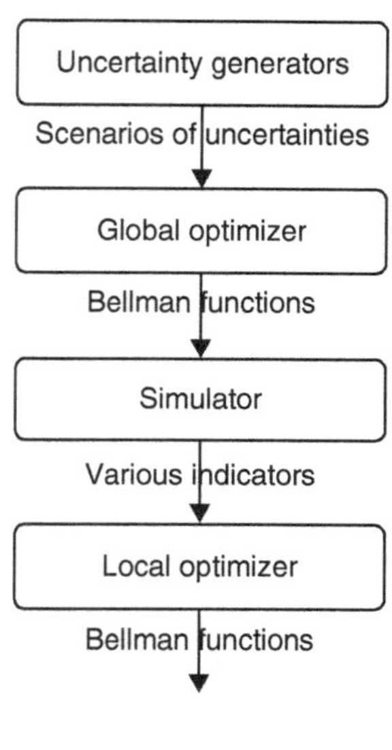

Fig. 1 Mid-term power generation management tools

reservoirs, depending on their characteristics (duration to empty the reservoir, possibility of pumping). Uncertainties are represented by a Markov chain built from uncertainties scenarios.

The "Simulator" simulates the supply and demand balance for each scenario of uncertainties. For each scenario, it minimizes the global cost of generation (time-step by time-step or day by day or week by week) using the Bellman functions calculated by the "Global optimizer." The "Simulator" also uses aggregated hydraulic reservoirs.

The "Local optimizer" calculates Bellman functions for each large hydraulic reservoir in France (around 60). As it cannot calculate Bellman functions in dimension 60, it uses scenarios of marginal costs calculated by the simulator to calculate a Bellman function for each hydraulic system one by one. A hydraulic system can comprise one or several reservoirs in series or parallel.

In the end, mid-term generation management tools give the indicators of the supply and demand balance (thanks to the simulator) and a Bellman function for each large reservoir in France (thanks to the "Local optimizer").

2.4 Mid-term Power Generation Management Tools as an Approximated Dynamic Programming Method

The use of the tools described above enables to compute Bellman functions, although 60 reservoirs are accounted for. It can be seen as an implementation of an approximated dynamic programming method.

This method could be very efficient if the three aggregated reservoirs were equivalent to the 60 ones. It nevertheless remains a very good method to compute Bellman functions in very high dimension.

3 The Simulator

3.1 The Former Simulator and Its Limitations

The former simulator, called COMPAS, minimized the production cost time-step by time-step, using a heuristic. The results obtained were rather good even if they were not wholly optimal.

Indeed, the heuristic mainly chose units according to their costs. Using a more expensive unit, which can provide secondary or tertiary system reserve, can actually be overall cheaper. Moreover, this heuristic could not consider starting costs. Finally, unit commitment was done time-step by time-step, as if the future (4 h later) were totally unknown.

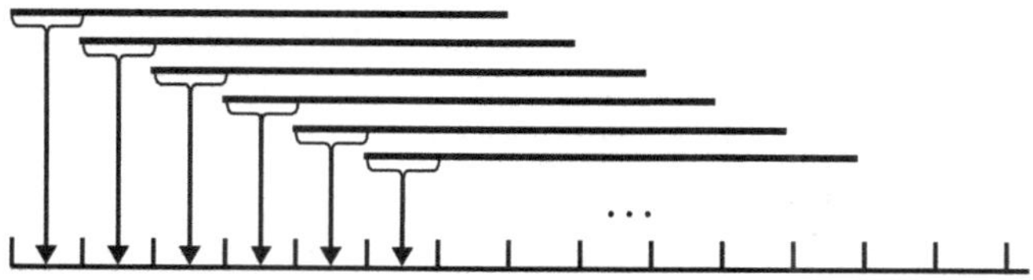

Fig. 2 Optimization on 7 days, slipping of 1 day

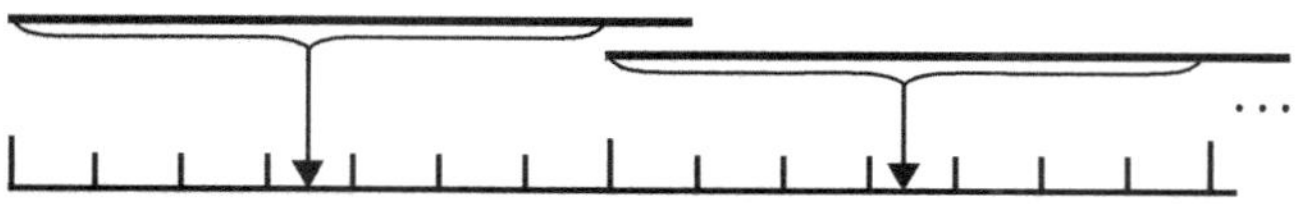

Fig. 3 Optimization on 8 days, slipping of 7 days

3.2 Horizon of Simulation

As for COMPAS, the new simulator simulates the demand and supply balance on a 2–5 year period, with 6 time-steps each day. But instead of doing it time-step by time-step, it can achieve it on a day by day or week by week basis. Actually, weather forecasts are very good on a weekly level; therefore, a demand forecast is possible without too much error on a weekly level. This would lead to the following algorithm: for each scenario, unit commitment is optimized for a week (from day number 1 to day number 7), the first day decisions are the only ones stored, then unit commitment is optimized from day number 2 to day number 8 (Fig. 2).

This method requires as many optimizations as there are simulated days, each optimization dealing with 7 days. Because of calculation duration, the following methods were implemented:

- Just one optimization a week
- Each optimization deals with 8 days.

Optimizing just on 8 days and slipping 7 days (instead of optimizing on 7 days and slipping 7 days) avoids switching too many generation units off at the end of the week (Fig. 3).

3.3 The Simulator Modeling

The following entities and constraints are considered by the simulator:

- *Coupling constraints:* For each time-step, three coupling constraints are allowed for: demand constraint, secondary system reserve constraint, and tertiary system reserve constraint.
- *Conventional thermal power units:* A maximal capacity of generation and possible provisions for secondary and tertiary system reserves are represented. Some

dynamic constraints are permitted: minimum up and down times. Several types of cost are used: running cost (proportional to the production), constant cost (to be paid whenever the unit is working), and starting cost (to be paid whenever the unit starts). Availability is different for each scenario.

- *Nuclear thermal power units:* Nuclear thermal power units are represented as conventional thermal power units with some extra features. Between two consecutive stops for uranium refilling, nuclear fuel must be managed as a limited stock.
- *Hydraulic units:* As described earlier, hydraulic reservoirs are aggregated in three reservoirs.
- *Peak day option (EJP or Effacement Jour de Pointe):* Some clients choose a special tariff called EJP. They usually have a lower price than other clients in exchange for a more expensive power price 22 days a winter. During these 22 days, clients who chose the EJP fare will reduce their consumption of power. EDF is to choose the days the price is more expensive. EDF has a stock of 22 days that must be placed at the best moments. To optimize this stock, the simulator uses a Bellman function calculated by the optimizer.
- *Market:* A spot market is considered. It represents the amounts EDF can buy or sell a day ahead.

Compared to COMPAS, modeling improvements mainly concern the following items:

- Representation of some dynamic constraints and starting costs for thermal units
- Unit commitment management foreseeing the next week and not just one time-step.

Those two points lead to a modeling closer to a short-term one, even if many differences remain.

3.4 MIP Resolution

3.4.1 Modeling

The problem described above can easily be formulated as a mixed integer linear programming (MIP). Binary variables are used to model the on/off state of generation units and the fact that they start. All other variables are continuous.

The writing of the MIP was rather natural except for two items:

- *Minimum up and down times:*
  Some notations:
  $\beta(c,t)$: binary variable giving the state of the generation unit c at time-step t (0: off, 1: on)
  $\delta(c,t)$: binary variable indicating that the unit switches on between time-step t-1 and t

If a generation power unit must work at least N time-steps, this constraint can be written like this: If the unit switches on between time-step t-1 and t ($\delta(c,t) = 1$), it must remain on for time-steps t to t+N-1 ($\beta(c,t) = \beta(c,t+1) = \cdots = \beta(c,t+N-1) = 1$). This can be formulated in a linear way:

$$\forall t, \forall t' \in \{t, t+1, \ldots, t+N-1\}, \beta(c,t') \geq \delta(c,t).$$

This expression requires a large number of constraints and leads to a huge calculation duration. That is why the next formulation was preferred:

$$\forall t', \beta(c,t') \geq \sum_{t=t'-N+1}^{t'} \delta(c,t).$$

This constraint can be seen as, "before a time-step where a unit is off, there cannot be any start for a minimal duration of work." This formulation reduces the number of constraints and adds a cut for the MILP. The same type of constraints can be written for minimal duration of stop constraint:

$$\forall t', \beta(c,t') + \sum_{t=t'+1}^{t'+N} \delta(c,t) \leq 1.$$

This constraint can be seen as, "after a time-step where a unit is on, there cannot be any start for a minimal duration of stop."

- *Taking into account the Bellman function:*
As seen in the Sect. 2.3, the "Global Optimize" calculates Bellman functions for three independent aggregated reservoirs. For each reservoir and each time step, the Bellman function depends on the level of the stock and on the level of complementary stock. The complementary stock, that is only virtual, is obtained by summing the capacity of stock and the maximum capacity of generation of all other stocks. This approach is very efficient on our problem and was introduced by Turgeon in 1980. So, for each reservoir, we consider Bellman functions with two dimensions (main stock and complementary stock). Moreover, if we suppose that the level of complementary stock is almost constant during a time-step, for each aggregated reservoir, Bellman functions calculated by the Global optimiser are piecewise linear. This assumption will be discussed later in this section. The Bellman functions correspond to a benefit and are therefore concave (Fig. 4).
In a simulator, the mathematical program for time-step t should schematically be the following (bounds and details are not indicated):

$$min \left(\sum_{c=1}^{C} \Delta(t).cost(c).p(c,t) - \sum_{r=1}^{R} B_r(x(r,t)) \right)$$

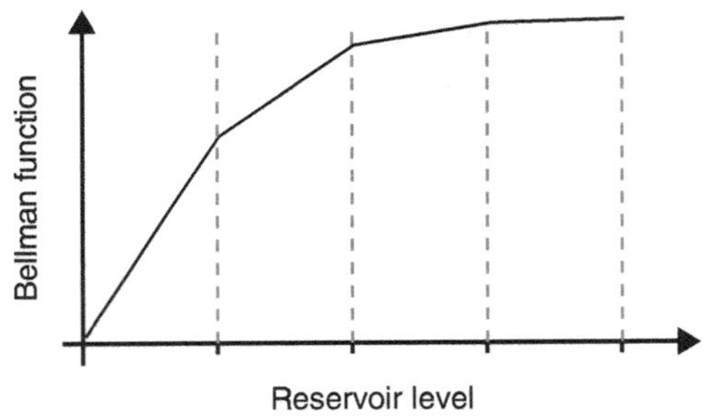

Fig. 4 Piecewise linear Bellman function

subject to:

$$\begin{cases} \sum_{c=1}^{C} p(c,t) + \sum_{r=1}^{R} p(r,t) = d(t) \\ \forall r, x(r,t) = x(r,t-1) + \Delta(t).(a(r,t) - p(r,t)) \end{cases}$$

where
$\Delta(t)$: duration of time-step t (in hours)
$d(t)$: demand to satisfy at time-step t (in MW)
$cost(c)$: cost of generation for the thermal power unit c (in /MWh)
$B_r()$: Bellman function of the aggregated reservoir r (in)
$a(r,t)$: inflows for the aggregated reservoir r during time-step t (in MW)
$x(r,t-1)$: level of the aggregated reservoir r at the beginning of time-step t (in MWh)
$x_{max}(r)$: maximum level of the aggregated reservoir r (in MWh)

Variables:
$p(c,t)$: production of the thermal unit c at time-step t (in MW)
$p(r,t)$: production of the aggregated reservoir r at time-step t (in MW)
$x(r,t)$: level of the aggregated reservoir r at the end of time-step t (in MWh)

The objective function is equivalent to the next one:

$$min\left(\sum_{c=1}^{C} \Delta(t).cost(c).p(c,t) + \sum_{r=1}^{R} B'_r(x(r,t)).p(r,t)\right),$$

where B'_r is the derivate of B_r with respect to the level of the reservoir.

In COMPAS (the former simulator), this objective function was simplified assuming that x(r,t) is not very different from x(r,t-1), which is the level of stock at the beginning of time-step t, and also that level of complementary stock is not different between t-1 and t. With these assumptions, we were able to deduce in which linear piece of the Bellman function x(r,t) will probably be. Then, we were able to replace the previous problem with the next one:

$$min\left(\sum_{c=1}^{C}\Delta(t).cost(c).p(c,t)+\sum_{r=1}^{R}B'_r(x(r,t-1)).p(r,t)\right)$$

subject to:

$$\begin{cases}\sum_{c=1}^{C}p(c,t)+\sum_{r=1}^{R}p(r,t)=d(t)\\ \forall r, x(r,t)=x(r,t-1)+\Delta(t).(a(r,t)-p(r,t))\end{cases}$$

It is obvious that this problem is linear (as $B'_r(x(r,t-1))$ is known when solving the problem at time-step t).
At first, we imagined that the way COMPAS takes Bellman functions into account could be used in the new simulator. The new simulator optimizes several time-steps together. With COMPAS' formulation, the problem between time-steps t_0 and t_1 would be written like this:

$$min\sum_{t=t_0}^{t_1}\left(\sum_{c=1}^{C}\Delta(t).cost(c).p(c,t)+\sum_{r=1}^{R}B'_r(x(r,t_0-1)).p(r,t)\right)$$

subject to:

$$\begin{cases}\forall t\in\{t_0,\ldots,t_1\}, \sum_{c=1}^{C}p(c,t)+\sum_{r=1}^{R}p(r,t)=d(t)\\ \forall t\in\{t_0,\ldots,t_1\}, \forall r, x(r,t)=x(r,t-1)+\Delta(t).(a(r,t)-p(r,t))\end{cases}$$

But this formulation supposes that the level of stock is almost constant between t0 and t1. This is actually false for some small reservoirs that can be emptied or filled in 2 or 3 days. Besides, the complementary stock is always big enough to admit the fact that it is almost constant between t0 and t1 because the involved duration is less than a week. That is why we came back to the original formulation, using the Bellman function and not its derivative. Thanks to the concavity of the Bellman function, we managed to write this mathematical problem as a linear problem without binary or integer variables:

$$min\sum_{t=t_0}^{t_1}\left(\sum_{c=1}^{C}\Delta(t).cost(c).p(c,t)+\sum_{r=1}^{R}\sum_{s=1}^{S}U(r,t_1,s).xb(r,t_1,s)\right)$$

subject to:

$$\begin{cases}\forall r, \sum_{s=1}^{S}xb(r,t_1,s)=x(r,t_1)\\ \forall r, \forall s, 0\le xb(r,t_1,s)\le x_{max}(r)\\ \forall t\in\{t_0,\ldots,t_1\}, \sum_{c=1}^{C}p(c,t)+\sum_{r=1}^{R}p(r,t)=d(t)\\ \forall t\in\{t_0,\ldots,t_1\}, \forall r, x(r,t)=x(r,t-1)+\Delta(t).(a(r,t)-p(r,t))\end{cases}$$

where $U(r, t_1, s)$ is the slope of the Bellman function sth linear piece at time-step t_1. S is the number of linear pieces for the Bellman function of the reservoir r at time step t_1.

These formulations are not new but allowed EDF to take into account minimum up and down times and piecewise linear Bellman functions in a very effective way in a tool used every day with a strong constraint on calculation duration.

3.4.2 Comparison Between Solvers

We have compared three linear solvers to solve the MIP formulation of the problem: Cplex 10, Xpress-MP 2006B.2, and COIN-OR on Sun UltraSparc III and IV. The tolerance on optimality (mipgap) was set to 10^{-3} (i.e., 0.1%). The set of problems solved has around 8,500 instances. On these problems Cplex was faster than the two other solvers, by 9% on ultraSparc III and by 18% on ultraSparc IV against Xpress-MP. The performance of COIN-OR was very far from the other two, and it was close to 60 times slower.

3.4.3 Some Results

Figure 5 shows how EDF mixes all forms of energy during a winter week (from Sunday to Saturday) to meet the clients' requirements under the best possible economic. Nuclear units produced as much as they could and even more than the demand to fill water reservoirs during the night. The thermal power stations (coal, fuel oil, and gas) and hydroelectric power plants supply additional power to meet peaks in demand. In some cases, EDF also buys or sells electricity on day-ahead market (like Powernext).

The curve below illustrates a typical use of weekly reservoir management. We can clearly see the intensive pumping during the weekend and the night and the use of water to produce electricity during the peak of each workday (Fig. 6).

The curves below show the impact of taking starting cost into account. Here are the uses of a particular coal power plant with (black one) and without starting costs (grey one). In the first graph, the unit is not stopped at night to avoid paying the starting cost again. In the second graph, the plant is not started because of the starting cost.

These illustrate the fact that a more precise model does have impact on some results (Fig. 7).

3.5 *Heuristics*

We have tried two types of heuristics. The first one (H1) is based on COMPAS' heuristic (called H0). H0 begins with a heuristic choice of the thermal power stations

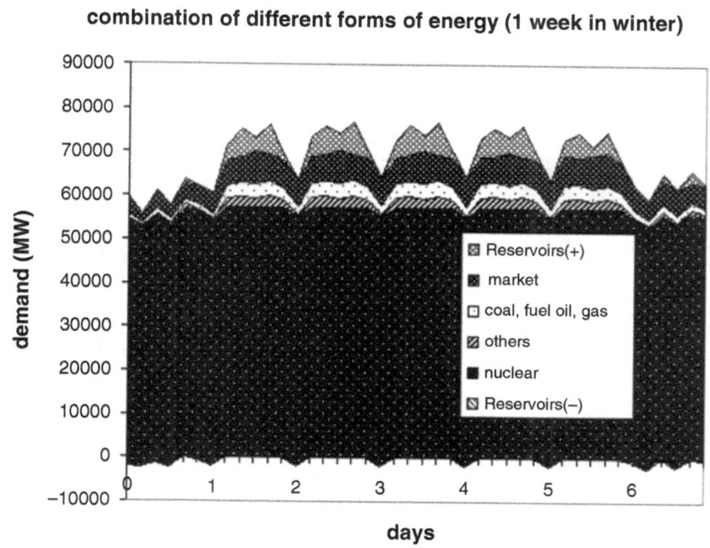

Fig. 5 Combination of different forms of energy during 1 week

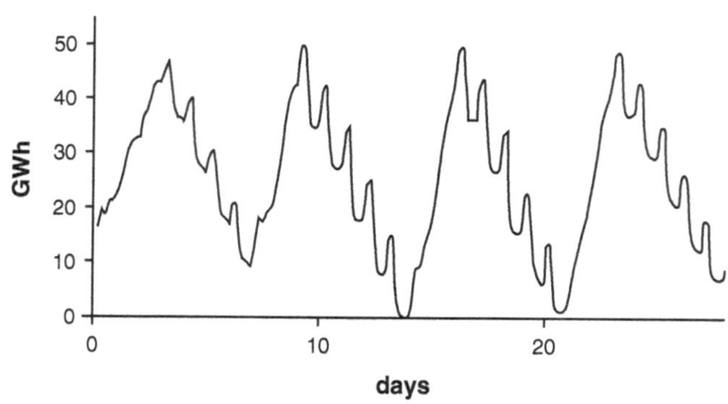

Fig. 6 Typical use of weekly reservoir management on 4 weeks

that are in use at a specific time-step, and then it allocates a part of the total production needed for each power plant. H1 corrects the solution given by H0 to take all the constraints into account. The second method (H2) does a heuristic branch and bound to solve the MIP.

3.5.1 First Heuristic H1

Initially, the first heuristic (H0) solves an independent problem for each time-step. But, we have to rework the solution to take all the constraints into account; heuristic (H0) with these improvements is called (H1) as described in Fig. 8.

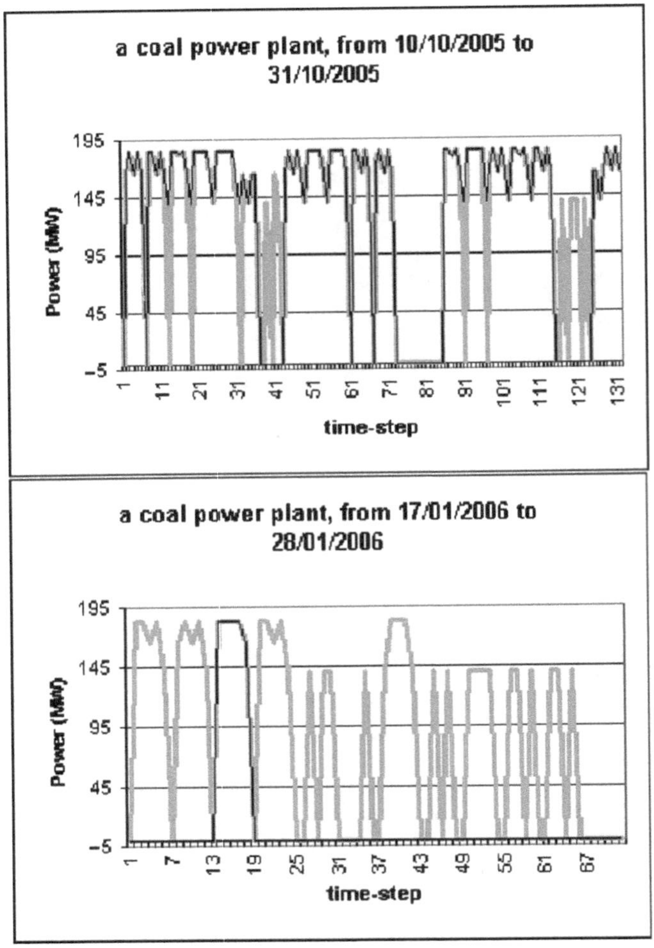

Fig. 7 Impact of taking starting costs into account

Especially, weekly reservoirs are really misused. Therefore, we add a preliminary algorithm to fix this management issue all over the week considering the demand variation. The principle consists in pumping up to the maximum capacities during the periods with low demands and up to the maximum producing capacities at demand peaks.

We have also added a last process called "Solution correction" to take the not yet considered constraints into consideration, like minimum operation time or starting costs. For example, we endeavor to avoid stopping a plant for a short duration and prefer keeping it producing, so as not to pay start up cost. Conversely, we also try to avoid starting a plant just for a short duration and when it cannot be avoided, we compare the sum of start up cost and running cost. Thus production is swapped to increase the feasibility. Then, production is swapped to decrease the total cost.

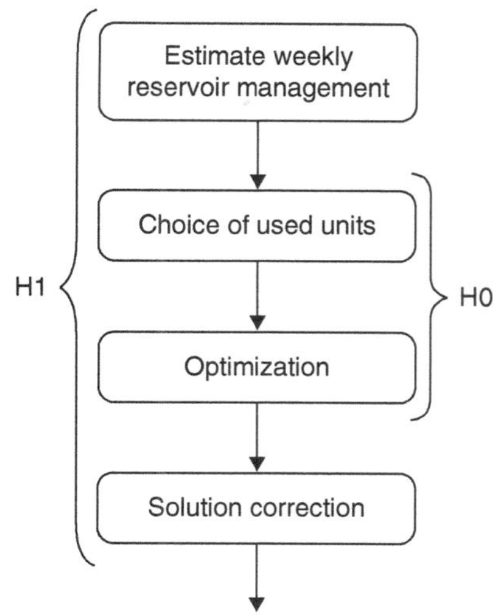

Fig. 8 Description of first type of heuristic (H1)

The choice of units to be used is mainly but not solely based on running cost. Especially, being able to provide system reserve is crucial.

The "optimization" step is equivalent to solving a linear program that can be seen as a shortest path problem.

If we consider N power plants, the linear program is for each time-step described below.

The variables are as follows:
$u(i)$: the production of the unit i at time-step (in MW)
$v(i)$: the participation of secondary system reserve at time-step (in MW)
$w(i)$: the participation of tertiary system reserve at time-step (in MW)

The data are the following:
D: the demand at time-step (in MW)
D': the demand of secondary system reserve at time-step (in MW)
D'': the demand of tertiary system reserve at time-step (in MW)
$P_{min}(i)$: the minimum power of the unit i at time-step (in MW)
$P_{max}(i)$: the maximum power of the unit i at time-step (in MW)
$V_{max}(i)$: the maximum participation of secondary system reserve (in MW)

The linear program is

$$min \sum_{i=1}^{N} cost(i).u(i)$$

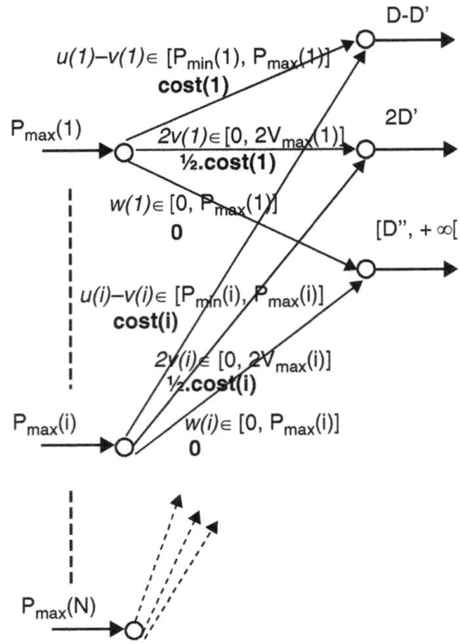

Fig. 9 Equivalent graph

subject to:

$$\begin{cases} \sum_{i=1}^{N} u(i) = D \\ \sum_{i=1}^{N} v(i) = D' \\ \sum_{i=1}^{N} w(i) = D'' \\ \forall i \in [1, N], u(i) \geq 0 \\ \forall i \in [1, N], v(i) \geq 0 \\ \forall i \in [1, N], w(i) \geq 0 \\ \forall i \in [1, N], u(i) - v(i) \geq P_{min}(i) \\ \forall i \in [1, N], u(i) + v(i) + w(i) \leq P_{max}(i) \\ \forall i \in [1, N], v(i) \leq V_{max}(i) \end{cases}$$

This linear program can be seen as a shortest path problem and is described with Fig. 9.

On each edge, the bold value is the cost, the italic one is the flow value (these are unknown) and the normal one is the admissible interval for the flow.

3.5.2 Second Heuristic H2

The second heuristic uses the MIP formulation of the problem. But, we only solve linear relaxation, then set all binary variables (H2.1). We have also tried to set a subset of binary variables and iterate the process until we obtain an integer solution (H2.2) (Fig. 10).

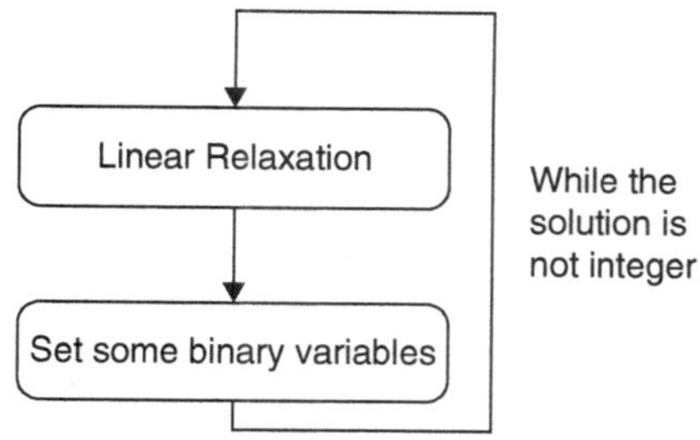

Fig. 10 Description of the heuristic (H2.2)

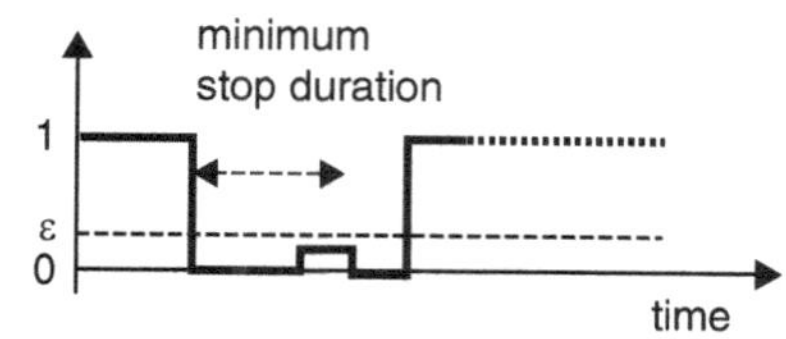

Fig. 11 Linear relaxation solution

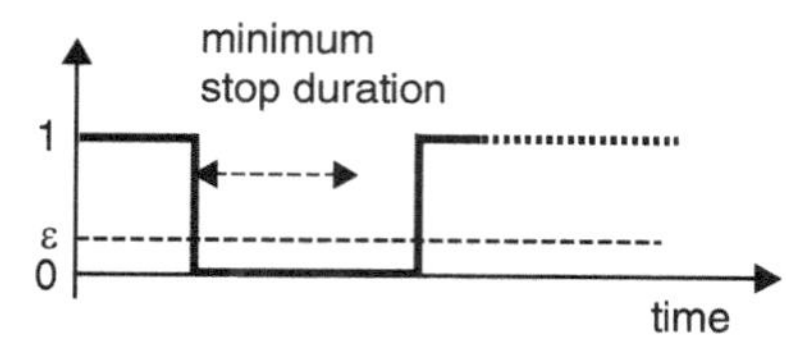

Fig. 12 Solution with β settled on

The problem here is to settle on variables that respect all the constraints without degrading the objective function. The relevant variables are $\beta(c,t)$ and $\delta(c,t)$ described above. Knowing β gives the value of δ. So we only work on β. Therefore, the problem is to know when the plants are on or off. When a plant is used under a level ε during a period longer than the minimum down time, we establish $\beta = 0$ for all the relevant period (Figs. 11 and 12).

The level ε is either a constant (H2.1) or it can change for each iteration (H2.2). Moreover, (H1.1) also establishes those β that are greater than ε and (H2.2) also establishes those β that are greater than $1-\varepsilon$.

3.6 Calculation Duration Comparison

As this tool must be used every day with 500 scenarios and as each scenario requires the resolution of 200 MIP (one for each week of the period of study), calculation duration are really critical.

Table 1 shows the average gap, maximum gap, and average CPU time for each heuristic, optimal resolution, and a resolution with a relative mipgap tolerance fixed to 10^{-3}.

These calculation durations seem short. Nevertheless, for 500 scenarios, they become huge (more than 15 h for the MIP 0.1% Cplex), which requires the use of distributed computing and many Cplex licences to meet total calculation duration

Table 1 Calculation duration and gap

Algorithm	Average gap	Max gap	Average CPU time (s)
MIP Cplex	0.00%	0.00%	168.77
MIP 0.1% Cplex	0.04%	0.09%	112.08
LP H2.1 Cplex	0.09%	0.18%	70.70
LP H2.1 Coin	0.04%	0.11%	265.49
LP H2.2 Cplex	0.11%	0.26%	91.65
H1	1.22%	7.24%	22.02
H0	3.67%	13.16%	5.22

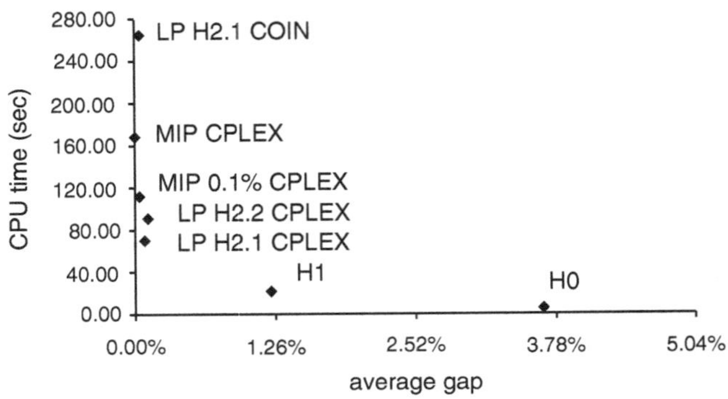

Fig. 13 Calculation duration and gap

less than half an hour. Moreover, EDF wants to run the simulator with 50,000 scenarios once a week to compute accurate risk indicators. Every tiny decrease of the calculation duration per scenario is crucial (Fig. 13).

Intuitive logic is often respected: the faster the algorithm, the bigger the gap until the best known solution (except COIN and heuristic H2.2). COIN is only 3.8 times slower than CPLEX. Considering that COIN is free, its use is conceivable depending on CPU time constraints.

Among all the methods presented above, future users prefer the CPLEX MIP resolution with a relative mipgap tolerance fixed to 10^{-3}, which is quite accurate (0.1% guaranteed) and not too slow. Moreover, a MIP formulation is more upgradeable than a heuristic formulation, especially if EDF wants to improve the modeling with new constraints.

4 Conclusion

Thanks to the new simulator, mid-term unit management has been improved (with better modeling and better optimization) and gives more accurate results. Among all the methods presented above, future users prefer the CPLEX MIP resolution with a relative mipgap tolerance fixed to 10^{-3}, which is quite accurate (0.1% guaranteed), easily upgradeable, and not too slow (even if distributed computing was required to meet calculation duration constraints). But some limitations still remain. The simulator modeling has been improved and is now closer to short-term unit commitment modeling, but there are still differences. For instance, gradient or primary system reserve constraints are not considered in the mid-term simulator. Considering the different goal of the two simulators, a small modeling gap can be accepted. Moreover, as global optimizer modeling has not been improved, now there is a gap between optimizer and simulator modeling. This leads to non-optimal results in the simulator. We will now focus on this issue to go on improving EDF mid-term power management.

Further Reading

EDF Modelling System

Carpentier P, Gohen G, Culioli J-C, Renaud A (1996) Stochastic optimization of unit commitment: a new decomposition framework. IEEE Trans Power Syst 11(2):1067–1073

Dubost L, Gonzalez R, Lemarchal C (2003) A primal-proximal heuristic applied to the unit commitment problem. INRIA: Report no 4978, October 2003

Dubost L, Gonzalez R, Lemarchal C (2005) A primal-proximal heuristic applied to the French Unit-commitment problem. Math Program A B 104(1):129–151

Lederer P, Torrion P, Bouttes J (1984) Overall control of an electricity supply and demand system: A global feedback for the French system. In: System modeling and optimization proceedings of the 11th IFIP conference Copenhagen, Denmark, 25-29 July 1983. Springer, Berlin, pp. 609–617

Lemarchal C, Sagastizabal C, Pellegrino F, Renaud A (1996) Bundle methods applied to the unit commitment problem. System modelling and optimization, pp. 396–402. Chapman & Hall, London

Renaud A (1993) Daily generation management at Electricite de France: from planning towards real time. IEEE Trans Automat Contr 38(7):1080–1093

Other References

Arroyo J, Conejo A (2004) Modeling of start-up and shut-down power trajectories of the thermal units. IEEE Trans Power Syst 19(3):1562–1568

Borghetti A, Frangioni A, Lacalandra F, Lodi A, Martello S, Nucci CA, Trebbi A (2001) Lagrangian relaxation and tabu search approaches for the unit commitment problem. In: Saraiva JT, Matos MA (eds) Proceedings IEEE 2001 Powerteck Porto Conference, vol. 3, Paper no PSO5-397, 2001

Fournier J (2006) Théorie des graphes et applications. Hermès - Lavoisier, Paris

Frangioni A, Gentile C, Lacalandra F (2007) Hybrid lagrangian-MILP approaches for unit commitment problems. Research Report R. 668, Istituto di Analisi dei Sistemi ed Informatica "A. Ruberti" - C.N.R.

Frangioni A, Gentile C, Lacalandra F (2009) Tighter approximated MILP formulations for unit commitment problems. IEEE Trans Power Syst 24(1):105–113

Frangioni A, Gentile C, Lacalandra F. Solving unit commitment problems with general ramp contraints. Int J Electr Power Energ Syst (To appear)

Gruhl J, Schweppe F, Ruane M (1975) Unit commitment scheduling of electric power systems. In: Fink LH, Carlsen K (eds) System engineering for power: Status and prospects. Henniker, NH

Nemhauser G, Wolsey L (1999) Integer and combinatorial optimization. Wiley, New York

Nilsson O, Sjelvgren D (1996) Mixed-integer programming applied to short-term planning of a hydro-thermal system. IEEE Trans Power Syst 11(1):281–286

Turgeon A (1980) Optimal operation of multireservoir power systems with stochastic inflows. Water Resources Res 16:275–283

Varoquaux W (1996) Calcul économique et électricité. Presses Universitaires de France, Paris

Large Scale Integration of Wind Power Generation

Pedro S. Moura and Aníbal T. de Almeida

Abstract In a scenario of large scale penetration of renewable production from wind and other intermittent resources, it is fundamental that the electric systems have appropriate means to compensate the effects of the variability and randomness of the wind power availability. This concern was traditionally met by the promotion of the wind resource studies and in the identification of solutions based on reversible hydropower dams. However, in the electric system planning, other options deserve to be evaluated. This chapter evaluates the methods and technologies that can be used to minimize the intermittence, such as grid integration, technical distribution of the generators, geographic distribution of the generators, improved forecasting techniques, power plants providing operational and capacity reserve, interconnection with other grid systems, curtailment of intermittent technology, distributed generation, complementarily between renewable sources, energy storage, demand-side management, and demand-side response.

Keywords Demand response · Demand-side management · Energy storage · Grid integration · Hydro power · Renewable Generation · Solar power · Wind power

1 Wind Intermittence

Wind energy has characteristics that differ from conventional energy sources. If the contribution of this production vector in energy terms is not a cause of concern, the power balance, and therefore the impact in the supply security, needs attention due to the intermittent and random character of this production option. Wind capacity is installed to generate energy with negligible CO_2 emissions, but its contribution to meet peak load growth requirements is limited.

A.T. de Almeida (✉)
Department of Electrical and Computer Engineering, University of Coimbra, Portugal
e-mail: adealmeida@isr.uc.pt

S. Rebennack et al. (eds.), *Handbook of Power Systems I*, Energy Systems,
DOI 10.1007/978-3-642-02493-1_5,

Wind power cannot fully replace the need for a variety of "capacity resources," which are dispatchable generators that are available to be used when needed to meet peak load. Wind power must be considered an energy resource, but not a peak capacity resource, because only a small fraction of total wind capacity has a high probability of running consistently. Wind is used when it is available, and if it has some capacity value for reliability operation planning purposes, then that should be viewed as a bonus.

The output of wind power is driven by environmental conditions mainly outside the control of the generators or the system operators. Since the wind is determined by random meteorological processes, it is inherently variable. Supply of power from wind turbines is stochastic in nature and the actual power is more or less proportional to the third power of the wind velocity. The wind output varies seasonally between summer and winter (Hamacher et al. 2004), and the variations are also present on shorter time scales, namely on hourly basis (Holttinen 2005). Unlike conventional capacity, wind-generated electricity cannot be reliably dispatched or perfectly forecasted, and exhibits significant temporal variability.

In addition to being variable, wind power production is also a challenge to accurately predict on the time scales of interest to day ahead and for long-term planning of the system adequacy. It is possible to forecast, if in a concrete zone, the average wind power density for the whole year; however it is impossible to precisely forecast the days or the hours with wind (Schneller 2004). Electric energy production from a large wind generation facility over a period of time – months, years, or even the life of the project – can be estimated accurately enough, but over shorter time frames, however, production is less predictable.

The variability also decreases as the time scale decreases. The second and minute variability of large scale wind power is generally small, whereas the variability over several hours can be large even for distributed wind power. For time scales from several hours to day-ahead, forecasting of wind power production is crucial.

The large-scale variability can be divided on the following types (Table 1) (Holttinen 2007):

- Very fast variations (second-minute level) of distributed wind power are low.
- The largest hourly step changes recorded from regional distributed wind power range from ±10 to ±35% depending on region size and how dispersed the wind power plants are.
- Wind power production can vary a lot in longer time scales, above 4 h.

Several extreme ramp rates were recorded during storms (Holttinen 2007):

- Denmark – 2,000 MW (83% of capacity) decrease in 6 h or 12 MW (0.5% of capacity) in a minute on 8th January 2005.
- North Germany – over 4,000 MW (58% of capacity) decrease within 10 h, extreme negative ramp rate of 16 MW/min (0.2% of capacity) on 24th December 2004.
- Ireland – 63 MW in 15 mins (approximately 12% of capacity at the time), 144 MW in 1 h (approximately 29% of capacity) and 338 MW in 12 h (approximately 68% of capacity).

Table 1 Extreme variations of large scale regional wind power, as % of installed capacity (Holttinen 2007)

Region	Region size (km^2)	Number of sites	10–15 min max dec	10–15 min max inc	1 h max dec	1 h max inc	4 h max dec	4 h max inc	12 h max dec	12 h max inc
Denmark	300×300	>100			−23%	+20%	−62%	+53%	−74%	+79%
West Denmark	200×200	>100			−26%	+20%	−70%	+57%	−74%	+84%
East Denmark	200×200	>100			−25%	+36%	−65%	+72%	−74%	+72%
Ireland	280×480	11	−12%	+12%	−30%	+30%	−50%	+50%	−70%	+70%
Portugal	300×800				−16%	+13%	−34%	+23%	−52%	+43%
Germany	400×400	>100	−6%	+6%	−17%	+12%	−40%	+27%		
Finland	400×900	30			−15%	+16%	−41%	+40%	−66%	+59%
Sweden	400×900	56			−17%	+19%	−40%	+40%		
US Midwest	200×200	3	−34%	+30%	−39%	+35%	−58%	+60%	−78%	+81%
US Texas	490×490	3	−39%	+39%	−38%	+36%	−59%	+55%	−74%	+76%
US Midwest+OK	1200×1200	4	−26%	+27%	−31%	+28%	−48%	+52%	−73%	+75%

- Portugal – 700 MW (60% of capacity) decrease in 8 h on 1st June 2006.
- Spain – 800 MW (7%) increase in 45 min (ramp rate of 1,067 MW/h, 9% of capacity), and 1,000 MW (9%) decrease in 1 h and 45 min (ramp rate −570 MW/h, 5% of capacity). Generated wind power between 25 and 8,375 MW have occurred (0.2% and 72% of capacity), in a single year.
- Texas, USA – loss of 1,550 MW of wind capacity at the rate of approximately 600 MW/hr over a 2 h and 30 min period on February 24 2007.

2 Impact in the Power System

Integrating large amounts of wind energy into the electric generation mix requires some special considerations. Beyond the variability, a lot of wind generation occurs in hours when energy use is low. The uncontrollable nature of wind makes it less valuable to system operators than dispatchable power. The variability and uncertainty of wind energy production require that power system operators take measures to manage its delivery. These measures may increase the cost incurred to balance the system and maintain reliability. Substantial amounts of wind generation in a utility system can increase the demand for the various ancillary services. Previous results also reveal a diminishing benefit as wind power penetration increases (Kennedy 2005).

In 2005, the impact of meeting 100% of western Denmark's annual electrical energy requirement from wind energy was analyzed (Pedersen (2005)). The results of the study demonstrated that the system could absorb about 30% energy from wind without any wasted production. However, when wind share reaches 50%, the excess wind energy starts to grow considerably. With the hypothesis of total energy demand of 26 terawatt-hours (TWh) generated by wind power, 8 TWh of the wind generation would be surplus because it would be generated during periods of low consumption. In that case, the electricity cost will double.

As the wind penetration level increases, three factors lower the economic value of wind power (DeCarolis and Keith 2005):

- Increasing wind generation usually displaces capacity from increasingly lower cost plants
- Operational losses due to repeated plant starts or partial plant loading
- Unnecessary wind energy, which cannot be absorbed, due to operational constraints or excess production

Two penetration limits can be defined (Grubb and Meyer 1993):

- When the marginal fuel savings have dropped by one-quarter (economic target)
- When the marginal fuel savings have been halved (maximum credible penetration level)

However, the wind penetration level can be defined using several different measures: energy penetration, capacity penetration, and instantaneous penetration DOE (2008):

- *Energy penetration*: ratio of the energy delivered from the wind generation to the total energy delivered
- *Capacity penetration*: ratio of the wind plant capacity to the peak load
- *Instantaneous penetration*: ratio of the wind plant output to the load at a specific time, or over a short period of time

Traditional system planning techniques are oriented toward ensuring enough capacity levels at all times. However, most of them have flexibility for interconnection and operation (choice of interconnection voltage, operation as a price-taker in a spot market, and limited curtailment) and can receive additional energy resources. Even in a small scale the undispatchable wind energy imposes costs at the grid operation, increasing the costs while the wind production enlarges. The system reliability will also decreases due to an increasing variation in the system.

The variability can occur in different time scales, which range from seconds to days and have impact in the correspondent grid operation time scale (Fig. 1). In variations from seconds to minutes, system regulation will be affected since in this timeframe generation automatically responds to deviations in load or to the net load wind balance, allowing operators to maintain system balance. Load following is a longer period capability (from 10 min to few hours) that includes both capacity and energy services, incorporating the morning load pick-up and evening load drop-off. The "scheduling" and "unit-commitment" processes ensure that sufficient generation will be available when needed over several hours or days ahead of the real time schedule (Milligan et al. 2006).

Figure 2 shows the increased load-following requirements due to wind on an electrical system (Zavadil et al. 2004). More high-ramp requirements and a general reduction in small-ramp requirements can be observed in the case with wind. Following load requirements subtracted by the wind generation creates a larger variability in the magnitude of the load change between two adjacent hours. Thus wind

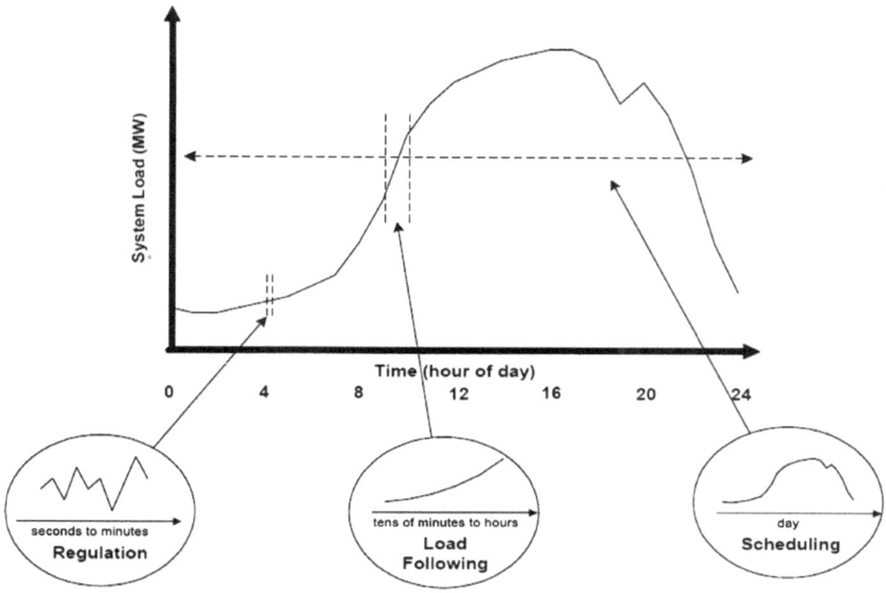

Fig. 1 Time scales for grid operation (Milligan et al. 2006)

generation requires that the system has more active load-following generation capability, or more load-management capability to compensate the variability of the wind and the load. Recent wind integration studies have demonstrated that the variations of most importance and cost are those in the hourly and daily timeframe related to the ancillary services of load following and unit commitment.

3 Options for Managing Intermittency

The connection of wind turbines to the electricity grid can potentially affect supply reliability and power quality due to the unpredictable fluctuations in wind power output (Chompoo-inwai et al. 2005; Ackermann 2005; Georgilakis 2008). In a scenario of large scale penetration of renewable production from wind and other intermittent resources, it is fundamental that the electric system has appropriate means to compensate the effects of the variability and randomness of the wind power availability. This concern was traditionally addressed by the promotion of the wind resource studies and by the identification of solutions based on reversible hydropower dams (de Almeida et al. 2005). However, in the electric system planning, other options deserve to be evaluated in deregulated markets.

The traditional planning methods are centered on reliability and capacity planning offered by the units that make up the generation mix. Incorporating wind energy into power system planning and operation will need new methodologies.

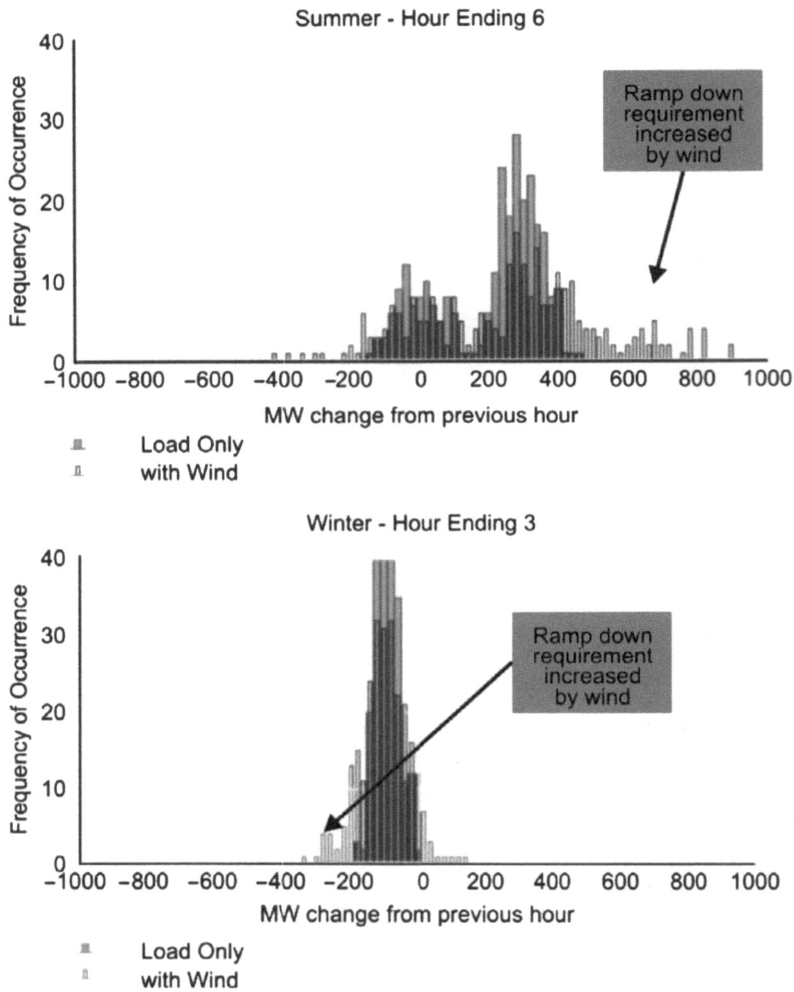

Fig. 2 Impact of wind on load-following requirements (Zavadil et al. 2004)

Wind power offers additional energy planning reserves to the system, but because wind is not a capacity resource, it does not require 100% backup to ensure replacement capacity when the wind is not blowing.

The intermittency of wind energy can be reduced by some techniques:

- Grid integration
- Technical distribution of the generators
- Geographic distribution of the generators
- Improved forecasting techniques

The last three techniques can be grouped as aggregation and distribution methods. These techniques have as aim the increasing of the predictability of the production and the substantial reduction of the global variations.

However, although those improvements bring benefits, several periods of low wind production and substantial variations will remain. Thus tools to respond to short- to medium-term and long-term variability will be necessary, managing the operational and capacity reserve, respectively. For large scale integration of wind power, the provision of flexible capacity reserve will be of crucial importance. To achieve that aim several options are possible:

- Power plants providing operational and capacity reserve
- Interconnection with other grid systems
- Curtailment of intermittent technology
- Distributed generation
- Complementarily between renewable sources
- Energy storage
- Demand-side management
- Demand-side response

All the options have as aim balancing demand and supply continuously and backing up other capacity shortage.

3.1 Wind Power Forecasting

The forecasting of wind power output plays a major role in the short run operation of wind power in the electricity-generation system. However, this forecast is never a certainty and forecast errors will occur. Apart from that, the technology is also relatively new and the information about wind power is not based on the same amount of experience as for conventional technologies. The inflexibility, variability, and relative unpredictability of wind power as a means for electricity production are the most obvious barriers to an easy integration and widespread application of wind power.

The short-term forecasting of wind power production is still a recent power system tool when compared to load forecasting. For wind power, the level of accuracy will not be as high as for load forecasting. The shape of the wind energy production can be predicted in large part of the time, but large divergences can occur in the timing of and in the winds amplitude (Giebel et al. 2003). Wind electricity production will always fluctuate with weather conditions, but the more precise the forecasting and modeling becomes, the smaller will be the error margin in forecasting, and thus the lower the requirements can become for operational reserves and balancing energy. This is reflected in Fig. 3, which contrasts the gap between simple "persistence forecasting" and "perfect forecasting" and the impact on required operational reserve (Holttinen 2005).

The "capacity factor" of wind farms is usually low (25–40% – depending on location) because the wind speed is highly variable. However, it should be recognized

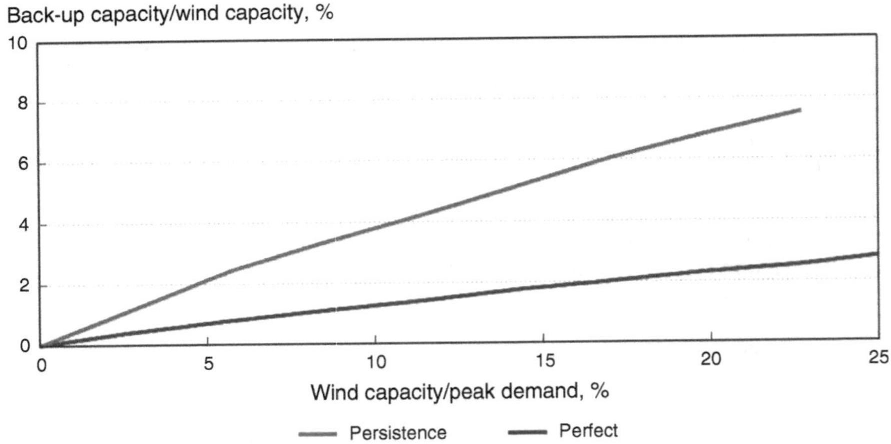

Fig. 3 Reducing added back-up for wind IEA (2005)

that better wind forecasting can reduce the need for operational reserve to a minimum, because wind turbines do have very few unexpected outages and need less maintenance than traditional power plants.

The accurate forecasting of wind power have an increasing importance when the wind power penetration grows, enabling the real time operation and commitment of generation. For wind penetrations lesser than 5%, wind forecasting is generally assumed as not needed, but as wind penetration rises, wind forecasting increasingly adds value to wind power. System operators can significantly reduce the uncertainty of wind output by using wind forecasts that incorporate meteorological data to predict wind production. An incorrect forecast can largely affect the system operation, because an under forecast might result in an over commitment of generation and an over forecast may result in an under commitment of generation, and the reliability of the power system may be affected.

To have an efficient unit commitment and scheduling, it will be necessary to develop accurately day-ahead and near real-time forecasts of wind power to support real-time operations. Forecasting allows operators to anticipate wind generation levels and adjust the remainder of generation units leading to economic benefits (a perfect wind forecast would reduce annual variable production costs by $125 million in the USA (Piwko et al. 2005)). Advanced forecasting systems can also help to warn the system operator if extreme wind events are likely so that the operator can implement a defensive system posture if needed.

When the forecasts are combined for larger areas, the accuracy level is increased (Fig. 4). The error for day-ahead forecasts in control area is below 10%, but for a single wind power plant it is between 10 and 20% (Holttinen 2007). For comparison, prediction errors of consumption are generally in the region of 1–5%. The level of accuracy also improves when the forecast horizon decreases and when different forecasting models are combined (Fig. 5).

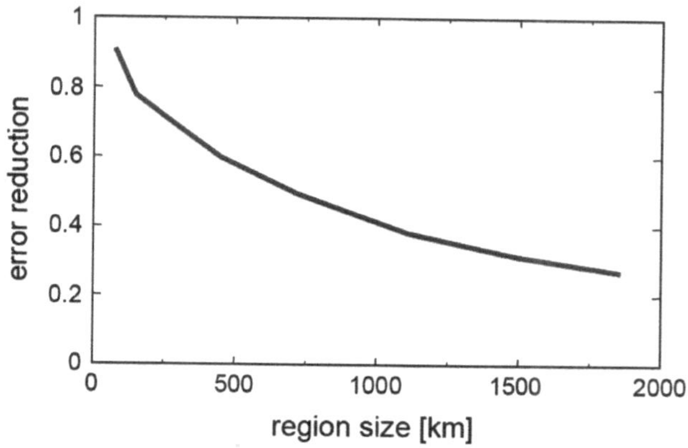

Fig. 4 Decrease of the forecast error prediction for aggregated wind power production due to spatial smoothing effects (Holttinen 2007)

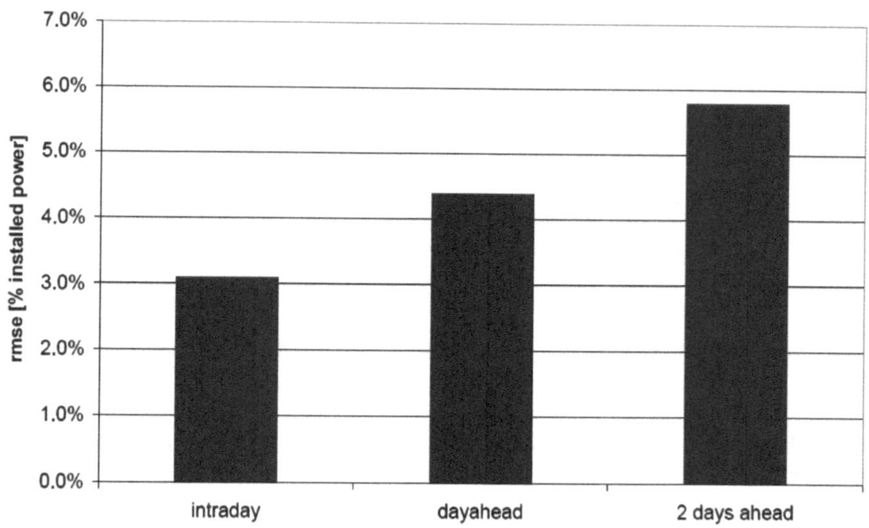

Fig. 5 Increasing forecast error as forecast time horizon increases (Holttinen 2007)

3.2 *Aggregation and Distribution*

An analysis of individual wind turbines or single wind farms is not relevant to determine the impact of the wind power integration. The objective is not to provide a mitigation solution for each individual turbine, but to all the system, and the global impact will not be the sum of all individual impacts. The larger is the number of wind turbines operating in a given area, the lesser is their total production variability.

As a general tendency, as more wind turbines are operating in a given period, lower is the production variability (Table 2). Similarly, as more wind turbines are

Table 2 Wind power step change average magnitude and standard deviation (Std) values as a function of an increasing number of aggregated wind turbines in a large wind plant in the Midwest of the US (Holttinen 2007)

		14 turbines		61 turbines		138 turbines		250+ turbines	
		(kW)	(%)	(kW)	(%)	(kW)	(%)	(kW)	(%)
1 s	Average	41	0.4	172	0.2	148	0.1	189	0.1
1 s	Std	56	0.5	203	0.3	203	0.2	257	0.1
1 min	Average	130	1.2	612	0.8	494	0.5	730	0.3
1 min	Std	225	2.1	1038	1.3	849	0.8	1486	0.6
10 min	Average	329	3.1	1658	2.1	2243	2.2	3713	1.5
10 min	Std	548	5.2	2750	3.5	3810	3.7	6418	2.7
1 h	Average	736	7.0	3732	4.7	582	6.4	12755	5.3
1 h	Std	1124	10.7	5932	7.5	10032	9.7	19213	7.9

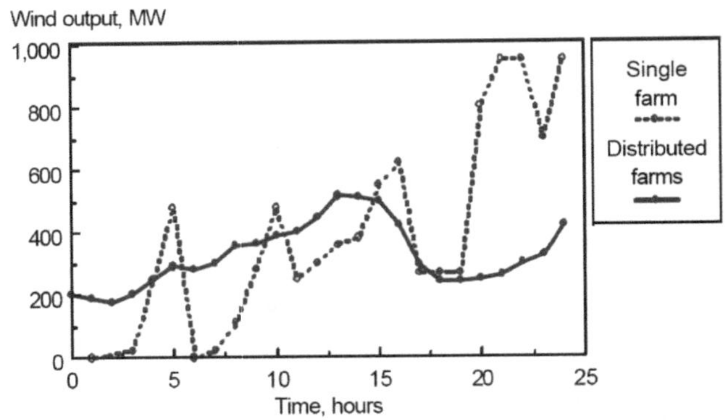

Fig. 6 The smoothing effects of geographical dispersion (MacDonald 2003)

installed across larger geographic areas, the total wind generation becomes more predictable and less variable. In a wind power region, the abrupt loss of all wind power will be less probable than that in a single turbine, due to the spatial variations of wind from turbine to turbine. On a larger degree the same occurs on a system due to the spatial variations between wind power plants. Just as aggregation of consumer loads in an integrated electricity network smooths the total demand, the aggregation of wind plant smooths wind fluctuations. It is a simple statistical phenomenon: the larger is the integrated grid, the more pronounced is this effect.

Figure 6 describes a hypothetical situation of a 1,000 MW wind capacity, comparing the concentration in a single wind farm with the distribution between several wind farms. A considerable reduction both in the size and in the volatility of output variations can be observed. During a 24 h period, the output from a 1,000 MW wind farm might fluctuate from zero to close to its rated output, but the output from 1,000 MW of distributed wind would only vary between 200 and 500 MW, approximately (MacDonald 2003).

In larger areas the correlation among wind generators between the produced energy in the wind plants will be lower, causing the smoothing effect in the global

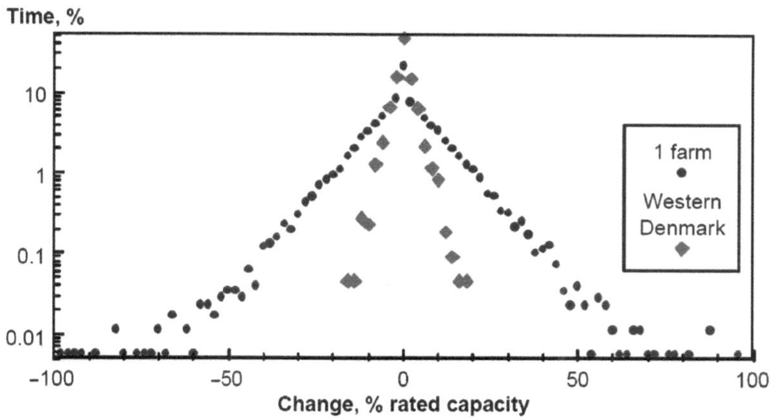

Fig. 7 The smoothing effects of geographical dispersion: a single wind farm of 5 MW, and all the wind plants in Western Denmark (MacDonald 2003)

wind power production. The correlation coefficient falls to around 0.5 at a distance of 300 km and to 0.2 at 800 km (Milborrow 2001). The number of hours with zero output will also decrease to larger areas, because one wind power plant can have zero output for more than 1,000 h during a year, but the total wind power in a large area is always more than zero.

Figure 7 compares the fluctuations in all the wind plants in Western Denmark with those verified in a single wind farm of 5 MW. The maximum one-hourly power swing from the wind farms installed in Western Denmark (1,860 MW) was 18% of installed capacity (335 MW), and for 47% of the time it was less than 2% of installed capacity (37 MW) (MacDonald 2003).

When the time scale is lower the correlation will also be smaller, reducing the variation of the global wind power production (Fig. 8). Wind production changes very little over short time periods, but when the time period increases from seconds to hours, the output variability enlarges due to changes in weather patterns. In a given period, when more wind turbines are operating, the production variability during that period will be lower. In summary, the size of swings in output from wind farms and the volatility of average output are significantly reduced through geographical aggregation.

Wind power variability and the smoothing effect can be quantified using the standard deviation (Fig. 9) of the time series for variations (Holttinen 2007). The determination of the magnitude and frequency of occurrence of changes in the net load on the system during the time frames of interest (seconds, minutes, and hours) will be very important. The analysis will help determine the additional requirements on the balance of the generation mix and should be performed both before and after the addition of wind power to the system.

Just like demand, the fluctuations of wind energy are statistically random and thus can or cannot correlate with movements of other variables. Milborrow takes

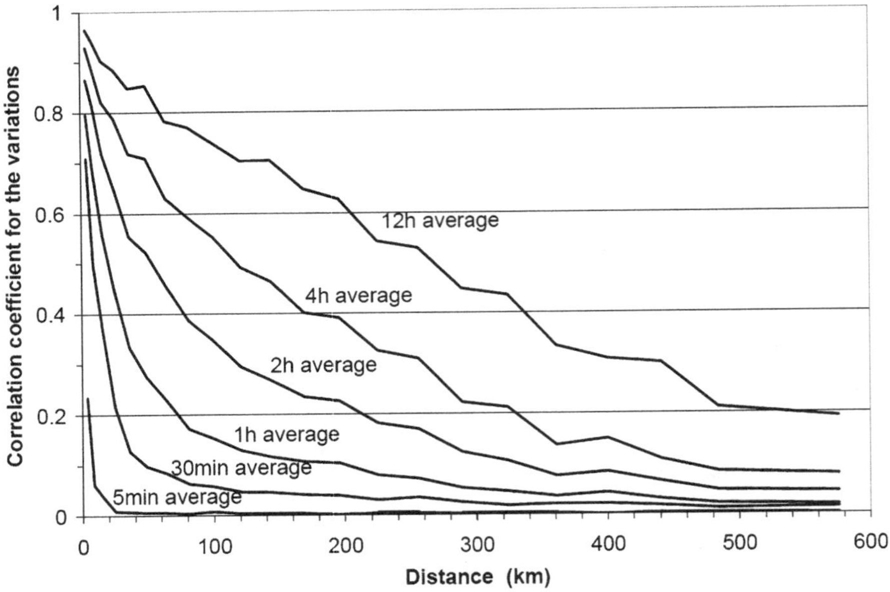

Fig. 8 Correlation of variations for different time scales in Germany (Ernst 1999)

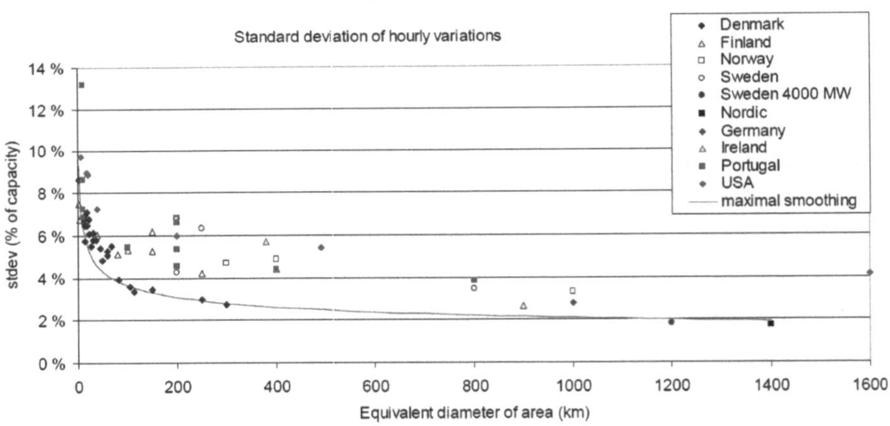

Fig. 9 Reduction in standard deviation of hourly variations of wind power (Holttinen 2007)

two important conclusions analyzing the German and Danish grid (Milborrow 2004) (Fig. 10):

- The maximum hourly swing in output power from distributed wind rarely, if ever, exceeds 20% of the installed capacity of the wind plant. The standard deviation of the hourly swings is 3%.
- The maximum measured change in output per minute from 2,400 MW of wind in western Denmark is about 6 MW.

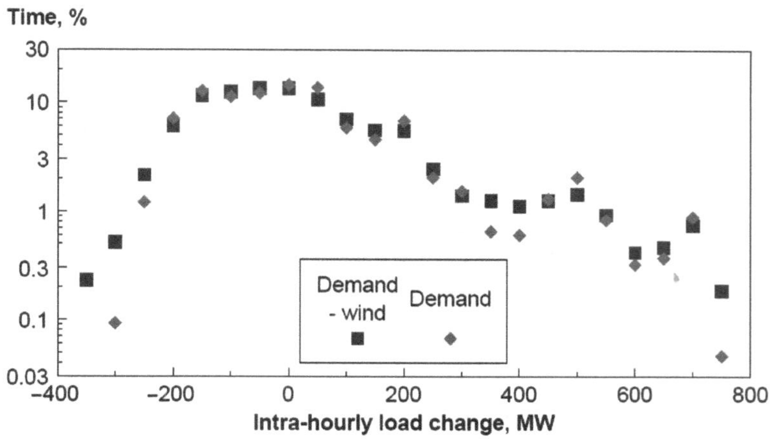

Fig. 10 Intra-hourly load changes in western Denmark, with and without 20% wind (Milborrow 2004)

3.3 Interconnection with Other Grids

The interconnection with other grids enables the export of energy in times of wind power production in excess and the import when the production is reduced. One example is the good interconnections between the Danish grid and German, Norway and Sweden that makes possible the high wind power penetration in Denmark. When the wind energy production is in surplus, Denmark exports the energy to the other countries (but at low price) and import in situations of low wind power production. An interesting possibility is the use of the Norwegian hydro-power as reserve capacity to the Danish wind power.

3.4 Power Plants Providing Reserve

The use of other power plants to provide operational and capacity reserve is the most traditional method to the integration of intermittent power. The power plants that can provide such systems must be flexible and with short response times to make possible the fast repositions of the lost wind capacity. Hydropower is one of the possible technologies that presents most advantages. In the fossil fuel power plants, open-cycle gas turbines (OCGT) or steam-fired power plants, like coal and oil, running at below full-capacity, can be used to such objective. The major disadvantage of such method is the high associated cost, due to the required extra capacity, that operates only in situations of sudden reduction of wind energy production. Other disadvantage is the greenhouse gases emission associated to this kind of power plants.

3.5 Curtailment of Wind Farms

To ensure system stability and control, a minimum level of conventional generation must be maintained in operation, even in periods of low demand. In systems with large level of nuclear power, the problem is larger, due to the high minimum load capacity. In such situations the curtailment of wind power can be used to reduce the overall system integration costs.

The curtailment is made by constraining the output of a group of wind generators, shutting down some or all the turbines. This will result in a loss of energy production and in economic losses. The costs also include the time taken for the wind farm to become fully operational following grid curtailment.

Other situation that can require the curtailment of the wind power is when the transmission and distribution capacity is congested near the wind farm. This situation is clearly more frequent in offshore wind farms, because large wind farms can grant a similar level of ancillary services to that of conventional generators, switching off a few wind turbines for operational reserve or running them at reduced output.

3.6 Distributed Generation

The use of other types of distributed generation can provide several benefits in the network, such as alleviating congestion, reducing transmission losses, and providing ancillary services. Thus, distributed generation can help in the wind power integration, providing reserve necessities as a substitute of conventional power plants. However, wind power is normally a form of distributed generation and has the same requirements for the grid connection. Several technologies of distributed generation can also have intermittency problems, like the wind power (e.g., photovoltaic power).

3.7 Complementarity Between Renewable Sources

In a way similar to wind power, hydropower and solar power are also intermittent resources, due to the dependence of meteorological conditions. However, the variations do not occur at the same time to the three renewable sources and can be mutually compensated (Moura and de Almeida 2008).

To evaluate the intermittence and complementarity from renewable energy production, climate data over several decades was collected. The collected information includes the global solar radiation (monthly average), the wind velocity (monthly average), and the monthly water inflow in dams. The locations for data collection were selected based on the approximation between the collected data and the annual variation of the wind power, solar photovoltaic, and hydropower in all the country.

In each variable, a 50 years' time series was collected. With the collected data, a mathematical model was developed to generate random years, enabling the study of the sources complementarity for a large number of years. Using the climate model, a series of 500 years was generated for the three variables. Wind velocity presents high variations relatively to the average year, with impact in the yearly variation curve shape (Fig. 11).

Also the daily variation of the wind velocity presents a large variation and unpredictability; however, it has a higher availability in the hours of higher energy consumption (Fig. 12).

As it can be observed in (Fig. 13), the solar radiation has small fluctuations relatively to the average year, and additionally, the yearly variation curve does not change in shape.

The daily variation of the solar radiation also presents a constant shape, with the advantage of the concentration of the availability in the hours of higher energy consumption (Fig. 14).

The hydro inflow presents huge variations relatively to the average year and large unpredictability (Fig. 15).

Figure 16 shows the average monthly values for each variable (wind velocity, solar radiation, and water inflow). As it can be observed, the solar radiation is higher between May and September, the opposite occurring with the wind velocity and water inflow. Thus, the solar radiation has the maximum value in July and the minimum in December. Both the wind velocity and water inflow has the maximum value in February, registering the minimum in September and August, respectively.

The wind velocity and the water inflow have average variations along the year with a very similar course, having the two curves at a high correlation (0.98). The solar radiation varies almost inversely relatively to the wind velocity and the water

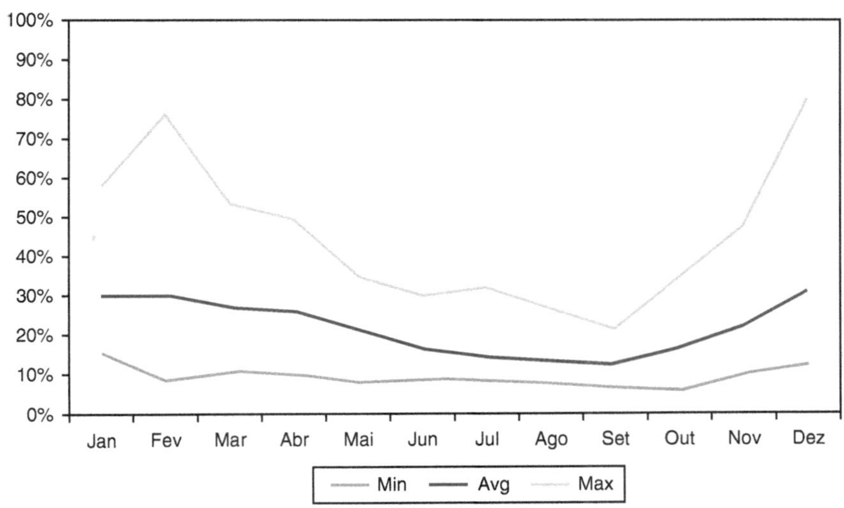

Fig. 11 Monthly variation of the wind velocity in Portugal

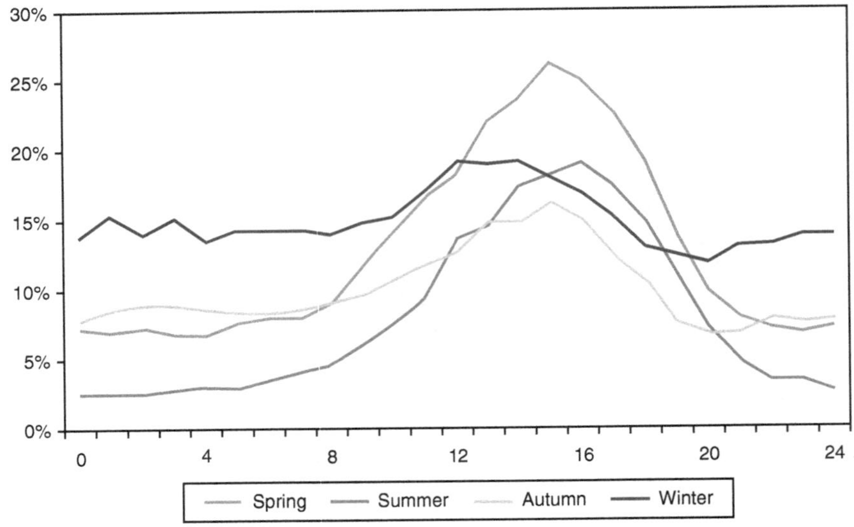

Fig. 12 Average hourly values of the wind velocity in Portugal

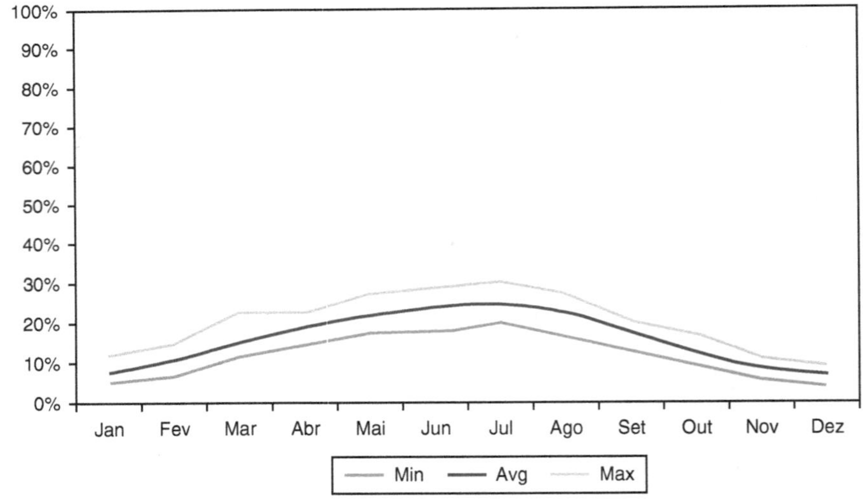

Fig. 13 Monthly variation of the solar radiation in Portugal

flow (correlation of –0.7 and –0.66, respectively). This observation indicates that the complementarity between solar energy and the pair wind power/hydropower is high. Solar energy can then be used to face the seasonal variations of wind power. The hydropower is not complementary to wind power, but due to its similar variations is the ideal means to store the excess wind energy to cope with the intermittence, using the storage, dispatchable power, and dynamic response capacities.

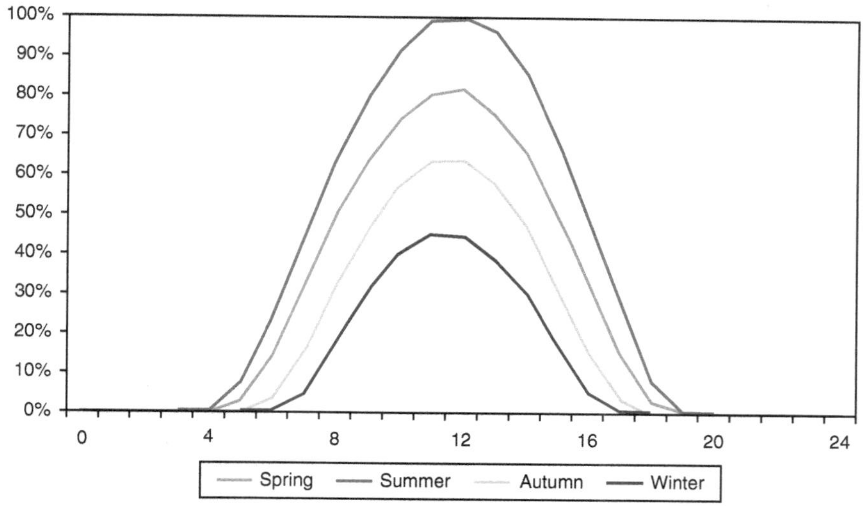

Fig. 14 Average hourly values of the solar radiation in Portugal

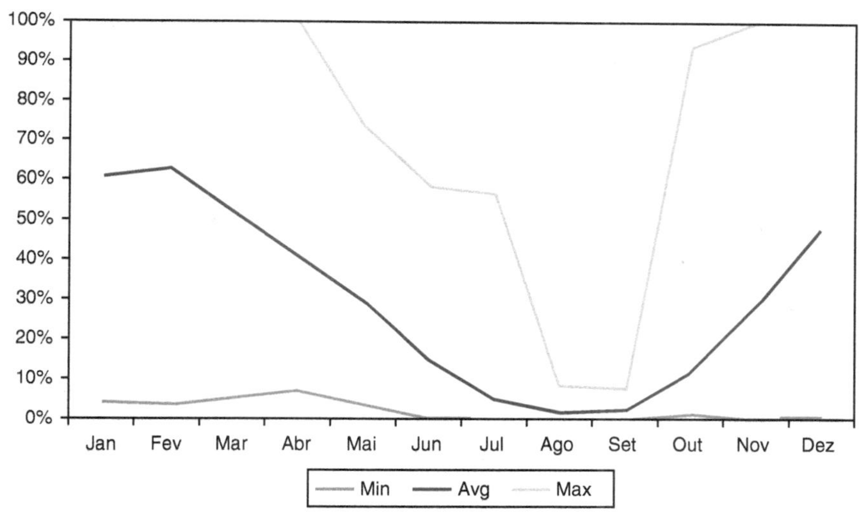

Fig. 15 Monthly variation of the water inflows in Portugal

Also other dispatchable energy technologies, such as biomass, can have a positive contribution, reducing the intermittent power requirements.

3.8 Demand-Side Management

Rather than attempting to match power generation to consumer demand, the philosophy of load management takes action to vary the load to match the power available.

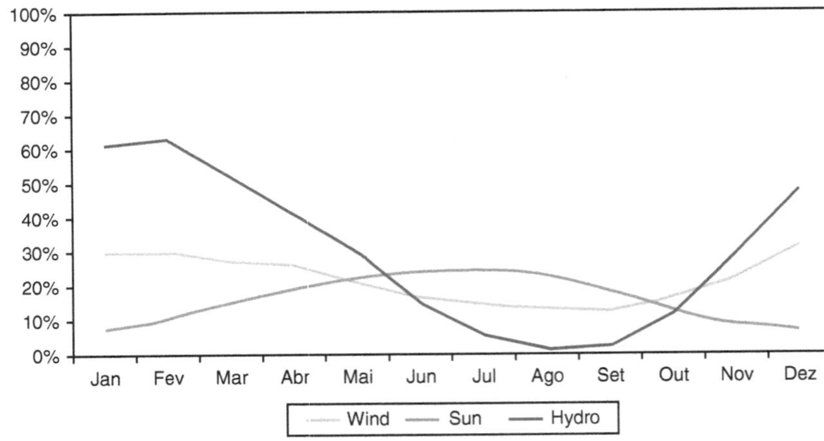

Fig. 16 Yearly variation of the wind, solar, and hydro in Portugal

Through the proper application of demand-side management (DSM) technologies (Kushler et al. 2003), it is possible to reduce the need of new installed intermittent power to achieve the renewable penetration targets.

The most critical situations due to high penetration of wind power occur in periods with high energy consumption. Thus, the DSM technologies have a major role in avoiding critical situations due to intermittent power, mainly the technologies with impact in the peak load reduction. One serious problem is the summer days with high temperatures (elevated consumptions in air conditioning) with reduced wind velocities; therefore, the DSM technologies in space conditioning are the most important.

As an example, the impact of the European Union Energy Services Directive in Portugal, which has the target of achieving a consumption reduction of 9%, between 2008 and 2016, is shown in Fig. 17. Several demand-side management technologies were considered to accomplish such objective, trying to achieve a larger impact with a minimal cost. The aggregated impact in the load diagrams of the selected technologies in the residential, services, and industrial sectors was determined.

The application of DSM measures will reduce the investment needs to integrate intermittent power and will lead to a large reduction on the peak power, which can prevent the most dangerous situations of unbalanced power.

3.9 Demand Response

Other technology that can perform a major role in the integration of renewable intermittent power is demand response (DR) (IEA 2003). With such technologies it is possible to directly or indirectly force a consumption reduction in critical situations, during a short time.

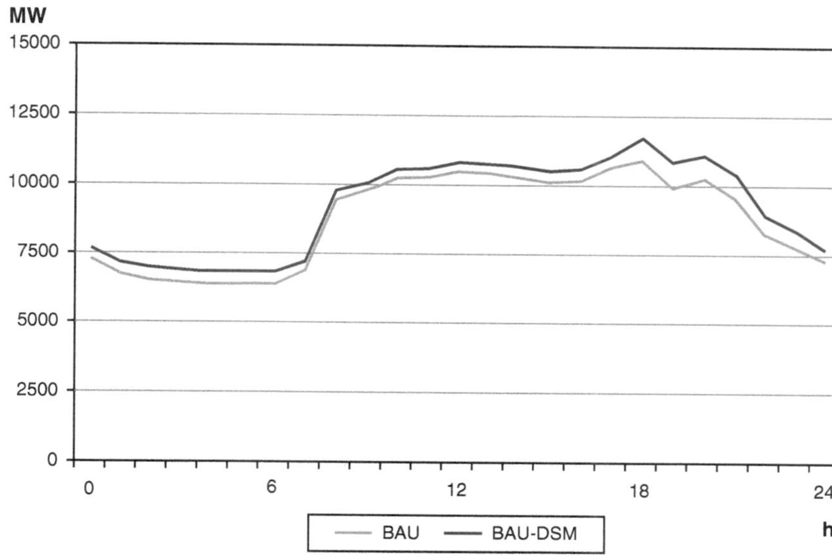

Fig. 17 Impact of the DSM in the portuguese load diagram (January 2016)

In the past, the electric system has been planned and operated under the supposition that the supply system must meet all customer's energy use, and that is not possible to control the demand. However, that supposition is starting to change due to the creation of opportunities for customers to manage their energy use in response to signals (prices or load contracts).

The idea behind DR is that if the marginal peak load price is higher than the value that a consumer gets out of the services derived from the electricity, he would be willing to modify the demand, if paid the peak price or slightly less instead. A grid operator can obtain an economic benefit by paying a customer to reduce the consumption instead of paying a power producer to supply more output, because in peak periods, the production cost can be very high.

Traditionally, the DR technologies were typically used to attend to economical concerns. However, nowadays they can be used to improve the system reliability, reducing instantaneously the energy consumption to prevent the most unbalanced situations, like the problems that result from the large space conditioning consumption on days with reduced wind velocity. As more customers practice automated price-responsive demand or automatically receive and respond to directions to increase or decrease their electricity use, system loads will be able to respond to, or manage, variability from wind power production.

Considering the portuguese load diagram with the impact of DSM measures to 2016, the DR impact was showed in Fig. 18. With a control of 5% of the peak load, it is possible to reduce it and make an identical control on unbalanced situations, when abrupt reductions of wind velocity occurs.

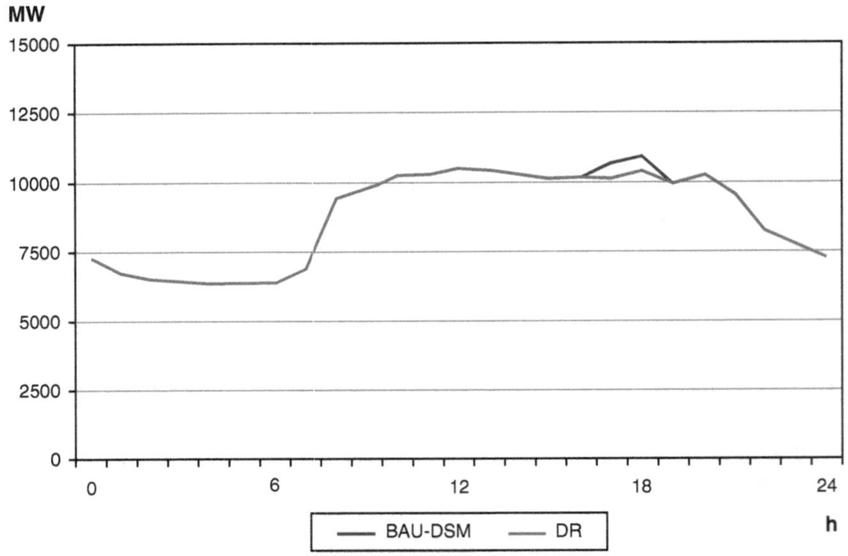

Fig. 18 Impact of the DR in the portuguese load diagram (January 2016)

3.10 Energy Storage

Energy storage has crucial importance in the electric sector, because the energy demand has high hourly, daily, and seasonal variations. Additionally, the energy generation from renewable energy sources has significant variations, either in the short term (few seconds) or in the long term (hourly, daily, and seasonal). Energy storage is an appropriated option to make possible the large-scale integration of intermittent renewable sources.

Energy storage in electric energy generation systems enables the adjustment between the energy production and demand (Lund 2005). The produced energy by intermittent renewable sources can be transferred in time to be released in low production or high consumption times. Storage technology has the advantage of generally not using fossil fuel generation, and so storage facilities do not directly contribute to greenhouse gas production. One disadvantage of the energy storage is the inherent loss due to the energy conversion (about 75–80% efficiency).

The storage devices do not need to be located in the wind farms and can be installed in any point of the grid. That choice enables the support to the integration of intermittent energy and congestion mitigation. Several energy storage technologies can mitigate over-generation problems absorbing the surplus energy in few seconds. Each technology have its response rate, varying from a few seconds to some minutes, but all can quickly connect to the system and ramp up to add load to the system. Hydro storage facilities, whether in the form of pumped-hydro or hydro reservoirs, have played a key role to provide several grid balancing services. Large pump hydropower plants can be switched from generation mode to pumping mode

Technology	Hydropower	Compressed Air	Electrochemical Batteries	Flow Cell	Flywheel	Hydrogen Storage
	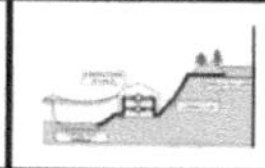			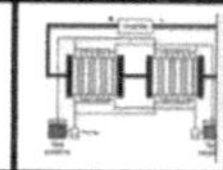	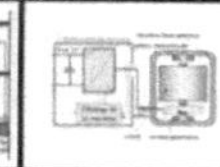	
Energy Density	1 kWh/m³ (360 m of fall)	12 kWh/m³ (cavern at 100 bar)	Acid 33 kWh/t Lithium 100 kWh/t	33 kWh/m³	1 - 5 Wh/kg	300 - 600 Wh/kg (200 - 350 bar)
Capacity	1000 - 100000 MWh	100 - 1000 MWh	0.1 - 40 MWh	10 - 100 MWh	1 - 10 kWh	NA
Power	100 - 1.000 MW	100 - 1.000 MW	0,1 - 10 MW	1 - 15 MW	1 - 50 kW	50 kW – 1MW
Efficiency	65 - 80%	50% (natural gas)	70%	70%	80 - 90%	32 - 50%
Initial Cost (€/kWh)	70 - 150	50 - 80	116 - 860	65 - 400	300 - 250000	2000 - 5500

Fig. 19 Energy storage technologies applicable in integration of intermittent sources (Multon et al. 2004)

within few minutes, storing the energy produced in excess by the wind power plants and releasing such energy when the production decreases. Those plants have a high potential for large-scale electricity storage, fast response times, and reduced operating costs. Other example is the plug-in hybrid vehicles, which can buy and sell energy to the grid.

Some storage devices can provide regulation services and frequency control and can also help with ramping issues by quickly absorbing excess energy when wind generation ramps up and it can deliver energy when wind generation ramps down. Storage systems that can rapidly inject power into the system or add a block of load could mitigate some of the ramp problems and allow other resources to be dispatched and catch-up with the ramp. Hydropower can ensure such requirement because of having a fast ramp rates and can keep the maximum power for several hours. Also the flywheels can ramp in a few seconds and maintain the production until 15 min.

Ancillary services such as regulation and load following capacity need to be increased, in the case of a high level of intermittent sources. The traditional way to ensure such services is hydropower. However, other technologies can be used, such as NAS (sodium–sulphur) batteries or flywheels. Storage systems that incorporate an inverter can also deliver reactive power, supporting the voltage regulation.

In the storage energy technologies (Fig. 19), only hydropower is being utilized from many years ago and is well-established in the market. The other storage technologies present noncompetitive costs so far (but decreasing) and reduced commercial availability. Certain storage systems such as flywheels, flow cells, and certain battery types could become viable. Another viable technology is compressed air depending on available locations, which is stored in geologic structures under the ground and released when necessary.

4 Solutions Adopted in the Existent Markets

Several European electricity markets are already dealing with large penetration of wind power and its intermittency. Countries such as Denmark, Germany, and Spain have large percentages of Wind Power on the grid and need to face several intermittency problems. The actual approach to enable the wind integration includes market-based approaches as well as technical solutions, which are the main subject of this paper.

The technical solutions to reduce the intermittence of the wind power are already widely used. Either the technical or the geographic distribution of the generators (Sect. 3.2) are used to reduce the intermittency, and studies of the wind power potential, in the different regions, are made helping the determination of the locals that can be used to minimize the overall variation of the wind power production. However, large imbalances remain and the uncertain part of the variability is left for the power system reserves. During the operating hours, the imbalance of wind power is added to all other imbalances in the power system, and they are jointly treated in the supply-demand balance settlement.

In what concerns the short-term reserves, the limited wind power variations are being well handled by the capabilities of the existing systems. As an example, in Spain there is no need to increase secondary reserve (short-term reserve available within 5 min) due to wind penetration (Eriksen et al. 2005).

The demand-side management (Sect. 3.8) and demand response (Sect. 3.9) technologies are broadly used, but so far not with the aim of the integration of renewable resources. However, the demand-side management technologies with larger impact in the peak hours (e.g., HVAC) have a major role in avoiding the most unbalanced situations, and therefore the use of such technologies deserves to be increased. Also the demand response technologies usually used due to economical concerns (mainly in the USA) will have a potential major role in the short-term reserves requirements, by instantly adjusting the consumption when critical events occur. The curtailment of the wind power (Sect. 3.5) is already used, mostly due to congestion reasons, mainly in the offshore plants in the North of Europe. However, this methodology is used only in critical situations due to the economical disadvantages.

The requirements on longer-term balancing reserves caused by wind power are inherent to the wind speed forecasting error. The forecast mainly focuses on predicting the power output of the aggregated wind power capacity in the control area for a time horizon of 1–48 h, thereby providing the basis for scheduling the required longer-term balancing power reserves (15 min–8 h). Hence, increasing wind power penetration usually requires more frequent usage of long-term reserves, because the predictability of wind power generation has a higher uncertainty than load forecasting (forecasting errors of 30% are possible over 24 h). However, new improved forecasting techniques (Sect. 3.1) are already used, enabling an increasing reduction of the error. So far European power systems have sufficient reserve capacity available to provide long-term reserve capacity, and there has been no need for new plants to be constructed just for that purpose (Eriksen et al. 2005).

The use of other power plants to provide operational and capacity reserve (Sect. 3.4) was the most used method for the integration of intermittent power. Nowadays such method is already used, but the use of fossil fuel power plants (open-cycle gas turbines or steam-fired power plants) is decreasing, due to the environmental and economic disadvantages. Distributed generation technologies (Sect. 3.6) are already used in substitution of the conventional power plants and such utilization expected an increase in the next years (e.g., cogeneration, microturbines, and fuel cells). However, hydropower remains the most used technology to provide operational and capacity reserve due to the dynamic response and energy storage capacities. Other energy storage technologies (e.g., different kinds of batteries, electrochemical flow cells, flywheels, and the plug-in hybrid vehicles) are already used, but only at small scale level, due to the so far noncompetitive costs and reduced commercial availability (Sect. 3.10). With the expected costs reduction, such technologies may have a larger implementation in the near future to address the intermittence reduction.

Other methodology that can help the integration of renewable resources is the analysis of the complementarity between renewable sources (Sect. 3.7). Nowadays, a few experiences have been made at a small scale level. In areas where there is potential to exploit several renewable resources, the complementarity between all the renewable sources deserves to be analyzed. Such process will help the selection process of the renewable technologies to install and will minimize the storage requirements. This approach will become more attractive due to the expected costs reduction of such forms of production in the next years (e.g., photovoltaic costs are now converging to grid parity).

The interconnection between electric grids is the most widely used methodology (Sect. 3.3) to ensure power systems reliability, including the integration of intermittent generation. The increased interconnection capabilities allow the use of the described technologies and methodologies outside the borders of only one country, enabling the creation of markets to deal with the intermittence. Also at the country level the market mechanisms make easier the application of the technologies and methodologies for balancing the intermittence. At present, several balancing service markets already operate in Europe and North America. However, to achieve the objective of creation of larger markets of balancing services, harmonization of regulations and technical issues will be needed, because the rules for electric markets are often made at a national level.

The markets mechanisms can be real time energy markets, ancillary services markets, or traditional regulation. As general tendency, supply-demand balancing is a task for the grid operator, but the market mechanism through which the associated financial transactions take place varies between countries. Possible examples can be indicated:

- In the Nordic liberalized electricity market, balancing is the responsibility of so-called balance responsible players (BRPs). Wind energy goes through BRPs and payments for deviations in schedule have to be paid according to the contracts for balancing that the producers have made. The balancing power can be bought locally or from the regulating power market operated by the Nordel TSOs.

- In Germany the TSOs take care of the balancing in their control areas with a market for primary and secondary reserves.
- In Ireland balancing is taken care of by the TSO via the "top-up" and "spill" imbalance market that all other operators use.
- In Spain, balancing is taken care of also by the TSO, and the imbalances are charged at a fixed tariff or with a market option.

5 Conclusion

To face the wind power intermittence several options can be considered. In first place, the intermittence can be reduced by using project techniques, such as grid integration, geographic distribution of the generators, and improved forecasting techniques. However, in large scale integration of wind power, periods of large intermittence will remain.

To face such intermittence, the most traditional solutions were based on back-up power plants providing operational and capacity reserve, as well as the interconnection with other grid systems. Other options can be used, such as the curtailment of intermittent generation or the use of distributed dispatchable generation. However, the options that can have the greatest impact are complementarity between renewable sources, demand-side management, demand response, and energy storage.

The option to mix complementary energy sources like wind power, solar power, and hydropower will mitigate the problems, when comparing with concentrating in only one source of renewable energy. More important is the utilization of dispatchable technologies of renewable generation, such as biomass, that can compensate the fluctuations of the other sources. As a complement, energy storage technologies can be used. DSM and DR can also have a major role, either by reducing the needs of new intermittent capacity or by adjusting the consumption in real time to face production variations.

References

Ackermann T (2005) Wind power in power systems. Wiley, Chichester

Chompoo-inwai C et al (2005) System impact study for the interconnection of wind generation and utility system. IEEE Trans Ind Appl 41(1):163–168

de Almeida A, Moura P, Marques A, Almeida J (2005) Multi-impact evaluation of new medium and large hydropower plants in Portugal centre region. Renew Sustain Energ Rev 9(2):149–167

DeCarolis J, Keith D (2005) The costs of wind's variability: is there a threshold?. Electricity J 18:69–77

DOE (2008) 20% Wind Energy by 2030, Increasing Wind Energy's Contribution to US Electricity Supply, United States Department of Energy DOE/GO-102008-2567, May 2008

Eriksen PB, Ackermann T, Abildgaard H, Smith P, Winter W, Graca JR (2005) System operation with high wind penetration. IEEE Power Energ 3:65–74

Ernst B (1999) Analysis of wind power ancillary services characteristics with German 250 MW wind data, p 38. NREL Report No. TP-500-26969
Georgilakis PS (2008) Technical challenges associated with the integration of wind power into power systems. Renew Sustain Energ Rev 12(3):852–863
Giebel G, Brownsword R, Kariniotakis G (2003) The state-of-the-art in short-term prediction of wind power. A literature overview. EU project ANEMOS (ENK5-CT- 2002-00665)
Grubb M, Meyer N (1993) Wind energy: resources, systems, and regional strategies. In: Johansson TB, Burnham L (eds) Renewable energy: sources for fuels and electricity. Island Press, Washington, DC
Hamacher T, Haase T, Weber H, Dweke J (2004) Integration of large scale wind power into the grid, Power-Gen Europe, 2004
Holttinen H (2005) Hourly wind power variations in the nordic countries. Wind Energ 8(2):73–195
Holttinen H (2007) Design and operation of power systems with large amounts of wind power, State-of-the-art report, International Energy Agency, IEA WIND Task 25, 2007
IEA (2003) Demand response in liberalised electricity markets, International Energy Agency, 2003
IEA (2005) Management Options and strategies for variability of wind power and other renewables, International Energy Agency, 2005
Kennedy S (2005) Wind power planning: assessing long-term costs and benefits. Energ Policy 33(13):1661–1675
Kushler M et al (2003) Using energy efficiency to help address electric systems reliability. Energy 28(4):303–317
Lund H (2005) Large-scale integration of wind power into different energy systems. Energy 30(13):2402–2412
MacDonald M (2003) Intermittency literature survey, Annex 4 to the Carbon Trust and DTI Renewables Network Impact Study, 2003
Milborrow D (2001) Penalties for intermittent sources of energy. Working Paper for PIU Energy Review, 2001
Milborrow D (2004) Assimilation of wind energy into the Irish electricity network. Sustainable Energy, Ireland, Dublin
Milligan M, Parsons B, Smith JC, DeMeo E, Oakleaf B, Wolf K, Schuerger M, Zavadil R, Ahlstrom M, Nakafuji D (2006) Grid impacts of wind variability: recent assessments from a variety of utilities in the United States. Report No. NREL/CP-500-39955. National Renewable Energy Laboratory (NREL), Golden, CO, 2006
Moura P, de Almeida A (2008) Large scale integration of wind power generation, Clean Technology 2008, CTSI Clean Technology and Sustainable Industries Conference and Trade Show, Boston, USA, 1–5 July, 2008
Multon B, Robin G, Erambert E, Ben Ahmed H (2004) Colloque Energie lectrique: besoins, enjeux,technologies et applications, Stockage de l'nergie dans les applications stationnaires, SATIE, Belfort, 18 June 2004
Pedersen J (2005) System and Market Changes in a Scenario of Increased Wind Power Production. Report No. 238389. Energinet.dk, Fredericia, Denmark, 2005
Piwko R, Xinggang B, Clark K, Jordan G, Miller N, Zimberlin J (2005) The effects of integrating wind power on transmission system planning, reliability, and operations. Schenectady, NY, Prepared for The New York State Energy Research and Development Authority (NYSERDA) by Power Systems Energy Consulting, General Electric International, Inc, 2005
Schneller C (2004) Integrating wind power into the power supply system, Intelligent Policy Options, 2004
Zavadil R, King J, Xiadon L, Ahlstron M, Lee B, Moon D, Finley C et al (2004) Wind Integration Study – Final Report. Prepared for Xcel Energy and the Minnesota Department of Commerce. Knoxville, TN, EnerNex and WindLogics, 2004

Optimization Models in the Natural Gas Industry

Qipeng P. Zheng, Steffen Rebennack, Niko A. Iliadis, and Panos M. Pardalos

Abstract With the surge of the global energy demand, natural gas plays an increasingly important role in the global energy market. To meet the demand, optimization techniques have been widely used in the natural gas industry, and has yielded a lot of promising results. In this chapter, we give a detailed discussion of optimization models in the natural gas industry, with the focus on the natural gas production, transportation, and market.

Keywords Gas market · Gas recovery · Gas transmission · Mixed integer nonlinear programming (MINLP) · Mixed integer programming (MIP) · Natural gas industry · Optimization

1 Introduction

Concerned about global warming and shortage of crude oil, people become more interested in natural gas, which is a relatively clean energy source and abundant in many places. Natural gas mainly consists of methane, and when burnt, it releases a fair amount of energy and less green house gases (e.g., CO_2) than oil and coal. As we can see from Fig. 1, the world gas consumption/production is linearly growing since 1980 from approximately 52,890 billion cubic feet to approximately 104,424 billion cubic feet in 2006, according to the US Department of Energy (Energy Information Administration 2008). Moreover, the natural gas consumption is expected to continue to grow linearly to approximately 153 trillion cubic feet in 2030, which is an average growth rate of about 1.6% per year (International Energy Outlook 2009).

In 2008, the residential use of natural gas accounted for 21%, the commercial use for 13%, the industrial use for 34%, the transportation for 3%, and the electric power

Q.P. Zheng (✉)
Department of Industrial & Systems Engineering, Center for Applied Optimization
University of Florida, Gainesville, FL 32611, USA
e-mail: zqp@ufl.edu

S. Rebennack et al. (eds.), *Handbook of Power Systems I*, Energy Systems,
DOI 10.1007/978-3-642-02493-1_6,

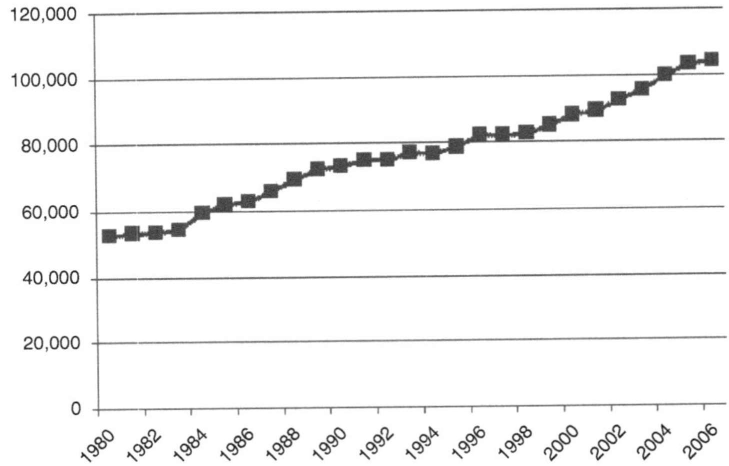

Fig. 1 World gas consumption in billion cubic feet (Energy Information Administration 2008)

production for 29% (AER 2009). The industrial sector is expected to remain the largest end-use sector for natural gas through 2030 with an expected share of 40% (International Energy Outlook 2009). The electric power generation from natural gas was the second largest consumer of natural gas after the industrial sector in 2006. The electricity generation accounted in 2006 for 32% of the world's total natural gas consumption. Because of the worldwide discussions/attempts to reduce green house gas emissions, the electricity generation via natural gas is expected to become even more important and its share of the world's total natural gas consumption is expected to increase to 35% in 2030 (International Energy Outlook 2009). Hence, natural gas remains an important source of energy for both the industrial and the electricity sectors.

This chapter discusses different optimization models in the natural gas industry. We focus on three different key applications: the natural gas production, the natural gas transportation, and the natural gas market. This chapter is organized in such a way that we start with the introduction of the problem itself, and then discuss a mathematical formulation of the problem and finally review solution techniques to solve these models. However, especially when well known algorithms, such as Branch & Cut, are used to solve the mathematical programs, we do not go into details but refer to the literature instead.

Section 2 discusses the optimization applications in gas recovery and production. Two problems are discussed in this section, the production scheduling problem and the maximal recovery problem. Section 3 focuses on gas transportation, where the network design problems and the optimal fuel cost problem are discussed. The natural gas market is discussed in Sect. 4, where both regulated and deregulated market models are considered. We conclude with Sect. 5.

2 Optimization in Gas Production (Recovery)

There is still a huge amount of natural gas reserves in the world: in 2009, the reserves were estimated at 6,254 trillion cubic feet; 69 trillion cubic feet above the estimate for 2008 (Worldwide Look at Reserves and Production 2008). This follows the general upwards trend of the world natural gas reserves over years. With a share of approximately 40.7%, the Middle East has the largest natural gas reserves of the world, followed by Eurasia with 32.2% and Africa with 7.8% (Worldwide Look at Reserves and Production 2008). On the country level, Russia has approximately 26.9% of the world's natural gas reserves and holds together with Iran (15.9%) and Qatar (14.3%) approximately 57% of the world's natural gas reserves, while the top 20 countries hold together 90.7% (Worldwide Look at Reserves and Production 2008). Interestingly, for most regions, the reserves-to-production rates are substantial, with an worldwide estimate of 63 years (BP 2008). Hence, natural gas production and recovery will continue to be an important task in the future.

Optimization models and techniques are applied extensively in natural gas recovery processes, such as production scheduling, placement of well head, gas recovery systems, or facilities designs. For a survey on gas and oil recovery and production, we refer the reader to Horne (2002). These optimization problems are computationally difficult to solve. One reason is that a huge number of parameters are subject to uncertainties. Other reasons are the nonlinear/nonsmooth/nonconvex functions and constraints, due to the properties of the gas production operations as explained in Beggs (1984). In the following, we discuss some specific optimization problems occurring in gas production.

2.1 *Production Scheduling Considering Well Placement*

Usually, a gas reservoir is accessed by drilling multiple wells on its surface. Also gas withdrawal from any of the wells will lead to pressure reductions at all wells drilled on the same reservoir. Then the pressure reductions will come back to decrease the withdrawal rate at every well for the next period. The optimal production scheduling problem is to find the optimal withdrawal rate at every drilled well at each time period while determining the well location at the same time.

2.1.1 Mixed Integer Linear Programming Formulation

Murray and Edgar (1978) formulate this problem as a mixed integer linear programming (MILP) problem. They try to determine the optimal well configuration (withdrawal rates) while satisfying the demand schedule without exceeding it. Drilling or not at a particular location, i, can be denoted by a binary variable, say, y_i. Hence, the drilling decision can only be made at particular locations i, which have to be identified beforehand. Also, use q_i^k to denote the withdrawal rate from well i at time period k. The interaction between withdrawal rates and pressures at all the wells can be delineated by the following gas flow equation:

$$\nabla k_g \nabla \Phi + q = \phi c_t \frac{\partial \Phi}{\partial t}, \tag{1}$$

where $\Phi = 2\int_0^p \frac{\rho}{z(\rho)\mu(\rho)} d\rho$. Including this constraint in a mathematical programming formulation leads to huge computational difficulties. However, as stated in Murray and Edgar (1978), this nonlinear constraint has a very good linearization substitute, called influence equations (Al-Hussainy 1967; Wattenbarger 1970). In these equations, the pressure drop at well i is a linear function of withdrawal flow rates from all drilled wells. This is defined by influence function matrices, Φ^k, $k = 1, \dots, m$, where Φ_{ij} denotes the pressure drop at well i for a unit flow at well j during time period k. The maximal profit problem can be formulated as follows:

$$\max \quad \sum_{k=1}^{m} \sum_{i=1}^{n} b_i^k q_i^k \tag{2}$$

$$\text{s.t.} \quad \sum_{j=1}^{n} \Phi_{ij}^k q_j^k = p_i^k, \qquad i = 1, \dots, n, k = 1, \dots, m, \tag{3}$$

$$\sum_{j=1}^{n} \Phi_{ij}^k q_j^k \le \bar{p}_i^k, \qquad i = 1, \dots, n, k = 1, \dots, m, \tag{4}$$

$$\sum_{k=1}^{l} \sum_{j=1}^{n} \Phi_{ij}^k q_j^k \le \hat{p}_i^l, \qquad i = 1, \dots, n, l = 1, \dots, m, \tag{5}$$

$$\sum_{j=1}^{n} q_j^k \le d^k, \qquad k = 1, \dots, m, \tag{6}$$

$$q_i^k \le M_i y_i, \qquad i = 1, \dots, n, \tag{7}$$

$$q_i^k \ge 0, \qquad i = 1, \dots, n, k = 1, \dots, m, \tag{8}$$

$$y_i \in \{0, 1\}, \qquad i = 1, \dots, n, \tag{9}$$

where, from well i during time period k, b_i^k is the benefit of one unit gas flow, p_i^k is the pressure reduction, and $\bar{p}_i^k$ is the maximal pressure reduction at period k. $\hat{p}_i^l$ is the maximal total pressure drop allowed from the initial time point to time period l and d^k is the demand at time k. M_i is a big number to bound the withdrawal flow rate if $y_i = 1$. Its objective function is the total benefit from the withdrawal of gas. Constraints (3) compute the pressure drop at each well location during every time period. Constraints (4) specify the upper bound by which the pressure can drop during a specific single period for each well location. Also there is an upper bound by which the pressure can drop during the period between the initial time point and the current time period, which is stated in constraints (5). Constraints (6) ensure that the total gas withdrawal from all wells does not exceed the demand at each time period. Constraints (7) show that only drilled wells can have a positive withdrawal flow rate. This results in a mixed integer programming (MIP) problem, which can be solved by well Branch & Bound or Branch & Cut techniques. We refer the reader to Horst et al. (2000); Locatelli and Thoai (2000); Nemhauser and Wolsey (1999); Wolsey (1998); Kallrath and Wilson (1997) for comprehensive discussions of these techniques.

Let us discuss now the drawbacks of the proposed model (2)–(9). The model does not include any other cost such as well drilling cost, it does not take into account the relationship between the profit coefficient b_i^k and the demand d^k, and it assumes that the operator can choose any flow rate without considering the concurrent wellhead pressure. Also, after the deregulation of the natural gas market, constraint (6) is not necessary and can be incorporated into the objective function instead. Furthermore, the different periods are intercorrelated to each other. For instance, the price of gas at time period t will affect the demand at the next time period $t + 1$ and vice versa. By incorporating all these factors, a nonlinear mixed integer programming problem can be formulated.

2.1.2 Nonlinear Programming Formulation

A multiple-stage nonlinear optimization problem is also proposed by Murray and Edgar (1978). They formulate a nonlinear problem for each time period taking into account the interactions between two consecutive stages. The objective function for each time period k incorporates more factors such as the well placement cost, compressor operating cost, compressor setup cost, and the price of gas, which is shown as follows:

$$f^k = \sum_{j=1}^{n} \left(Aq_j^k - C_w \frac{q_j^k}{q_j^k + \epsilon} - U_j^k q_j^k - C_j^k q_j^k + D_j^k\right), \tag{10}$$

where A is the price per unit gas flow and C_w is the setup cost of any well placement. Instead of using the binary variables y_i to denote whether a well is drill or not, this nonlinear programming formulation uses the term $\frac{q_j^k}{q_j^k+\epsilon}$ to approximate y_i, where ϵ is a small constant compared to the magnitude of gas withdrawal flow rates q_j^k, $j = 1, \ldots, n, k = 1, \ldots, m$. To be able to use this approximation, the magnitude of the flow rates are assumed to be known. U_j^k is the operating cost of the compressors for a unit flow of q_j^k. $-C_j^k q_j^k$ and D_j^k approximate the setup cost of a compressor at this location before time period k. Setting $D_j^k = C_j^k q_j^{k-1}$ makes the summation of these two terms equal to 0, which ensures that the compressor setup cost occur only once. For the nonlinear formulation, the deliverability equations are considered besides the constraints in the MIP formulation. The deliverability constraints specify the relationship between the withdrawal rate and the well head pressure, which is also approximated by linear functions and shown as follows:

$$q_j^k \leq e_j^1 + e_j^2 \rho_j^k, \qquad j = 1, \ldots, n, k = 1, \ldots, m, \tag{11}$$

where e_j^1 and e_j^2 are the linear coefficients and ρ_j^k is the bottom-hole pressure at well site j after time period k.

Also a multi-stage based algorithm is proposed in Murray and Edgar (1978), in which all stages (time periods) are solved in a sequential order from 1 to m. We describe this algorithm as follows:

Step 1: Set up the problem: obtain parameters, e_j^1 and e_j^2, by some regression techniques; assume that no compressor is needed initially and set $U_j^1 = C_j^k = D_j^k = 0$; start from the first period problem.

Step 2: Solve the period k problem with an appropriate nonlinear programming algorithm, such as the gradient projection method (Rosen 1960).

Step 3: Examine the dual variables of the deliverability constraints. If none is positive, an optimal solution has been found for time period k, then go to Step 6. Otherwise, go to Step 4.

Step 4: If all positive dual variables are associated with deliverability constraints of the lowest feasible delivery pressure, an optimal solution is found for time period k, then go to Step 6. Otherwise, go to Step 5.

Step 5: Select the deliverability constraint with the largest associated dual variable, and then relax this constraint to the next lowest delivery pressure. Go to Step 2.

Step 6: By using the current period optimal solution, update the parameters in the next period problem. If $k = m$, terminate the whole program. Otherwise, set $k = k + 1$, and go to Step 2.

The drawback of the proposed model is that it does not consider all time periods together but considers them separately. Obviously, with this approach, an optimal solution to the practical problem cannot be obtained, as the interactions among all time periods are not taken into account.

2.2 *Total Gas Recovery Maximization: An Optimal Control Formulation*

To withdraw as much natural gas from a reservoir as possible, one option is to use waterflooding. This leads to the following immediate question. What is an optimal water injection rate with respect to different objectives, such as the maximal ultimate recovery, or the total revenues? A lot of models have been proposed for this problem. Mantini and Beyer (1979) proposed optimal control models to this system and defined several objective functions due to different aspects of the problem.

Now, suppose there are two wells drilled on the surface of the gas reservoir, one for gas recovery and one for water injection. Therefore, let $r(t)$ denote the withdrawal rate of gas, which is bounded by the maximum rate of gas extraction $r_m(t)$. Through the water injection, well water is injected into the reservoir at the nonnegative rate $s(t)$. This model assumes a constant g, which is the ratio of gas entrapped behind the injected water to the volume of water at any time. The model to maximize the ultimate gas recovery can then be stated as

$$\max \int_0^\infty r(t)dt \tag{12}$$

$$\text{s.t. } PV = NRT, \tag{13}$$

$$\frac{dV}{dt} = -s(t) - gs(t), \tag{14}$$

$$\frac{dN}{dt} = -r(t) - \frac{gs(t)P(t)}{RT}, \tag{15}$$

$$r_m(t) \geq r(t) \geq 0,$$

$$s(t) \geq 0,$$

where $P(t)$ and $V(t)$ are the pressure and volume of the gas reservoir, $N(t)$ is the amount of gas which is not entrapped at time t, R is the universal constant of gas, and T is the temperature. Constraint (13) is the ideal gas law, constraint (14) shows the entrapped gas equals to constant g times the volume of the water, while constraint (15) states that gas is entrapped at the current pressure in the reservoir and remains at the same pressure and has no effect on the reservoir. By introducing another variable $Q = P/RT$ and plugging constraint (13) into constraint (15), a more concise model can be obtained

$$\begin{aligned} \max \quad & \int_0^\infty r(t)dt \\ \text{s.t.} \quad & \frac{dV}{dt} = -(1+g)s(t), \\ & \frac{dQ}{dt} = \frac{-r(t) + P(t)s(t)}{V(t)}, \\ & r_m(t) \geq r(t) \geq 0, \\ & s(t) \geq 0. \end{aligned}$$

Mantini and Beyer (1979) also discuss several other objective functions. For example, the objective function to maximize the present worth value of the net revenues for internal rate of return, ρ, not equal to 0, is

$$\beta \int_0^\infty e^{-\rho t}[r(t) - \alpha s(t)]dt,$$

where α is the ratio of the water price (per cubic meter) to the gas price (per mole) and β is the gas price per mole. Owing to the presence of the differential equations, these problems are generally computationally difficult to solve. However, Mantini and Beyer established a very interesting theorem, characterizing the properties of (some) optimal solutions of the control variable $r(t)$ and $s(t)$. Let us re-state this theorem here.

Theorem 1. *(Mantini and Beyer 1979) The objective function* $\int_0^\infty r(t)dt$ *is maximized by any functions* $\hat{r}$ *and* $\hat{s}$ *such that,*

$$\int_0^{t_1} \hat{r}(t) = V_0(P_0 - P_c), \tag{16}$$

$$\int_0^{t_2} \hat{r}(t) = \frac{P_c(V_0 - V_c)}{1+g}, \tag{17}$$

$$\hat{r}(t) = 0, \forall t > t_2, \tag{18}$$

$$\hat{s}(t) = \begin{cases} 0, & 0 \leq t < t_1, \\ \frac{\hat{r}(t)}{P_c}, & t_1 \leq t \leq t_2, \\ 0, & t > t_2. \end{cases} \tag{19}$$

for t_1 *and* t_2 *are any numbers with* $0 < t_1 < t_2$*, where* P_0 *and* V_0 *are the initial pressure and volume, respectively, and gas recovery stops when* $P \leq P_c$ *or* $V \leq V_c$*.*

This theorem leads to the interesting statement that it is optimal to start the waterflooding when the first time P is lower than P_c; that is, the entrapped gas is at the lowest possible pressure. Although, in practice, this may not be valid for some specific gas wells due to discrepancies between modeling and reality.

3 Natural Gas Pipeline Network Optimization

Originally natural gas was treated as a byproduct of crude oil or coal mining and was spared. The flares in the mining field were usually natural gas (Tussing and Barlow 1984). Not until the introduction of pipelines did the natural gas become one of the major sources of energy. The earliest gas pipelines were constructed in the 1890s and they were not as efficient as those that we are using nowadays. The modern gas pipelines did not come into being until the second quarter of the twentieth century. Because of the properties of natural gas, pipelines were the only way to transport it from the production sites to the demanding places, before the concept of Liquefied Natural Gas (LNG). The transportation of natural gas via pipelines remains still very economical, but it is highly impractical across oceans. Although LNG market is burgeoning in high speed now, pipeline network remains the main transportation system for natural gas.

Gas pipelines play a major role in energy supply and security. The Nord Stream Gas Pipeline (NSGP) project, transporting Russian gas to Germany, is one of the recent large scale pipeline projects (Cameron 2007). The NSGP is planned as a twin-pipeline with a total capacity of 55 billion cubic meters per annum. The estimated investment cost is 4 billion €, financed by a joint venture of the three companies

JSC Gazprom, BASF AG, and E.ON AG. Not least, the decision to build the marine pipeline was driven politically, passing by Poland, Lithuania, Estonia, Belarus and Ukraine, to increase the natural gas supply security for Germany, mainly.

After the post war gas pipeline boom, a lot of research has been done in optimization applications to pipeline networks; for instance, how to setup the pipeline network, how to determine the optimal diameter of the pipelines, how to allocate compressor stations in the pipeline network, and what is the minimal fuel consumption of the network. Typically, the mathematical programming formulations of the pipeline optimization problems contain a lot of nonlinear/nonconvex/nonsmooth constraints and functions. The most common constraints are the so-called Weymouth panhandle equations, which relate the pressure and flow rate through a segment of pipeline (i, j). They read as follows

$$\text{sign}(f_{ij})f_{ij}^2 = p_i^2 - p_j^2, \qquad (i, j) \in A_p, \tag{20}$$

where f_{ij} is the flow rate of pipeline (i, j), p_i and p_j are the pressures at node i and j, respectively. Hence, the direction of the gas flow depends on the pressure difference of the two nodes i and j. Therefore, the nonsmooth function $\text{sign}(f_{ij})$ is needed.

Recently, more research is related to the network optimization of gas transmission, given the network structure other than the design of the network topology. One of the few papers dealing with the design of network topology is the one by Rothfarb et al. (1970), where the authors propose a tree generating algorithm to design the network topology.

3.1 Compressor Station Allocation Problem Considering Pipeline Configurations

Once a network topology is chosen, one problem is to determine the optimal configuration of the pipelines and the location of the compressor stations in this network. Because of the high setup cost and high maintenance cost, it is desirable to have the best network design with the lowest cost. This problem concerns a lot of variables: the number of compressor stations, which is an integer variable, the pipeline length between two compressor stations, the diameters of the pipelines, and the suction and discharge gas pressures at compressor stations. This problem is computationally very challenging since it includes not only nonlinear functions in both objective and constraints but, in addition, also integer variables.

A simple and typical network for this type of problem is shown in Fig. 2. Node s is the supply node where the gas is produced. Nodes a and b are the demand nodes where the gas is consumed. The trapezoids 1 through 6 denote the compressor stations. There are three branches: s to 3 is the first branch, 3 to a is the second branch, and 3 to b is the third branch.

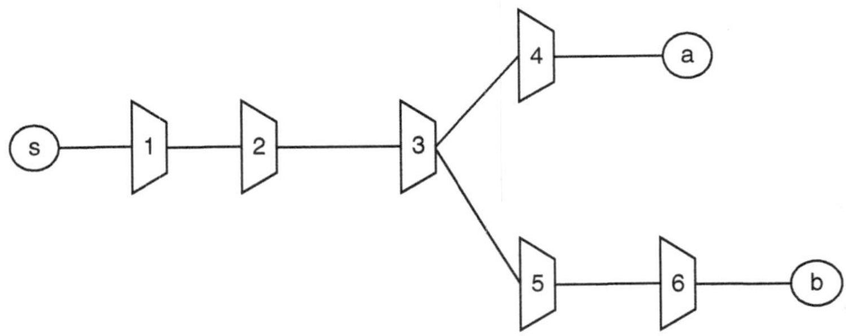

Fig. 2 A gas pipeline network configuration problem with three branches

Suppose there are at most n compressor stations to be set up, and n_1, n_2, and n_3 denote the number of compressor stations on branch 1, 2, and 3, respectively. For each pipeline segment i, there are five associated parameters: the flow rate f_i, the discharge pressure (from the upstream compressor) p_i^d, the suction pressure (from the downstream compressor) p_i^s, the diameter d_i, and the length l_i.

The formulation for the three branches problem by Edgar et al. (Edgar and Himmelblau 2001; Edgar et al. 1978) reads as follows

$$\min \quad \sum_{i=1}^{n}(O_y + C_c)\alpha\frac{T_s}{\eta_s}\frac{\gamma}{\gamma-1}\left[1-\left(\frac{p_i^d}{p_i^s}\right)^{\frac{z(\gamma-1)}{\gamma}}\right]+\sum_{i=1}^{n+1}C_l l_i d_i \tag{21}$$

$$\text{s.t.} \quad p_i^d \geq p_i^s, \qquad i = 1, \ldots, n, \tag{22}$$

$$p_i^d \leq K_i p_i^s, \qquad i = 1, \ldots, n, \tag{23}$$

$$\underline{p}_i^d \leq p_i^d \leq \bar{p}_i^d, \qquad i = 1, \ldots, n, \tag{24}$$

$$\underline{p}_i^s \leq p_i^s \leq \bar{p}_i^s, \qquad i = 1, \ldots, n, \tag{25}$$

$$\underline{l}_i \leq l_i \leq \bar{l}_i, \qquad i = 1, \ldots, n, \tag{26}$$

$$\underline{d}_i \leq p_i^d \leq \bar{p}_i^d, \qquad i = 1, \ldots, n, \tag{27}$$

$$f_i = A d_i^{\frac{8}{3}}\left[\frac{(p_i^d)^2-(p_i^s)^2}{l_i}\right]^{\frac{1}{2}} \qquad i = 1, \ldots, n, \tag{28}$$

$$\sum_{i=1}^{n_1} l_i + \sum_{i=n_1+1}^{n_1+n_2} l_i = L_1, \tag{29}$$

$$\sum_{i=1}^{n_1} l_i + \sum_{i=n_1+1}^{n_1+n_3} l_i = L_2, \tag{30}$$

where γ is the ratio of specific heats, T_s is the suction temperature, z is the gas compressibility factor, η_s is the efficiency factor, O_y and C_c are cost functions with respect to horsepower. The objective function (21) contains two parts, of which the first is the compressor station costs and the second is the maintenance costs of the pipeline segments. Constraints (22)–(27) are the upper and lower bounds on pressures, pipeline lengths, and diameters. L_1 and L_2 are the distances between the supply node and two demand nodes.

Model (21)–(30) can be solved by applying Branch and Bound techniques using reduced gradient nonlinear optimization method to solve the subproblem at each node in the Branch and Bound tree (Edgar and Himmelblau 2001; Edgar et al. 1978). Alternatively, a differential evolution method is proposed to solve this problem formulation by Babu et al. (2008). The drawback of this model is that it highly depends on the topology of the network.

3.2 *Least Gas Purchase Problem and Optimal Dimensioning of Gas Pipelines*

In the modern natural gas industry, the gas production companies are rarely affiliated with the gas transmission and distribution companies. Thus, for gas distribution companies, one problem is to determine the best flow rate and gas pressures in each pipeline by which the least cost on purchasing gas from producers is achieved. This problem can be formulated as an optimization problem with linear objective function and nonlinear/noconvex constraints.

Consider now Fig. 3. s_1 and s_2 are the supplies for source nodes 1 and 2, the set of which is denoted by N_s. Nodes 6 to 9 are demand nodes with demands $-s_i$, $i = 6, 7, 8, 9$. In this model, there are two kinds of arcs: those with compressor stations such as $(1, 4)$ and $(2, 4)$, which is denoted by A_c, and those without, which are also called pipeline arcs and denoted by A_p. Flows on arcs with compressors are directed such that $f_{ij} \geq 0, \forall (i, j) \in A_c$, and flows on pipeline arcs are undirected and the direction depends on the pressures of both ends of this arc.

A mathematical programming formulation can be stated as

$$\min \quad \sum_{i \in N_s} c_i s_i \tag{31}$$

$$\text{s.t.} \quad \sum_{j \in A_i^+} f_{ij} - \sum_{j \in A_i^-} f_{ji} = s_i, \qquad \forall i \in N, \tag{32}$$

$$\text{sign}(f_{ij}) f_{ij}^2 = C_{ij}(p_i^2 - p_j^2), \qquad \forall (i, j) \in A_p, \tag{33}$$

$$f_{ij}^2 \geq C_{ij}(p_i^2 - p_j^2), \qquad \forall (i, j) \in A_c, \tag{34}$$

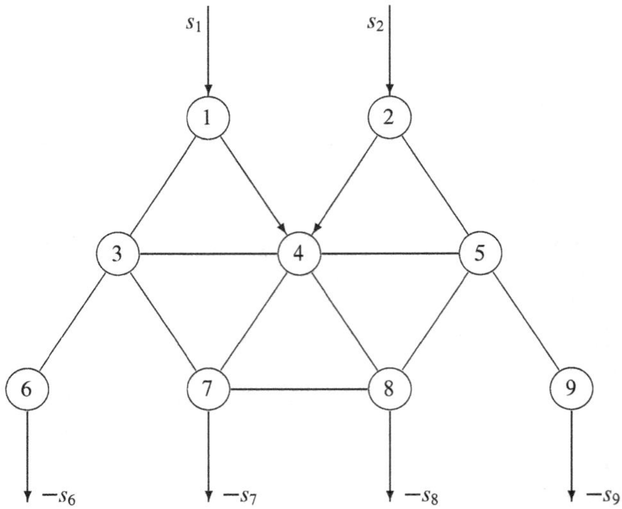

Fig. 3 Least cost problem network

$$\underline{s}_i \le s_i \le \bar{s}_i, \qquad \forall i \in N, \tag{35}$$

$$\underline{p}_i \le p_i \le \bar{p}_i, \qquad \forall i \in N, \tag{36}$$

$$f_{ij} \ge 0, \qquad \forall (i,j) \in A_c, \tag{37}$$

where p_i is the gas pressure at node i, c_i is the purchase cost per unit gas from supplier i, and C_{ij} is a coefficient for arc (i,j), which is determined by the length, diameter, and so on. A_i^+ denotes the set of arcs that are emanating from node i, while A_i^- denote the one of incoming arcs to node i.

The nonlinear constraints of the above model can be simplified by letting π_i substitute p_i^2. Then constraints (33), (34), and (36) can be replaced by

$$\text{sign}(f_{ij})f_{ij}^2 = C_{ij}(\pi_i - \pi_j), \forall (i,j) \in A_p,$$

$$f_{ij}^2 = C_{ij}(\pi_i - \pi_j), \forall (i,j) \in A_p,$$

$$\underline{\pi}_i \le \pi_i \le \bar{\pi}_i, \forall i \in N.$$

With this substitution, the "only" nonlinear functions left are $\text{sign}(f_{ij})$ and f_{ij}^2.

De Wolf and Smeers (2000) propose a piecewise linear programming algorithm to solve this problem, in which they construct a piecewise linear approximation to the nonlinear constraints and solve the relaxed problem by simplex algorithm extensions (De Wolf 1991). The performance of the algorithm depends highly on the choice of the initial point. It is crucial to have a good starting solution, which can be obtained by solving the following problem:

$$\min \quad \sum_{(i,j)\in A} \frac{|f_{ij}|f_{ij}^2}{3C_{ij}^2} \tag{38}$$

$$\text{s.t.} \quad \sum_{j\in A_i^+} f_{ij} - \sum_{j\in A_i^-} f_{ji} = s_i, \qquad \forall i \in N,$$

$$\underline{s}_i \le s_i \le \bar{s}_i, \qquad \forall i \in N.$$

The objective function (38) in this problem is the amount of mechanical energy consumed in the gas pipeline per unit time. Its KKT necessary conditions (see Bazaraa et al. (2006)) is equivalent to the constraints (32), (33), and (35). The KKT necessary point is a good approximation starting point, which does not take into account pressure bounds and the existence of compressors. The algorithm proposed by De Wolf and Smeers (2000) is as follows:

(o) **Initialization:** Let (f^0, p^0, s^0) be a vector of flows, pressures, and net supplies that satisfy constraints (32), (33), (34), (35), and (37). Replace the nonlinear function $\text{sign}(f_{ij})f_{ij}^2$ by a piecewise linear approximation including f_{ij}^0 as a breakpoint. Use f_{ij}^0 as starting point for the piecewise linear programming approach. Also set $k = 1$.

(i) **Iteration** k**:** Solve the approximation problem by the piecewise linear programming approach. Let (f^k, p^k, s^k) be the solution.

(ii) **Stopping rule:** Compute $\bar{f}_{ij}^k$ by the following equation:

$$\bar{f}_{ij}^k = \text{sign}(p_i^k - p_j^k)C_{ij}|(p_i^k)^2 - (p_j^k)^2|^{\frac{1}{2}}.$$

If the error $e_{ij}^k = \bar{f}_{ij}^k - f_{ij}^k$ is greater than a given tolerance, for example, 10^{-5}, then add $\bar{f}_{ij}^k$ as a new discretization point and return to step (i). Otherwise stop and the incumbent solution is optimal.

It can be noticed that the optimal objective function value of problem (31) is a function of the diameters of the pipelines, say, $Q(D)$, because the parameter C_{ij} of pipeline (i, j) is a function of the diameter, where $C_{ij} = K_{ij}D_{ij}$ and K_{ij} is a coefficient. If the network structure and the length of each pipeline are fixed, the investment problem is to find the best pipeline diameters that achieve the lowest investment cost, including both the gas purchase cost $Q(D)$ and the pipeline construction cost $C(D)$. They are given as

$$C(D) = \sum_{(i,j)\in A} = (k_G D_{ij}^2 + k'_G D_{ij} + k''_G)l_{ij},$$

where l_{ij} is the length of pipeline (i, j). Then the investment problem becomes

$$\min \quad C(D) + Q(D) \tag{39}$$
$$\text{s.t.} \quad D_{ij} \ge 0, \qquad \forall (i, j) \in A,$$

which is a bilevel programming problem. The second part of the cost function, $Q(D)$, is nonconvex/nondifferential and has an implicit domain. De Wolf and

Smeers (1996) propose how to get one generalized subgradient, as in the next proposition.

Proposition 1. *Denote by* f^*, s^*, π^* *an optimal solution of the operations problem (31). Let* w^*_{ij} *be an optimal value of the dual variable associated to constraint (33). Then*

$$(\ldots, w^*_{ij} 5K^2_{ij} D^4_{ij}, \ldots) \in \partial Q(D), \tag{40}$$

where $\partial Q(D)$ *is the generalized subdifferential.*

The investment problem (39) can be solved by a bundle method, which performs well for nondifferential optimization problems. By using a bundle method, we do not need to know the explicit domain of the objective function. Hence it is a good fit for the investment problem because the objective function domain is implicit. At each step, it only needs the value of the objective function and one of the generalized subgradient, which can be computed by (40). The dual variables, w_{ij}, can be obtained while solving the operations problems by using simplex algorithm extensions. Readers may find more comprehensive discussions of the bundle method in Hiriart-Urruty and Lemarechal (1993).

3.3 Minimum Fuel Consumption Problem

To let the consumer receive an acceptable withdrawal rate of gas, the pipeline needs to maintain a certain pressure. This is achieved by adding compressor stations in the network. One well known problem is the minimal fuel cost problem due to the fuel consumption of compressor stations, which are usually considered as special arcs in the network of this type of models. The minimal fuel cost problem has been widely discussed in the literature; see for instance Peretti and Toth (1982); Rios-Mercado (2002); Wu et al. (2000); Rios-Mercado et al. (2006); Chebouba et al. (2009); Goldberg (1983).

A typical gas pipeline network is shown in Fig. 4. Node s is the source node, and t, p, and q are the demand nodes. Arc (j,t) is an ordinary pipeline arc, arcs (i,j), (k,p), and (s,q) are compressor station arcs. In each compressor station (i,j), there are C_{ij} compressors, and the pressures at i and j are denoted by p_i and p_j, respectively. Let A' denote the set of compressor station arcs, A'' denote the set of ordinary pipe arcs, and V denote the node set. Then the minimal fuel cost problem can be stated as

$$\min \sum_{(i,j)\in A'} g_{ij}(x_{ij}, p_i, p_j) = \sum_{(i,j)\in A'} \frac{\frac{x_{ij} Z_i RT_i}{\omega}[(\frac{p_j}{p_i})^{\omega} - 1]}{\mu_{ij}} \tag{41}$$

$$\text{s.t.} \sum_{j\in A_i^+} x_{ij} - \sum_{j\in A_i^-} x_{ji} = b_j, \quad \forall i \in V \tag{42}$$

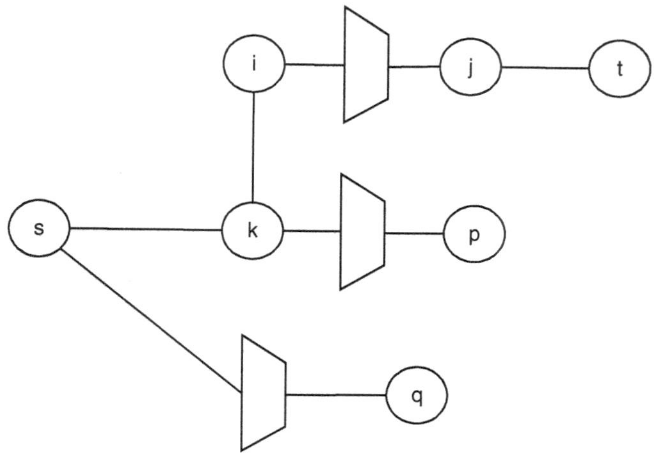

Fig. 4 A gas pipeline network

$$p_i^2 - p_j^2 = R_{ij}x_{ij}^2, \quad \forall (i,j) \in A'' \tag{43}$$

$$0 \leq x_{ij} \leq u_{ij}, \quad \forall i,j \in A \tag{44}$$

$$p_i^L \leq p_i \leq p_i^U, \quad \forall i \in V \tag{45}$$

$$(\frac{x_{ij}}{n_{ij}}, p_i, p_j) \in \mathcal{D}_{ij}, \quad \forall (i,j) \in A' \tag{46}$$

$$n_{ij} \in 0, 1, 2, \ldots, N_{ij}, \quad \forall (i,j) \in A', \tag{47}$$

where p_i^L and p_i^U are the lower and upper bounds on the pressure of node i. At each compressor station (i,j), u_{ij} is the capacity, $N_{i,j}$ is the total number of compressor, and x_{ij} and n_{ij} are the gas flow rate and number of compressor in use, respectively. Also there are several other related parameters for (i,j): z_i is the gas compressibility factor, T_i is the gas temperature, μ_{ij} is the compressor adiabatic efficiency, and R_{ij} is a gas constant. The most complicated constraint is (46) in which $\mathcal{D}_{ij}$ is the feasible domain of compressor station (i,j) as for variable triplet $(\frac{x_{ij}}{n_{ij}}, p_i, p_j)$. The feasible domain is stated below by the set of equations,

$$\frac{h_{ij}}{s_{ij}^2} = A_H + B_H(\frac{q_{ij}}{s_{ij}}) + C_H(\frac{q_{ij}}{s_{ij}})^2 + D_H(\frac{q_{ij}}{s_{ij}})^3 \tag{48}$$

$$\mu_{ij} = \frac{C_E(\frac{q_{ij}}{s_{ij}})^2 + B_E(\frac{q_{ij}}{s_{ij}}) + A_E}{100} \tag{49}$$

$$S_{min} \leq s_{ij} \leq S_{max} \tag{50}$$

$$Surge \leq \frac{q_{ij}}{s_{ij}} \leq Stonewall \tag{51}$$

$$h_{ij} = \frac{Z_i RT_i}{\omega}[(\frac{p_j}{p_i})^{\omega} - 1] \tag{52}$$

$$q_{ij} = Z_i RT_i \frac{x_{ij}}{p_i n_{ij}}. \tag{53}$$

In the above equations, q_{ij} denote the flow through the compressor unit, s_{ij} denote the speed of the compressor(s), and A_H, B_H, C_H, D_H, C_E, B_E, A_E are the compressor unit's constants.

This problem is very difficult to solve, and its solution algorithms are highly dependent on the topology of underlying network. Most of the algorithms for this problem are based on dynamic programming (Rios-Mercado 2002; Rios-Mercado et al. 2006; Peretti and Toth 1982) and gradient search approaches (Wu et al. 2000). Also meta heuristic approaches have been conducted, such as ant colony optimization (Chebouba et al. 2009) or genetic algorithms (Goldberg 1983).

4 Natural Gas Market Models

Government regulation over the gas industry dates back to the early days of natural gas usage. At the first glance, this seams to be reasonable, as government and the public are the main users of natural gas, and investments in the natural gas industry are tremendous. Not until the 1980s began the deregulation of this industry to improve both equity and efficiency of the natural gas market. Between the original producers and end users, there exists a variety of participants, each of which acts to optimize its own benefits. Under different government policies, a lot of natural gas market models are proposed. In this section we discuss optimization models of both a regulated and a deregulated gas market.

4.1 Reallocation Problem in a Regulated Natural Gas Market

O'Neil et al. (1979) propose a model on how to allocate gas to users with different priorities under the government regulations when encountering a gas shortage emergency. In this model there are multiple gas transmission systems among which any two systems are not necessarily connected physically. All users are divide into nine categories with priorities 1 through 9. The transportation network is composed of two types of arcs and nodes: the physical arcs and nodes, which really exist in practice – denoted by A_{phy} and N_{phy}, respectively – and the pseudo counterparts, which are for convenience of modeling – denoted by A_{pseudo} and N_{pseudo}, respectively. Let K_w be the set of users who withdraw gas from gas system w. This model also includes the panhandle constraints (20) for each of the pipeline arcs. However, instead of using the actual nonlinear constraints, this model incorporates two linearized approximation constraints in each iteration, which read as

$$-\epsilon_{ij} \leq -f_{ij} + \alpha_i p_i - \alpha_j p_j \leq \epsilon_{ij}, \quad \forall (i,j),$$

$$\epsilon_{ij}^1 \leq p_i - p_j \leq \epsilon_{ij}^2, \quad \forall (i,j),$$

where ϵ, ϵ_{ij}^1, and ϵ_{ij}^2 are parameters determined at each iteration through

$$\begin{aligned}
\epsilon_{ij} &= \alpha_1 |f_{ij}^{new}|, \\
\epsilon_{ij}^1 &= (1-\alpha_1)(p_i^{new} - p_j^{new}), \\
\epsilon_{ij}^2 &= (1+\alpha_1)(p_i^{new} - p_j^{new}), \\
\alpha_1 &= \max\{\alpha(\gamma_2)^m, \frac{1}{2}\delta_2\},
\end{aligned}$$

with the positive constants $\alpha, \gamma_2, \delta_2$.

The allocation algorithm proposed by O'Neil et al. (1979) is as follows:

Step 0: Allocate the minimum amounts that all users must receive. If no feasible solution exists, then stop; no allocation exists under the specified parameters.

Step 1: Allocate gas according to the priorities within each transporter's system, starting with priority 1 and proceeding in ascending order of priority.

Step 2: Determine if priorities 1 through 5 are satisfied. If so, go to step 4. Otherwise, fix the lower (6 through 9) priority users, in pipelines with a shortage in any higher priority, at their lower bounds.

Step 3: Allocate gas according to the priorities within the entire system.

Step 4: Incorporate the linearized nonlinear constraints and find the optimal solution minimizing the amount transferred between systems, as in the optimization problem (54)–(64).

The linear programming formulation used in the allocation problem (O'Neil et al. 1979) can be stated as

$$\min \quad \sum_{(i,j)\in I} |f_{ij}| + \sum_{(i,j)\in S} f_{ij} \tag{54}$$

$$\text{s.t.} \quad \sum_{j\in A_i^+} f_{ij} - \sum_{j\in A_i^-} f_{ji} = s_i - \sum_{k\in K}\sum_{l=0}^{9} d_{ikl}, \quad \forall i \in N, \tag{55}$$

$$\sum_{i\in N}\sum_{k\in K} d_{ikl} + u_l = \bar{d}_l, \quad l = 0,\ldots,9, \tag{56}$$

$$\sum_{l=1}^{5}\sum_{k\in K_w}\sum_{i\in N} d_{ikl} + r_w = g_w, \quad \forall w \in W, \tag{57}$$

$$-\epsilon_{ij} \leq -f_{ij} + \alpha_i p_i - \alpha_j p_j \leq \epsilon_{ij}, \quad \forall (i,j) \in A_{ps}, \tag{58}$$

$$\epsilon_{ij}^1 \leq p_i - p_j \leq \epsilon_{ij}^2, \quad \forall (i,j) \in A_{vc}, \tag{59}$$

$$0 \leq s_i \leq \bar{s}_i, \quad \forall i \in N, \tag{60}$$

$$\underline{d}_{ikl} \leq d_{ikl} \leq \bar{d}_{ikl}, \qquad \forall i \in N, k \in K, l = 0, \ldots, 9 \tag{61}$$

$$\underline{p}_i \leq p_i \leq \bar{p}_i, \qquad \forall i \in N, \tag{62}$$

$$u_l \geq 0, \qquad l = 0, 1, \ldots, 9, \tag{63}$$

$$r_w \geq 0, \qquad \forall w \in W, \tag{64}$$

where s is the supply, d is the demand, u is the slack variable for the demand of each priority, and r is the slack variable for the demand of priority 1 through 5. In constraints (58), $-f_{ij} + \alpha_i p_i - \alpha_j p_j$ is the linearized version of the panhandle equation, where α_i and α_j are the coefficients of the first order Taylor series expansion. A_{ps} and A_{vc} denote the pipeline arc set and the compressor arc set, respectively. The objective function is the amount of gas transferred between two systems, I is the set of physical arcs that connect two systems, and S is the set of pseudo arcs that realize swapping by allowing flow into redistribution node. This is one of the earliest mathematical models describing the natural gas market under regulation.

4.2 Deregulated Natural Gas Market Models

In North America, before the 1980s, the natural gas market had been greatly regulated by the government since the 1930s. In the regulated market, there were primarily four participants: the gas producers, the gas pipeline companies, local gas distribution companies, and customers. The relationship of these participants is shown in Fig. 5, where producers sold gas to pipeline companies, and pipeline companies sold the gas to local gas distribution companies, and then local distribution companies sold the gas to various customers, such as industrial, commercial,

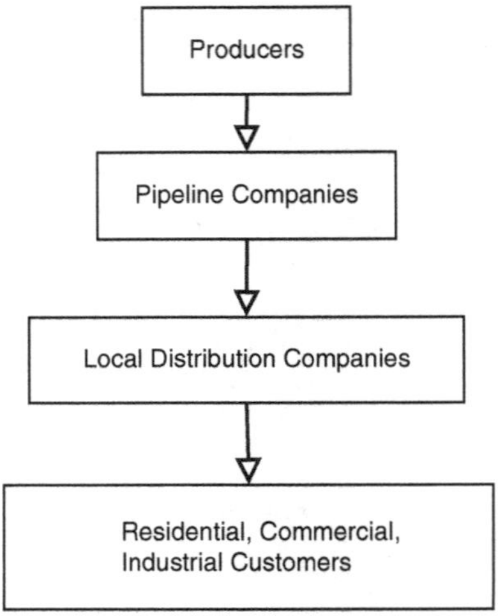

Fig. 5 Participants relationship in regulated gas market

and residential customers. In this regulated market, gas prices in each of the above transactions are tightly regulated by Federal and State governments as pipeline companies and local distribution companies had monopolies in the gas market. Since the mid 1980s, a series of deregulation policies have been announced. These polices encourage pipeline companies to switch from their traditional role as owners of natural gas by allowing producers and buyers to bypass the pipeline companies, in that the buyers can transport their own gas through the pipeline system by paying some fees.

The deregulation of the gas market not only changed the roles of the former participants but also helped to create more participants, such as the gas marketing companies. Many models have been proposed for the deregulated gas market, especially for North America and Europe. Optimal purchasing strategies considering storage, contract, spot prices, peak day demands local distribution companies under North America gas market conditions have been studied by Avery et al. (1992). A model based on generalized network to provide optimal strategies for the marketing companies and local distribution companies and a system, GRIDNET, to store all the dealt information were proposed by Brooks and Neill (2003) and Brooks (2003). The Natural Gas Transmission and Distribution Module (NGTDM) is an important model of the North American gas market, which is a submodule of the US Department of Energy's National Energy Modeling System (NEMS) and can be found in Energy Information Administration (2003). The Gas System Analysis Model (GSAM) is another North American gas market model, which tries to maximize the social welfare function to get the equilibrium, see for instance Gabriel et al. (2003). One of the most recent North American gas market models is the Mixed Complementarity-based Equilibrium Model of Natural Gas Markets, see Gabriel et al. (2005).

Gabriel et al. (2005) consider six types of participants: the pipeline operators, the production operators, the marketers/shippers, the storage reservoir operators, the peak gas operators, and the customers. Each participant is trying to minimize cost or maximize profit for itself. For the sake of simplicity, this model assumes only linear relationship within each problem faced by a participant. Hence, every participant faces a linear programming problem. Because natural gas is a highly seasonal product, the model specifies three seasons in each year, which are denoted by $s = 1, 2, 3$. Every year has index $y \in Y$.

- $s = 1$: low demand season, Apr–Oct
- $s = 2$: high demand season, Nov, Dec, Feb, Mar
- $s = 3$: peak demand season, Jan

In this formulation, pipeline gas is available for all three seasons, and gas is injected to storage reservoir in season 1 and extracted in season 2 and 3, and peak gas is used only in the peak season.

The operator of pipeline a is trying to maximize its own profit by solving the following problem

$$\max \quad \sum_{y \in Y} \sum_{s=1}^{3} days_s \tau_{asy} f_{asy} \tag{65}$$

$$\text{s.t.} \quad f_{asy} \le \bar{f_a}, \qquad \forall s, y, \tag{66}$$

$$f_{asy} \ge 0, \qquad \forall s, y, \tag{67}$$

where $days_s$ is the number of days in season s, τ_{asy} and f_{asy} are the prices and flow rates, respectively, of pipeline a in season s of year y. Constraints (66) are the upper bound constraints of the flows. τ_{asy} are the equilibrium show prices determined by the optimization problems of the other participants. Other than τ_{asy}, there are some other conditions relating this pipeline operator problem to the other pipelines and other kinds of participants. These conditions are usually called market-clearing conditions. The corresponding market-clearing conditions for the gas pipeline operator problem reads

$$days_1 f_{a1y} = \sum_{r \in R(n_1(a))} days_1 g_{ary} + \sum_{m \in M(n_1(a))} days_1 h_{am1y} \tau_{a1y} \text{ free} \quad \forall y \in Y, \tag{68}$$

$$days_s f_{asy} = \sum_{m \in M(n_s(a))} days_s h_{amsy} \tau_{asy} \text{ free} \quad s = 2, 3, \forall y \in Y. \tag{69}$$

These two market-clearing conditions state that all the supplies equal all the demands. g_{ary} is the flow rate of gas to storage operator r from the producers of season 1 through arc a, and h_{amsy} is the gas flow rate from producers of season s to marketer m through arc a.

The production operator's problem, for production company $c \in C$ at node $n \in N$, is to maximize its profit by solving the following problem:

$$\max \quad \sum_{y \in Y} \sum_{s=1}^{3} days_s \left(\pi_{nsy} q_{csy} - c_c^{pr} q_{csy} \right) \tag{70}$$

$$\text{s.t.} \quad q_{csy} \le \bar{q}_c, \qquad \forall s, y, \tag{71}$$

$$\sum_{y \in Y} \sum_{s=1}^{3} days_s q_{csy} \le prod_c \tag{72}$$

$$q_{csy} \ge 0, \qquad \forall s, y, \tag{73}$$

where π_{nsy} and c_c^{pr} are the price of gas sold by the production company and cost to produce one unit of gas, respectively, for company c, and q_{csy} is the production rate of the company in season s of year y. Constraints (71) specify the upper bounds of the production rate in each period, and constraints (72) give the total production capacity for the whole planning horizon. Except this optimization problem, the coupling conditions for the production company c at node n are as follows:

$$\sum_{c \in C(n)} days_1 q_{c1y} = \sum_{a \in A_n^+} \Big(\sum_{r \in R(n_1(a))} days_1 g_{ary} + \sum_{m \in M(n_1(a))} days_1 h_{am1y} \Big), \quad \pi_{n1y} \text{ free} \quad \forall y \in Y, \tag{74}$$

$$\sum_{c \in C(n)} days_s q_{csy} = \sum_{a \in A_n^+} \sum_{m \in M(n_s(a))} days_s h_{asy}, \quad \pi_{asy} \text{ free} \quad s = 2, 3, \forall y \in Y. \tag{75}$$

The storage reservoir operator's problem, the marketer's problem, and the peak gas operator's problem are all described in the same way, first the linear programming problem and then the market-clearing conditions. Since all operator's problems are linear programming problems, the KKT conditions are necessary and sufficient. Combining all the KKT conditions and market-clearing conditions of every operator's problem, we then get a linear complementarity problem (LCP), which is a special case of nonlinear complementarity problem (NCP) or variational inequality problem (VI). Gabriel et al. (2005) proved that there exists a solution of the system and the prices are unique in this case. For more details about LCP, NCP, and VI, we refer the reader, for instance, to Cottle et al. (1992); Murty (1988); Facchinei and Pang (2003); Luo et al. (1996).

Also a lot of models for the European gas markets have been proposed. A stochastic Stackelberg–Nash–Cournot equilibrium model for natural gas producers are proposed by De Wolf and Smeers (1997). Breton and Zaccour (2001) propose a duopoly producer model. A recent European gas market model similar to the model in Gabriel et al. (2005) is GASTALE proposed by Boots et al. (2004).

4.3 Optimization in the Energy System Combining Natural Gas System and Electricity System

Natural gas is widely used in electricity production. Because combined-cycle plants are highly efficient and have less damage to the environment, more and more power plants of this type are build around the world. Hence the electricity and the gas system are now highly correlated. Here we discuss some related optimization applications regarding this relationship.

4.3.1 Electricity System Reliability Study using Natural Gas Transmission Network Modeling

Because of the increasing number of combined-cycle power plants being built, electricity production relies more and more on the amount of gas the power plants can get. However, the electricity plants are not the only users of natural gas; see Sec. 1.

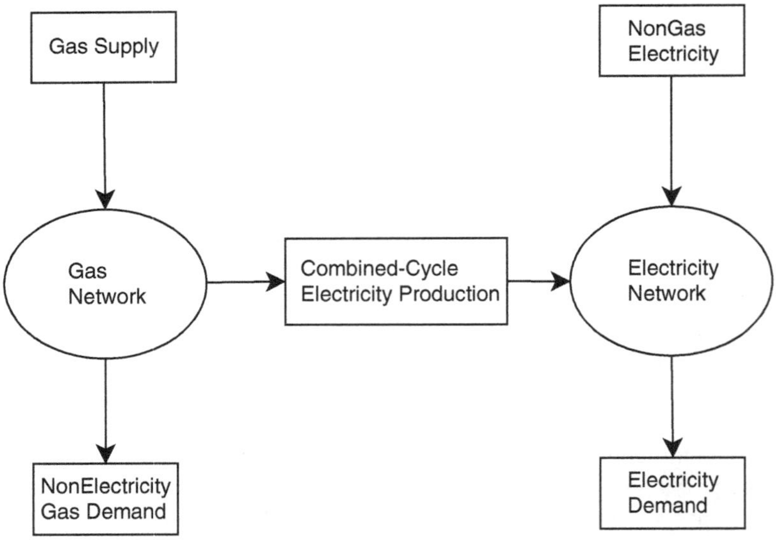

Fig. 6 Relationship between gas network and electricity network

To perform a reliability analysis of the electricity system, it is important to study the maximal amount of gas that the gas network can supply to the electricity plants. The relation between gas network and electricity network is shown in Fig. 6. Munoz et al. (2003) studied the problem of the maximal gas supply the electricity system can receive, taking into account the other gas users, the pipeline capacity, and the production capacity. The formulation is very similar to the gas pipeline operations problem (31). Instead of minimizing the gas purchase cost as in (31), this problem maximizes the total electricity that can be produced by using gas from the gas system. It can be formulated as

$$\max \quad \sum_{i \in N_e} A_i e_i + B_i e_i^2 + C_i e_i^3 \tag{76}$$

$$\text{s.t.} \quad \sum_{j \in A_i^+} f_{ij} - \sum_{j \in A_i^-} f_{ji} = s_i - d_i - e_i, \qquad \forall i \in N, \tag{77}$$

$$\text{sign}(f_{ij}) f_{ij}^2 = C_{ij}(p_i^2 - p_j^2), \qquad \forall (i,j) \in A_p, \tag{78}$$

$$f_{ij}^2 \geq C_{ij}(p_i^2 - p_j^2), \qquad \forall (i,j) \in A_c, \tag{79}$$

$$\underline{s}_i \leq s_i \leq \bar{s}_i, \qquad \forall i \in N, \tag{80}$$

$$\underline{p}_i \leq p_i \leq \bar{p}_i, \qquad \forall i \in N, \tag{81}$$

$$\underline{d}_i \leq d_i \leq \bar{d}_i, \qquad \forall i \in N, \tag{82}$$

$$\underline{e}_i \leq e_i \leq \bar{e}_i, \qquad \forall i \in N, \tag{83}$$

$$f_{ij} \geq 0, \qquad \forall (i,j) \in A_c, \tag{84}$$

where the objective function is a polynomial function of withdrawal of gas from the gas network. e_i is the gas withdrawal to produce electricity. d_i is the demand not related to electricity production. A_i^+ denotes the set of arcs that are emanating from node i, while A_i^- denotes the set of incoming arcs to node i.

Munoz et al. (2003) solve the above problem in two phases. First, by dropping all nonlinear constraints, a mixed integer linear programming problem is obtained and then solved, where the integer variables denote the directions of flows in the pipeline segments. Second, by knowing the directions of flows from the phase I problem, a nonlinear problem is solved. However, two theoretical questions still remain in the correctness of optimality obtained by the method. First, it remains unanswered whether the solution from phase I will ensure the phase II problem to be feasible. Second, it is not true that the second phase problem is a convex problem for which a simple counterexample can easily be constructed, such as the single pipeline segment problem.

4.3.2 Optimization in Natural Gas Contracts

Many electricity production plants use a lot of sources among which natural gas is a very reliable alternative to meet the high electricity demand. The optimization of fuel contracts for a hydro-based power system is a very good example. In hydro power systems, precipitation varies from season to season. For the low precipitation seasons, the plants need to buy gas to generate electricity.

Let us now discuss a model that deals with the optimal dispatch strategy while considering the particular specifications of gas supply contracts as in Chabar et al. (2006). This model assumes a take-or-pay contract, which is widely adopted, especially in Europe. If a take-or-pay contract is signed, specifying a monthly amount and a total annual amount, then at least $X\%$ of the monthly amount has to be bought every month and at least $Y\%$ of the contracted annual amount for the year has to be bought. Hence, there might be some gas excess based on contracts of this type. Two reservoirs are added into this model to accommodate the situations where excess gas exists. All excesses of gas not consumed monthly are stored in the gas reservoir A, the difference between the annual take-or-pay amount and the sum of all monthly take-or-pay amounts of the year is stored in reservoir B. Also, one of the gas contract provisions state that the gas purchased at any time point cannot "stay in the reservoir" or actually held by the gas provider by more than N time periods, which means that if any amount of gas stays in the reservoir for more than N time periods, it will have to be discarded. GD_t is used to denote the amount of gas discarded at time t. Figure 7 shows how the model, based on reservoirs, deals with the contract provisions.

Also the maintenance schedule is modeled by reservoirs. A fictitious remaining-hours reservoir is assigned to every power unit for each maintenance cycle. For a three power units three cycles problem, there will be nine reservoirs. The length of each kind of cycle is shown in Table 1. For each power unit, the reservoirs are filled with the amount of remaining hours of operation until next maintenance.

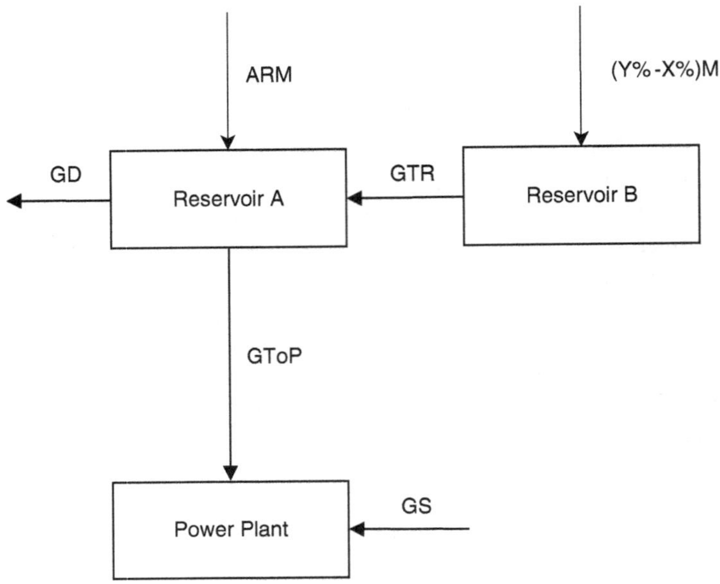

Fig. 7 Gas contracts modeled by reservoirs

Table 1 Maintenance cycle length

Cycle	Frequency (h)	Average duration (days)	Cost (MMR$)
Combustor	8,000	7	3.5
Hot path circuit	24,000	14	10
Major maintenance	48,000	21	20

The capacity of each reservoir is the length of the cycle. As the unit operates, all reservoirs for that unit are reduced by the quantity of the elapsed hours. After maintenance, the fictitious maintenance reservoir is filled to its capacity.

Considering also the maintenance scheduling of the thermal plant, a dynamic programming formulation of the problem, for a given stage and price, is proposed by Chabar et al. (2006):

$$FBF_t^k(VA_t, VB_t, \{VH_t^{i,j}, i = 1, \ldots, n, j = 1, \ldots, m\}, \pi_t^k) \tag{85}$$

$$= \max \quad RI_t + \sum_{s=1}^{S} p_{t+1}(k,s) FBF_{t+1}^s(VA_{t+1}, VB_{t+1},$$

$$\{VH_{t+1}^{ij}, i = 1, \ldots, n, j = 1, \ldots, m\}, \pi_{t+1}^k) \tag{86}$$

$$\text{s.t.} \quad VA_{t+1} = VA_t + ARM_t - GToP_t + GTR_t - GD_t, \tag{87}$$

$$VB_{t+1} = VB_t - GTR_t \tag{88}$$

$$VH_{t+1}^{i,j} = VH_t^{i,j}(1 - x_t^{i,j}) + \overline{VH}^j x_t^{i,j} - \gamma EG_t^i, \quad i = 1,\ldots,n, j = 1,\ldots,m, \tag{89}$$

$$\sum_{i=1}^{n} \psi_t^i EG_t^i = H_c(CToP_t + GToP_t + \nabla G_t), \tag{90}$$

where VA_t and VB_t are the volume of gas in reservoirs A and B, respectively, and $VH_t^{i,j}$ is the "volume" of remaining hours of operation that the unit i has until the next maintenance of cycle j. $GToP_t$ is the amount of gas actually used to generate electricity, and GS_t is the amount of gas purchased or sold to the gas spot market. ARM_t is the amount of gas purchased from the gas distributor and should be bounded below by $X\%M$. GTR_t is the amount of gas transfer from A to B, and GD_t is the amount of gas discarded when it is in the reservoir more than the maximum storage time, N. n and m are the total number of power units and total number of maintenance cycles, respectively. RI_t is the immediate revenue in stage t. π_t^s is the spot price in stage t of scenario s. $p_{t+1}(k,s)$ is the transition probability of the spot price of scenario k in stage t to the spot price of scenario s in stage $t+1$. $x_t^{i,j}$ is the binary decision variable associated with the schedule of maintenance of cycle j for unit i at stage t. $\overline{VH}^j$ is the maximum capacity of the reservoir of remaining hours of operation until the next maintenance of cycle j. EG_t^j is the energy generated by unit i at stage t. γ is an inverse coefficient of the power unit, ψ_t^i is conversion factor from MMBTU to MWh of unit i at stage t, and H_c is the heat rate of the gas.

Constraints (87)–(89) are the fictitious reservoir balance constraints and (90) is the transformation from gas to electricity. Except constraints (87)–(90), there are also a lot of other constraints, such as gas consumption priority constraints, maximum and minimum gas consumption constraints, maintenance constraints, constraints related to the mechanism implemented for the modeling of the contracts and so on. For this problem, each stage is a mixed integer linear programming problem. And the whole problem is solved by using stochastic dual dynamic programming, first proposed by Pereira and Pinto (1991).

Also the natural gas market can be modeled as a natural gas value chain. The primary component is natural gas in this chains. Various market models are proposed and utilized in reality at different stages along this value chain, for example, production, transportation and processing, storage, import terminals and markets, wholesale and retail markets. Please refer to the natural gas value chapter by Tomasgard et al. and Midthun (2007) for more details about market models within the natural gas value chain.

5 Conclusion

This chapter discusses various optimization models occurring in the natural gas industry, focusing on three aspects: production, transportation, and market. As we can see, the natural gas industry is a complex system and in great need of optimization techniques to improve performance. Especially, the nonlinear and nonconvex nature of the problems makes it computationally challenging to find good solutions. We observe that linearization techniques are the common methods to tackle these nonconvex functions, often reducing the problem to a (series) of linear or mixed integer liner programming problems. With the computational power of computers increasing over the last decade, the use of meta-heuristics has become more and more popular, especially for problems that cannot be handled with the current MINLP solvers either due to the size of the problem or due to the degeneracy.

The deregulation of the gas market introduced additional modeling aspects and computational challenges: various (additional) stochastic elements have been added to the "classical" problems. This underlying structure of the problems cannot be ignored by any serious model and we expect that future research will focus on stochastic models and, especially, on new techniques on how to solve these (large-scale) practical problems when also integer, nonconvex, and nonlinear functions are present.

References

Al-Hussainy R (1967) Transient flow of ideal and real gases through porous media. PhD thesis, Texas A&M University, College Station

Annual Energy Review (AER) (2009) Technical Report DOE/EIA-0384(2008), US Department of Energy, Energy Information Administration, 26 June 2009

Avery W, Brown GG, Rosenbranz JA, Wood RK (1992) Optimization of purchase, storage and transmission contracts for natural gas utilities. Oper Res 40(3):446–462

Babu BV, Angira R, Chakole PG, Syed Mubeen JH (2008) Optimal design of gas transmission network using differential evolution. http://discovery.bits-pilani.ac.in/discipline/chemical/BVb/RevisedBabRakPalMub%20CIRAS-2003.pdf

Bazaraa M, Sherali HD, Shetty CM (2006) Nonlinear programming, 3rd edn. Wiley, New York

Beggs HD (1984) Gas production operations. Oil Gas Consultants International Inc., Tulsa, Oklahoma

Boots MG, Rijkers FAM, Hobbs BF (2004) Modeling the role of trading companies in the downstream European gas market: A successive oligopoly approach. Energ J 25(3):73–102

BP (2008) BP Statistical Review of World Energy 2008. London, UK, June 2008

Breton N, Zaccour Z (2001) Equilibria in an asymmetric duopoly facing a security constraint. Energ Econ 25:457–475

Brooks RE (2003) Optimizing complex natural gas models. http://rbac.com/Articles/tabid/63/Default.aspx

Brooks RE, Neill CP (2003) Natural gas operations optimizing system. http://rbac.com/Articles/tabid/63/Default.aspx

Cameron F (2007) The north stream gas pipeline project and its strategic implications. Briefing Note for The European Parliament's committee on Petitions, December 2007

Chabar RM, Pereira MVF, Granville S, Barroso LA, Iliadis N (2006) Optimization of fuel contracts management and maintenance scheduling for thermal plants under price uncertainty. In Proceedings of the 2006 Power Systems Conference Expo (PSCE 06), October, pp. 923–930

Chebouba A, Yalaoui F, Smati A, Amodeo L, Younsi K, Tairi A (2009) Optimization of natural gas pipeline transportation using at colony optimization. Comput Oper Res 36(6):1916–1923

Cottle RW, Pang JS, Stone RE (1992) Linear complementarity problem. Academic Press, NY

De Wolf D, de Bisthoven OJ, Smeers Y (1991) The simplex algorithm extended to piecewise linearly constrained problems I: The method and an implementation, CORE DP No. 9119, Universite Catholique de Louvain, Belgium
De Wolf D, Smeers Y (1996) Optimal dimensioning of pipe networks with application to gas transmission networks. Oper Res 44:596–608
De Wolf D, Smeers Y (1997) A stochastic version of a Stackelberg Nash-Cournot equilibrium model. Manag Sci 43(2):190–197
De Wolf D, Smeers Y (2000) The gas transmission problem solved by an extension of the simplex algorithm. Manag Sci 46:1454–1465
Edgar TF, Himmelblau DM (2001) Optimization of chemical processes. McGraw-Hill, New York
Edgar TF, Himmelblau DM, Bickel TC (1978) Optimal design of gas transmission networks. Soc Petrol Eng J 30:96–104
Energy Information Administration (2003) The national energy modeling system: an overview, natural gas transmission and distribution module. http://www.eia.doe.gov/oiaf/aeo/overview/nat_gas.html
Facchinei F, Pang JS (2003) Finite-dimensional variational inequalities and complementarity problems, vol. I and II. Springer, New York
Gabriel SA, Manik J, Vikas S (2003) Computational experience with a large-scale, multi period, spatial equilibrium model of the North America natual gas system. Networks Spatial Econ 3:97–122
Gabriel SA, Kiet S, Zhuang J (2005) A mixed complementarity-based equilibrium model of natural gas markets. Oper Res 53(5):799–818
Goldberg DE (1983) Computer-aided gas pipeline operation using genetic algorithms and rule learning. PhD thesis, University of Michigan
Hiriart-Urruty JB, Lemarechal C (1993) Convex analysis and minimization algorithms Springer, Berlin
Horne RN (2002) Optimization applications in oil and gas recovery. In: Handbook of applied optimization, pp. 808–813. Oxford University Press, New York
Horst R, Pardalos PM, Thoai NV (2000) Introduction to global optimization, 2nd edn. Kluwer, The Netherland
International Energy Outlook 2009 Technical Report DOE/EIA-0484(2009), US Department of Energy, Energy Information Administration, 27 May 2009. Chapter 3 – Natural Gas
Kallrath J, Wilson JM (1997) Business Optimization using Mathematical Programming. MacMillan Business
Locatelli M, Thoai NV (2000) Finite exact branch-and-bound algorithms for concave minimization over polytopes. J Global Optim 18:107–128
Luo ZQ, Pang JS, Ralph D (1996) Mathematical programs with equilibrium constraints. Cambridge University Press, London
Mantini LA, Beyer WA (1979) Optimization of natural gas production by waterflooding. Appl Math Optim 5:101–116
Midthun KT (2007) Optimization models for liberalized natural gas markets. PhD thesis, Norwegian University of Science and Technology, 2007
Munoz J, Jimenez-Redondo N, Perez-Ruiz J, Barquin J (2003) Natural gas network modeling for power systems reliability studies. 2003 IEEE Bologna PowerTech Conference, 23–26 June, Bologna, Italy, 2003
Murray JE, Edgar TF (1978) Optimal scheduling of production and compression in gas fields. SPE J Petrol Technol 30:109–116
Murty KG (1988) Linear complementarity, linear and nonlinear programming. Helderman. http://ioe.engin.umich.edu/people/fac/books/murty/linear_complementarity_webbook/
Nemhauser GL, Wolsey LA (1999) Integer and combinatorical optimization. Wiley, New York
O'Neil RP, Williard M, Wilkins B, Pike R (1979) A mathematical programming model for allocation of natural gas. Oper Res 27(5):857–873
Pereira MVF, Pinto LMVG (1991) Multi-stage stochastic optimization applied to energy planning. Math Program 52:359–375

Peretti A, Toth P (1982) Optimization of a pipeline for the natural gas transport. Eur J Oper Res 11:247–254

Rios-Mercado (2002) Natural gas pipleline optimization. In: Handbook of applied optimization. Oxford University Press, New York, pp. 813–826

Rios-Mercado RZ, Kim S, Boyd EA (2006) Efficient operation of natural gas transmission systems: a network-based heuristic for cyclic structures. Comput Oper Res 33:23–51

Rosen JB (1960) The gradient projection method for nonlinear programming. Part I. linear constraints. SIAM J 22:181–217

Rothfarb B, Frank H, Rosenbaum DM, Steiglitz K, Kleitman DJ (1970) Optimal design of offshore natural-gas pipeline systems. Oper Res 18:992–1020

Tussing AR, Barlow CC (1984) The natural gas industry: evolution, structure, and economics. Ballinger Publishing Company, Cambridge, MA

US Department of Energy, Energy Information Administration (2008) International Energy Annual 2006, 25 September 2008

Wattenbarger RA (1970) Maximizing seasonal withdrawal from gas storage reservoir. SPE J Petrol Technol 22:994–998

Wolsey LA (1998) Integer programming. Wiley, New York

Worldwide Look at Reserves and Production (2008) Oil Gas J 106(48):22–23

Wu S, Rios-Mercado RZ, Boyd EA, Scott LR (2000) Model relaxations for the fuel cost minimization of steady-state gas pipeline networks. Math Comput Model 31:197–220

Integrated Electricity–Gas Operations Planning in Long-term Hydroscheduling Based on Stochastic Models

B. Bezerra, L.A. Barroso, R. Kelman, B. Flach, M.L. Latorre, N. Campodonico, and M. Pereira

Abstract The integration of natural gas and electricity sectors has increased sharply in the last decade as a consequence of combined cycle natural gas thermal power plants. In some countries such as Brazil, gas-fired generation has been a major factor in the overall growth of natural gas consumption. When related to the operations planning, in some hydrothermal systems, a national system operator dispatches these gas-fired plants (along with other thermal sources such as coal, oil, and nuclear) in conjunction with the country's hydroelectric plants by using a production-costing model based on stochastic programming. The algorithm determines the optimal hydro-to-thermal energy production ratio on the basis of the expected benefit of reducing thermal plant generation over a large number of hydrological scenarios, along a planning horizon of some years. This means that the optimal scheduling decision today depends on the assumptions about future load growth and future entrance of new generation capacity. Stochastic dynamic programming models are extensively used. However, the hydrothermal scheduling models usually do not take into account the possibility of future fuel supply constraints, either in production or in transportation. The assumption of fuel supply adequacy is felt to be reasonable for the more mature markets such as coal and oil. However, because of the fast growth of the natural gas market, it is possible that demand outpaces supply or transportation investments. Indications that gas-related constraints could be relevant were observed in New England, in the US, and in Brazil in 2004, where several megawatt of combined-cycle generation could not be dispatched when needed due to constraints in pipeline capacity. The objective of this work is to present a methodology for representing the natural gas supply, demand, and transportation network in the stochastic hydrothermal power scheduling model. Application of the integrated electricity–gas scheduling model is illustrated in case studies, with realistic configurations of the 90 GW Brazilian system.

L.A. Barroso (✉)
PSR, Rua Voluntarios da Patria 45, Botafogo, Rio de Janeiro, Brazil
e-mail: luiz@psr-inc.com

S. Rebennack et al. (eds.), *Handbook of Power Systems I*, Energy Systems,
DOI 10.1007/978-3-642-02493-1_7, © Springer-Verlag Berlin Heidelberg 2010

Keywords Hydroelectric-thermal power generation · Natural gas industry · Stochastic dual dynamic programming

1 Introduction

The integration of natural gas and electricity sectors was intensified in the last decade as a consequence of a widespread construction of new gas-fired power plants, both combined-cycle and single-cycle. Several countries in South America, Europe, Asia, and the US have built a lot of gas-fired generation to have a more economical and clean resource than the standard coal and oil-fired resources.

On the operational side, the integration between gas-fired and hydroplants in hydro-based systems is usually not straightforward (due to the low cost of hydro), as opposed to the case of thermal systems.

In some hydro-based systems, specially those in South America (such as Brazil, Chile, Argentina, etc.), both hydro and thermal plants are dispatched by the country's national system operator on the basis of a production-costing model. This model determines the optimal hydro-to-thermal energy production ratio based on the expected benefit of reducing thermal plant generation over a large number of hydrological scenarios, along a planning horizon of 5 year, and on the stochastic dynamic programming (SDP) techniques, such as the stochastic dual dynamic programming (SDDP) (Pereira et al. 1998). This means that the optimal scheduling decision today depends on assumptions about future load growth and future entrance of new generation capacity. However, the hydrothermal scheduling model does not take into account the possibility of future fuel supply constraints, either in production or in transportation. The assumption of fuel supply adequacy is felt to be reasonable for the more mature markets such as coal and oil. However, owing to the fast growth of the natural gas market, it is possible that demand outpaces supply or transportation investments.

This actually has been observed in some countries. Brazil, for example, built over 7,000 MW of gas-fired generation in the last 5 years. These gas-fired plants, along with other thermal sources such as coal, oil, and nuclear energy, correspond to 15% of the country's installed capacity; the major source of power production being hydroelectric power. A first indication that gas-related constraints could be relevant was observed in January 2004, when 800 MW of combined-cycle generation (out of a total capacity of 1,200 MW) could not be dispatched due to constraints in pipeline capacity. In the same vein, the ISO New England commanded the dispatch of about 3,000 MW of gas-fired generation in 2004, which turned out to be unavailable due to lack of natural gas.[1]

[1] ISO NEW ENGLAND INTERNAL REPORT on natural gas shortage, available at: http://www.iso-ne.com/special_studies/Interim_Report_on_January_14_-_16_2004_Cold_Snap/

Coordinating these two sectors, especially of hydrosystems, is a critical issue. If gas production and transportation constraints are ignored, the scheduler may be optimistic with respect to the firm capacity of the thermal plants, and jeopardize the supply reliability: hydro-reservoirs may get depleted faster today on the basis of the availability of future gas-fired generation, which may not occur.

The objective of this work is to present a methodology for representing the natural gas supply, demand, and transportation network in the stochastic hydrothermal power scheduling model used to schedule real hydropower systems. This will be done in two steps. The first step consists in developing a model to examine the feasibility of the gas-based generation resulting from a hydrothermal scheduling tool that does not take into account any constraints related to the supply and transportation of natural gas. The objective of the model is to allocate the natural gas supply to meet the total demand in each node of the gas network, while minimizing the amount of gas for use by the power sector that is rationed. The obtained results reveal that it is imperative to jointly represent both the electricity and the natural gas sectors so as not to compromise with the security of supply for either one of them. This is accomplished in the second step, which consists in explicitly introducing the natural gas constraints into the dynamic programming recursion of the energy planning model. Gas demand in each node is given by the sum of non-power gas consumption forecasts plus gas consumption factors for the gas-fired power plants; gas production in each node is represented by minimum and maximum production levels – for example, if the gas field is associated with oil production. Finally, fuel transportation is modeled both through pipelines and through liquified natural gas (LNG) supply.

An application illustrating the proposed methodology will be done using the 90 GW Brazilian hydrosystem as an example. The Brazilian system provides good case-studies for the methodology because it has a large-scale hydrosystem, but on the other hand, the gas sector is developing at aggressive growth rates and gas-fired plants account for an important share of overall thermoelectric resources. Therefore, it concentrates several characteristics and challenges that are of interest to several other power systems worldwide.

This work is organized as follows: Sect. 2 presents an overview of the electricity and gas sectors in Brazil. Section 3 describes and motivates the main issues in the energy–gas integration in the country. Section 4 presents a procedure that has been developed for assessing the feasibility of the schedules of the gas-fired power plants. Section 5 presents the integrated representation of the electricity–gas sectors in a hydrothermal scheduling model and Sect. 6 concludes.

2 Overview of Electricity and Gas Sectors

Brazil is the largest electricity market in South America, accounting for 40% of the continent's energy consumption. As mentioned in the Introduction, the country is hydro-dominated: 85% of the 90 GW installed capacity and more than 90% of

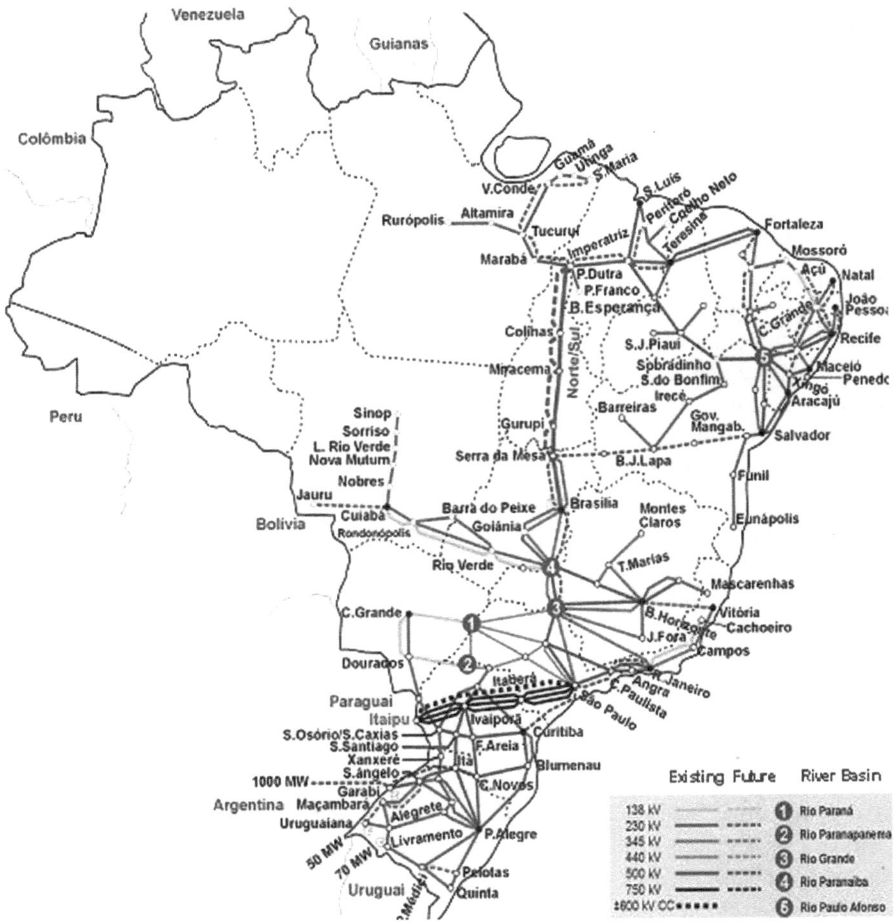

Fig. 1 Power transmission network (source ONS)

the electricity production (average of 44 GW) comes from hydropower. Thermal generation includes nuclear power, coal, diesel, biomass and, more recently, natural gas plants. The country is fully interconnected at the bulk power level by a 80,000 km meshed high-voltage transmission network, shown in Fig. 1. The direct international interconnections are the back-to-back links with Argentina (2,200 MW) and smaller interconnections with Uruguay and Venezuela.

On the natural gas side, Brazil has proven gas reserves of 320 bcm (Barroso et al. 2005; IEA 2003). The country also has a natural gas production[2] of about 27 MMm^3 per day available to the market, mostly associated with the exploration of oil. Since 1999, up to 30 MMm^3 per day of imported natural gas has been flowing into the

[2] This number excludes reinjection, E&P consumption, and flares and losses

country through pipelines from Bolivia and Argentina. In 2003, the discovery of a large offshore natural gas field (Santos field), capable of more than doubling the country's reserves, was announced.

In contrast with Argentina and Chile, Brazil's gas market is relatively undeveloped. One of the reasons is that there is no market for space heating, which is an important factor in the other countries.

Figure 2 shows the gas pipelines and the areas of exploration and production. There are three separate systems: the largest comprises the South and Southeast regions; coastal cities from the Northeast form the country's second natural gas system; the third system is in the Amazon region.

Finally, a natural gas law that regulates pipeline access and other topics is currently being discussed in Congress.

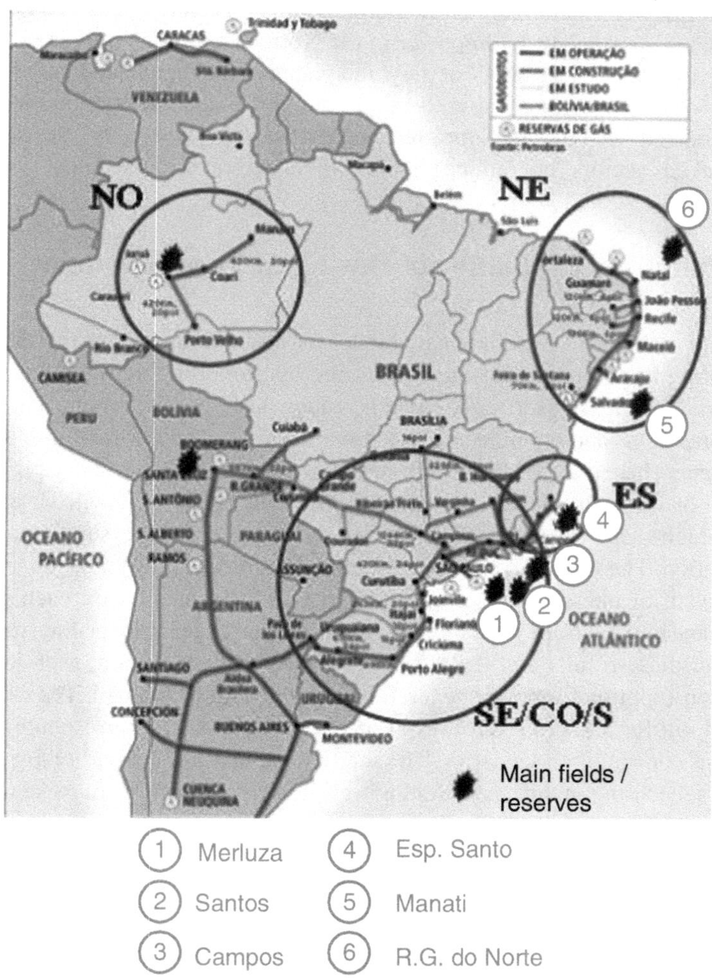

Fig. 2 Natural gas network (source ANP & PSR)

3 Electricity–Natural Gas Integration Issues

As mentioned previously, Brazil has 7,000 MW of gas-fired plants. Their potential gas consumption is quite significant: if dispatched simultaneously, the gas-fired plants would use 35 MMm^3 per day of gas, about the same amount as the entire "non-power" gas demand. Also as mentioned previously, the thermal plants' dispatch depends on the hydrological conditions: if the system is "wet," the entire electricity load can be met with hydro-generation alone.

In other words, power-related gas consumption is both large and stochastic. This creates a complex problem for investment decisions in new gas fields and in new pipelines, which may be either excessive or insufficient, depending on hydrological conditions. Although take or pay contracts can alleviate part of the financial uncertainty, a mismatch between gas supply and demand can have significant consequences for power scheduling. One example of this mismatch happened in January 2004, when a shortage of hydropower in the Northeast of Brazil made ONS command the dispatch of 1,200 MW of gas-fired plants of the region, and only a third of this (400 MW) was delivered due to gas production and transportation constraints. This episode showed the need for greater coordination between the electricity and the natural gas sectors' operations planning. This will be discussed next.

4 Probabilistic Evaluation of Gas-Fired Plant Schedules

We initially developed a probabilistic model for evaluating whether the sum of gas consumption requirements resulting from the hydrothermal dispatch and of non-power gas consumption forecasts could be adequately supplied by the existing and planned gas fields and pipeline network. Figure 3 shows the information flowchart.

The upper shaded area shows the first step of the process: the use of a production-costing tool for hydrothermal scheduling based on SDDP – it will be discussed next – which dispatches the power system for a given electric supply x demand configuration. The main driver of uncertainty is hydrology.

The result of interest is a set of power generation scenarios for each gas-fired power plant in each stage and simulated hydrological scenario for the study horizon. From these results and the consumption rates of each plant, a projection of the gas consumption for power generation is immediately obtained. The simulation is carried out for a set of hydrological scenarios,[3] yielding a corresponding set of natural gas consumption scenarios. The shaded area in the lower part of Fig. 3 represents the scheduling of the gas sector and verifies the "feasibility" of these scenarios under the gas sector point of view.

Each step will be discussed next.

[3] The hydrological scenarios are produced by a streamflow periodic autoregressive stochastic model, which bases its samples on a log-normal probability distribution. We use a sample of 80 hydrological scenarios.

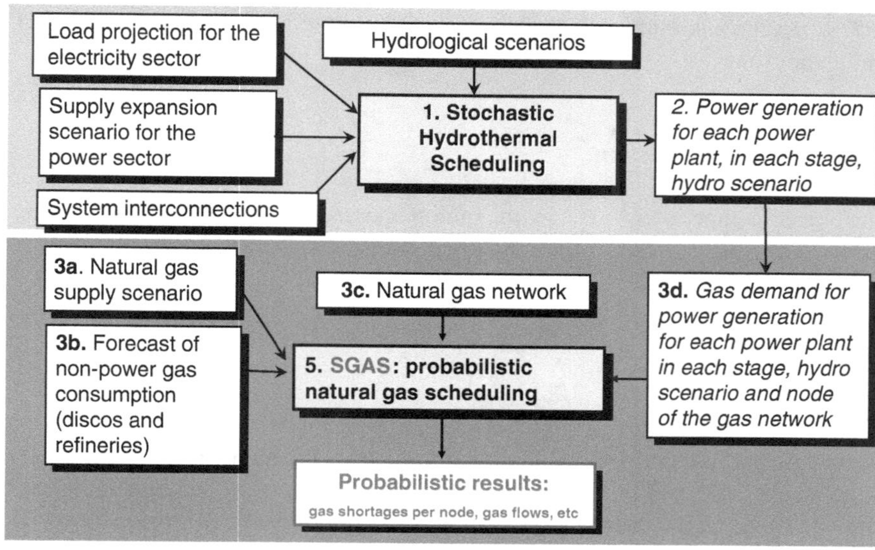

Fig. 3 Data flow procedure

4.1 Stochastic Hydrothermal Scheduling

Systems with considerable share of hydropower, such as Brazil, Colombia, Norway, and New Zealand, have been using hydrothermal scheduling tools for at least two decades. The objective of hydrothermal scheduling is to determine an operation strategy of a hydrothermal system, which for each stage of the planning period produces generation targets for each plant (hydro-releases and thermal production). This strategy should minimize the expected value of the operation cost along the period, including fuel cost and penalties for failure of load supply.

Hydroplants are dispatched on the basis of their marginal water values, which are computed by a multi-stage stochastic optimization methodology, SDDP, which approximates the cost-to-go functions by a set of linear inequalities, known as Benders' cuts, avoiding the well-know curse of dimensionality of traditional SDP models. The major advantage is the possibility to represent hydroplants individually. The SDDP approach is reviewed in great details in Annex, and it has been applied to the scheduling of large-scale power systems in more than 30 countries, including detailed modeling of system components and transmission networks (Granville et al. 2003).

However, as mentioned previously, the implementation of the SDDP algorithm in the majority of hydro-based countries as a dispatch model does not consider the gas supply–transportation constraints.

A simplified formulation of the one-stage problem solved in the SDDP recursion is shown next; further details can be found in Pereira et al. (1998); Granville et al.

(2003); Pereira and Pinto (1984); Pereira and Pinto (1985); Gorenstin et al. (1992)[4] and in the Annex.

4.1.1 Objective Function

The objective function is given by the minimization of thermal costs and rationing, plus a term that represents the cost-to-go function (also known as "future cost function (FCF)").

$$\alpha_t(v_t, a_{t-1}) = \text{Min} \sum_{k=1}^{K} \sum_{j \in J} c_j \times g_{tki} + c_\delta \times \delta + \alpha_{t+1}(v_{t+1}, a_t) \qquad (1)$$

where k is the indexes' load block in the stage, K is the number of load blocks, j is the indexes' thermal plants, J is the set of thermal plants, c_j is the operating cost of plant j, g_{tki} is the energy produced by the thermal plant j (decision variable), c_δ is the generic representation of operating constraint violation cost, δ is the violation amount (decision variable), v_{t+1} is the final storage vector at stage t (decision variable), and a_t is the lateral inflow vector at stage t.

The FCF is expressed as a scalar variable subject to linear inequalities (Benders' cuts), which are determined according to the SDDP algorithm:

$$\begin{aligned} &\alpha_{t+1}(v_{t+1}, \alpha_t) = \alpha \\ &\text{s.t. } \alpha \geq w_{tp} + \sum_{i \in I} \lambda_{tvip} v_{t+1,i} + \sum_{i \in I} \lambda_{taip} a_{ti}, \quad p = 1, \ldots, \text{p} \end{aligned} \qquad (2)$$

where α is the scalar variable that represents expected future operating cost, P is the indexes' segments of the piecewise FCF, w_{tp} is the constant term of pth segment, λ_{tvip} is the plant i's final storage coefficient in the pth segment, λ_{taip} is the plant i's lateral inflow coefficient in the pth segment, and P is the number of segments in the piecewise FCF.

4.1.2 Water Balance Equations

The *water balance* equation represents the coupling between successive stages: the reservoir storage v_{t+1} at stage $t+1$ is equal to the initial storage v_t minus outflow volumes (turbined variable u_t and spilled variable s_t) plus inflow volumes (lateral inflow a_t plus releases from immediately upstream plants belonging to set U), all in stage t, for all hydroplants in set K belonging to set M(i).

[4] Application of Stochastic Dual DP and Extensions to Hydrothermal Scheduling – PSR TR 012/99 – available at http://www.psr-inc.com

$$v_{t+1,i} = v_{ti} + a_{ti} - \varepsilon(v_{ti}) - \sum_{k=1}^{K} [u_{tki} + s_{tki}] + \sum_{m \in M(i)} \sum_{k=1}^{K} [u_{tkm} + s_{tkm}] \quad \text{for } i \in I \quad (3)$$

where i is the indexes hydroplants, i is the set of hydroplants, $\underline{M}(i)$ is the set of upstream plants immediately upstream of plant i, $v_{t+1,i}$ is the final storage of i in stage t (decision variable), v_{ti} is the initial storage of i in stage t, a_{ti} is the lateral inflow to plant i, $\varepsilon(v_{ti})$ is the vaporated volume from reservoir i, u_{tki} is the urbined outflow volume of plant i along stage t in load block k (decision variable), and s_{tki} is the pilled outflow volume of plant i along stage t in load block k (decision variable).

4.1.3 Bounds on Storage, Turbined Volumes, and Thermal Generation Variables

$$\underline{v}_i \leq v_{ti} \leq \overline{v}_i \quad \text{for } i \in I \quad (4)$$

$$u_{tki} \leq \overline{u}_{ti} \quad \text{for } i \in I;\ k = 1, \ldots, K \quad (5)$$

$$\underline{g}_{tkj} \leq g_{tkj} \leq \overline{g}_{tkj} \quad \text{for } j \in J;\ \text{for } k = 1, \ldots, K \quad (6)$$

4.1.4 Load Balance Equation

The *load supply* equation relates total thermal and hydro-generation to system load D_{tk} (MWh). The hydro-generation for unit i is given by the product of its production coefficient ρ_i (MWh m^{-3}) and its turbined outflow u_{tki}, resulting in

$$\sum_{i \in I} g_{hki} + \sum_{j \in J} g_{tkj} = D_{tk} \quad \text{for } k = 1, \ldots, K \quad (7)$$

4.2 Probabilistic Gas Scheduling Model

A gas network consists of supply nodes, where the gas is injected into the system; demand nodes, where gas flows out of the system due to thermal power or non-thermal use; and intermediate nodes. A pipeline is represented by an arc linking the nodes. When modeling gas pipelines for short-term scheduling studies, the gas flow through pipelines depends on the pressure difference between the entry and exit nodes; also, nonlinear expressions relate flow limits with the pressure in the pipeline (Mercado 2002; Wolf and Smeers 2000; Munoz et al. 2002; Mello and Ohishi 2004). For the purposes of the present study – long-term planning, with monthly steps – a linear network flow model was felt to be adequate. In this sense, the following constraints are modeled.

4.2.1 Gas Production and Flow Limits

Local production sources may be available at each node of the gas system. Operational constraints may impose daily minimum and maximum limits, represented by the following set of equations:

$$\underline{P}_{\mathrm{tn}} \leq P_{\mathrm{tn}} \leq \bar{P}_{\mathrm{tn}} \quad \text{for } n \in \mathrm{N} \tag{8}$$

where P_{tn} is the gas production at node n (decision variable), stage t, and the pair $\{P_{\mathrm{tn}},\ \bar{P}_{\mathrm{tn}}\}$ is, respectively, the minimum and maximum production limits at node n, stage t. Finally N is the set of gas nodes.

The nodes of the gas system are interconnected by pipelines. Each pipeline can be characterized by its maximum and minimum flow limits under equilibrium (steady state) conditions, originating the following constraints:

$$\underline{f}_{\mathrm{tnl}} \leq f_{\mathrm{tnl}} \leq \bar{f}_{\mathrm{tnl}} \quad \text{for } n, l \in \mathrm{N} \tag{9}$$

where f_{tnl} is natural gas flow in the pipeline (decision variable) that connects nodes n and l and the pair $\{f_{tnl},\ \bar{f}_{tnl}\}$ is, respectively, the minimum and maximum flow limit between nodes n and l.

4.2.2 Gas Balance Equations

At each stage, the sum of the demands at each node must be equal to the sum of the supply – either locally produced or imported through the pipelines – and of the deficit – in case there is not enough natural gas to completely fulfill the demand. For each node of the gas system, we have

$$P_{\mathrm{t}}(n) + \sum_{l \in \Omega(n)} [1 - w_{\mathrm{tln}}] f_{\mathrm{tln}} - \sum_{l \in \Omega(n)} f_{\mathrm{tnl}} + \sum_{k \in D(n)} \delta_{\mathrm{tk}} + \sum_{j \in T(n)} \delta'_{\mathrm{tj}} = \sum_{k \in D(n)} d_{\mathrm{tk}} + \sum_{j \in T(n)} \phi_{\mathrm{tj}} g^{*}_{\mathrm{tj}} \quad \text{for } n \in \mathrm{N} \tag{10}$$

where $\Omega(n)$ is the set of nodes of the gas system connected to node n, $T(n)$ is the set of thermal plants associated to node n of the gas system, and $D(n)$ is the set of non-thermoelectric demands at node n of the gas system (distribution companies, refineries, and others). The parameters are w_{tnl} for the loss factor of the pipeline connecting nodes n and l and ϕ_{tj} for the gas consumption conversion factor for thermal plant j, and d_{tk} is the non-electric natural gas demand k. The generation of the gas-fired plant j, $g_{\mathrm{tj}}{}^{*}$ is also known in this context, as it is obtained from the hydroscheduling simulation.

The decision variables of the problem are the following: (a) scheduling of gas supply sources; (b) scheduling of gas flows in the pipelines; and (c) deficits of natural gas for non-electrical demand k (δ_{tk}) and the deficit of natural gas for thermal power plant j ($\delta_{tj}{}'$). The deficit costs for the natural gas non-electrical demand k and the electrical demand j are represented ib the objective function by c_k and $c_j{}'$, respectively.

4.2.3 Objective Function

The objective function is to minimize the natural gas rationings costs, thus:

$$\text{Min} \sum_k c_k \delta_{tk} + \sum_j c'_j \delta'_{tj} \tag{11}$$

4.3 Case Study

The probabilistic evaluation scheme will be illustrated with basis on the (publicly available) power system configuration of the Brazilian Monthly Operations Plan ("PMO") for December 2005–December 2009. As shown in Fig. 3, the stochastic operational policy for 2005/2009 was calculated (with five additional years as a buffer to prevent depletion at the end of the period) using the SDDP hydrothermal dispatch algorithm described previously. Monthly steps were used, with three demand blocks in each step. Once the hydrothermal operational policy was calculated, the system operation was simulated for a set of hydrological scenarios, resulting in energy production schedules for each gas-fired power plant, for each month, and for each hydrological scenario.

Next, these energy production schedules were transformed into gas consumption schedules through the use of efficiency factors for each power plant. Finally, these gas schedules were added to the non-power gas consumption forecasts at the appropriate consumption nodes.

Table 1 shows the gas supply projections, including production increase in local fields and imports. Figure 4 shows the pipeline network for the South–Southeast region. A similar procedure was applied for the Northeast network (remember that the gas networks are not integrated yet).

Finally, the non-power gas consumption was estimated for each sector (industrial, automotive, commercial, residential, and co-generation), in addition to Petrobras (Brazil's oil and gas company) internal consumption in refineries and fertilizer plants. Figure 5 compares total supply and demand for the years of study. We see that the gas consumption from thermal plants is crucial for the demand x supply balance: if the thermal plants are not dispatched at all along the year (zero consumption of power-related gas), supply exceeds demand; at the other extreme, if the thermal plants are 100% dispatched along the year (base-loaded), supply cannot match demand. Given that the thermal plant dispatch depends, as seen previously, on hydrological conditions and on the overall supply s demand balance of the electricity

Table 1 Gas supply projection available to market

(MMm3 per day)	2006	2007	2008	2009
South/Southeast				
Campos	14.4	14.9	15.5	15.0
Merluza + Lagosta	1.2	1.9	1.9	1.9
Gasbol	30.0	30.0	30.0	30.0
TSB	0.0	0.0	0.0	0.0
Santos	0.0	0.0	12.0	12.0
Total	45.6	46.8	59.4	58.9
Espírito Santo				
Total	4.4	6.6	10.0	10.0
Northeast				
Total	14.2	15.4	14.4	13.4
Brazil				
Total	64.2	68.8	83.8	82.3

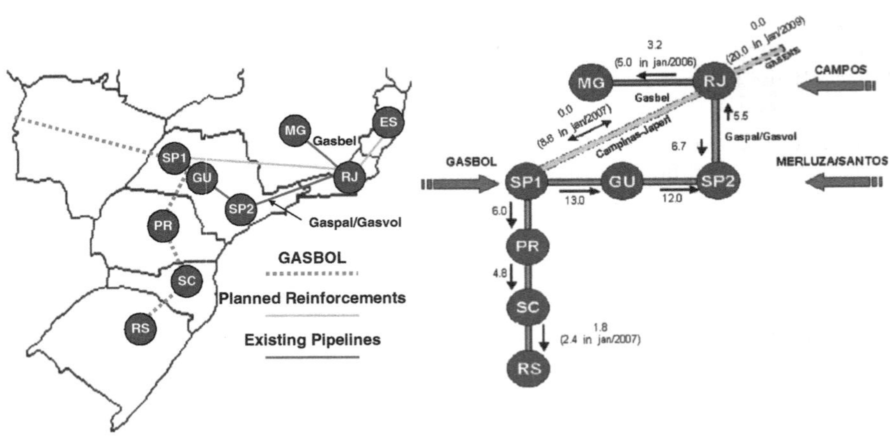

Fig. 4 South/Southeast gas network

sector, the question is then to assess the likelihood and severity of the gas supply shortfalls.

Figure 6 shows the frequency of gas supply shortfalls in volumes higher than 5% of the gas-to-power demand. Figure 7 shows the cumulative duration curve of the gas volumes shortfall, expressed in average MW (in other words, assuming that the supply of non-power demand has priority over the supply of power-related consumption). We see in Fig. 6 that, in 2007, 19% of the scenarios had shortfalls; in turn, Fig. 7 shows that the severity of the shortfalls is concentrated in fewer scenarios, which is consistent with the skewed probability distribution of droughts ("wet" scenarios are more likely than dry scenarios).

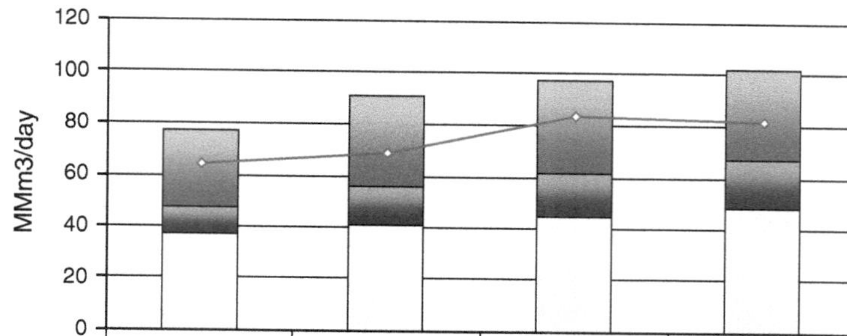

	2006	2007	2008	2009
Thermal Cons	30.0	34.3	35.7	35.7
Petrobras Cons	10.5	15.3	17.0	18.4
LDC	36.9	41.0	44.7	48.3
Supply	64.2	68.8	83.8	82.3
Deficit	(16.0)	(26.3)	(13.7)	(20.1)

LDC: Local Distribution Companies (non-power consumption)

Fig. 5 Gas supply × demand balance

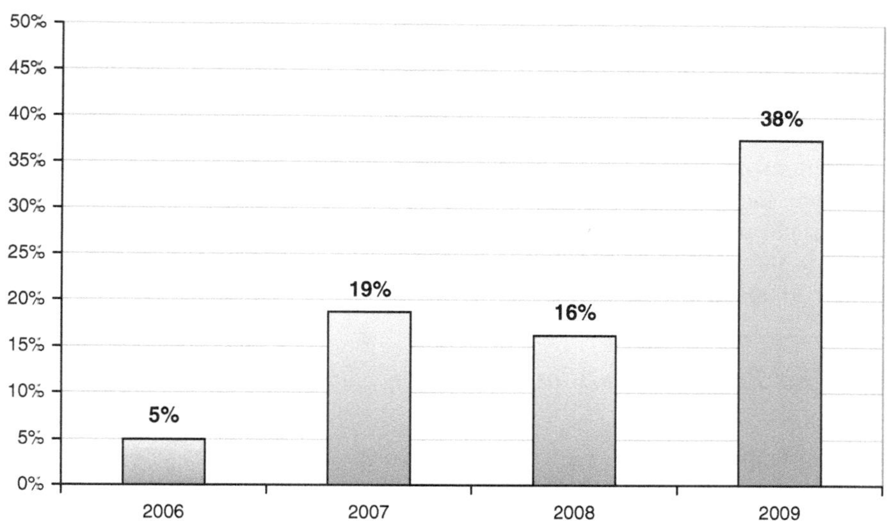

Fig. 6 Gas deficit probability

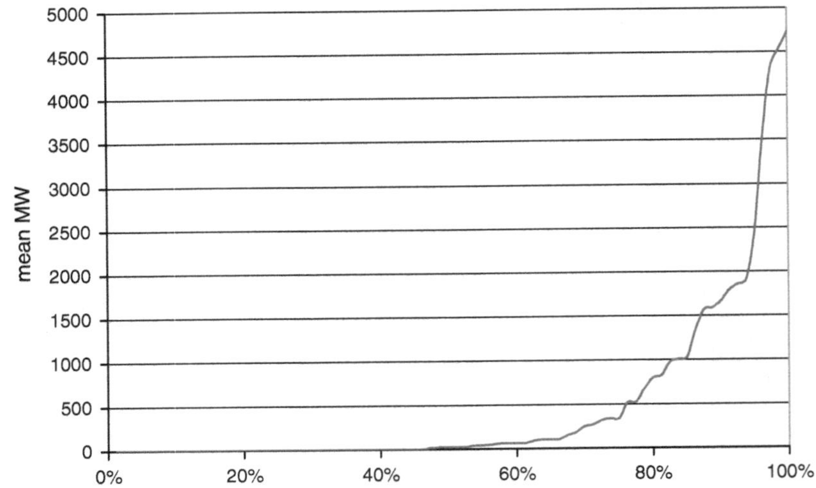

Fig. 7 Gas deficit distribution in 2007

5 Integrated Electricity–Gas Modeling in HydroScheduling Models

The previous study showed that the probability of dispatch failures of gas-fired plants due to fuel supply problems could be significant. Given that the hydrothermal dispatch model did not "know" about this possibility when calculating the water value of the hydroelectric plants, this means that the hydrothermal dispatch is not fully optimized: the system reservoirs will be depleted faster than expected, thus increasing the risks of energy deficits or of dispatching more expensive thermal plants such as fuel oil and diesel. One clear possibility for improving this situation is to incorporate the gas supply equations and constraints into the stochastic hydrothermal model, as described next.

5.1 Gas Pipeline Equations

The set of equations (8)–(10) is added to the one-stage presented problem formulation above. The only change lies in (10): thermal generation values $g_{tj}{}^*$ were known values in problems (8)–(11) and are decision variables g_{tj} here. The modified equation becomes

$$P_{tn} + \sum_{l \in \Omega(n)} [1 - w_{tln}] f_{tln} - \sum_{l \in \Omega(n)} f_{tnl} + \sum_{k \in D(n)} \delta_{tk} + \sum_{j \in T(n)} \delta'_{tj} - \sum_{j \in T(n)} \phi_{tj} g_{tj} = \sum_{k \in D(n)} d_{tk} \quad \text{for } n \in \mathrm{N} \qquad (12)$$

5.2 Case Study

The integrated electricity–gas hydrothermal scheduling was applied for the same electricity–gas configuration and data of the previous analysis. Also, as in the previous study, we gave more priority for the non-power gas supply than for the gas-fired generation, in case of fuel shortages.

Figure 8 shows the yearly short-run marginal cost (SRMC) of electricity (averaged over all months, load levels, and hydrological scenarios) of the Southeast system for two situations: unrestricted gas supply and supply constraints. We see that the fuel supply constraints had an important effect on electricity costs. Figure 9 shows the distribution of yearly SRMC over the hydrological scenarios, again for the fuel-constrained and unconstrained cases. We see that fuel constraints did not

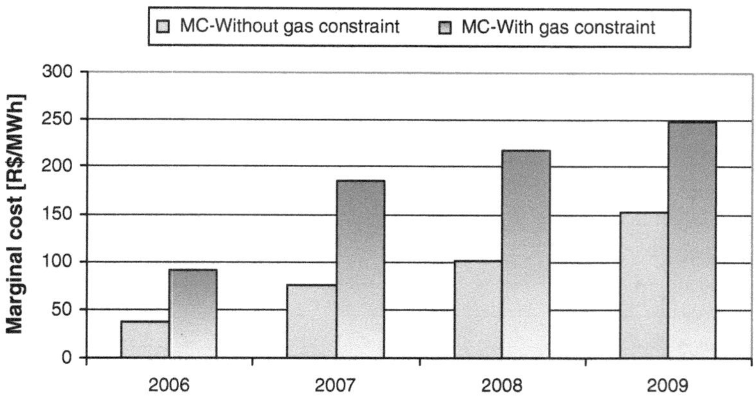

Fig. 8 Annual SRMC – Southeast region

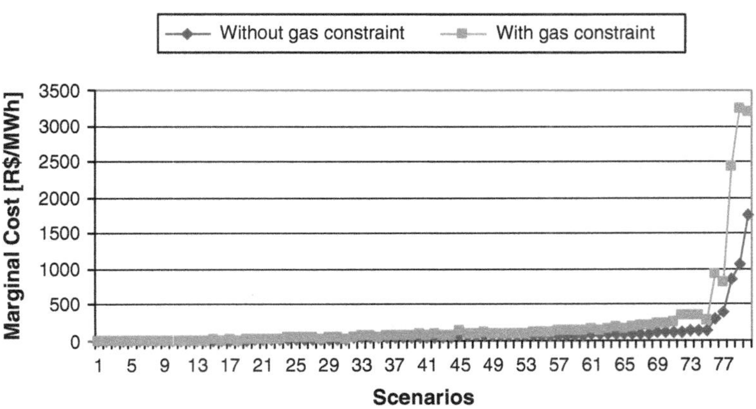

Fig. 9 Distribution of the Southeast system marginal cost in 2008

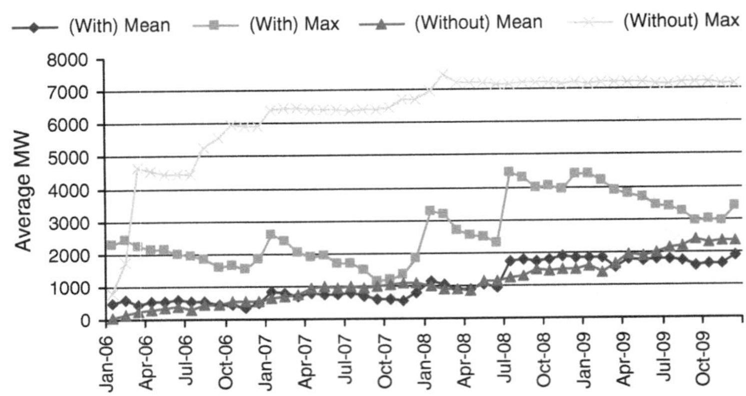

Fig. 10 Mean and maximum gas-fired generation

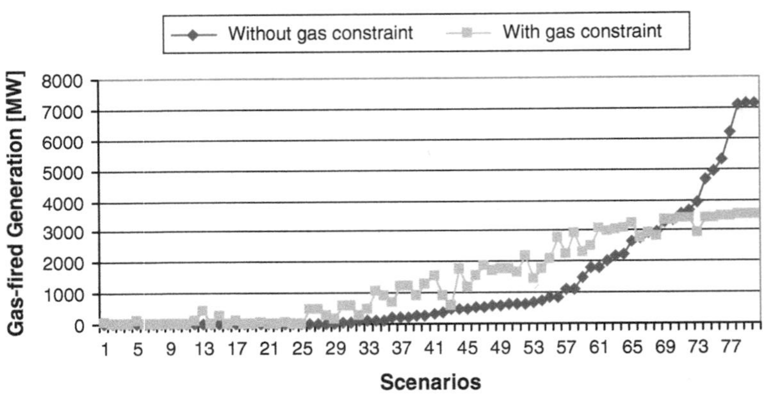

Fig. 11 Distribution of the gas-fired thermal power generation in 2008

affect electricity prices in most hydrological scenarios, which are "wet" and do not require thermal generation. However, they had a large impact on the remaining dry scenarios.

The impact of the natural gas constraints to the gas-fired thermal generation is shown in Figs. 10 and 11. Figure 10 compares the maximum and mean total generation for both cases – with and without the constraints. The maximum generation of the case with no constraints is nearly twice as much as the one with constraints. The mean generation, in turn, is more similar in both cases. The likely reason for this is that, when constraints are included in the policy calculation of the SDDP algorithm, there is a tendency for the occurrence of preventive thermal generation to compensate for a smaller firm power availability caused by the gas constraints. The end effect is not a big change in the *mean* generation, but rather in the tails of the distribution. In other words, the consideration of the gas constraints will result in less generation for the critical scenarios, but a higher generation for the moderate scenarios. This effect can be clearly seen in Fig. 11. Notice that the case with the generation constraints has much less amplitude then the other (values are sorted following the results of the simulation without gas constraints).

The impact of gas supply constraints on electricity prices could be alleviated by other measures, which can also be evaluated by the integrated gas–electricity scheduling model. One possibility is to transform the gas-fired plants into bi-fuel plants (the other fuel being diesel oil). Another possibility is to negotiate interruptible (flexible) gas contracts with industry, which would switch to an alternative fuel or even decrease production in case the gas-fired plants were dispatched. These alternatives bring more flexibility to the electricity–gas market and should be evaluated in a future work.

6 Conclusions

The vigorous growth of the natural gas market in hydro-dominated countries poses special challenges for planning and operations scheduling of both the electricity and the gas sectors due to the substantial oscillation in power-related gas consumption when hydrological conditions vary from "wet" to "dry." In this paper, we examined two alternatives for coordinating these sectors. In the first one, power dispatch assumes that there are no fuel constraints and produces a (stochastic) gas consumption schedule, which is added to the non-power gas consumption forecasts, all to be managed by the gas dispatch. In the second alternative, power and gas are dispatched jointly. It is shown that both alternatives can be modeled by stochastic optimization techniques, and their application is illustrated in the case studies based on realist data from the Brazilian power system. It should be noted that this type of modeling and analysis, introduced by these authors, are being the basis of the studies and evaluations currently carried out by Brazilian authorities on this subject in the country.

References

Barroso LA, Flach B, Kelman R, Bezerra B, Binato S, Bressane JM, Pereira MV (2005) Integrated gas-electricity adequacy planning in Brazil: technical and economical aspects. Proceedings of the IEEE PES general meeting, San Francisco

OECD, International Energy Agency (2003) South American gas – daring to tap the bounty, OECD/IEA, Paris, 253 p

Mercado RR (2002) Natural gas pipeline optimization. Pardalos PM, Resende MGC (eds) Handbook of applied optimization Oxford university Press, New York

Wolf D, Smeers Y (2000) The gas transmission problem solved by an extension of the simplex algorithm. Manage Sci 46(11):1454–1465

Munoz J, Redondo NJ, Ruiz JP (2002) Natural gas network modeling for power systems reliability studies. IEEE PES-Summer Meeting. Chicago

Mello OD, Ohishi T (2004) Natural gas transmission for thermoelectric generation problem. IX Simpósio de Especialistas em Planejamento da Operação e Expansão Elétrica, Maio
Pereira MV, Campodónico N, Kelman R (1998) Long-term hydro scheduling based on stochastic models. Proceedings of EPSOM Conference, Zurich, Available at http://www.psr-inc.com
Granville S, Oliveira GC, Thomé LM, Campodónico N, Latorre ML, Pereira M, Barroso LA (2003) Stochastic optimization of transmission constrained and large scale hydrothermal systems in a competitive framework. Proceedings of the IEEE general meeting, Toronto
Pereira MVF, Pinto LMVG (1984) Operation planning of large-scale hydrothermal systems. Proceedings of the 8th PSCC, Helsinki, Finland
Pereira MVF, Pinto LMVG (1985) Stochastic optimization of multireservoir hydroelectric system – a decomposition approach. Water Resour Res 21(6)
Gorenstin BG, Campodónico NM, Costa JP, Pereira MVF (1992) Stochastic optimization of a hydrothermal system including network constraints. IEEE Transactions on PAS 7(2):791–797
Read E, George J (1990) Dual dynamic programming for linear production/inventory systems. Comput Math Appl 19(11):29–42

Annex: Hydroscheduling and the SDDP Algorithm

The objective of hydrothermal scheduling is to determine the sequence of hydro releases, which minimizes the expected thermal operation cost (given by fuel cost plus penalties for rationing) along the planning horizon. Nevertheless, the availability of this hydroenergy is limited by reservoir storage capacity. Therefore, there is a relationship between the operative decision in a given stage and the future consequences of this decision. For example, if the stored hydroelectric energy is used today, and a drought occurs, it may be necessary to use expensive thermal generation in the future or even interrupt the energy supply. If, on the other hand, reservoir levels are kept high through a more intensive use of thermal generation today, and high inflows occur in the future, reservoirs may spill and waste energy, resulting in increased operation costs. Figure 12 illustrates this decision tree.

In contrast with thermal systems, whose long-term operation is decoupled in time, hydrosystem operation is *coupled* in time, and so a decision today affects operating costs in the future. Also, since future inflows are unknown and difficult to forecast, the scheduling of hydrothermal systems is essentially stochastic.

The scheduling problem is decomposed into several one-stage subproblems, where the objective is to minimize the sum of *immediate* and *future* operating costs, where the tradeoff between *immediate* and *future* operating costs is illustrated in Fig. 13.

The immediate cost function (ICF) is related to thermal generation costs in the present stage. The more the stored water is used for energy production, the cheaper the ICF will be today, since less thermal generation is needed to meet the load. However, using more water today leaves less storage for future use. So, in terms of the final storage, the ICF increases for higher final storage values. In turn, the FCF is associated with the *expected* thermal generation expenses from the next stage to the end of the study period. We see that the FCF decreases with final storage, as more water becomes available for future use.

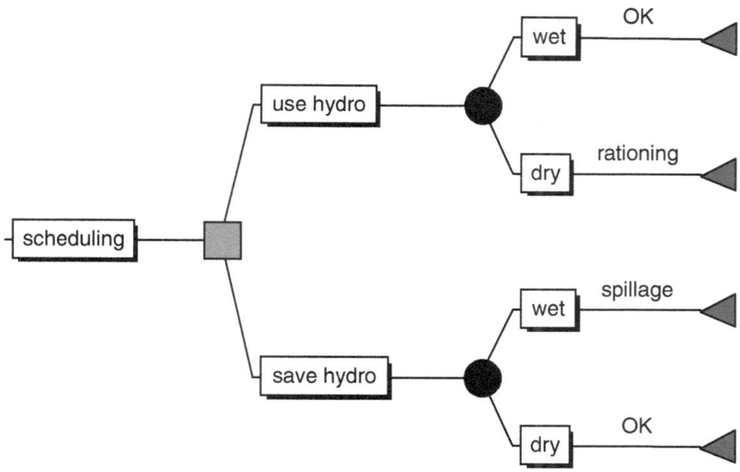

Fig. 12 Decision process for hydrothermal systems

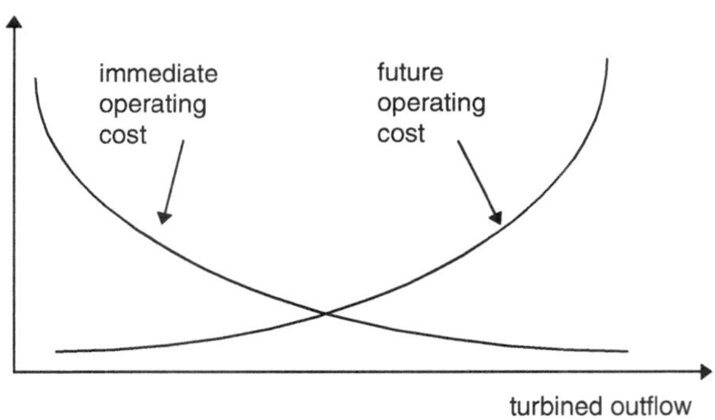

Fig. 13 Immediate and future operation costs as a function of final storage

Conceptually, the FCF can be obtained by simulating the system operation in the future for different starting values of initial storage and calculating the operation costs. If the capacity is relatively small, as in the Spanish or Norwegian system, the impact of a decision is diluted in several months. If the capacity is substantial, as in Brazil, the simulation horizon may reach up to 5 years. The simulation must take into account the *variability* of inflows to reservoirs, which fluctuate seasonally, regionally, and from year to year. Inflow forecasts are generally inaccurate, in particular when it comes from rainfall, not snowmelt. Therefore, inflows are usually modeled as a multivariate stochastic process, which preserves relevant serial and spatial dependencies observed in the past. As a consequence, FCF calculation has to be carried out on a *probabilistic* basis, using a large number of hydrological scenarios.

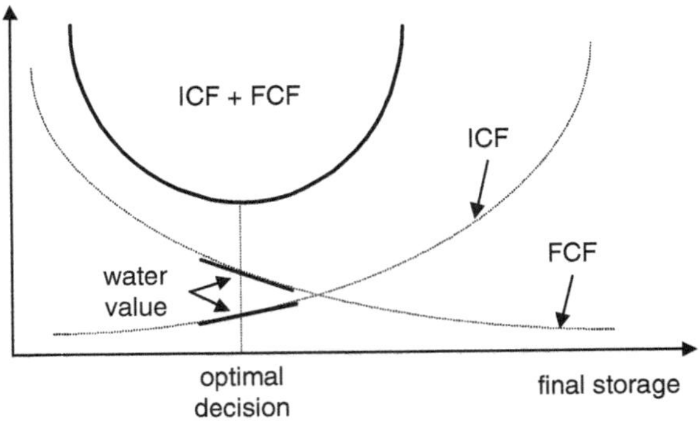

Fig. 14 Optimal hydroscheduling

The optimal use of stored water corresponds to the point that minimizes the sum of immediate and future costs. As shown in Fig. 14, this is also where the derivatives of ICF and FCF (in absolute value) with respect to storage are equal. These derivatives are known as water *values*. The optimal hydro-dispatch is at the point that equalizes immediate and future water values.

Formulation of One-Stage Hydrothermal Dispatch

The immediate cost is given by the thermal operating costs in stage t, $\sum_{j \in J} c_j g_{tj}$, where J denotes the set of thermal plants, c is the vector of thermal unit operating costs, and variable g_t (MWh) is the vector of thermal generations in stage t. In turn, the future cost is represented by the function $\alpha_{t+1}(v_{t+1})$, where variable v_{t+1} is the vector of reservoir levels in stage $t+1$. Let us consider independent inflow scenarios. Given the initial storage vector v_t, the objective of the one-stage hydroscheduling problem is to minimize the sum of immediate and future discounted operating costs (β is the discount factor):

$$z_t(v_t) = \min \sum_{j \in J} c(j) g_t(j) + \beta\, \alpha_{t+1}(v_{t+1}) \qquad (13)$$

Plant operation is modeled through the following constraints. The *water balance* equation (shown in Fig. 15) represents the coupling between successive stages: the reservoir storage v_{t+1} at stage $t + 1$ is equal to the initial storage v_t minus outflow volumes (turbined variable u_t and spilled variable s_t) plus inflow volumes (lateral inflow a_t plus releases from immediately upstream plants belonging to set U), all in stage t, for all hydroplants in set I:

$$v_{t+1,i} = v_t i - u_t i - s_{ti} + a_{ti} + \sum_{m \in U(i)} [u_{tm} + s_{tm}], \quad \text{for } i \in I \qquad (14)$$

The *load supply* equation relates total thermal and hydro-generation to system load d_t (MWh). The hydro-generation for unit i is given by the product of its production

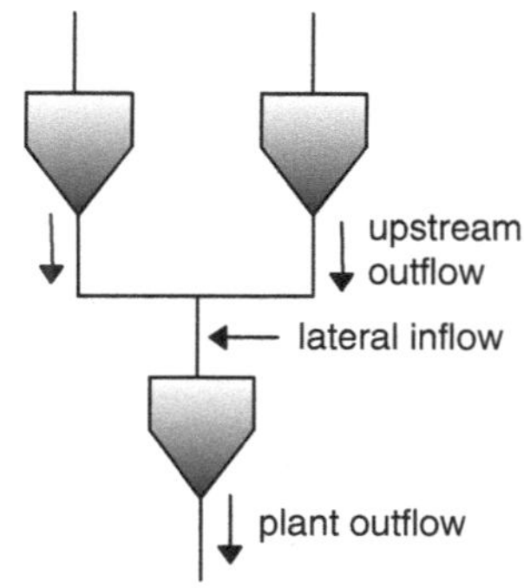

Fig. 15 Reservoir water balance

coefficient ρ_i (MWh m^{-3}) and its turbined outflow u_{ti}, resulting in

$$\sum_{i\in I} \rho_i u_{ti} + \sum_{j\in J} g_{tj} = d_t \tag{15}$$

Finally, there are bounds on thermal generation (g^{max}), maximum storage (v^{max}), and turbine capacity (u_{max}) for each hydroplant:

$$g_{tj} \le g_{max}, \quad j \in J \tag{16}$$

$$v_{t+1,i} \le v_i^{max}, \quad i \in I \tag{17}$$

$$u_t(i) \le u_i^{max}, \quad i \in I \tag{18}$$

For simplicity, network constraints are not represented in the above formulation. These constraints are not coupled in time and are expressed as linearized power flow equations with transmission limits.

Calculation of Future Cost Function

The FCF calculation is naturally the key aspect of the state-space scheme. In theory, $\alpha_{t+1}(v_{t+1})$ could be calculated by *simulating* system operation in the future for different starting values of initial storage and calculating the operating costs, as illustrated in Fig. 16.

However, this "brute force" approach has the same computational drawbacks as the explicit stochastic formulation. Therefore, the FCF in each stage is calculated through a more efficient SDP recursion: reservoir levels are discretized, and starting from the last stage T, problem (13–18) is solved assuming the first storage level for each reservoir. Since we are at the last stage, the FCF is zero. Because of inflow uncertainty, the hydroscheduling problem is successively solved for N different inflow scenarios and the expected operation cost is calculated as the mean of the costs over the N scenarios. For each remaining storage states in stage T, repeat the calculation of expected operation costs and interpolate in order to produce the FCF $\alpha_T(v_T)$ for stage $T-1$. This process is then repeated for all states in stages $T-1$, $T-2$, etc. Note that the objective in those stages is to minimize immediate

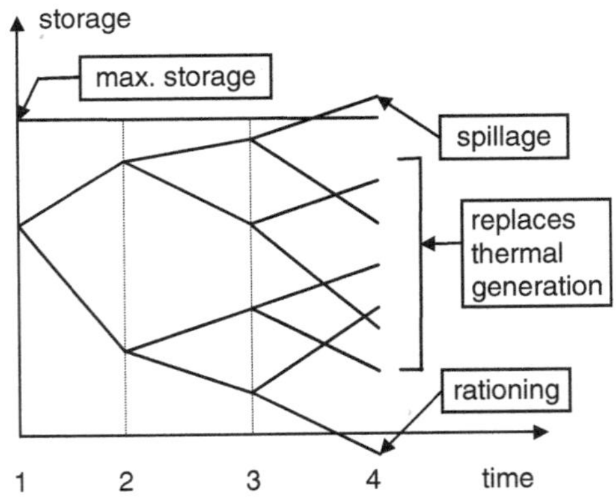

Fig. 16 "Brute force" FCF calculation

operation plus expected future cost, given by previously calculated FCF. The final result of the SDP scheme outlined above is the set of FCFs $\alpha_{t+1}(v_{t+1})$ for each stage t.

The procedure can be depicted as follows:
Initialize the end-of-horizon FCF $\alpha_{T+1}(v_T) \leftarrow 0$

for $t = T, T-1, \ldots, 1$
for each storage value $v_t = v_t^1, \ldots, v_t^m, \ldots, v_t^M$
for each inflow scenario $a_t = a_t^1, \ldots, a_t^k, \ldots, a_t^K$
solve the one-stage problem for initial storage v_t^m and inflow a_t^k:

$$\alpha_t^k(v_t^m) = \text{Min}\, c_t(u_t) + \alpha_{t+1}(v_{t+1})$$

subject to

$$v_{t+1} = v_t^m - u_t - s_t + a_t^k$$
$$v_{t+1} \leq \bar{v}$$
$$u_t \leq \bar{u}$$

next
calculate the expected operation cost over all inflow scenarios:

$$\alpha_t(v_t^m) = \sum_{k=1}^{K} p_k \times \alpha_t^k(v_t^m)$$

next
create a complete FCF $\alpha_t(v_t)$ for the previous stage by interpolation on the discrete values $\{\alpha_t(v_t^m),\ m = 1, \ldots, M\}$
next

However, because of the discretization, the SDP computational effort increases exponentially with the number of reservoirs, the well-known "curse of dimensionality" of DP. Therefore, it is not practical for systems with many reservoirs.

For this reason, it has become necessary to develop computationally feasible state-space schemes. The traditional approach, still adopted in many countries, has been to reduce system dimensionality by the aggregating system reservoirs into one reservoir that represents the energy production capability of the cascade. This scheme is in some cases coupled with the use of partial dynamic programming schemes (typically, calculation of separate FCFs for each basin).

More recently, an approach based on the analytical representation of the FCF, known as SDDP, has been applied in several countries in South and Central America, plus in the United States, New Zealand, Spain, and Norway.[5] The SDDP scheme does not require discretization of the state space and, as a consequence, alleviates the computational requirements of the stochastic DP recursion. It will be described next.

The Dual Dynamic Programming Scheme

The stochastic dual DP scheme (DDP) proposed independently by Pereira and Pinto (1985) and Read and George (1990) is based on the observation that the FCF can be represented as a piecewise linear function, and so there is no need to create an interpolated table. Furthermore, the slope of the FCF around a given point can be analytically obtained from the one-stage dispatch problem (13–18).

The last-stage dispatch problem is shown below (note that the FCF in this stage is zero):

$$z_T = \min \sum\nolimits_{j \in J} c_j g_{Tj} \tag{19}$$

$$v_{T+1,i} = v_{T,i} - u_{Ti} - s_{Ti} + a_{Ti} + \sum\nolimits_{m \in U(i)} [u_{Tm} + s_{Tm}], \quad i \in I \tag{20}$$

$$\sum\nolimits_{i \in I} \rho_i u_{Ti} + \sum\nolimits_{j \in J} g_{Tj} = d_T \tag{21}$$

$$g_{Tj} \le g_j^{max}, \quad j \in J \tag{22}$$

$$v_{T+1,i} \le v_i^{max}, \quad i \in I \tag{23}$$

$$u_{Ti} \le u_i^{max}, \quad i \in I \tag{24}$$

The Lagrange multiplier vector π_h associated with the water balance equation (20), also known as the *water value*, represents the derivative of z_T with respect to a variation in the initial storage v_T, and corresponds to the slope of FCF for stage $T-1$. Figure 17 shows the calculation of the operation cost and FCF slopes for each state in stage T. It can be seen that the FCF for stage $T-1$ corresponds to the *piecewise* cost surface produced by taking the linear segment with the highest cost value in each state (the convex hull).

The dispatch problem for stage $T-1$ is now

$$z_{T-1}(v_{T-1}) = \min \sum\nolimits_{j \in J} c_j g_{Tj} + \beta\, \alpha_T \tag{25}$$

---

[5] A related scheme, called constructive dynamic programming, has been applied to the Australian system.

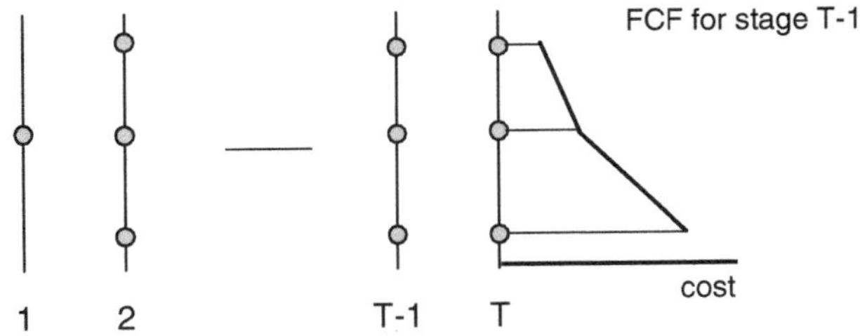

Fig. 17 Calculation of piecewise FCF for stage T-1

Fig. 18 Piecewise linear FCF for stage T

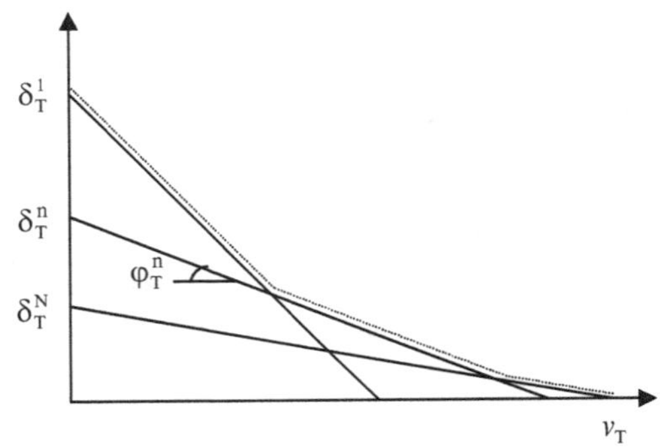

subject to

$$v_{Ti} = v_{T-1,i} - u_{T-1,i} - s_{T-1,i} + a_{T-1,i} + \sum\nolimits_{m \in U(i)} [u_{T-1,m} + s_{T-1,m}], \quad i \in I \quad (26)$$

$$\sum_{i \in I} \rho_i u_{T-1,i} + \sum\nolimits_{j \in J} g_{T-1,j} = d_{T-1} \quad (27)$$

$$g_{T-1,j} \leq g_j^{max}, \quad j \in J \quad (28)$$

$$u_{T-1,i} \leq u_i^{max}, \quad i \in I \quad (29)$$

$$\alpha_T \geq \varphi_T^n v_T + \delta_T^n, \quad n = 1, \ldots, N \quad (30)$$

The FCF is represented by the scalar variable α_T and the N linear constraints $\{\alpha_T \geq \varphi_T{}^n v_T + \delta_T{}^n\}_{n=1,N}$, where N is the number of linear segments. A shown in Fig. 18, these inequalities represent the piecewise characteristic of this function.

Therefore, in general, for each stage t, the FCF is represented by the scalar variable α_t and N linear constraints.

Backward Recursion and Lower Bound Calculation

The recursive calculation of the piecewise linear FCF is very similar to the standard SDP scheme. To take into account that future inflows are unknown, consider K inflow scenarios. The backward recursion scheme is shown below:

Set N equal to M, the number of initial storage values;

initialize the FCF for stage T as zero: $\delta^n{}_{T+1}$ *and* $\varphi^n{}_{T+1}$ *are null, for* $n = 1, \ldots, N$;

for $t = T, T-1, \ldots, 1$

for each storage value v_t^m, $m = 1, \ldots, M$

for each inflow scenario a_t^k, $k = 1, \ldots, K$

solve the one-stage scheduling problem for initial storage v_t^m *and inflow* a_t^k:

$$z_t^k\left(v_t^m\right) = \min \sum_{j\hat{I}J} c_j g_{tj} + \beta\, \alpha_{t+1}$$

$$v_{t+1,i} = v_{ti}^m - u_{ti} s_{ti} a_{ti}^k + \sum\nolimits_{m\hat{I}U(i)} [u_{tm} + s_{tm}], \quad i \in I$$

$$\sum\nolimits_{i\hat{I}I} \rho_i u_{ti} + \sum\nolimits_{j\hat{I}J} g_{tj} = d_t$$

$$g_{tj} \leq g_j^{max}, \quad j \in J$$

$$u_{ti} \leq u_i^{max}, \quad i \in I$$

$$\alpha_{t+1} \geq \varphi_{t+1}^n v_{t+1} + \delta_{t+1}^n, \quad n = 1, \ldots, N$$

end;

calculate the volume coefficients and constant term for the mth linear segment of FCF in the previous stage by taking averages over all scenarios (scenario k at stage t has conditional probabilility p_{kt}*)*:

$$\varphi_t^m = \sum_k p_{kt} \pi_{ht}^k$$

$$\delta_t^m = \sum_k p_{kt} a_t^k (v_t^m) - \varphi_t^m v_t^m$$

end;

end.

At first sight, there are no substantial differences between the Dual DP procedure in and the traditional DP scheme. Note, however, that the traditional scheme had to create a new FCF table in each stage by *interpolation* of the discrete values $\left\{\alpha_t\left(v_t^m\right)\right\}$. As a consequence, the required number of points in the table for a system of I hydroplants is at least equal to the 2^I combinations of extreme points (full/empty). In the Dual DP scheme, the piecewise linear segments can be used to *extrapolate* the FCF values, that is, it not necessary to use all combinations of points to obtain a complete (although approximate) FCF. Moreover, if a smaller number of initial storage values is used, a smaller number of linear segments will be generated. As seen in Fig. 18, the resulting FCF, which is based on the maximum value over all segments, will then be a *lower bound* to the "true" function.

As a consequence, the FCF for the first stage is a lower bound $\underline{z}$ to the optimal solution of the hydrothermal scheduling problem:

$$Z^L = Z_1(v_1)$$

Forward Simulation and Upper Bound Calculation

If we use the FCF produced by the backward recursion scheme, an upper bound to the optimal solution of the hydrothermal scheduling problem can be obtained by Monte Carlo simulation of system operation. (Read and George (1990) uses a complete representation of the piecewise linear function; his approach was limited to two reservoirs.) This is due to the fact that the only FCF that can result in the optimal operation cost is the optimal function itself; all others, by definition, will have higher operation costs. The simulation scheme is shown below.

Define inflow scenarios a_t^m, $m = 1, \ldots, M$ *for all stages* $t = 1, \ldots, T$;
for each inflow scenario a_t^m, $m = 1, \ldots, M$
initialize storage value for stage 1 as $v_t^m = v_1$;
for $t = 1, \ldots, T$
solve the one-stage scheduling problem (19–24);
calculate the total operating cost z^m *for scenario m as the sum of all immediate thermal costs along the study period*;
end;
end.

An upper bound for the expected operation cost is estimated as the mean total cost over all scenarios:

$$z^U = M^{-1} \sum_{m=1,M} Z^m$$

This estimator is unbiased, converging to the population value. Because of the sampling variation, there is an uncertainty around the "true" expected value. A 95% confidence interval $[z^U - 1.96\sigma,\ z^U + 1.96\sigma]$ can be derived by estimating the variance of the estimator as

$$\sigma^2 = M^{-1} å_{m=1,M}(z^m - z^U)^2$$

Optimality Check and New Iteration

Optimality is achieved when the lower bound z^L is within the confidence interval of the upper bound. Note that, because of sampling variation, the lower bound may exceed the upper bound mean estimate z^U.

If the lower bound is outside the confidence interval, the backward recursion step described previously is repeated with an additional set of storage values. These values are produced by the forward simulation step. Note that all linear constraints produced along the iterative process are retained, since the piecewise FCF is given by the convex hull. Therefore, the representation of the FCF is gradually improved along the process until convergence is achieved.

Recent Progress in Two-stage Mixed-integer Stochastic Programming with Applications to Power Production Planning

Werner Römisch and Stefan Vigerske

Abstract We present recent developments in two-stage mixed-integer stochastic programming with regard to application in power production planning. In particular, we review structural properties, stability issues, scenario reduction, and decomposition algorithms for two-stage models. Furthermore, we describe an application to stochastic thermal unit commitment.

Keywords Decomposition algorithms · Discrepancy · Mixed-integer · Two-stage · Scenario reduction · Stability · Stochastic programming · Unit commitment

1 Introduction

Since its beginnings in the late 1980s, mixed-integer stochastic programming has undergone a considerable development both in theory and in computations. We refer to the excellent overviews in Louveaux and Schultz (2003), Schultz (2003), Sen (2005), and to the very comprehensive bibliography (van der Vlerk 1996).

The aim of this paper is to look at some of the more recent developments that bear further potential for applications to power systems modeling and optimization. First, we mention new results on structures and approximating the underlying probability distribution (Eichhorn and Römisch 2007; Römisch and Vigerske 2008), and as a consequence of the latter, on scenario reduction in two-stage mixed-integer stochastic programs (Henrion et al. 2008, 2009). Another line of work deals with the consequences of replacing the (traditional) expectation functional in the objective by risk functionals on structural properties and algorithms (see Eichhorn and Römisch (2005); Schultz and Tiedemann (2006), e.g.). Much work was directed to algorithmic issues and, in particular, to decomposition schemes (Alonso-Ayuso et al. 2003; Ahmed et al. 2004; Carøe and Schultz 1999; Dentcheva and Römisch 2004;

W. Römisch (✉)
Humboldt University, 10099 Berlin, Germany
e-mail: romisch@math.hu-berlin.de

S. Rebennack et al. (eds.), *Handbook of Power Systems I*, Energy Systems,
DOI 10.1007/978-3-642-02493-1_8, © Springer-Verlag Berlin Heidelberg 2010

Escudero et al. 2007; Lulli and Sen 2004; Ntaimo and Sen 2008; Schultz et al. 1998; Sen and Higle 2005; Sen and Sherali 2006), where much is due to the pioneering work of Sen and his coworkers.

In the following, we review some of the recent work. We start with a review of structural properties, discuss stability issues, methods for scenario reduction, and decomposition algorithms. As an illustration, we finally discuss an application to the stochastic unit commitment problem in power production planning.

2 Models and Structural Properties

Stochastic programs with mixed-integer recourse arise as deterministic equivalents of linear programs containing a random parameter vector ξ (varying in Ξ) and being of the form

$$\min\{\langle c,x\rangle \mid x \in X,\ T(\xi)x \geq h(\xi)\},$$

where X is a closed subset of $\mathbb{R}^m$, $c \in \mathbb{R}^m$, the (technology) matrix $T(\xi)$ and the vector $h(\xi)$ may depend on ξ. Given a realization of ξ, a possible violation of $h(\xi) - T(\xi)x \leq 0$ is compensated by the recourse cost $\langle q_1(\xi), y_1(\xi)\rangle + \langle q_2(\xi), y_2(\xi)\rangle$, where the pair $(y_1(\xi), y_2(\xi))$ with integral y_2 satisfies the constraint $W_1 y_1 + W_2 y_2 \leq h(\xi) - T(\xi)x$. Here, the cost coefficients $q_1(\xi)$ and $q_2(\xi)$ may depend on ξ. The modeling idea consists in adding the expected recourse cost $\mathbb{E}(\langle q_1(\xi), y_2(\xi)\rangle + \langle q_2(\xi), y_2(\xi)\rangle)$ to the original cost $\langle c,x\rangle$ and in minimizing the total cost with respect to (y_1, y_2). This leads to the stochastic program with mixed-integer recourse

$$\min\left\{ \int_{\Xi} f_0(x,\xi)dP(\xi) \;\middle|\; x \in X \right\}, \tag{1}$$

where the function f_0 is given by

$$f_0(x,\xi) := \langle c,x\rangle + \Phi(q(\xi), h(\xi) - T(\xi)x) \quad ((x,\xi) \in \mathbb{R}^m \times \Xi), \tag{2}$$

Φ is the infimum function of a mixed-integer linear program

$$\Phi(u,t) := \inf\{\langle u_1, y_1\rangle + \langle u_2, y_2\rangle \mid y_1 \in \mathbb{R}^{m_1}, y_2 \in \mathbb{Z}^{m_2}, W_1 y_1 + W_2 y_2 \leq t\} \tag{3}$$

for all pairs $(u,t) \in \mathbb{R}^{m_1+m_2} \times \mathbb{R}^r$, Ξ is a polyhedron in $\mathbb{R}^s$, W_1 and W_2 are (r, m_1)- and (r, m_2)-matrices, respectively, $q(\xi) \in \mathbb{R}^{m_1+m_2}$, $h(\xi) \in \mathbb{R}^r$, the (r,m)-matrix $T(\xi)$ are affine functions of $\xi \in \mathbb{R}^s$, and P is a probability distribution on the set Ξ (shortly $P \in \mathcal{P}(\Xi)$). Since the decisions x and $y(\xi)$ are made before and after the realization of ξ, they are called first and second stage decisions, respectively.

The following conditions are imposed to have the model (1) well-defined:
(C1) The matrices W_1 and W_2 have only rational elements.

(C2) For each pair $(x, \xi) \in X \times \Xi$ it holds that $h(\xi) - T(\xi)x \in \mathcal{T}$, where

$$\mathcal{T} := \{t \in \mathbb{R}^r \mid \exists y = (y_1, y_2) \in \mathbb{R}^{m_1} \times \mathbb{Z}^{m_2} \text{ such that } W_1 y_1 + W_2 y_2 \le t\}.$$

(C3) For each $\xi \in \Xi$ the recourse cost $q(\xi)$ belongs to the dual feasible set

$$\mathcal{U} := \left\{u = (u_1, u_2) \in \mathbb{R}^{m_1+m_2} \;\middle|\; \exists z \in \mathbb{R}^r_- \text{ such that } W_1^\top z = u_1,\ W_2^\top z = u_2\right\}.$$

(C4) $P \in \mathcal{P}_2(\Xi)$, that is, $P \in \mathcal{P}(\Xi)$ and $\int_\Xi \|\xi\|^2 P(d\xi) < +\infty$.

Condition (C2) means that a feasible second stage decision always exists (*relatively complete recourse*). Both (C2) and (C3) imply $\Phi(u,t)$ to be finite for all $(u,t) \in \mathcal{U} \times \mathcal{T}$. Clearly, it holds $(0,0) \in \mathcal{U} \times \mathcal{T}$ and $\Phi(0,t) = 0$ for every $t \in \mathcal{T}$. With the convex polyhedral cone

$$\mathcal{K} := \{t \in \mathbb{R}^r \mid \exists y_1 \in \mathbb{R}^{m_1} \text{ such that } t \ge W_1 y_1\} = W_1(\mathbb{R}^{m_1}) + \mathbb{R}^r_+,$$

one obtains the representation

$$\mathcal{T} = \bigcup_{z \in \mathbb{Z}^{m_2}} (W_2 z + \mathcal{K}). \tag{4}$$

The two extremal cases are (a) W_1 has rank r implying $\mathcal{K} = \mathbb{R}^r = \mathcal{T}$ (*complete recourse*) and (b) $W_1 = 0$ (*pure integer recourse*) leading to $\mathcal{K} = \mathbb{R}^r_+$.

In general, the set $\mathcal{T}$ is connected (i.e., there exists a polygon connecting two arbitrary points of $\mathcal{T}$) and condition (C1) implies that $\mathcal{T}$ is closed. If, for each $t \in \mathcal{T}$, $Z(t)$ denotes the set

$$Z(t) := \{z \in \mathbb{Z}^{m_2} \mid \exists y_1 \in \mathbb{R}^{m_1} \text{ such that } W_1 y_1 + W_2 z \le t\},$$

the representation (4) implies that $\mathcal{T}$ may be decomposed into subsets of the form

$$\mathcal{T}(t_0) := \{t \in \mathcal{T} \mid Z(t) = Z(t_0)\} = \bigcap_{z \in Z(t_0)} (W_2 z + \mathcal{K}) \setminus \Big(\bigcup_{z \in \mathbb{Z}^{m_2} \setminus Z(t_0)} (W_2 z + \mathcal{K})\Big) \tag{5}$$

for every $t_0 \in \mathcal{T}$. In general, the set $Z(t_0)$ is finite or countable, but condition (C1) implies that $Z(t_0)$ in the intersection in (5) may be replaced by a single element of $\mathcal{T}$ and $\mathbb{Z}^{m_2} \setminus Z(t_0)$ in the union by a finite subset of $\mathbb{Z}^{m_2}$, respectively (see Bank et al. 1982, Lemmas 5.6.1 and 5.6.2). Hence, if (C1) is satisfied, there exist countably many elements $t_i \in \mathcal{T}$ and $z_{ij} \in \mathbb{Z}^{m_2}$ for j belonging to a finite subset N_i of $\mathbb{N}$, $i \in \mathbb{N}$, such that

$$\mathcal{T} = \bigcup_{i \in \mathbb{N}} \mathcal{T}(t_i) \quad \text{with} \quad \mathcal{T}(t_i) = (t_i + \mathcal{K}) \setminus \bigcup_{j \in N_i} (W_2 z_{ij} + \mathcal{K}). \tag{6}$$

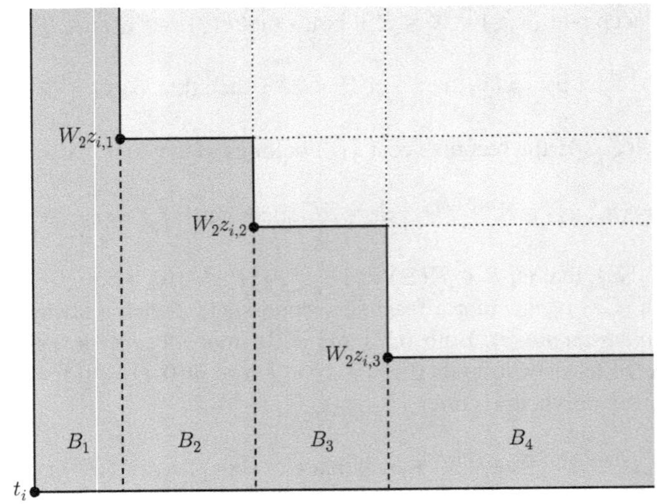

Fig. 1 Illustration of $\mathcal{T}(t_i)$ (see (6)) for $W_1 = 0$ and $r = 2$, i.e., $\mathcal{K} = \mathbb{R}^2_+$, with $N_i = \{1, 2, 3\}$ and its decomposition into the sets B_j, $j = 1, 2, 3, 4$, whose closures are convex polyhedral (*rectangular*)

The sets $\mathcal{T}(t_i)$, $i \in \mathbb{N}$, are nonempty and connected (even star-shaped cf. Bank et al. 1982, Theorem 5.6.3), but nonconvex in general (see the illustration in Fig. 1). If for some $i \in \mathbb{N}$ the set $\mathcal{T}(t_i)$ is nonconvex, it can be decomposed into a finite number of subsets of $\mathcal{T}(t_i)$ whose closures are convex polyhedra with facets parallel to suitable facets of $W_1(\mathbb{R}^{m_1})$ or of $\mathbb{R}^r_+$ (see Fig. 1). By renumbering all such subsets (for every $i \in \mathbb{N}$) one obtains countably many subsets B_j, $j \in \mathbb{N}$, of $\mathcal{T}$, which form a partition of $\mathcal{T}$. Since the sets $Z(t)$ of feasible integer decisions do not change if t varies in some B_j, the function $\Phi(u, \cdot)$ is Lipschitz continuous (with modulus not depending on j) on B_j for every $j \in \mathbb{N}$ and every fixed $u \in \mathcal{U}$.

Now, let (C1)–(C3) be satisfied. Then the function Φ is lower semicontinuous and the function $(u, t) \mapsto \Phi(u, t)$ from $\mathcal{U} \times \mathcal{T}$ to $\mathbb{R}$ has the (convex) polyhedral continuity regions $\mathcal{U} \times B_j$, $j \in \mathbb{N}$. More precisely, the estimate

$$|\Phi(u,t) - \Phi(\tilde{u},\tilde{t})| \le L(\max\{1, \|t\|, \|\tilde{t}\|\}\|u - \tilde{u}\| + \max\{1, \|u\|, \|\tilde{u}\|\}\|t - \tilde{t}\|) \quad (7)$$

holds for all pairs $(u, t), (\tilde{u}, \tilde{t}) \in \mathcal{U} \times B_j$ and some constant $L > 0$. For proofs and further details the interested reader is referred to Bank et al. (1982, Chap. 5.6).

Next, we consider the integrand

$$f_0(x, \xi) = \langle c, x\rangle + \Phi(q(\xi), h(\xi) - T(\xi)x)$$

for all pairs $(x, \xi) \in X \times \Xi$ and study the continuity properties and growth behavior of $f_0(x, \cdot)$ on Ξ for fixed $x \in X$. The properties of Φ imply that, for every $x \in X$, there exists a partition $\{\Xi_{x,j}\}_{j\in\mathbb{N}}$ of Ξ given by

$$\Xi_{x,j} = \{\xi \in \Xi \mid h(\xi) - T(\xi)x \in B_j\} \qquad (j \in \mathbb{N}). \tag{8}$$

Furthermore, the function $f_0(x,\cdot)$ (on Ξ) satisfies the properties

$$|f_0(x,\xi) - f_0(x,\tilde{\xi})| \leq \hat{L} \max\{1, \|\xi\|, \|\tilde{\xi}\|\}\|\xi - \tilde{\xi}\| \quad (x \in X,\ \xi, \tilde{\xi} \in \Xi_{x,j}), \tag{9}$$
$$|f_0(x,\xi)| \leq C \max\{1, \|x\|\} \max\{1, \|\xi\|^2\} \quad (x \in X,\ \xi \in \Xi), \tag{10}$$

with some positive constants $\hat{L}$ and C. Because of (10), condition (C4) implies the existence of the integral in (1). We note that $f_0(x,\cdot)$ is globally Lipschitz continuous on $\Xi_{x,j}$ if the recourse cost $q(\xi)$ does not depend on ξ. It is even globally Lipschitz continuous on Ξ if only $q(\xi)$ depends on ξ. In both cases, $|f_0(x,\cdot)|$ grows only linearly with $\|\xi\|$ and a finite first order moment of P, that is, $P \in \mathcal{P}_1(\Xi)$ (instead of (C4)), implies the existence of the integral.

Since the objective function of (1) is lower semicontinuous if the conditions (C1)–(C4) are satisfied, solutions of (1) exist if X is closed and bounded. If the probability distribution P has a density, the objective function of (1) is continuous but nonconvex in general. If the support of P is finite, the objective function is piecewise continuous with a finite number of polyhedral continuity regions. The latter is illustrated by Fig. 2, which shows the expected recourse function

$$x \mapsto \int_{\Xi} \Phi(q, h(\xi) - Tx)dP(\xi) \qquad (x \in [0,5]^2),$$

with $r = s = 2, h(\xi) = \xi, m_1 = 0, W_1 = 0, m_2 = 4, q = (-16, -19, -23, -28)^\top$, the matrices

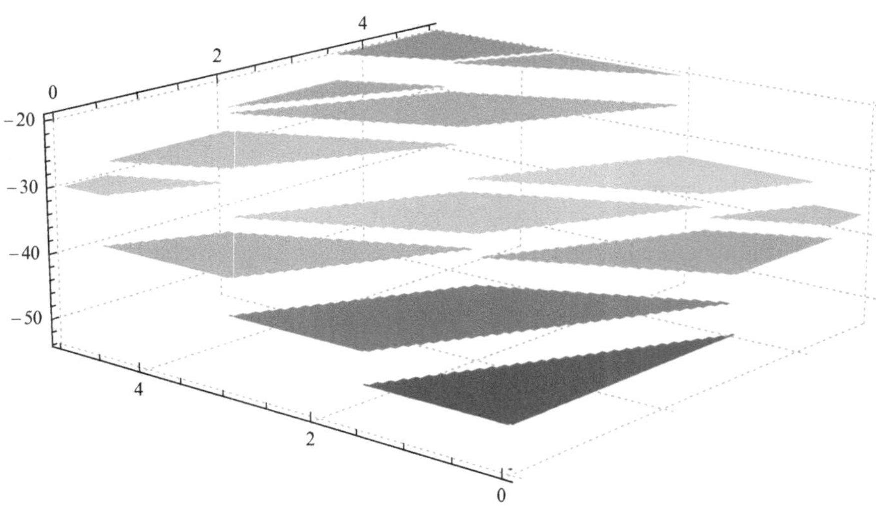

Fig. 2 Illustration of an expected recourse function with pure 0–1 recourse, random right-hand side, and discrete uniform probability distribution

$$W_2 = \begin{pmatrix} 2 & 3 & 4 & 5 \\ 6 & 1 & 3 & 2 \end{pmatrix} \quad \text{and} \quad T = \begin{pmatrix} 2/3 & 1/3 \\ 1/3 & 2/3 \end{pmatrix},$$

and binary restrictions on the second stage variables as in Schultz et al. (1998), but with a uniform probability distribution P having a smaller finite support than in Schultz et al. (1998), namely, $\text{supp}\,(P) = \{5, 10, 15\}^2$.

3 Stability

In this section, we review stability results for mixed-integer two-stage stochastic programs (1), that is, results on the dependence of their solutions and optimal values on the underlying probability distribution P. Such results also provide information on how an underlying probability distribution should be approximated such that approximate solutions and optimal values get close to the original ones.

In this context it is well known that the behavior of the (first stage) solution set

$$S(P) := \left\{ x \in X \,\middle|\, \int_{\Xi} f_0(x,\xi) P(d\xi) = v(P) \right\}$$

with respect to changes of P requires knowledge on the growth of the objective function

$$x \mapsto F_P(x) := \mathbb{E}(f_0(x,\xi)) = \int_{\Xi} f_0(x,\xi) P(d\xi)$$

near $S(P)$. Here, $v(P)$ denotes the infimum of the objective function or the optimal value, that is,

$$v(P) := \inf\left\{ \int_{\Xi} f_0(x,\xi) P(d\xi) \,\middle|\, x \in X \right\}.$$

However, the growth behavior of F_P depends essentially on properties of the underlying probability distribution P. The situation is different for optimal values $v(P)$. Their behavior with respect to changes of P depends essentially on structural properties of the function f_0, which are well studied (cf. Sect. 2).

It is shown in Römisch and Vigerske (2008) that the following distances of probability distributions are important for mixed-integer two-stage stochastic programs:

$$\zeta_{\ell,\mathcal{B}}(P,Q) := \sup\left\{ \left| \int_B f(\xi) P(d\xi) - \int_B f(\xi) Q(d\xi) \right| \,\middle|\, f \in \mathcal{F}_\ell(\Xi), B \in \mathcal{B} \right\}, \tag{11}$$

where $\ell \in \{1,2\}$ and $\mathcal{B}$ is a set of convex polyhedra, which contains the closures of $\Xi_{x,j}$, $j \in \mathbb{N}$, $x \in X$ (see (8)), and $\mathcal{F}_\ell(\Xi)$ contains all functions $f : \Xi \to \mathbb{R}$ such that

$$|f(\xi)| \leq \max\{1, \|\xi\|^\ell\} \text{ and } |f(\xi) - f(\tilde{\xi})| \leq \max\{1, \|\xi\|^{\ell-1}, \|\tilde{\xi}\|^{\ell-1}\}\|\xi - \tilde{\xi}\|$$

holds for all $\xi, \tilde{\xi} \in \Xi$. While the set $\mathcal{F}_\ell(\Xi)$ of functions has its origin in property (9) of the integrand f_0, but depends on the specific structure of the second stage program only with respect to $\ell \in \{1, 2\}$, the class $\mathcal{B}$ of convex polyhedra strongly depends on that structure.

If the conditions (C1)–(C4) are satisfied and X is closed and bounded, there exists a constant $L > 0$ such that the estimate

$$|v(P) - v(Q)| \leq L\varphi_P(\zeta_{\ell,\mathcal{B}}(P, Q)) \tag{12}$$

holds for every $Q \in \mathcal{P}_\ell(\Xi)$ with $\ell \in \{1, 2\}$ and $\ell = 2$ if ξ enters $q(\xi)$ and, in addition, $h(\xi)$ or $T(\xi)$. Here, the function φ_P is defined by $\varphi_P(0) = 0$ and

$$\varphi_P(t) := \inf_{R \geq 1} \left\{ R^{r+1}t + \int_{\{\xi \in \Xi \mid \|\xi\| > R\}} \|\xi\|^\ell P(d\xi) \right\} \qquad (t > 0).$$

The function characterizes the tail behavior of P and is continuous at $t = 0$. If P has a finite pth moment, that is, if $\int_\Xi \|\xi\|^p P(d\xi) < +\infty$, for some $p > \ell$, the estimate

$$\varphi_P(t) \leq C t^{\frac{p-\ell}{p+r-1}} \qquad (t \geq 0)$$

is valid for some constant $C > 0$, and if Ξ is bounded, the estimate (12) simplifies to

$$|v(P) - v(Q)| \leq L\zeta_{\ell,\mathcal{B}}(P, Q).$$

If the set $\Xi \subset \mathbb{R}^s$ belongs to $\mathcal{B}$, we obtain from (11) by choosing $B := \Xi$ and $f \equiv 1$, respectively,

$$\max\{\zeta_\ell(P, Q), \alpha_\mathcal{B}(P, Q)\} \leq \zeta_{\ell,\mathcal{B}}(P, Q) \tag{13}$$

for all $P, Q \in \mathcal{P}_\ell(\Xi)$. Here, ζ_ℓ and $\alpha_\mathcal{B}$ denote the ℓth order Fortet–Mourier metric (see Rachev (1991, Sect. 5.1)) and the polyhedral discrepancy

$$\zeta_\ell(P, Q) := \sup \left\{ \left| \int_\Xi f(\xi)P(d\xi) - \int_\Xi f(\xi)Q(d\xi) \right| \;\middle|\; f \in \mathcal{F}_\ell(\Xi) \right\}, \tag{14}$$

$$\alpha_\mathcal{B}(P, Q) := \sup_{B \in \mathcal{B}} |P(B) - Q(B)|, \tag{15}$$

respectively. Hence, convergence of probability distributions with respect to $\zeta_{\ell,\mathcal{B}}$ implies their weak convergence, convergence of ℓth order absolute moments, and convergence with respect to the polyhedral discrepancy $\alpha_\mathcal{B}$. For bounded Ξ, the technique in Schultz (1996, Proposition 3.1) can be employed to obtain

$$\zeta_{\ell,\mathcal{B}}(P, Q) \leq C_s \alpha_\mathcal{B}(P, Q)^{\frac{1}{s+1}} \qquad (P, Q \in \mathcal{P}(\Xi)) \tag{16}$$

for some constant $C_s > 0$. In view of (13) and (16), the metric $\zeta_{\ell,\mathcal{B}}$ is stronger than $\alpha_{\mathcal{B}}$ in general, but in case of bounded Ξ, both distances metrize the same convergence on $\mathcal{P}(\Xi)$.

For more specific models (1), improvements of the stability estimate (12) may be obtained by exploiting specific recourse structures, that is, by using additional information on the shape of the sets B_j, $j \in \mathbb{N}$, and on the behavior of the function Φ on these sets. This may lead to stability estimates with respect to distances that are (much) weaker than $\zeta_{\ell,\mathcal{B}}$. For example, if $W_1 = 0$, Ξ is rectangular, T is fixed and some components of $h(\cdot)$ coincide with some of the components of ξ, and the closures of $\Xi_{x,j}$, $x \in X$, $j \in \mathbb{N}$, are rectangular subsets of Ξ, that is, belong to

$$\mathcal{B}_{\text{rect}} := \{I_1 \times I_2 \times \cdots \times I_s \mid \emptyset \neq I_j \text{ is a closed interval in } \mathbb{R},\ j = 1,\ldots,s\}, \tag{17}$$

then the stability estimate (12) is valid with respect to $\zeta_{\ell,\mathcal{B}_{\text{rect}}}$. As shown in Henrion et al. (2009), convergence of a sequence of probability distributions with respect to $\zeta_{\ell,\mathcal{B}_{\text{rect}}}$ is equivalent to convergence with respect to both ζ_ℓ and $\alpha_{\mathcal{B}_{\text{rect}}}$. If, in addition to the previous assumptions, q is fixed and Ξ is bounded, the estimate (12) is valid with respect to the rectangular discrepancy $\alpha_{\mathcal{B}_{\text{rect}}}$ (see also Schultz 1996, Sect. 3).

4 Scenario Reduction

A well known approach for solving two-stage stochastic programs computationally consists in replacing the original probability distribution by a discrete distribution based on a finite number of scenarios. Let P be such a discrete distribution with scenarios ξ^i and probabilities p_i, $i = 1,\ldots,N$. The corresponding stochastic programming model is of the form

$$\min\left\{\langle c, x\rangle + \sum_{i=1}^{N} p_i(\langle q_1(\xi^i), y_1^i\rangle + \langle q_2(\xi^i), y_2^i\rangle) \;\middle|\; \begin{array}{l} W_1 y_1^i + W_2 y_2^i \le h(\xi^i) - T(\xi^i)x \\ y_1^i \in \mathbb{R}^{m_1},\, y_2^i \in \mathbb{Z}^{m_2},\, i = 1,\ldots,N, \\ x \in X \end{array}\right\}.$$

It may turn out that the computing times for solving the resulting mixed-integer linear programs are not acceptable. In such a case one might wish to reduce the number of scenarios entering the stochastic program. In Dupačová et al. (2003) and Heitsch and Römisch (2007), a stability-based approach for scenario reduction in two-stage models without integrality requirements is developed. This approach suggests to look at stability results for optimal values and to use the corresponding distance of probability distributions for determining discrete distributions based on a smaller and prescribed number of scenarios as best approximations of P. According to the stability estimate (12) in Sect. 3, the distances $\zeta_{\ell,\mathcal{B}}$ or $\alpha_{\mathcal{B}}$ (if Ξ is bounded) appear as the right choice, where $\mathcal{B}$ is a set of convex polyhedra that depend on the structure of the stochastic program (1).

In Henrion et al. (2008), the scenario reduction approach is elaborated for $\alpha_{\mathcal{B}}$ and for a relevant set $\mathcal{B}$ of convex polyhedra. The numerical results show that the complexity of scenario reduction algorithms increases if $\mathcal{B}$ gets more involved. To avoid this effect, the distance $\zeta_{\ell,\mathcal{B}_{\text{rect}}}$ or, equivalently,

$$\mathbf{d}_\lambda(P,Q) := \lambda\alpha_{\mathcal{B}_{\text{rect}}}(P,Q) + (1-\lambda)\zeta_\ell(P,Q) \tag{18}$$

for some $\lambda \in (0,1)$ is considered in this section and in Henrion et al. (in preparation).

Let Q_J denote a probability distribution whose support $\operatorname{supp}(Q_J)$ contains the following subset of $\{\xi^1,\dots,\xi^N\}$:

$$\operatorname{supp}(Q_J) = \left\{\xi^i \mid i \in \{1,\dots,N\}\setminus J\right\} \quad \text{and} \quad J \subset \{1,\dots,N\}.$$

Let q_i $(i \notin J)$ denote the probability of scenario ξ^i of Q_J. Now, the aim is to determine Q_J such that the distance $\mathbf{d}_\lambda(P,Q_J)$ is minimal, that is, for arbitrary subsets J of $\{1,\dots,N\}$, we are interested in

$$D_J := \min\left\{\mathbf{d}_\lambda(P,Q_J) \;\middle|\; q_i \geq 0,\, i \notin J,\, \sum_{i\notin J} q_i = 1\right\}. \tag{19}$$

In the following, we show that D_J can be computed as optimal value of a linear program. To this end, we assume without loss of generality that $J = \{n+1,\dots,N\}$, that is, $\operatorname{supp}(Q_J) = \{\xi^1,\dots,\xi^n\}$ for some $1 \leq n < N$. We consider the system of index sets

$$\mathcal{I}_{\mathcal{B}_{\text{rect}}} := \{I(B) := \{i \in \{1,\dots,N\} \mid \xi^i \in B\} \mid B \in \mathcal{B}_{\text{rect}}\}$$

and obtain the following representation of the rectangular discrepancy:

$$\alpha_{\mathcal{B}_{\text{rect}}}(P,Q_J) = \sup_{B\in\mathcal{B}_{\text{rect}}} |P(B) - Q_J(B)| = \max_{I\in\mathcal{I}_{\mathcal{B}_{\text{rect}}}} \left|\sum_{i\in I} p_i - \sum_{j\in I\cap\{1,\dots,n\}} q_j\right|, \tag{20}$$

$$= \min\left\{t_\alpha \;\middle|\; \begin{array}{l} -\sum_{j\in I\cap\{1,\dots,n\}} q_j \leq t_\alpha - \sum_{i\in I} p_i,\ I \in \mathcal{I}_{\mathcal{B}_{\text{rect}}} \\ \sum_{j\in I\cap\{1,\dots,n\}} q_j \leq t_\alpha + \sum_{i\in I} p_i,\ I \in \mathcal{I}_{\mathcal{B}_{\text{rect}}} \end{array}\right\} \tag{21}$$

Since the set $\mathcal{I}_{\mathcal{B}_{\text{rect}}}$ may be too large to solve the linear program (21) numerically, we consider the system of reduced index sets

$$\mathcal{I}^*_{\mathcal{B}_{\text{rect}}} := \{I(B) \cap \{1,\dots,n\} \mid B \in \mathcal{B}_{\text{rect}}\}$$

and the quantities

$$\gamma^{I^*} := \max\left\{ \sum_{i\in I} p_i \;\middle|\; I \in \mathcal{I}_{\mathcal{B}_{\text{rect}}}, I \cap \{1,\dots,n\} = I^* \right\}$$

$$\gamma_{I^*} := \min\left\{ \sum_{i\in I} p_i \;\middle|\; I \in \mathcal{I}_{\mathcal{B}_{\text{rect}}}, I \cap \{1,\dots,n\} = I^* \right\}$$

for every $I^* \in \mathcal{I}^*_{\mathcal{B}_{\text{rect}}}$. Since any such index set I^* corresponds to some left-hand side of the inequalities in (21), γ^{I^*} and γ_{I^*} correspond to the smallest right-hand sides in (21). Hence, the rectangular discrepancy may be rewritten as

$$\alpha_{\mathcal{B}_{\text{rect}}}(P, Q_J) = \min\left\{ t_\alpha \;\middle|\; \begin{array}{l} -\sum_{j\in I^*} q_j \le t_\alpha - \gamma^{I^*},\ I^* \in \mathcal{I}^*_{\mathcal{B}_{\text{rect}}} \\ \sum_{j\in I^*} q_j \le t_\alpha + \gamma_{I^*},\ I^* \in \mathcal{I}^*_{\mathcal{B}_{\text{rect}}} \end{array} \right\}. \tag{22}$$

Since the number of elements of $\mathcal{I}^*_{\mathcal{B}_{\text{rect}}}$ is at most 2^n (compared to 2^N in $\mathcal{I}_{\mathcal{B}_{\text{rect}}}$), passing from (21) to (22) indeed drastically reduces the maximum number of inequalities and may make the linear program (22) numerically tractable.

Because of duality arguments, the Fortet–Mourier distance $\zeta_\ell(P, Q_J)$ (see (14)) allows the representation as linear program (cf. Heitsch and Römisch 2007)

$$\zeta_\ell(P, Q_J) = \inf\left\{ \sum_{i=1}^{N}\sum_{j=1}^{n} \eta_{ij}\hat{c}_\ell(\xi^i, \xi^j) \;\middle|\; \eta_{ij} \ge 0,\ \begin{array}{l} \sum_{i=1}^{N} \eta_{i,j} = q_j,\ j = 1,\dots,n \\ \sum_{j=1}^{n} \eta_{i,j} = p_i,\ i = 1,\dots,N \end{array} \right\},$$

where $c_\ell(\xi, \tilde{\xi}) := \max\{1, \|\xi\|^{\ell-1}, \|\tilde{\xi}\|^{\ell-1}\}\|\xi - \tilde{\xi}\|$ for all $\xi, \tilde{\xi} \in \Xi = \{\xi^1, \dots, \xi^N\}$ and $\hat{c}_\ell$ denotes the reduced costs

$$\hat{c}_\ell(\xi, \tilde{\xi}) := \inf\left\{ \sum_{k=1}^{K} c_\ell(\xi^{i_{k-1}}, \xi^{i_k}) \;\middle|\; K \in \mathbb{N}, i_k \in \{1,\dots,N\}, \xi^{i_0} = \xi, \xi^{i_K} = \tilde{\xi} \right\}.$$

Hence, extending the representation (22) of $\alpha_{\mathcal{B}_{\text{rect}}}$, we obtain the following linear program for determining D_J and the probabilities q_j, $j = 1, \dots, n$, of the discrete reduced distribution Q_J,

$$D_J = \min\left\{ \lambda\, t_\alpha + (1-\lambda) t_\zeta \;\middle|\; \begin{array}{l} t_\alpha, t_\zeta \ge 0,\ q_j \ge 0,\ \sum_{j=1}^{n} q_j = 1, \\ \eta_{ij} \ge 0, i = 1,\dots,N,\ j = 1,\dots,n, \\ t_\zeta \ge \sum_{i=1}^{N}\sum_{j=1}^{n} \hat{c}_\ell(\xi^i, \xi^j)\eta_{ij}, \\ \sum_{j=1}^{n} \eta_{ij} = p_i,\ i = 1,\dots,N, \\ \sum_{i=1}^{N} \eta_{ij} = q_j,\ j = 1,\dots,n, \\ -\sum_{j\in I^*} q_j \le t_\alpha - \gamma^{I^*},\ I^* \in \mathcal{I}^*_{\mathcal{B}_{\text{rect}}} \\ \quad\sum_{j\in I^*} q_j \le t_\alpha + \gamma_{I^*},\ I^* \in \mathcal{I}^*_{\mathcal{B}_{\text{rect}}} \end{array} \right\}. \tag{23}$$

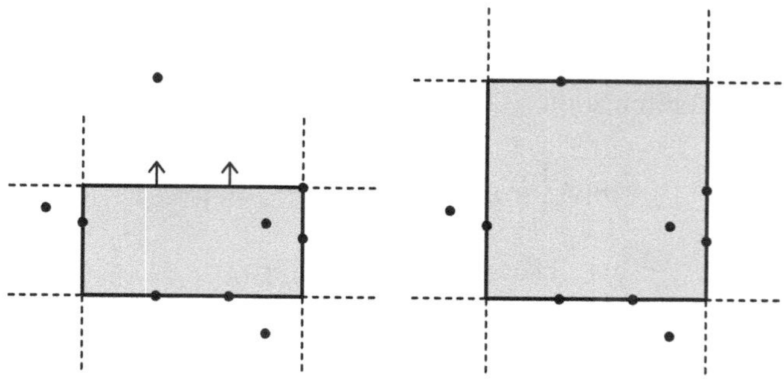

Fig. 3 Non-supporting rectangle (*left*) and supporting rectangle (*right*). The dots represent the remaining scenarios $\xi^1, \ldots, \xi^n$ for $s = 2$

While the linear program (23) can be solved efficiently by available software, the determination of the index set $\mathcal{I}^*_{\mathcal{B}_{\text{rect}}}$ and the coefficients γ^{I^*}, γ_{I^*} is more intricate.

It is shown in Henrion et al. (2008, in preparation, Sect. 3) that the parameters $\mathcal{I}^*_{\mathcal{B}_{\text{rect}}}$ and γ^{I^*}, γ_{I^*} can be determined by studying the set $\mathcal{R}$ of supporting rectangles. A rectangle B in $\mathcal{B}_{\text{rect}}$ is called *supporting* if each of its facets contains an element of $\{\xi_1, \ldots, \xi_n\}$ in its relative interior (see also Fig. 3). On the basis of $\mathcal{R}$, the following representations are valid according to Henrion et al. (2008, Prop. 1 and 2):

$$\mathcal{I}^*_{\mathcal{B}_{\text{rect}}} = \bigcup_{B \in \mathcal{R}} \left\{ I^* \subseteq \{1, \ldots, n\} \;\middle|\; \cup_{j \in I^*} \{\xi^j\} = \{\xi^1, \ldots, \xi^n\} \cap \operatorname{int} B \right\},$$

$$\gamma^{I^*} = \max \left\{ P(\operatorname{int} B) \;\middle|\; B \in \mathcal{R}, \cup_{j \in I^*} \{\xi^j\} = \{\xi^1, \ldots, \xi^n\} \cap \operatorname{int} B \right\},$$

$$\gamma_{I^*} = \sum_{i \in \underline{I}} p_i \quad \text{where } \underline{I} := \left\{ i \in \{1, \ldots, N\} \;\middle|\; \min_{j \in I^*} \xi^j_l \leq \xi^i_l \leq \max_{j \in I^*} \xi^j_l,\ l = 1, \ldots, s \right\}$$

for every $I^* \in \mathcal{I}^*_{\mathcal{B}_{\text{rect}}}$. Here, $\operatorname{int} B$ denotes the interior of the set B.

An algorithm is developed in Henrion et al. (in preparation) that constructs recursively l-dimensional supporting rectangles for $l = 1, \ldots, s$. Computational experiments show that its running time grows linearly with N, but depends on n and s via the expression $\binom{n+1}{2}^s$. Hence, while N may be large, only moderately sized values of n given s are realistic.

Since an algorithm for computing D_J is now available, we finally look at determining a scenario index set $J \subset \{1, \ldots, N\}$ with cardinality $\# J = n$ such that D_J is minimal, that is, at solving the combinatorial optimization problem

$$\min\{D_J \mid J \subset \{1, \ldots, N\}, \# J = n\}, \tag{24}$$

which is known as n-median problem and as NP-hard. One possibility is to reformulate (24) as mixed-integer linear program and to solve it by standard software.

Since, however, approximate solutions of (24) are sufficient, heuristic algorithms like *forward selection* are of interest, where u_k is determined in its kth step such that it solves the minimization problem

$$\min\left\{D_{J^{[k-1]}\setminus\{u\}} \;\middle|\; u \in J^{[k-1]}\right\},$$

where $J^{[0]} = \{1,\dots,N\}$, $J^{[k]} := J^{[k-1]} \setminus \{u_k\}$ $(k = 1,\dots,n)$, and $J^{[n]} := \{1,\dots,N\} \setminus \{u_1,\dots,u_n\}$ serves as approximate solution to (24). Recalling that the complexity of evaluating $D_{J^{[k-1]}\setminus\{u\}}$ for some $u \in J^{[k-1]}$ is proportional to $\binom{k+1}{2}^s$ shows that even the forward selection algorithm is expensive.

Hence, heuristics for solving (24) became important, which require only a low number of D_J evaluations. For example, if P is a probability distribution on $[0,1]^s$ with independent marginal distributions P_j, $j = 1,\dots,s$, such a heuristic can be based on Quasi–Monte Carlo methods (cf. Niederreiter (1992)). The latter provide sequences of equidistributed points in $[0,1]^s$, which approximate the uniform distribution on the unit cube $[0,1]^s$. Now, let n Quasi–Monte Carlo points $z^k = (z_1^k,\dots,z_s^k) \in [0,1]^s$, $k = 1,\dots,n$, be given. Then we determine

$$y^k := \left(F_1^{-1}(z_1^k),\dots,F_s^{-1}(z_s^k)\right) \qquad (k = 1,\dots,n),$$

where F_j is the (one-dimensional) distribution function of P_j, that is,

$$F_j(z) = P_j((-\infty, z]) = \sum_{i=1,\,\xi_j^i \le z}^{N} p_i \qquad (z \in \mathbb{R})$$

and $F_j^{-1}(t) := \inf\{z \in \mathbb{R} \mid F_j(z) \ge t\}$ $(t \in [0,1])$ its inverse $(j = 1,\dots,s)$. Finally, we determine u_k as the solution of

$$\min_{u \in J^{[k-1]}} \|\xi^u - y^k\|$$

and set again $J^{[k]} := J^{[k-1]} \setminus \{u_k\}$ for $k = 1,\dots,n$, where $J^{[0]} = \{1,\dots,N\}$. Figure 4 illustrates the results of such a Quasi–Monte Carlo based heuristic and Fig. 5 shows the discrepancy $\alpha_{\mathcal{B}_{\mathrm{rect}}}$ for different n and the running times of the Quasi–Monte Carlo based heuristic.

5 Decomposition Algorithms

When the size of an optimization problem becomes intractable for standard solution approaches, a decomposition into small tractable subproblems by relaxing certain coupling constraints is often a possible resort. The task of the decomposition

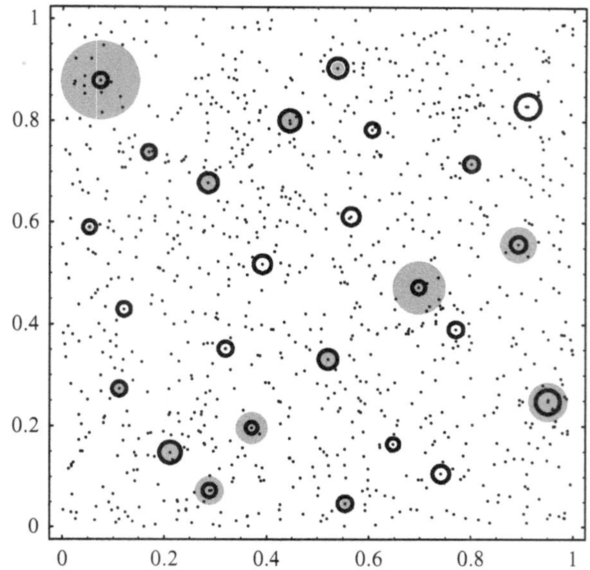

Fig. 4 $N = 1{,}000$ samples ξ^i from the uniform distribution on $[0, 1]^2$ and $n = 25$ points ξ^{u_k}, $k = 1, \ldots, n$, obtained via the first 25 elements z_k, $k = 1, \ldots, n$, of the *Halton sequence* (in the bases 2 and 3) (see Niederreiter 1992, p. 29). The probabilities q_k of ξ^{u_k}, $k = 1, \ldots, n$, are computed for the distance $\mathbf{d}_\lambda$ with $\lambda = 1$ (*gray balls*) and $\lambda = 0.9$ (*black circles*) by solving (23). The diameters of the circles are proportional to the probabilities q_k, $k = 1, \ldots, 25$

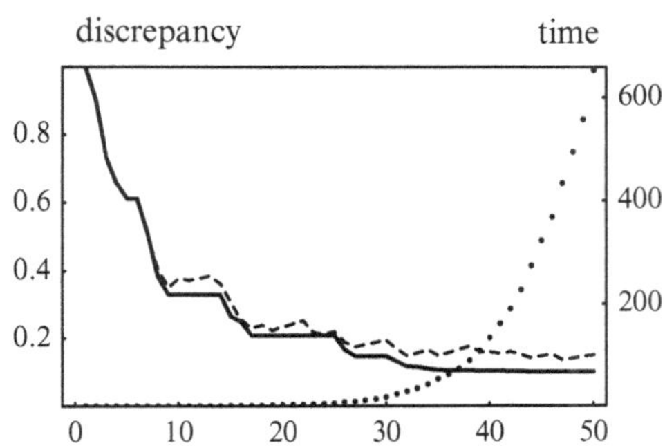

Fig. 5 Distance $\alpha_{\mathcal{B}_{\text{rect}}}$ between P (with $N = 1{,}000$) and equidistributed QMC-points (*dashed*), QMC-points, whose probabilities are adjusted according to (23) (*bold*), and running times of the QMC-based heuristic (in seconds)

algorithm is then to coordinate the search in the subproblems in a way that their solutions can be combined into one that is feasible for the overall problem and has a "good" objective function value. Often, the algorithm also provides a certified lower bound on the optimal value, which allows to evaluate the quality of a found solution. Since, on the one hand, mixed-integer stochastic programs easily reach a size that is intractable for standard solution approaches but, on the other hand, are also very structured, many decomposition algorithms have been developed (Louveaux and Schultz 2003; Schultz 2003; Sen 2005). In the following, we discuss some of them in more detail.

Let us assume that the set of first stage feasible solutions X is given in the form

$$X = \{x \in \mathbb{Z}^{m_0} \times \mathbb{R}^{m-m_0} \mid Ax \le b\},$$

where m_0 denotes the number of first stage variables with integrality restrictions, A is a (r_0, m)-matrix, and $b \in \mathbb{R}^{r_0}$. Further, we denote by

$$\bar{X} := \{x \in \mathbb{R}^m \mid Ax \le b\}$$

the *linear relaxation* of X. We recall the *value function* (3),

$$\Phi(u,t) = \inf\{\langle u_1, y_1\rangle + \langle u_2, y_2\rangle \mid y_1 \in \mathbb{R}^{m_1}, y_2 \in \mathbb{Z}^{m_2}, W_1 y_1 + W_2 y_2 \le t\},$$

and define the *expected recourse function* of model (1) by

$$\Psi(x) := \int_{\Xi} \Phi(q(\xi), h(\xi) - T(\xi)x) P(d\xi) \qquad (x \in \bar{X}).$$

For continuous ($m_2 = 0$) stochastic programs, the Benders' decomposition is an established method (Van Slyke and Wets 1969; Birge 1985). It decomposes the decision on the first stage from the recourse decisions on the second stage by replacing the value function $\Phi(u,t)$ in (1) by an explicit approximation based on supporting hyperplanes. Unfortunately, Benders' decomposition relies heavily on the convexity of the value function $t \mapsto \Phi(u,t)$. Thus, in the view of Sect. 2, it cannot be directly applied to the case where discrete variables are present.

However, there are several approaches to overcome this difficulty. One of the first is the integer L-shaped method (Laporte and Louveaux 1993), which assumes that the first stage problem involves only binary variables. This property is exploited to derive linear inequalities that approximate the value function $\Phi(u,t)$ pointwise. While the algorithm makes only moderate assumptions on the second stage problem, its main drawback is the weak approximation of the value function due to lacking first-order information about the value function. Thus, the algorithm might enumerate all feasible first stage solutions to find an optimal solution.

A cutting-plane algorithm is proposed in Carøe and Tind (1997). Here, the deterministic equivalent of (1) is solved by improving its linear relaxation with lift-and-project cuts. Decomposition arises here in two ways. First, the linear relaxation (including additional cuts) is solved by Benders' decomposition. Second, lift-and-project cuts are derived scenario-wise. Further, in case of a fixed technology matrix $T(\cdot) \equiv T$, cut coefficients that have been computed for one scenario can also be reused to derive cuts for other scenarios. This algorithm can be seen as a predecessor of the dual decomposition approach presented in Sen and Higle (2005). While the cuts in Carøe and Tind (1997) include variables from both stages, Sen and Higle

(2005) extend the Benders' decomposition approach to the mixed-integer case by sequentially convexifying the value function $\Phi(u,t)$. It is discussed in detail in Sect. 5.2.

In Klein Haneveld et al. (2006) it is observed that, even though the value function $\Phi(u,t)$ might be nonconvex and difficult to handle, under some assumptions on the distribution of ξ, the expected recourse function $\Psi(x)$ can be convex. Starting with simple integer recourse models and then extending to more general classes of problems, techniques to compute tight convex approximations of the expected recourse function by perturbing the distribution of ξ are developed in a series of papers (Klein Haneveld et al. 2006; van der Vlerk 2004, 2005). We sketch this approach in more detail in Sect. 5.1.

In the case that the second stage problem is purely integer ($m_1 = 0$), the value function $\Phi(u,t)$ has the nice property to be constant on polyhedral subsets of $\mathcal{U} \times \mathcal{T}$. Therefore, in case of a finite distribution, also the expected recourse function $\Psi(x)$ is constant on polyhedral subsets of $\bar{X}$. This property allows to reduce the set X to a finite set of solution candidates that can be enumerated (Schultz et al. (1998)). Since the expected recourse function $\Psi(x)$ has to be evaluated for each candidate, many similar integer programs have to be solved. In Schultz et al. (1998) a Gröbner basis for the second stage problem is computed once in advance (which is expensive) and then used for evaluation of $\Psi(x)$ for every candidate x (which is then cheap).

Another approach based on enumerating the sets where $\Psi(x)$ is constant is presented in Ahmed et al. (2004). Instead of a complete enumeration, here a branch-and-bound algorithm is applied to the first stage problem to enumerate the regions of constant $\Psi(x)$ implicitly. Branching is thereby performed along lines of discontinuity of $\Psi(x)$, thereby reducing its discontinuity in generated subproblems.

While all approaches discussed so far explore the structure of the value or expected recourse function in some way, Lagrange decomposition is a class of algorithms where decomposition is achieved by relaxation of problem constraints. By moving certain coupling restrictions from the set of constraints into the objective function as penalty term, the problem decomposes into a set of subproblems, each of them often much easier to handle than the original problem. This relaxed problem then yields a lower bound onto the original optimal value, which is further improved by optimization of the penalty parameters. Since, in general, a solution of the relaxed problem violates the coupling constraints, heuristics and branch-and-bound approaches are applied to obtain good feasible solutions of the original problem. While there are several alternatives to choose a set of coupling constraints for relaxation, each one providing a lower bound of different quality (Dentcheva and Römisch 2004), in general, scenario and geographical decomposition are the preferred strategies (Carøe and Schultz 1999; Nowak and Römisch 2000). In scenario decomposition, nonanticipativity constraints are relaxed so that the problem decomposes into one deterministic subproblem for each scenario. We discuss this approach in more detail in Sect. 5.3. In geographical decomposition, model-specific constraints are relaxed, which leads to one subproblem for each component of the model. Even though each subproblem then corresponds to a stochastic program itself, its structure often allows to develop specialized algorithms to solve them very efficiently. Similarly, the modelers' knowledge can be explored to make solutions from the relaxed problem feasible for the original problem. Geographical decomposition is demonstrated for a unit commitment problem in Sect. 6.

5.1 Convexification of the Expected Recourse Function

In a simple integer recourse model, the second stage variables are purely integer ($m_1 = 0$) and are partitioned into two sets $y^+, y^- \in \mathbb{Z}^s_+$ with $2s = m_2$. The cost-vector $q(\xi) \equiv (q^+, q^-)$ and technology matrix $T(\xi) \equiv T$ are fixed, $r = 2s$, $h(\xi) \equiv \xi$, and the value function takes the form

$$\Phi(q(\xi), h(\xi) - T(\xi)x) = \inf\left\{ \langle q^+, y^+\rangle + \langle q^-, y^-\rangle \;\middle|\; \begin{array}{r} y^+ \geq \xi - Tx, \\ y^- \geq -(\xi - Tx) \\ y^+, y^- \in \mathbb{Z}^s_+ \end{array} \right\}.$$

The simple structure of the value function allows to write the expected recourse function in a separable form,

$$\Psi(x) = \sum_{i=1}^{s} q_i^+ \mathbb{E}[\lceil \xi_i - (Tx)_i \rceil^+] + q_i^- \mathbb{E}[\lfloor \xi_i - (Tx)_i \rfloor^-],$$

where $\lceil \alpha \rceil$ denotes the smallest integer that is at least α, $\lfloor \alpha \rfloor$ is the largest integer that is at most α, $\alpha^+ = \max(0, \alpha)$, and $\alpha^- = \min(0, \alpha)$. Thus, it is sufficient to consider one-dimensional functions of the form $Q(z) := q^+ \mathbb{E}_\zeta[\lceil \zeta - z \rceil^+] + q^- \mathbb{E}_\zeta[\lfloor \zeta - z \rfloor^-]$ (with ζ a random variable).

In Klein Haneveld et al. (2006), convex approximations of $Q(z)$ are derived from a piecewise linear function in the points $(z, Q(z))$, $z \in \alpha + \mathbb{Z}$, where $\alpha \in [0, 1)$ is a parameter. Further, if ζ has a continuous distribution, then the approximation of $Q(z)$ can be realized as expected recourse function of a continuous simple recourse model,

$$Q_\alpha(z) = q^+ \mathbb{E}_{\zeta_\alpha}[(\zeta_\alpha - z)^+] + q^- \mathbb{E}_{\zeta_\alpha}[(\zeta_\alpha - z)^-] + \frac{q^+ q^-}{q^+ + q^-},$$

where ζ_α is a discrete random variable with support in $\alpha + \mathbb{Z}$ (Klein Haneveld et al. 2006).

The results in Klein Haneveld et al. (2006) are extended to derive convex approximations of the expected recourse function for models of the form (1), where $m_1 = 0$, $h(\xi) \equiv \xi$, $q(\xi) \equiv q$, and $T(\xi) \equiv T$ are fixed (van der Vlerk 2004). Further, the parameter α can be chosen such that the derived convex approximation underestimates the original expected recourse function. Since this convex underestimator is at least as good as an LP-based underestimator (obtained by relaxing the integrality condition on y) and even yield the convex hull of $\Psi(x)$ in the case that T is unimodular, it can be utilized to derive lower bounds in a Branch-and-Bound search for a solution of (1).

Another extension of the methodology from Klein Haneveld et al. (2006) considers mixed-integer recourse models where $r = 1$ and the value function is semi-periodic, c.f. van der Vlerk (2005).

5.2 Convexification of the Value Function

From now on we assume that the random vector ξ has only finitely many outcomes ξ^i with probability $p_i > 0$, $i = 1, \dots, N$. Thus, we can write the expected recourse function as

$$\Psi(x) = \sum_{i=1}^{N} p_i \Phi(q(\xi^i), h(\xi^i) - T(\xi^i)x) \qquad (x \in \bar{X}).$$

As discussed in Sect. 2, the nonconvexity of the function $\Phi(u, t)$ forbids a representation by supporting hyperplanes as used in a Benders' decomposition. However, while in the continuous case ($m_2 = 0$) the hyperplanes are derived from dual feasible solutions of the second stage problem, it is conceptually possible to carry over these ideas to the mixed-integer case by introducing (possibly nonlinear) dual price functions (Tind and Wolsey 1981). Indeed, Chvátal and Gomory functions are sufficiently large classes of dual price functions that allow to approximate the value function $\Phi(u, t)$ (Blair and Jeroslow 1982). These dual functions can be obtained from a solution of (3) with a branch-and-bound or Gomory cutting plane algorithm (Wolsey 1981). In Carøe and Tind (1998) this approach is used to carry over the Benders' decomposition algorithm for two-stage linear stochastic programs to the mixed-integer linear case by replacing the hyperplane approximation of the expected recourse function by an approximation based on dual price functions. Although Carøe and Tind (1998) do not discuss how the master problem with its (nonsmooth and nonconvex) dual price functions can be solved, the series of papers (Sen and Higle 2005; Ntaimo and Sen 2008; Sen and Sherali 2006; Ntaimo and Sen 2005, 2008) show that a careful construction of dual price functions combined with a convexification step based on disjunctive programming allows to implement an efficient Benders' decomposition for mixed-integer two-stage stochastic programs.

We consider the following *master problem* obtained from (1) by replacing the value functions $x \mapsto \Phi(q(\xi^i), h(\xi^i) - T(\xi^i)x)$ by approximations $\Theta_i : \mathbb{R}^m \to \mathbb{R}$:

$$\min \left\{ \langle c, x \rangle + \sum_{i=1}^{N} p_i \Theta_i(x) \;\middle|\; x \in X \right\}, \tag{25}$$

where each function $\Theta_i(\cdot)$, $i = 1, \ldots, N$, is given in the form

$$\Theta_i(x) := \max\{\min\{\eta_1(x), \ldots, \eta_k(x)\} \mid (\eta_1(\cdot), \ldots, \eta_k(\cdot)) \in C_i\}, \qquad x \in \bar{X},$$

and a tuple $\eta := (\eta_1(\cdot), \ldots, \eta_k(\cdot)) \in C_i$ consists of k (where k is allowed to vary with η) affine linear functions $\eta_j(\cdot) : \mathbb{R}^m \to \mathbb{R}$, $j = 1, \ldots, k$. The tuple η takes here the role of an optimality cut in Benders' decomposition for the continuous case. That is, each $\eta \in C_i$ is constructed in a way such that for all $x \in X$

$$\Phi(q(\xi^i), h(\xi^i) - T(\xi^i)x) \geq \eta_j(x) \text{ for at least one } j \in \{1, \ldots, k\}. \tag{26}$$

Hence, we have $\Phi(q(\xi^i), h(\xi^i) - T(\xi^i)x) \geq \Theta_i(x)$ and the optimal value of problem (25) is a lower bound to the optimal value of (1). Before discussing the construction of the tuples η, we shortly discuss an algorithm to solve problem (25).

5.2.1 Solving the Master Problem

Note that problem (25) can be written as a disjunctive mixed-integer linear problem:

$$\min\left\{ \langle c, x\rangle + \sum_{i=1}^{N} p_i\theta_i \;\middle|\; \begin{array}{l} x \in X, \\ \theta_i \geq \eta_1(x) \vee \ldots \vee \theta_i \geq \eta_k(x),\ \eta \in C_i,\ i = 1, \ldots, N \end{array} \right\}. \tag{27}$$

Problem (27) can be solved by a branch-and-bound algorithm (Ntaimo and Sen 2008). To this end, assume that for each tuple $\eta \in C_i$ an affine linear function $\bar{\eta}(\cdot) : \mathbb{R}^m \to \mathbb{R}$ is known, which underestimates each $\eta_j(\cdot)$, $j = 1, \ldots, k$, on X, that is, $\eta_j(x) \geq \bar{\eta}(x)$ for $j = 1, \ldots, k$ and $x \in X$. $\bar{\eta}(\cdot)$ allows to derive a linear relaxation of problem (27):

$$\min\left\{ \langle c, x\rangle + \sum_{i=1}^{N} p_i\theta_i \;\middle|\; x \in \bar{X},\ \theta_i \geq \bar{\eta}(x),\ \eta \in C_i,\ i = 1, \ldots, N \right\}. \tag{28}$$

Let $(\hat{x}, \hat{\theta})$ be a solution of (28). If $\hat{x}$ is feasible for (27), then an optimal solution for (27) has been found. Otherwise, $\hat{x}$ either violates an integrality restriction on a variable x_j, $j = 1, \ldots, m_0$, or a disjunction $\theta_i \geq \min\{\eta_1(x), \ldots, \eta_k(x)\}$ for some tuple $\eta \in C_i$ (with $k > 1$) and some scenario i. In the former case, that is, $\hat{x}_j \notin \mathbb{Z}$, two subproblems of (28) are created with additional constraints $x_j \leq \lfloor \hat{x}_j \rfloor$ and $x_j \geq \lceil \hat{x}_j \rceil$, respectively. In the latter case, the tuple η is partitioned into two tuples $\eta' = (\eta_1(\cdot), \ldots, \eta_{k'}(\cdot))$ and $\eta'' = (\eta_{k'+1}(\cdot), \ldots, \eta_k(\cdot))$, $1 \leq k' < k$, and corresponding linear underestimators $\bar{\eta}'(\cdot)$ and $\bar{\eta}''(\cdot)$ are computed (where $\bar{\eta}' = \eta_1$ if $k' = 1$ and $\bar{\eta}'' = \eta_k$ if $k' = k - 1$), and two subproblems where the tuple $\eta \in C_i$ is replaced by η' and η'', respectively, are constructed. Next, the same method is applied to each subproblem recursively. The first feasible solution for problem (27) is stored as "incumbent solution." In the following, new feasible solutions replace the incumbent solution if they have a better objective value. If a subproblem is infeasible or the value of its linear relaxation exceeds the current incumbent solution, then it can be discarded from the list of open subproblems. Since in each subproblem the number of feasible discrete values for a variable x_j or the length of a tuple $\eta \in C_i$ is reduced with respect to the ascending problem, the algorithm can generate only a finite number of subproblems and thus terminates with a solution of (27).

5.2.2 Convexification of Disjunctive Cuts

A linear function $\bar{\eta}(\cdot)$ in (28) that underestimates $\min\{\eta_1(\cdot),\ldots,\eta_k(\cdot)\}$ can be constructed by means of disjunctive programming (Balas 1998; Sen and Higle 2005): For a fixed scenario index i and a tuple $\eta \in C_i$, an inequality $\theta \geq \bar{\eta}(x)$ is valid for the feasible set of (27) if it is valid for $\bigcup_{j=1}^{k}\{(x,\theta) \in \mathbb{R}^{m+1} \mid x \in \bar{X}, \theta \geq \eta_j(x)\}$. That is, we require

$$\bar{\eta}(x) \leq \eta_j(x) \qquad \text{for all } x \in \bar{X}, \quad j = 1,\ldots,k. \tag{29}$$

We write $\bar{\eta}(x) = \bar{\eta}_0 + \langle \bar{\eta}_x, x\rangle$ and $\eta_j(x) = \eta_{j,0} + \langle \eta_{j,x}, x\rangle$ for some $\bar{\eta}_0, \eta_{j,0} \in \mathbb{R}$ and $\bar{\eta}_x, \eta_{j,x} \in \mathbb{R}^m$, $j = 1,\ldots,k$. Then (29) is equivalent to

$$\begin{aligned}\bar{\eta}_0 - \eta_{j,0} &\leq \min\{\langle \eta_{j,x} - \bar{\eta}_x, x\rangle \mid x \in \mathbb{R}^m,\ Ax \leq b\}\\ &= \max\{\langle \lambda_j, b\rangle \mid \lambda_j \in \mathbb{R}^{r_0}_{-},\ A^\top \lambda_j = \eta_{j,x} - \bar{\eta}_x\}.\end{aligned}$$

Therefore, choosing $\lambda_j \in \mathbb{R}^{r_0}_{-}$ and $\bar{\eta}_x \in \mathbb{R}^m$ such that $A^\top\lambda_j + \bar{\eta}_x = \eta_{j,x}$ and setting $\bar{\eta}_0 := \eta_{j,0} + \min\{\langle \lambda_j, b\rangle | j = 1,\ldots,k\}$ yields a function $\bar{\eta}(x)$ that satisfies (29).

Sen and Higle (2005) note that, given an extreme point $\hat{x}$ of $\bar{X}$, the linear underestimator $\bar{\eta}(\cdot)$ can be chosen such that $\bar{\eta}(\hat{x}) = \min\{\eta_1(\hat{x}),\ldots,\eta_k(\hat{x})\}$. Thus, if only extreme points of $\bar{X}$ are feasible for (1), then it is not necessary to branch on disjunctions η to solve (27). This is the case, for example, if all first stage variables are restricted to be binary.

5.2.3 Approximation of $\Phi(u,t)$ by Linear Optimality Cuts

The simplest way to construct a tuple η with property (26) is to derive a supporting hyperplane for the linear relaxation of $\Phi(u,t)$, which we denote by

$$\bar{\Phi}(u,t) := \min\{\langle u_1, y_1\rangle + \langle u_2, y_2\rangle \mid y_1 \in \mathbb{R}^{m_1},\ y_2 \in \mathbb{R}^{m_2},\ W_1y_1 + W_2y_2 \leq t\}. \tag{30}$$

It is well known that $\bar{\Phi}(u,t)$ is piecewise linear and convex in t. Thus, if, for fixed $(\hat{u},\hat{t}) \in \mathcal{U} \times \mathcal{T}$, $\hat{\pi}$ is a dual solution of (30), we obtain the inequality $\bar{\Phi}(\hat{u},t) \geq \bar{\Phi}(\hat{u},\hat{t}) + \langle \hat{\pi}, t - \hat{t}\rangle = \langle \hat{\pi}, t\rangle$ $(t \in \mathcal{T})$. Letting $\hat{u} = q(\xi^i)$ and $\hat{t} = h(\xi^i) - T(\xi^i)\hat{x}$

for a fixed scenario ξ^i and first stage decision $\hat{x} \in \bar{X}$, we obtain

$$\bar{\Phi}(q(\xi^i), h(\xi^i) - T(\xi^i)x) \geq \langle \hat{\pi}, h(\xi^i) - T(\xi^i)x \rangle =: \eta_1(x). \tag{31}$$

Since $\Phi(u,t) \geq \bar{\Phi}(u,t)$, (31) yields the *optimality cut* $\eta := (\eta_1(\cdot))$ (i.e., $k = 1$). Because of the polyhedrality of $\bar{\Phi}(u,t)$, a finite number of such cuts for each scenario is sufficient to obtain an exact representation of $\bar{\Phi}(u,t)$ in the master problem (27).

5.2.4 Approximation of $\Phi(u,t)$ by Lift-and-Project

However, to capture the nonconvexity of the original value function $\Phi(u,t)$, *nonconvex optimality cuts* are necessary, that is, tuples η of length $k > 1$. For the case that the discrete variables in the second stage are all of binary type, the following method is proposed in Sen and Higle (2005): Let $\bar{x} \in X$ be a feasible solution of the master problem (25), let $(\bar{y}_1^i, \bar{y}_2^i)$ be a solution of the relaxed second stage problem (30) for $u = q(\xi^i)$ and $t = h(\xi^i) - T(\xi^i)\hat{x}$, $i = 1, \ldots, N$. If $\bar{y}_2^i \in \mathbb{Z}^{m_2}$ for all $i = 1, \ldots, N$, then a linear optimality cut (31) is derived, c.f. (31). Otherwise, let $j \in \{1, \ldots, m_2\}$ be an index such that $0 < \bar{y}_{2,j}^i < 1$. We now seek for inequalities $\langle \pi_1^i, y_1 \rangle + \langle \pi_2^i, y_2 \rangle \geq \pi_0^i(x)$, $i = 1, \ldots, N$, which are valid for (3) for all $x \in \bar{X}$, but cut off the solution $\bar{y}^i$ from (30) for at least one scenario i with fractional $\bar{y}_{2,j}^i$. That is, we search for inequalities that are valid for the disjunctive sets

$$\left\{ y \in \mathbb{R}^{m_1+m_2} \;\middle|\; \begin{array}{r} W_1 y_1 + W_2 y_2 \leq t, \\ y_{2,j} \leq 0 \end{array} \right\} \cup \left\{ y \in \mathbb{R}^{m_1+m_2} \;\middle|\; \begin{array}{r} W_1 y_1 + W_2 y_2 \leq t, \\ -y_{2,j} \leq -1 \end{array} \right\}, \tag{32}$$

where $t = h(\xi^i) - T(\xi^i)\bar{x}$, $i = 1, \ldots, N$. Observe that points with fractional $y_{2,j}$ are not contained in (32). With an argumentation similar to the derivation of $\bar{\eta}(\cdot)$ before, it follows that, for fixed x, valid inequalities for (32) are described by the system

$$W_1^\top \lambda_{1,1}^i = \pi_1^i, \qquad W_1^\top \lambda_{2,1}^i = \pi_1^i, \tag{33a}$$

$$W_2^\top \lambda_{1,1}^i + e_j \lambda_{1,2}^i = \pi_2^i, \qquad W_2^\top \lambda_{2,1}^i - e_j \lambda_{2,2}^i = \pi_2^i, \tag{33b}$$

$$\langle h(\xi^i) - T(\xi^i)x, \lambda_{1,1}^i \rangle \geq \pi_0^i(x), \quad \langle h(\xi^i) - T(\xi^i)x, \lambda_{2,1}^i - \lambda_{2,2}^i \rangle \geq \pi_0^i(x), \tag{33c}$$

$$\lambda_{1,1}^i \in \mathbb{R}_-^r, \lambda_{1,2}^i \in \mathbb{R}_-, \qquad \lambda_{2,1}^i \in \mathbb{R}_-^r, \lambda_{2,2}^i \in \mathbb{R}_-, \tag{33d}$$

where $e_j \in \mathbb{R}^{m_2}$ is the jth unit vector. Observe further that the coefficients in (33a) and (33b) (i.e., W_1, W_2, e_j) are scenario independent. Thus, it is possible to use common cut coefficients $(\pi_1, \pi_2) \equiv (\pi_1^i, \pi_2^i)$ for all scenarios, thereby reducing the computational effort to the solution of a single linear program (Sen and Higle 2005):

$$\max \left\{ \sum_{i=1}^{N} p_i(\pi_0^i(\bar{x}) - \langle \pi_1, \bar{y}_1^i \rangle - \langle \pi_2, \bar{y}_2^i \rangle) \;\middle|\; \begin{array}{l} \lambda_{1,1}, \lambda_{2,1} \in \mathbb{R}_-^r,\ \lambda_{1,2}, \lambda_{2,2} \in \mathbb{R}_-, \\ \pi_1 \in \mathbb{R}^{m_1},\ \pi_2 \in \mathbb{R}^{m_2},\ \pi_0^i(\bar{x}) \in \mathbb{R}, \\ W_1^\top \lambda_{1,1} = \pi_1, W_2^\top \lambda_{1,1} + e_j \lambda_{1,2} = \pi_2, \\ W_1^\top \lambda_{2,1} = \pi_1, W_2^\top \lambda_{2,1} - e_j \lambda_{2,2} = \pi_2, \\ \langle h(\xi^i) - T(\xi^i)\bar{x}, \lambda_{1,1} \rangle \geq \pi_0^i(\bar{x}), \\ \langle h(\xi^i) - T(\xi^i)\bar{x}, \lambda_{2,1} \rangle - \lambda_{2,2} \geq \pi_0^i(\bar{x}), \\ \|\pi_1\|_\infty \leq 1, \|\pi_2\|_\infty \leq 1, |\pi_0^i(\bar{x})| \leq 1, \\ i = 1, \ldots, N \end{array} \right\}$$

The objective function of this simple recourse problem maximizes the average violation of the computed cuts by $(\bar{y}_1^i, \bar{y}_2^i)$. The functions $\pi_0^i(\cdot)$, $i = 1, \ldots, N$, with $\langle \pi_1, y_1 \rangle + \langle \pi_2, y_2 \rangle \geq \pi_0^i(x)$ for all $x \in \bar{X}$ is derived from a solution of this LP as

$$\pi_0^i(x) := \min\{\langle h(\xi^i) - T(\xi^i)x, \lambda_{1,1} \rangle, \langle h(\xi^i) - T(\xi^i)x, \lambda_{2,1} \rangle - \lambda_{2,2}\}. \tag{34}$$

Adding these new cuts to (30) for $u = q(\xi^i)$ and $t = h(\xi^i) - T(\xi^i)\bar{x}$, $i = 1, \ldots, N$, yields the updated second stage linear relaxations

$$\min \left\{ \langle q_1(\xi^i), y_1 \rangle + \langle q_2(\xi^i), y_2 \rangle \;\middle|\; \begin{array}{r} W_1 y_1 + W_2 y_2 \leq h(\xi^i) - T(\xi^i)\bar{x} \\ -\langle \pi_1, y_1 \rangle - \langle \pi_2, y_2 \rangle \leq -\pi_0^i(\bar{x}) \\ y_1 \in \mathbb{R}^{m_1}, y_2 \in \mathbb{R}^{m_2} \end{array} \right\}. \tag{35}$$

A dual solution (μ, μ_0) of (35) can then be used to derive the inequality

$$\Phi(q(\xi^i), h(\xi^i) - T(\xi^i)x) \geq \langle \mu, h(\xi^i) - T(\xi^i)x \rangle - \mu_0 \pi_0^i(x).$$

However, the nonconvexity of the right-hand side $\pi_0^i(x)$ yields a nonconvex optimality cut $\eta := (\eta_1(\cdot), \eta_2(\cdot))$, where

$$\begin{aligned} \eta_1(x) &:= \langle \mu - \mu_0 \lambda_{1,1}, h(\xi^i) - T(\xi^i)x \rangle, \\ \eta_2(x) &:= \langle \mu - \mu_0 \lambda_{2,1}, h(\xi^i) - T(\xi^i)x \rangle + \mu_0 \lambda_{2,2}. \end{aligned}$$

In a next iteration, when the second stage problems are revisited with a different first stage solution $\bar{x}$, the updated relaxation (35) takes the place of the original relaxation (30). Since the functions $\pi_0^i(\cdot)$ are known, the right-hand side of the added cut in (35) is updated when $\bar{x}$ changes.

5.2.5 Approximation of $\Phi(u, t)$ by Branch-and-Bound

For the general case where the discrete second stage variables can also be of integer type, the second stage problem (3) can be solved by a (partial) branch-and-bound algorithm and a (nonlinear) optimality cut η can be derived from the dual solutions of the linear programs in each leaf of the branch-and-bound tree (Sen and Sherali 2006): Let $\bar{x} \in X$ be again a feasible point to problem (25) and fix a scenario i. Assume that (3) with $\bar{u} = q(\xi^i)$ and $\bar{t} = h(\xi^i) - T(\xi^i)\bar{x}$ is (partially) solved by a branch-and-bound algorithm. Denote by $\mathcal{Q}$ the set of terminal nodes of the generated branch-and-bound tree. For any node $q \in \mathcal{Q}$, let $y_{2,l}^q$ and $y_{2,u}^q$ denote the vectors that define lower and upper bounds on the y_2 variables in the subproblem at node q. Then the LP relaxation of (3) for node $q \in \mathcal{Q}$ is given as

$$\min\left\{ \langle \bar{u}_1, y_1\rangle + \langle \bar{u}_2, y_2\rangle \;\middle|\; y \in \mathbb{R}^{m_1+m_2}, W_1 y_1 + W_2 y_2 \le \bar{t}, \begin{array}{l} y_2 \le y_{2,u}^q, \\ -y_2 \le -y_{2,l}^q \end{array} \right\}. \tag{36}$$

We assume that subproblems have been pruned if they are infeasible or their lower bound exceeds a known upper bound. Thus, all terminal nodes are associated with a feasible LP relaxation. The dual problem to (36) is

$$\max\left\{ \langle \mu, \bar{t}\rangle + \langle \pi_u, y_{2,u}^q\rangle - \langle \pi_l, y_{2,l}^q\rangle \;\middle|\; \begin{array}{ll} \mu \in \mathbb{R}_-^r, & W_1^\top \mu = \bar{u}_1 \\ \pi_l, \pi_u \in \mathbb{R}_-^{m_2}, & W_2^\top \mu + \pi_u - \pi_l = \bar{u}_2 \end{array} \right\}, \tag{37}$$

where we assume that a dual variable $\pi_{l,j}$, $\pi_{u,j}$ is fixed to 0 if the corresponding bound $y_{2,l,j}^q$, $y_{2,u,j}^q$ is $-\infty$ or $+\infty$, respectively, $j = 1, \ldots, m_2$. Based on a dual solution $(\mu^q, \pi_l^q, \pi_u^q)$ of (37), a supporting hyperplane of each nodes LP value function can be derived, c.f. (31). Since the branch-and-bound tree represents a partition of the feasible set of (3), it allows to state a disjunctive description of the function $t \mapsto \Phi(\bar{u}, t)$ by combining the LP value function approximations in all nodes $q \in \mathcal{Q}$:

$$\Phi(\bar{u}, t) \ge \langle \mu^q, t\rangle + \langle \pi_u^q, y_{2,u}^q\rangle - \langle \pi_l^q, y_{2,l}^q\rangle \quad \text{for at least one } q \in \mathcal{Q}. \tag{38}$$

This result directly translates into a nonlinear optimality cut $\eta := (\eta_1(\cdot), \ldots, \eta_q(\cdot))$ by letting $\eta_q(x) := \langle \mu^q, h(\xi^i) - T(\xi^i)x\rangle + \langle \pi_u^q, y_{2,u}^q\rangle - \langle \pi_l^q, y_{2,l}^q\rangle$.

5.2.6 Full Algorithm

We can now state a full algorithm for the solution of (1):

1. Solve the master problem (27) by branch-and-bound. If it is infeasible, then (1) is infeasible. Otherwise, let $(\bar{x}, \bar{\theta})$ be a solution of (27).
2. Solve (3) for each scenario $i = 1, \ldots, N$. Let $\phi_i := \Phi(q(\xi^i), h(\xi^i) - T(\xi^i)\bar{x})$ be the optimal value of (3) for the first stage decision $\bar{x}$ in scenario i.
3. For scenarios i where $\phi_i > \bar{\theta}_i$, derive an optimality cut η of the value function $x \mapsto \Phi(q(\xi^i), h(\xi^i) - T(\xi^i)x)$ either via linearization of $\bar{\Phi}(u, t)$ (see (31)), via lift-and-project (Sect. 5.2.4), or from a (partial) branch-and-bound search (Sect. 5.2.5). Add η to C_i in (27).

4. If no new tuples η have been constructed, that is, the master problem has not been updated, then finish: $\bar{x}$ is an optimal solution to (3). Otherwise, go back to 1.

Some remarks are in order:

- At the beginning, the sets C_i are empty, that is, no information about the value function $\Phi(u,t)$ is available in (27). Thus, (27) should be solved either with the variables θ_i removed or bounded from below by a known lower bound on $\Phi(u,t)$.
- In the first iterations, when almost no information about $\Phi(u,t)$ is available, it is unnecessary to solve the master problem (27) and the second stage problems (3) to optimality. Instead, at first it is more efficient to ignore the integrality conditions and to construct a representation of the LP value function $\bar{\Phi}(u,t)$ by a usual Benders' decomposition. Later, partial solves of (27) and the introduction of nonlinear optimality cuts η into (27) based on lift-and-project or partial branch-and-bound searches should be performed to capture the nonconvexity of $\Phi(u,t)$ in the master problem. Finally, to ensure convergence, first and second stage problems need to be solved to optimality, see also Sen and Higle (2005) and Ntaimo and Sen (2008).

5.2.7 Extension to Multistage Problems

While the algorithms discussed so far allow an efficient extension of the Benders' decomposition to two-stage mixed-integer stochastic programs, a further extension to the multistage case seems possible. Although in the two-stage case we have a nonconvex value function only in the first stage, in the multistage setting we are faced with such a function in each node of the scenario tree other than the leaves. That is, the master problems in each node before the last stage are of the form (27). Approximation of the value function of such a master problem then requires to take the nonlinear optimality cuts, which approximate the value functions of successor nodes, into account. For that matter, we have seen how such a master problem can be solved by branch-and-bound (Sect. 5.2.1, Ntaimo and Sen 2008) and how an optimality cut can be derived from a (partial) branch-and-bound search (Sect. 5.2.5, Sen and Sherali 2006).

However, the efficiency of such an approach might suffer under the large number of disjunctions that are induced from optimality cuts on late stages into the master problems on early stages. That is, while in the two-stage case the disjunctions in (27) are caused only by integrality constraints on the second stage, in the multistage setting we have to deal with disjunctions that are induced by disjunctions on succeeding stages. Therefore, solving a fairly large mixed-integer multistage stochastic program to optimality with this approach seems questionable.

Nevertheless, an interesting application are multistage problem that can be solved efficiently only by a temporal decomposition, for example, stochastic programs with recombining scenario trees (Küchler and Vigerske 2007). For the latter, the recombining nature of the scenario tree leads to coinciding value functions, a property that can be explored by a nested Benders' decomposition. Therefore, an extension to the mixed-integer case by application of the ideas discussed in this section seems promising.

5.3 Scenario Decomposition

Consider the following reformulation of (1) where the first stage variable x is replaced by one variable x^i for each scenario $i = 1, \dots, N$, and an explicit *nonanticipativity constraint* is added:

$$\min \quad \sum_{i=1}^{N} p_i(\langle c, x^i\rangle + \langle q_1(\xi^i), y_1(\xi^i)\rangle + \langle q_2(\xi^i), y_2(\xi^i)\rangle) \tag{39a}$$

$$\text{such that} \quad x^i \in X, \quad y_1^i \in \mathbb{R}^{m_1}, \quad y_2^i \in \mathbb{Z}^{m_2}, \qquad i = 1, \dots, N, \tag{39b}$$

$$T(\xi^i)x^i + W_1 y_1^i + W_2 y_2^i \leq h(\xi^i), \qquad i = 1, \dots, N, \tag{39c}$$

$$x^1 = x^2 = \dots = x^N. \tag{39d}$$

Problem (39) decomposes into scenario-wise subproblems by relaxing the coupling constraint (39d) (Carøe and Schultz 1999). The violation of the relaxed constraints is then added as a penalty into the objective function. That is, each subproblem has the form

$$D_i(\lambda) := \min \left\{ L_i(x^i, y^i; \lambda) \;\middle|\; \begin{array}{c} x^i \in X, \; y_1^i \in \mathbb{R}^{m_1}, \; y_2^i \in \mathbb{Z}^{m_2}, \\ T(\xi^i)x^i + W_1 y_1^i + W_2 y_2^i \leq h(\xi^i), \end{array} \right\}, \tag{40}$$

where $\lambda := (\lambda^1, \dots, \lambda^N) \in \mathbb{R}^{mN}$ is the *Lagrange multiplier* and

$$L_i(x^i, y^i; \lambda) := p_i(\langle c, x^i\rangle + \langle q_1(\xi^i), y_1(\xi^i)\rangle + \langle q_2(\xi^i), y_2(\xi^i)\rangle + \langle \lambda^i, x^i - x^1\rangle),$$

$i = 1, \dots, N$. For every choice of λ, a lower bound on (39) is obtained by computing

$$D(\lambda) := \sum_{i=1}^{N} D_i(\lambda). \tag{41}$$

That is, to compute (41), the deterministic problem (40) is solved for each scenario. To find the best possible lower bound, one now searches for an optimal solution to the *dual problem*

$$\max\{D(\lambda) \mid \lambda \in \mathbb{R}^{mN}\}. \tag{42}$$

The function $D(\lambda)$ is a piecewise linear concave function for which subgradients can be computed from a solution of (40). Thus, solution methods for the nonsmooth convex optimization problem (42) use a bundle of subgradients of $D(\lambda)$ to find promising values of λ (Kiwiel 1990).

The primal solutions (x^i, y^i), $i = 1, \ldots, N$, of (40), associated with a solution of (42), yield in general not a feasible solution to the original problem. To regain the relaxed nonanticipativity constraint, heuristics are employed, which, for example, select for x an average or a frequently occurring value among the x^i and then possibly resolve each second stage problem to ensure feasibility.

To find an optimal solution to (39), a branch-and-bound algorithm can be employed. Here, nonanticipativity constraints are insured by branching on the first stage variables. Since the additional bound constraints on x^i become part of the constraints in (40), the lower bound (42) improves by a branching operation.

An alternative for solving the dual problem (42) by a bundle method is proposed in Lulli and Sen (2004): As shown in Carøe and Schultz (1999), the problem (42) is equivalent to the primal problem

$$\min \quad \sum_{i=1}^{N} p_i(\langle c, x^i\rangle + \langle q_1(\xi^i), y_1(\xi^i)\rangle + \langle q_2(\xi^i), y_2(\xi^i)\rangle) \tag{43a}$$

$$\text{such that} \quad (x^i, y^i) \in \operatorname{conv}\left\{(x, y_1, y_2) \;\middle|\; \begin{array}{c} x \in X,\ y_1 \in \mathbb{R}^{m_1},\ y_2 \in \mathbb{Z}^{m_2}, \\ T(\xi^i)x + W_1 y_1 + W_2 y_2 \le h(\xi^i) \end{array}\right\}, \tag{43b}$$

$$i = 1, \ldots, N,$$

$$x^1 = x^2 = \ldots = x^N. \tag{43c}$$

This problem is solved by a column generation approach, which constructs an inner approximation of the convex hull in (43b). Feasible solutions for the original problem are obtained by application of branch-and-bound.

For problems where all first stage variables are restricted to be binary, Alonso-Ayuso et al. (2003) propose to relax both nonanticipativity and integrality constraints. Thereby, each scenario is associated with a branch-and-bound tree that enumerates the integer feasible solutions to the scenario's subproblem (i.e., the feasible set of (40)). Since each branch-and-bound fixes first stage variables to be either 0 or 1, a coordinated search across all n branch-and-bound trees allows to select feasible solutions from each subproblem that satisfy the nonanticipativity constraints. If also continuous variables are present in the first stage, Escudero et al. (2007) propose to "cross over" to a Benders' decomposition whenever the coordinated branch-and-bound search yields solutions which binary first stage variables satisfy the nonanticipativity constraints and second stage integer variables are fixed.

6 Application to Stochastic Thermal Unit Commitment

We consider a power generation system comprising thermal units and contracts for delivery and purchase, and describe a model for its cost-minimal operation under uncertainty in electrical load and in prices for fuel and electricity. Contracts for delivery and purchase of electricity are regarded as special thermal units. It is assumed that the time horizon is discretized into uniform (e.g., hourly) intervals. Let T and I denote the number of time periods and thermal units, respectively. For thermal unit i in period t, $u_{it} \in \{0,1\}$ is its *commitment* decision (1 if on, 0 if off) and x_{it} its *production*, with

$$u_{it}x_{it}^{\min} \le x_{it} \le x_{it}^{\max}u_{it} \qquad (i=1,\dots,I,\ t=1,\dots,T), \tag{44}$$

where $x_{it}^{\min}$ and $x_{it}^{\max}$ are the minimum and maximum capacities. Additionally, there are *minimum up/down-time requirements*: when unit i is switched on (off), it must remain on (off) for at least $\bar{\tau}_i$ ($\underline{\tau}_i$, resp.) periods, that is,

$$u_{i\tau} - u_{i,\tau-1} \le u_{it} \qquad (\tau = t-\bar{\tau}_i+1,\dots,t-1), \tag{45}$$

$$u_{i,\tau-1} - u_{i\tau} \le 1-u_{it} \qquad (\tau = t-\underline{\tau}_i+1,\dots,t-1), \tag{46}$$

for all $t=1,\dots,T$ and $i=1,\dots,I$. Let U_i denote the set of all pairs (x_i,u_i) satisfying the constraints (44), (45), and (46) for all $t=1,\dots,T$. The basic system requirement is to meet the electrical load d_t during all time periods $t=1,\dots,T$, that is,

$$\sum_{i=1}^{I} x_{it} \ge d_t \qquad (t=1,\dots,T). \tag{47}$$

The expected total system cost is given by the sum of expected startup and operating costs of all thermal units over the whole scheduling horizon, that is,

$$\mathbb{E}\left(\sum_{t=1}^{T}\sum_{i=1}^{I}(C_{it}(x_{it},u_{it}) + S_{it}(u_i))\right). \tag{48}$$

The *fuel cost* C_{it} for operating thermal unit i is assumed to be piecewise linear convex (concave for purchase contracts) during period t, that is,

$$C_{it}(x_{it},u_{it}) := \max_{l=1,\dots,\bar{l}} \{a_{ilt}x_{it} + b_{ilt}u_{it}\}$$

with cost coefficients a_{ilt}, b_{ilt}. The startup cost of unit i depends on its downtime; it may vary between a maximum cold-start value and a much smaller value when the unit is still relatively close to its operating temperature. This is modeled by the *startup cost*

$$S_{it}(u_i) := \max_{\tau=0,\dots,\tau_i^c} c_{i\tau}\left(u_{it} - \sum_{\kappa=1}^{\tau} u_{i,t-\kappa}\right),$$

where $0 = c_{i0} \le \cdots \le c_{i\tau_i^c}$ are cost coefficients, τ_i^c is the cool-down time of unit i, $c_{i\tau_i^c}$ is its maximum cold-start cost, $u_i := (u_{it})_{t=1}^{T}$, and $u_{i\tau} \in \{0,1\}$ for $\tau = 1-\tau_i^c,\dots,0$ are given initial values.

It is assumed that the stochastic input process $\xi = \{\xi\}_{t=1}^{T}$ is given by

$$\xi_t := (a_t, b_t, c_t, d_t) \quad (t = 1, \dots, T)$$

or by some of its components. Furthermore, it is assumed that $\xi_1, \dots, \xi_{t_1}$ (i.e., the input data for the first time period for which reliable forecasts are available), and thus, the (first stage) decisions $\{(x_{it}, u_{it}) \mid t = 1, \dots, t_1, i = 1, \dots, I\}$ are deterministic.

Minimizing the expected total cost (48) such that the operational constraints (44), (45), (46), and (47) are satisfied represents a two-stage (linear) mixed-integer stochastic program with (random) second stage decision $\{(x_{it}, u_{it}) \mid t = t_1 + 1, \dots, T, i = 1, \dots, I\}$.

In many cases it is possible to derive a model for the probability distribution P of ξ via time series analysis based on historical data (see, e.g., Eichhorn et al. (2005); Sen et al. (2006)). Sampling from P together with applying scenario reduction (see Sect. 4) then leads to a finite number of scenarios $\xi^j = (a_t^j, b_t^j, c_t^j, d_t^j)$ with probabilities p_j, $j = 1, \dots, N$, for the stochastic process ξ and to the corresponding decision scenarios (x_i^j, u_i^j) (for unit i). The scenario-based unit commitment problem then reads

$$\min \left\{ \sum_{j=1}^{N} \sum_{t=1}^{T} \sum_{i=1}^{I} p_j (C_{it}^j(x_{it}^j, u_{it}^j) + S_{it}^j(u_i^j)) \;\middle|\; \begin{array}{l} (x_i^j, u_i^j) \in U_i,\ i = 1, \dots, I, \\ \sum_{i=1}^{I} p_{it}^j \geq d_t^j,\ t = 1, \dots, T, \\ j = 1, \dots, N \end{array} \right\}, \tag{49}$$

where C_{it}^j and S_{it}^j denote the cost functions for scenario j.

Since the optimization problem (49) contains only NT (unit) coupling constraints while the number $2NTI$ of decision variables is typically (much) larger, *geographical decomposition* based on Lagrangian relaxation of the coupling constraints (47) seems to be promising. The Lagrangian function is of the form

$$\begin{aligned} L(x, u; \lambda) &= \sum_{j=1}^{N} \sum_{t=1}^{T} p_j \left(\sum_{i=1}^{I} (C_{it}^j(x_{it}^j, u_{it}^j) + S_{it}^j(u_i^j)) + \lambda_t^j (d_t^j - \sum_{i=1}^{I} x_{it}^j) \right) \\ &= \sum_{j=1}^{N} \sum_{t=1}^{T} p_j \left(\sum_{i=1}^{I} (C_{it}^j(x_{it}^j, u_{it}^j) + S_{it}^j(u_i^j) - \lambda_t^j x_{it}^j) + \lambda_t^j d_t^j \right), \end{aligned}$$

which leads to the dual function

$$D(\lambda) = \inf_{(x,u)} L(x, u; \lambda) = \sum_{j=1}^{N} p_j \left(\sum_{i=1}^{I} D_{ij}(\lambda^j) + \sum_{t=1}^{T} \lambda_t^j d_t^j \right)$$

$$D_{ij}(\lambda) = \inf_{u_i} \sum_{t=1}^{T} (\inf_{x_{it}} (C_{it}^j(x_{it}, u_{it}) - \lambda_t x_{it}) + S_{it}^j(u_i))$$

decomposing into unit subproblems for every scenario ξ^j and, hence, λ^j. While the inner minimization (with respect to x_{it}) can be solved explicitly, the outer minimization (with respect to u_i) can efficiently be done by dynamic programming. The dual concave (nondifferentiable) maximization problem

$$\max\left\{D(\lambda) \mid \lambda \in \mathbb{R}_+^{NT}\right\} \tag{50}$$

can be solved by bundle subgradient methods (e.g., Kiwiel (1990)). If $(\bar{x}, \bar{u}, \bar{\lambda})$ is an (approximate) solution of (50), $D(\bar{\lambda})$ is a lower bound of the infimum of (49) but, in general, the (maximal) load constraints

$$\sum_{i=1}^{I} x_{it}^{\max} \bar{u}_{it}^j \geq d_t^j \qquad (t = 1, \dots, T, j = 1, \dots, N) \tag{51}$$

are violated for some scenarios j and some time intervals t, respectively. However, as shown in Bertsekas (1982, Sect. 5.6.1), the relative duality gap gets small if the number I of units is large. In many practical situations, this allows to apply simple Lagrangian heuristics (like Zhuang and Galiana (1988)) to modify $\bar{u}$ scenario-wise such that (51) is satisfied for every pair (t, j). After having the commitment decision $\bar{u}$ fixed, a final scenario-wise economic dispatch (van den Bosch and Lootsma 1987) leads to good primal solutions $(\bar{x}, \bar{u})$.

The approach can be extended to multistage models by requiring in addition that the decisions (x_t, u_t) in (49) depend only on $(\xi_1, \dots, \xi_t)$ (for $t > t_1$). We refer to the relevant work Carpentier et al. (1996), Gröwe-Kuska et al. (2002), Gröwe-Kuska and Römisch (2005), Nowak and Römisch (2000), Philpott et al. (2000), Sen et al. (2006), Takriti et al. (1996), Takriti et al. (2000).

Furthermore, instead of the expected total system cost, a mean-risk objective of the form

$$\gamma\rho(Y_{t_1}, \dots, Y_T) - (1-\gamma)\mathbb{E}(Y_T), \quad Y_t := -\sum_{\tau=1}^{t}\sum_{i=1}^{I}(C_{it}(x_{it}, u_{it}) + S_{it}(u_i))$$
$$(t = t_1, \dots, T)$$

may be considered, where $\gamma \in (0, 1)$ and ρ is a multiperiod risk functional (see Eichhorn et al. (2010)). In this way, risk management is integrated into unit commitment. If the risk functional is polyhedral (Eichhorn and Römisch 2005; Eichhorn et al. 2010), the scenario-based unit commitment model may be reformulated as mixed-integer linear program.

Extensions of the two-stage stochastic unit commitment model are discussed in Nürnberg and Römisch (2002) and Nowak et al. (2005), respectively. In Nürnberg and Römisch (2002), a planning model is described whose (deterministic) first stage and (stochastic) second stage decisions are given on the whole time horizon $\{1, \dots, T\}$. The first stage decisions are determined such that a transition from the first to the second stage and vice versa is always feasible and compatible. In Nowak et al. (2005), day-ahead trading at a power exchange is incorporated into unit commitment.

7 Conclusions

We reviewed recent progress in two-stage mixed-integer stochastic programming. First we reviewed structural properties of optimal value functions of mixed-integer linear programs from the literature and discussed conclusions for continuity properties of integrands in two-stage mixed-integer stochastic programs. If the probability distribution has finite support, the expected recourse function is piecewise continuous with a finite number of polyhedral continuity regions. When perturbing or approximating the underlying probability distribution, the optimal value function behaves continuous with respect to a discrepancy distance of the original and perturbed probability measures. This result allowed to extend the stability-based scenario reduction algorithm from Dupačová et al. (2003) and Heitsch and Römisch (2007) to the mixed-integer two-stage situation.

For solving a two-stage mixed-integer stochastic program, several decomposition algorithms are reviewed. First, methods to convexify the expected recourse function of simple and more complex integer-recourse models by perturbing the probability measure are discussed. This allows to obtain tight bounds on the original optimal value. Second, algorithms that decompose the stochastic program in a Benders' decomposition style are detailed. Here, the nonconvexity in the second-stage value functions is captured by nonlinear optimality cuts, which might make a solution of the master problem by branch-and-bound necessary. Further, scenario decomposition-based algorithms based on relaxation of nonanticipativity constraints are reviewed. In a Lagrangian decomposition, the subproblems are coupled via a dual problem, which comprises the maximization of a piecewise linear concave function.

Finally, a geographical Lagrangian decomposition method is illustrated on a stochastic thermal unit commitment problem. Here, the problem decomposed into one (stochastic mixed-integer) subproblem for each thermal unit. This allows to exploit the subproblems structure by specialized algorithms and to use a Lagrangian heuristic specialized for unit commitment problems.

Acknowledgements This work was supported by the DFG Research Center MATHEON *Mathematics for Key Technologies* in Berlin (http://www.matheon.de) and by the German Ministry of Education and Research (BMBF) under the grant 03SF0312E.

References

Ahmed S, Tawarmalani M, Sahinidis NV (2004) A finite branch and bound algorithm for two-stage stochastic integer programs. Math Program 100:355–377

Alonso-Ayuso A, Escudero LF, Teresa Ortuño M (2003) BFC, a branch-and-fix coordination algorithmic framework for solving some types of stochastic pure and mixed 0-1 programs. Eur J Oper Res 151:503–519

Balas E (1998) Disjunctive programming: Properties of the convex hull of feasible points. Discrete Appl Math 89:3–44; originally MSRR#348, Carnegie Mellon University (1974)

Bank B, Guddat J, Kummer B, Klatte D, Tammer K (1982) Non-linear parametric optimization. Akademie-Verlag, Berlin

Bertsekas DP (1982) Constrained optimization and Lagrange multiplier methods. Academic Press, NY

Birge JR (1985) Decomposition and partitioning methods for multistage stochastic programming. Oper Res 33(5):989–1007

Blair CE, Jeroslow RG (1982) The value function of an integer program. Math Program 23:237–273

Carøe CC, Schultz R (1999) Dual decomposition in stochastic integer programming. Oper Res Lett 24(1–2):37–45

Carøe CC, Tind J (1997) A cutting-plane approach to mixed 0–1 stochastic integer programs. Eur J Oper Res 101(2):306–316

Carøe CC, Tind J (1998) L-shaped decomposition of two-stage stochastic programs with integer recourse. Math Program 83(3A):451–464

Carpentier P, Cohen G, Culioli JC, Renaud (1996) A stochastic optimization of unit commitment: A new decomposition framework. IEEE Trans Power Syst 11:1067–1073

Dentcheva D, Römisch W (2004) Duality gaps in nonconvex stochastic optimization. Math Program 101(3A):515–535

Dupačová J, Gröwe-Kuska N, Römisch W (2003) Scenarios reduction in stochastic programming: An approach using probability metrics. Math Program 95:493–511

Eichhorn A, Heitsch H, Römisch W (2010) Stochastic optimization of electricity portfolios: Scenario tree modeling and risk management. In: Rebennack S, Pardalos PM, Pereira MVF, Illiadis NA (eds.) Handbook of Power Systems, vol. II. Springer, Berlin, pp. 405–432

Eichhorn A, Römisch W (2005) Polyhedral risk measures in stochastic programming. SIAM J Optim 16:69–95

Eichhorn A, Römisch W (2007) Stochastic integer programming: Limit theorems and confidence intervals. Math Oper Res 32:118–135

Eichhorn A, Römisch W, Wegner I (2005) Mean-risk optimization of electricity portfolios using multiperiod polyhedral risk measures. In: IEEE St. Petersburg PowerTech Proceedings

Escudero LF, Garín A, Merino M, Pérez G (2007) A two-stage stochastic integer programming approach as a mixture of branch-and-fix coordination and Benders decomposition schemes. Ann Oper Res 152:395–420

Gröwe-Kuska N, Kiwiel KC, Nowak MP, Römisch W, Wegner I (2002) Power management in a hydro-thermal system under uncertainty by Lagrangian relaxation. In: Greengard C, Ruszczyński A (eds.) Decision making under uncertainty: Energy and power, Springer, NY, pp. 39–70

Gröwe-Kuska N, Römisch W (2005) Stochastic unit commitment in hydro-thermal power production planning. In: Wallace SW, Ziemba WT (eds.) Applications of stochastic programming. MPS/SIAM series on optimization, SIAM, Philadelphia, pp. 633–653

Heitsch H, Römisch W (2007) A note on scenario reduction for two-stage stochastic programs. Oper Res Lett 35:731–738

Henrion R, Küchler C, Römisch W (2008) Discrepancy distances and scenario reduction in two-stage stochastic mixed-integer programming. J Ind Manag Optim 4:363–384

Henrion R, Küchler C, Römisch W (2009) Scenario reduction in stochastic programming with respect to discrepancy distances. Comput Optim Appl 43:67–93

Henrion R, Küchler C, Römisch W (in preparation) A scenario reduction heuristic for two-stage stochastic integer programs

Kiwiel KC (1990) Proximity control in bundle methods for convex nondifferentiable optimization. Math Program 46:105–122

Klein Haneveld WK, Stougie L, van der Vlerk MH (2006) Simple integer recourse models: convexity and convex approximations. Math Program 108(2–3B):435–473

Küchler C, Vigerske S (2007) Decomposition of multistage stochastic programs with recombining scenario trees. Stoch Program E-Print Series 9 www.speps.org

Laporte G, Louveaux FV (1993) The integer L-shaped method for stochastic integer programs with complete recourse. Oper Res Lett 13(3):133–142

Louveaux FV, Schultz R (2003) Stochastic integer programming. In: Ruszczyński A, Shapiro A (eds.) Stochastic programming, pp. 213–266; Handbooks in operations research and management science vol. 10 Elsevier,

Lulli G, Sen S (2004) A branch and price algorithm for multi-stage stochastic integer programs with applications to stochastic lot sizing problems. Manag Sci 50:786–796

Niederreiter H (1992) Random number generation and Quasi-Monte Carlo methods. SIAM, Philadelphia

Nowak MP, Römisch W (2000) Stochastic Lagrangian relaxation applied to power scheduling in a hydro-thermal system under uncertainty. Ann Oper Res 100:251–272

Nowak MP, Schultz R, Westphalen M (2005) A stochastic integer programming model for incorporating day-ahead trading of electricity into hydro-thermal unit commitment. Optim Eng 6:163–176

Ntaimo L, Sen S (2005) The million-variable “march” for stochastic combinatorial optimization. J Global Optim 32(3):385–400

Ntaimo L, Sen S (2008) A comparative study of decomposition algorithms for stochastic combinatorial optimization. Comput Optim Appl 40(3):299–319

Ntaimo L, Sen S (2008) A branch-and-cut algorithm for two-stage stochastic mixed-binary programs with continuous first-stage variables. Int J Comp Sci Eng 3:232–241

Nürnberg R, Römisch W (2002) A two-stage planning model for power scheduling in a hydro-thermal system under uncertainty. Optim Eng 3:355–378

Philpott AB, Craddock M, Waterer H (2000) Hydro-electric unit commitment subject to uncertain demand. Eur J Oper Res 125:410–424

Rachev ST (1991) Probability Metrics and the Stability of Stochastic Models. Wiley, Chichester

Römisch W, Schultz R (2001) Multistage stochastic integer programs: An introduction. In: Grötschel M, Krumke SO, Rambau J (eds.) Online optimization of large scale systems, Springer, Berlin, pp. 581–600

Römisch W, Vigerske S (2008) Quantitative stability of fully random mixed-integer two-stage stochastic programs. Optim Lett 2:377–388

Schultz R (1996) Rates of convergence in stochastic programs with complete integer recourse. SIAM J Optim 6:1138–1152

Schultz R (2003) Stochastic programming with integer variables. Math Program 97:285–309

Schultz R, Stougie L, van der Vlerk MH (1998) Solving stochastic programs with integer recourse by enumeration: a framework using Gröbner basis reductions. Math Program 83(2A):229–252

Schultz R, Tiedemann S (2006) Conditional value-at-risk in stochastic programs with mixed-integer recourse. Math Program 105:365–386

Sen S (2005) Algorithms for stochastic mixed-integer programming models. In: Aardal K, Nemhauser GL, Weismantel R (eds.) Handbook of discrete optimization, North-Holland Publishing Co. pp. 515–558

Sen S, Higle JL (2005) The C^3 theorem and a D^2 algorithm for large scale stochastic mixed-integer programming: set convexification. Math Program 104(A):1–20

Sen S, Sherali HD (2006) Decomposition with branch-and-cut approaches for two-stage stochastic mixed-integer programming. Math Progr 106(2A):203–223

Sen S, Yu L, Genc T (2006) A stochastic programming approach to power portfolio optimization. Oper Res 54:55–72

Takriti S, Birge JR, Long E (1996) A stochastic model for the unit commitment problem. IEEE Trans Power Syst 11:1497–1508

Takriti S, Krasenbrink B, Wu LSY (2000) Incorporating fuel constraints and electricity spot prices into the stochastic unit commitment problem. Oper Res 48:268–280

Tind J, Wolsey LA (1981) An elementary survey of general duality theory in mathematical programming. Math Program 21:241–261

van den Bosch PPJ, Lootsma FA (1987) Scheduling of power generation via large-scale nonlinear optimization. J Optim Theor Appl 55:313–326

van der Vlerk MH (2004) Convex approximations for complete integer recourse models. Math Program 99(2A):297–310

van der Vlerk MH (1996-2007) Stochastic integer programming bibliography. http://mally.eco.rug.nl/spbib.html

van der Vlerk MH (2005) Convex approximations for a class of mixed-integer recourse models; Ann Oper Res (to appear); Stoch Program E-Print Series

Van Slyke RM, Wets R (1969) L-Shaped Linear Programs with Applications to Optimal Control and Stochastic Programming. SIAM J Appl Math 17(4):638–663

Wolsey LA (1981) Integer programming duality: Price functions and sensitivity analysis. Math Program 20:173–195

Zhuang G, Galiana FD (1988) Towards a more rigorous and practical unit commitment by Lagrangian relaxation. IEEE Trans Power Syst 3:763–773

Dealing With Load and Generation Cost Uncertainties in Power System Operation Studies: A Fuzzy Approach

Bruno André Gomes and João Tomé Saraiva

Abstract Power systems are currently facing a change of the paradigm that determined their operation and planning while being surrounded by multiple uncertainties sources. As a consequence, dealing with uncertainty is becoming a crucial issue in the sense that all agents should be able to internalize them in their models to guarantee that activities are profitable and that operation and investment strategies are selected according to an adequate level of risk. Taking into account the introduction of market mechanisms and the volatility of fuel prices, this paper presents the models and the algorithms developed to address load and generation cost uncertainties. These models correspond to an enhanced approach regarding the original fuzzy optimal power flow model developed by the end of the 1990s, which considered only load uncertainties. The paper also describes the algorithms developed to integrate an estimate of active transmission losses and to compute nodal marginal prices reflecting such uncertainties. The developed algorithms use multiparametric optimization techniques and are illustrated using a case study based on the IEEE 24 bus test system.

Keywords Generation Cost Uncertainties · Load Uncertainties · Fuzzy Models · DC Optimal Power Flow · Multiparametric Linear Programming · Nodal Marginal Prices.

1 Introduction

Power systems were always affected by uncertainties. They were traditionally related with load growth, consumer response to demand-side options, longevity and performance of life-extended or converted plants, potential supply of renewable

B.A. Gomes (✉)
INESC Porto, Departamento de Engenharia Electrotécnica e Computadores,
Faculdade de Engenharia da Universidade do Porto, Campus da FEUP,
Rua Dr. Roberto Frias 378, 4200 465 Porto, Portugal
e-mail: bgomes@inescporto.pt

S. Rebennack et al. (eds.), *Handbook of Power Systems I*, Energy Systems,
DOI 10.1007/978-3-642-02493-1_9,

resources, measurement errors or forecast inaccuracy, unscheduled outages, technological developments, fuel prices, or even regulatory requirements. In fact, these uncertainties are related with variables of different nature, such as physical, technical, economic, regulatory, and political.

Nowadays, power systems are also facing new challenges since uncertainties are also due to the introduction of market mechanisms in the sector and with the growing consciousness about environmental concerns. Apart from these new concerns, some others clearly increased their relevance, namely the volatility of fuel prices, the difficulty in predicting the availability of several primary resources highly used in renewable energy resources (RER), as well as the difficulty in predicting demand evolution, for instance as a consequence of economic problems in wider geographical areas.

On the other hand, the climate change is also putting a new pressure in the sector. As an attempt to tackle this problem, in 2005 the European Union (EU) launched the Emissions Trading System (ETS) to trade CO_2 allowances to create incentives to reduce emissions. Since RER are carbon free, it was also implemented in some European Countries a Green Certificate market to induce investments in these technologies. More recently and since energy efficiency is also an essential element of the EU energy policy, some member states also launched White Certificates markets (WhC) to promote energy efficiency.

In case these market instruments work properly, it is expected a net electricity demand reduction and consequently a decrease in new generation capacity investment and in the share of some more carbon intense generators. It is also important to refer that the implementation of a more ambitious quota for RER together with the Green Certificate system will contribute to a reduction in the price of the allowances within EU ETS. As a result, market participants will face a net combined effect. In fact, they have to face uncertainties in fuel prices, in the cost of purchasing ETS allowances if the CO_2 emissions exceed the administratively fixed limit, opportunity costs for selling allowances in case they are not fully used, and finally, uncertainties on demand and on RER generation levels (Soderholm 2008).

As a conclusion, internalizing the environmental damages due to energy generation and consumption into prices is bringing to the power sector a more conscious and sustainable way to plan and operate the system. On the other hand, it also brings a huge number of uncertainties and challenges that can contribute to raise the cost of capital and change investment decisions. Accordingly, market participants should use models and adopt methods that are able to address all these uncertainties and integrate them into decision-making processes. This means that risk has to be adequately addressed if one wants to have a complete and consistent picture of what can be the future power system operation while ensuring that profits are large enough to compensate the cost of capital.

Following these ideas, this paper is structured as follows. After this Introduction, Sect. 2 briefly addresses methodologies available in the literature to integrate uncertainties in several power system studies. Section 3 presents the fundamental concepts of Fuzzy Set theory useful to fully understand the remaining sections. Section 4 details the original fuzzy optimal power flow (FOPF) problem and Sect. 5

describes the new fuzzy dc optimal power flow (NFOPF) problem and the developed solution algorithms. Finally, Sect. 6 presents results obtained from a case study based on the IEEE 24 bus Test System and Sect. 7 draws the most relevant conclusions.

2 Uncertainty Modeling in Power System Studies

When planning and operating a power system, it is necessary to perform power flow studies to assess and monitor the steady state security of the system. These studies are one of the most frequently performed network calculations, and there are several well developed techniques to make these studies very quickly and accurately. Traditionally, these methods had a deterministic nature in the sense they admit that all values and parameters were completely and fully known. However, since system parameters such as nodal loads and generation levels cannot be considered perfectly constant, they were soon developed models to address the presence of uncertainties affecting these values. In this context, in the literature there are models admitting different types of data, namely probabilistic, fuzzy, boundary, and interval arithmetic ones.

Probabilistic methods were the pioneering methodologies developed in this area. Given their subsequent development, they can be organized into classes, such as simulation, analytical, or a combination of both (Chun 2005). Allan and Al-Shakarchi (1976, 1977), Allan et al. (1974, 1976), and Borkowska (1974) describe the main concepts related with this problem as well as the initially developed algorithms using convolution techniques, the DC model, and different linearized versions of the AC power flow problem. In general, these approaches translate the uncertainties specified in data to the results of traditional power flows under the form of probabilistic distributions. In view of the linearizations adopted in several of them, it was soon realized that the results were affected by errors that would eventually be larger in the tails of the output distributions. Addressing this issue, Allan et al. (1981) uses a Monte Carlo simulation based technique to evaluate the accuracy of the results, and Allan and Leite da Silva (1981) and Leite da Silva and Arienti (1990) propose the use of several linearization points to build partial probability distributions that, at the end, are aggregated to provide the final outputs. This approach was conceived to reduce the errors in the tails of the output distributions.

In subsequent years, there were other published contributions that enhanced these models or contributed to give them an increased realism. Karakatsanis and Hatziargyriou (1994) and Leite da Silva et al. (1985) are just two examples of these enhancements when considering network outages and operation constraints used to constrain the power flow results. Finally, Zhang and Lee (2004) describe a new approach to the probabilistic power flow problem to build branch flow distributions to be used in transmission investment planning problems.

Apart from probabilistic power flow models, the literature also includes a few references addressing the integration of probabilistic data in optimal power flow (OPF) models, as it is the case of El-Hawary and Mbamalu (1989), Madrigal et al. (1998), and Verbic and Canizares (2006).

Departing from the idea in Allan and Leite da Silva (1981) in terms of using multiple points to linearize the power flow equations, in Dimitrivski and Tomsovic (2004) it was presented, for the first time, the boundary load flow algorithm as a methodology to integrate load uncertainties in power flow studies. However, following Dimitrivski and Tomsovic (2004), the method is computationally intensive and could occasionally fail in getting the correct solution if the function exhibits extreme changes.

Apart from the data having probabilistic character, there are situations in which the uncertainty has no random nature but it derives, for instance, from the incomplete characterization of the phenomenon under analysis or from insufficient information required to build probability distributions, as it happens in very low frequency phenomena. In other cases, uncertainty is related with vagueness in the sense that human language has an intrinsic subjective nature. In this sense, expressions such as "larger than," "close to," "more or less," or "approximately" are inherently vague and their use does not reflect an historic average of past values, but it reflects a subjective evaluation of each user.

Probability theory is not fully adequate to model this type of uncertainty. Since the 1980s, fuzzy set models have been developed and applied to power systems to provide a new framework to model the vague or the ill-defined nature of some phenomena, namely fuzzy power flows, FOPF, risk analysis and reinforcement strategies (Saraiva and Miranda 1993), generation planning (Muela et al. 2007), reliability models, fuzzy reactive power control, fuzzy dispatch, fuzzy clustering of load curves, and transient or steady state stability evaluation.

Regarding power flow problems, the first DC and AC models admitting that at least one generation and demand are modeled by fuzzy numbers are described in Miranda et al. (1990). As a result, voltages and branch flows are now modeled by fuzzy numbers displaying their possible behavior under the specified uncertainties.

Miranda and Saraiva (1992) and Saraiva et al. (1994) give a step forward in this area because they describe a Fuzzy DC OPF model admitting that, at least, one load is represented by a fuzzy number. As a result, generations, branch flows, and power not supplied (PNS) (representing the power that the system cannot attend) display fuzzy representations translating data uncertainty. Afterwards, this approach was integrated in a Monte Carlo simulation (Saraiva et al. 1996) to obtain estimates of the expected value of PNS reflecting fuzzy loads and the reliability life cycle of equipments modeled by probability based approaches. In this sense, the model in Saraiva et al. (1996) has a hybrid nature when aggregating fuzzy and probabilistic models. This Fuzzy DC OPF model was also used to identify the most adequate expansion plan so that the risk of not being able to meet the demand gets reduced while accommodating the inherent uncertainty (Saraiva and Miranda 1996). Finally, Gomes and Saraiva (2007) present the basic concepts related with the simultaneous modeling of generation cost and demand uncertainties in OPF studies.

As we mentioned previously, the interval arithmetic model (Wang and Alvarado 1992) represents another class of models within this field. In this contribution loads are represented by arithmetic intervals and the power flow problem is solved using linearized expressions. Since uncertainties are represented by intervals, it is not possible to consider qualitative information on the phenomenon under analysis or even knowledge related with some kind of repetitive events. To overcome this conceptual problem, Chaturvedi et al. (2006) presents a distribution power flow model that uses a multi-point linearized procedure to find bounded intervals of loads that vary according to a probability distribution.

3 Fuzzy Set Basics

This section details some basic concepts of fuzzy set theory that are essential to fully understand the next sections. In the first place, a fuzzy set $\tilde{A}$ is a set of ordered pairs (1) in which the first element, x_1, is an element of the universe X under analysis and the second is the membership degree of that element to the fuzzy set, $\mu_{\tilde{A}}(x_1)$. These values measure the degree of compatibility of the elements of X with the proposition defining the fuzzy set meaning that $\mu_{\tilde{A}}(x)$ corresponds to a membership function that assigns a membership degree in [0.0; 1.0] to each element x.

$$\tilde{A} = \left\{(x_1; \mu_{\tilde{A}}(x_1)), x_1 \in X\right\} \tag{1}$$

Among all possible classes of fuzzy sets, a fuzzy number $\tilde{A}$ is a fuzzy set that is convex, and it is defined on the real line R such that its membership function is piecewise continuous. As an example, Fig. 1 reproduces the membership function of a trapezoidal fuzzy number. Its membership degree is maximum in $[A_2; A_3]$ and decreases from 1.0 to 0.0 from A_2 to A_1 and from A_3 to A_4. Once one knows that a fuzzy set belongs to this particular class, the shape of the membership function is known and so such a number is uniquely defined by the values A_1, A_2, A_3, and A_4. Therefore, a trapezoidal fuzzy number is usually denoted as $(A_1; A_2; A_3; A_4)$.

An α-level set or an α-cut of a fuzzy set $\tilde{A}$ is defined as the hard set A_α obtained from $\tilde{A}$ for each $\alpha \in [0.0; 1.0]$ according to (2). Taking the number in Fig. 1 as an example, the 0.0-cut is the interval $[A_1; A_4]$ and the 1.0-cut is given by the interval $[A_2; A_3]$. The central value of a fuzzy number is defined as the mean value of its 1.0-cut. Regarding the trapezoidal fuzzy number in Fig. 1, the central value of $\tilde{A}$, A^{ctr} is given by (3).

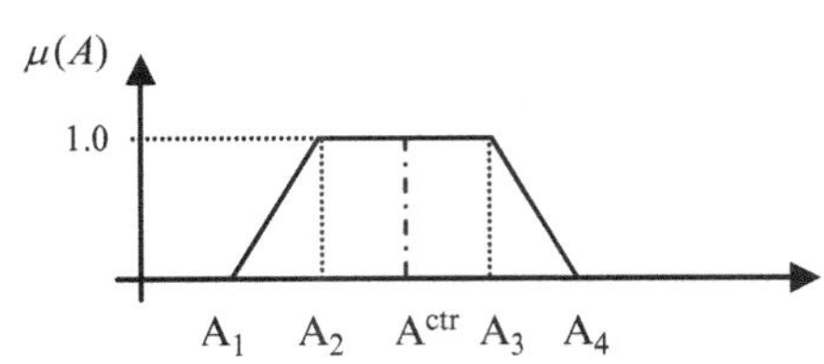

Fig. 1 Trapezoidal fuzzy number

$$A_\alpha = \{x_1 \in X : \mu_{\tilde{A}}(x_1) \geq \alpha\} \tag{2}$$

$$A^{ctr} = (A_2 + A_3)/2 \tag{3}$$

Finally, let us consider two fuzzy sets $\tilde{A}$ and $\tilde{B}$ represented by their membership functions $\mu_{\tilde{A}}(x)$ and $\mu_{\tilde{B}}(x)$. The literature describes several operators to perform the union of $\tilde{A}$ and $\tilde{B}$. The original contribution of Lotfi Zadeh (1965) defines the fuzzy union operator using (4). The membership grade of an element x in $\tilde{C} = \tilde{A} \cup \tilde{B}$ corresponds to the maximum of the membership grades of x in $\tilde{A}$ and in $\tilde{B}$. This ensures that the fuzzy set $\tilde{C}$ has the largest membership degrees when comparing them with the grades in $\tilde{A}$ and $\tilde{B}$.

$$\mu_{\tilde{C}}(x) = \max(\mu_{\tilde{A}}(x), \mu_{\tilde{B}}(x)) \tag{4}$$

4 Fuzzy Optimal Power Flow

A FOPF study can be defined as an optimization problem aiming at identifying the most adequate generation strategy driven by a generation cost function, and admitting that, at least, one load is represented by a fuzzy number. The original model described in Saraiva et al. (1994, 1996) uses the DC approach to represent the operation conditions of the network and includes power flow and generation limit constraints. In the first step, the algorithm developed to reflect load uncertainties in the results of this optimization exercise solves the deterministic DC-OPF problem (5–9). This problem is run considering that the fuzzy load in node k is represented by its central value, Pl_k^{ctr}, as defined in Sect. 3.

$$\min \mathrm{Z} = \sum c_k.Pg_k + G.\sum PNS_k \tag{5}$$

$$\text{s.t.} \sum Pg_k + \sum PNS_k = \sum Pl_k^{ctr} \tag{6}$$

$$Pg_k^{\min} \leq Pg_k \leq Pg_k^{\max} \tag{7}$$

$$PNS_k \leq Pl_k^{ctr} \tag{8}$$

$$P_b^{\min} \leq \sum a_{bk}.(Pg_k + PNS_k - Pl_k^{ctr}) \leq P_b^{\max} \tag{9}$$

In this model, Pg_k is the generation in bus k, c_k is the corresponding variable cost, PNS_k is the power not supplied in bus k, and G is the penalization specified for the power not supplied. On the other hand, $Pg_k^{\min}$, $Pg_k^{\max}$, $P_b^{\min}$, and $P_b^{\max}$ are the generation and branch flow limits, a_{bk} is the DC sensitivity coefficient of the flow in branch b regarding the injected power in bus k.

Once a feasible and optimal solution to this problem is identified, we integrate uncertainty parameters, λ_k, associated to each fuzzy load in the problem, leading to the condensed multiparametric problem (10–12).

$$\min Z = c^t.X \tag{10}$$

$$\text{s.t.A.X} = \text{b} + \text{b}'(\lambda_\text{k}) \tag{11}$$

$$\lambda_{k1} \leq \lambda_k \leq \lambda_{k4} \tag{12}$$

In this model, A is the coefficient matrix; X is the vector of the decision variables, namely generations; b is the right hand side vector; $b'(\lambda_k)$ is a vector of linear expressions on λ_k; and λ_{kI} and λ_{k4} are the minimum and maximum values of the load uncertainty on bus k. This means that they correspond to the extreme values of the 0.0-cut of the load membership function in bus k.

Constraints (12) define a hypervolume determined by the load uncertainty ranges, and in fact it encloses an infinite number of load combinations. Since the optimal solution of the crisp problem may not be feasible for all load combinations in this hypervolume, the algorithm proceeds identifying some vertices of this hypervolume. As a general rule, these vertices correspond to feasible solutions leading to maximum or minimum values of each basic variable in the initial solution or, alternatively, to vertices associated with unfeasible solutions. When this identification process is over, we have identified all vertices around the initial central value leading to individual extreme values of the basic variables or, instead, to non-feasible solutions. Afterwards, the algorithm proceeds by running a set of two consecutive parametric DC-OPF studies for each identified vertex. To clarify this procedure, let us admit a two fuzzy load system defined by (13) and (14) using the central values and a fuzzy deviation. For this case, Fig. 2 represents the rectangles enclosing all possible load combinations for the 0.0 and 1.0-cuts.

$$P_{l1} = P_{l1}^{ctr} + (\Delta_{11}; \Delta_{12}; \Delta_{13}; \Delta_{14})\,\text{MW} \tag{13}$$

$$P_{l2} = P_{l2}^{ctr} + (\Delta_{21}; \Delta_{22}; \Delta_{23}; \Delta_{24})\,\text{MW} \tag{14}$$

Taking Fig. 2 as reference and the vertex Y as an illustration, the first parametric study analyses load combinations along the segment OX considering that points O and X correspond, respectively, to $\alpha = 1.0$ and $\alpha = 0.0$ in the linear parametric optimization problem (15–17).

$$\min Z = c^t.X \tag{15}$$

$$\text{s.t. A.X} = \text{b} + \text{b}'(1.0 - \alpha) \tag{16}$$

$$0.0 \leq \alpha \leq 1.0 \tag{17}$$

When point X is reached, the optimal and feasible solution is given by (18):

$$\text{X}_\text{opt} = \text{B}^{-1}.(\text{b} + \text{b}'(1.0 - \alpha)) \tag{18}$$

The second parametric study is used to analyze load combinations along the segment XY, and its starting point corresponds to the solution obtained for point X in the first study. In this second study, points X and Y correspond to $\alpha = 1.0$ and $\alpha = 0.0$, respectively, and it is possible to obtain the solution for Y, given by (19).

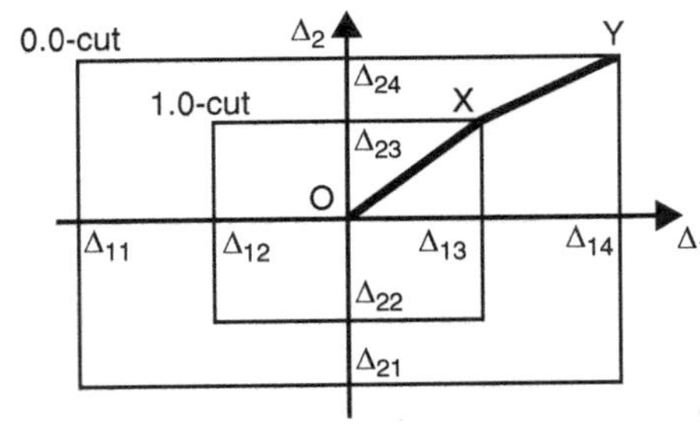

Fig. 2 0.0 and 1.0 cuts for a system with two trapezoidal fuzzy loads

$$X_{opt} = B^{-1}.(b + b' + b''(1.0 - \alpha)) \tag{19}$$

These parametric studies are formulated in such a way that it is possible to guarantee that the values obtained in the first one are assigned to 1.0 membership degree while the values obtained with the second one have a membership degree that corresponds to the value of α, and so it decreases from 1.0 to 0.0 membership degree (points X and Y, respectively). For each variable under analysis, it is therefore possible to build a membership function. Each of these functions is a partial one in the sense that each of them results from the analysis of only some combinations of the fuzzy loads. This means that when running the parametric problems to move the solution from O to X and then from X to Y, we are only analyzing the load combinations on the segments OX and XY. The final step corresponds to aggregate the partial membership functions obtained for each variable (branch flows, generation, and PNS) for each analyzed vertex. This aggregation is performed using the fuzzy union operator so that it is possible to capture the widest possible behavior of each branch flow, generation or PNS, underlying the specified load uncertainties. Although the literature describes many other operators, the fuzzy union operator as defined in Sect. 3 ensures that the final results are the widest possible, thus representing the possible operation of the system, that is, what may happen given the specified uncertainties.

5 New Fuzzy Optimal Power Flow Model

5.1 General Aspects

The original FOPF model described in Sect. 4 simplified the multiparametric problem (10)–(12) that described in an exact way the impact of load uncertainties in a number of parametric problems. This simplification, although convenient from the point of view of computational efficiency, would mean not to analyze in a systematic way all load combinations in the hypervolume described by (12), but only the ones lying on the segments departing from the central load combination (point O in Fig. 2) to the identified vertices (as point Y in Fig. 2). As a consequence, the final results for generations, branch flows, or PNS could be narrower than what they should be, that is, we were eventually not capturing the widest possible behavior of the system when reflecting load uncertainties. Apart from the above problem, we considered that it would be important to extend the original concept by translating to the results not only load uncertainties but also generation cost uncertainties, namely in view of the current volatility of fuel prices and also considering the development of market mechanisms in the sector.

As a result, the NFOPF enables obtaining more accurate solutions, as it adopts linear multiparametric optimization techniques and it allows addressing in a simultaneous way load and generation cost uncertainties. Considering load uncertainties as an example, the application of multiparametric techniques lead to the identification of a number of critical regions that effectively cover the uncertainty space corresponding to the hypervolume defined by (12). This ultimately means that this is a more realistic and accurate approach to address uncertainties in power system operation and planning.

The algorithms used to solve linear multiparametric problems were originally proposed in Gal (1979). Starting at the initial optimal and feasible solution of the deterministic optimization problem stated by (5)–(9), these algorithms identify critical regions in the uncertainty space, considering that their union covers the entire uncertainty space. When running this identification step, we admit that the problem integrates uncertainty parameters in the right hand side vector, λ_k, to model load uncertainties and uncertainty parameters in the cost function, Δ_k, to model generation cost uncertainties.

From a mathematical point of view, let B be an optimal and feasible basis, ρ the index for the corresponding set of basic variables, A the columns of the nonbasic variables in the Simplex tableau, C_0 the cost vector of the basic variables, and C^T the cost vector of the nonbasic variables. While analyzing load uncertainties, the solution obtained for the initial deterministic problem can lose its feasibility, that is, the set of constraints (20) expressing the feasibility condition can be violated. Similarly, when analyzing generation cost uncertainties, the solution obtained for the initial deterministic problem can lose its optimality, that is, the set of constraints (21) expressing the optimality condition can be violated.

$$B_\rho^{-1}.b(\lambda_k) = B_\rho^{-1}(b + b'(\lambda_k)) \geq 0 \tag{20}$$

$$C^T(\Delta_k)\text{-}C_0^T.B_\rho^{-1}.A = (c + c'(\Delta_k))\text{-}C_0^T.B_\rho^{-1}A \geq 0 \tag{21}$$

Using these two conditions, the solution algorithm starts at the optimal and feasible solution of the initial deterministic DC-OPF problem and it proceeds to find the set of other optimal and feasible solutions, provided they are valid in some region of the uncertainty space. These regions are called critical regions and each of them corresponds to a region in the uncertainty space where there is a basis B that is optimal and feasible. Their identification is conducted by pivoting over the initial basis as well as over all the new ones identified during the search process.

Since the dual solution does not depend on λ_k for right hand side parameterization, a critical region can be uniquely defined by the conditions in (20). Similarly, since the primal solution does not depend on Δ_k for cost parameterization, a critical region can be defined by the conditions given by (21). Apart from these conceptual aspects, two optimal and feasible basis, B_1 and B_2, are considered neighbor ones if and only if one can pass from B_1 to B_2 performing one dual pivot step in case of right hand side parameterization, one primal pivot step in case of cost vector parameterization, or one step of each type if we are addressing the simultaneous parameterization of cost and load uncertainties.

As a final comment, the ultimate objective to attain when solving a linear multiparametric problem is to find all possible optimal solutions, their corresponding optimal values, and critical regions, which can be defined as a closed nonempty polyhedron. These regions correspond to a set of linear inequalities in λ_k, Δ_k, or both. Mathematically, these constraints can be expressed as an equivalent set of nonredundant constraints, which in turn can be identified through a nonredundancy test for linear inequalities, as the one proposed by Gal (1979).

5.2 Integration of Load Uncertainties

The algorithm developed to solve the NFOPF problem is detailed in Fig. 3. It starts with the solution of the deterministic Fuzzy DC-OPF problem (5–9) considering the central values of the load fuzzy numbers. It should be mentioned that this formulation is general in the sense that it admits that some loads are represented by fuzzy numbers while the value of some others is deterministic. In this case, each fuzzy load is associated with a parameter λ_k modeling its uncertainty while the value of deterministic loads is fixed. After obtaining an optimal and feasible solution for this deterministic problem, the algorithm integrates in this initial problem the parameters used to model load uncertainties. This leads to the linear multiparametric optimization problem (22–26).

$$\min \mathrm{Z} = \sum \mathrm{c_k}.Pg_k + G.\sum PNS_k \tag{22}$$

$$\text{s.t.} \sum Pg_k + \sum PNS_k = \sum Pl_k^{ctr} + \sum \lambda_k \tag{23}$$

$$Pg_k^{\min} \leq Pg_k \leq Pg_k^{\max} \tag{24}$$

$$PNS_k \leq Pl_k^{ctr} + \lambda_k \tag{25}$$

$$P_b^{\min} \leq \sum a_{bk}.\left(Pg_k + PNS_k - \left(Pl_k^{ctr} + \lambda_k\right)\right) \leq P_b^{\max} \tag{26}$$

When solving this problem, it is possible that the optimal and feasible basis identified for the initial deterministic DC-OPF problem is no longer feasible in some regions of the uncertainty space represented by the hypervolume defined by (12). This means that for some combinations of the parameters λ_k, the optimal and feasible basis of the deterministic DC-OPF problem leads to negative values of some basic variables.

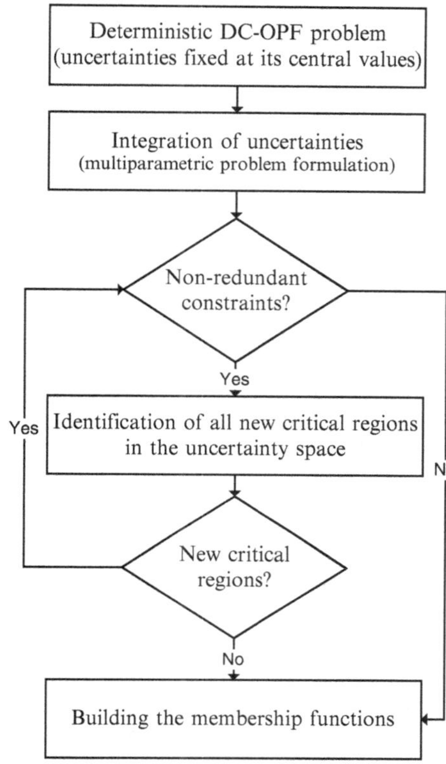

Fig. 3 Solution algorithm of the NFOPF integrating active load uncertainties

Following the algorithm in Fig. 3, we now use the feasibility condition (20) to identify new critical regions. Each of these regions is represented by the maximum and minimum excursion of each load parameter together with the set of nonredundant constraints obtained from the inequalities associated with the feasibility condition (20). If there are no nonredundant constraints, the algorithm stops. Otherwise, it performs a dual pivoting over the initial optimal and feasible basis to identify new critical regions. This process is repeated until no nonredundant constraints exist or until all identified critical regions correspond to already known ones. When this process is completed, all the uncertainty space is covered, and for all identified critical regions there is a feasible and optimal basis B of the problem (22–26).

To illustrate this procedure, let us consider Fig. 4. It displays the rectangles related with the 0.0 and 1.0-cuts for a two trapezoidal fuzzy load system. Lines a and b represent constraints, for instance, related with branch flow or generator capacity limits, point O represents the optimal and feasible solution of the initial deterministic DC-OPF problem, and the dashed lines define the rectangle representing the i th cut of the fuzzy load membership functions. In this case, the constraint represented by line a determines a basis change admitting that the process started in point O, that is, in region R_j. This means that the feasibility condition (20) including the load vector uncertainties is used to define the regions R_i and R_j.

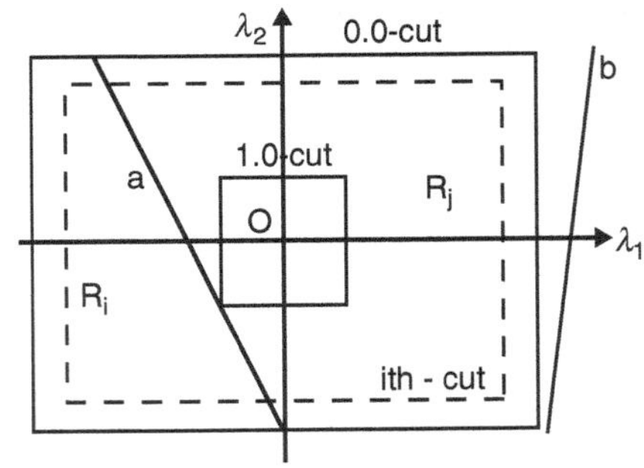

Fig. 4 Critical regions in the uncertainty space for a two fuzzy load system

In a systematic way, all critical regions are obtained by performing a dual pivoting over each of the optimal and feasible already identified basis.

To build the membership functions of the output variables (generations, branch flows, and PNS), it is important to identify their extreme possible behavior reflecting data uncertainty. The adoption of a linear formulation for the OPF problem, a DC OPF approach, ensures that the behavior of each output variable v is represented by a linear expression in function of the uncertain parameters. This means that the behavior of v is described by a linear function of the uncertainty parameters λ_1 and λ_2, say $v(\lambda_1, \lambda_2)$, in each identified critical region. If we want to capture the possible behavior of $v(\lambda_1, \lambda_2)$ in a critical region, we have to compute its minimum and maximum values in that region. This corresponds to solving the problem (27)–(30) once in terms of minimization and once as a maximization problem:

$$\min / \max \ \mathrm{f} = v(\lambda_1, \lambda_2) \tag{27}$$

$$\text{s.t.} \ k_{1i}.\lambda_1 + k_{2i}.\lambda_2 \leq b_i \tag{28}$$

$$\lambda_1^{\min\ ith-cut} \leq \lambda_1 \leq \lambda_1^{\max\ ith-cut} \tag{29}$$

$$\lambda_2^{\min\ ith-cut} \leq \lambda_2 \leq \lambda_2^{\max\ ith-cut} \tag{30}$$

This problem should be formulated for some α-cuts of the specified load uncertainties, meaning that each fuzzy load is discretized in a number of intervals, each one associated with an α-cut. Once this discretization is done, we reflect the uncertainties now represented by α-cuts in the output variables of the problem. Computation of the minimum and maximum values of v corresponds to minimizing and maximizing the linear expression describing v subjected to the constraints modeling the nonredundant conditions (28), where k_{1i} and k_{2i} are real numbers, together with the possible ranges of the input uncertainties regarding the ith cut under analysis (29) and (30). Constraints (28) define the critical region under analysis and they result from the process of identification of these regions, that is, from the application of the feasibility condition (20). After minimizing and maximizing v for the selected cuts, it is possible to build the membership function of v in the critical region in analysis. Once all regions are analyzed, the final membership function of v is obtained applying the fuzzy union operator (as defined in Sect. 3) to the partial membership functions obtained for v.

5.3 Integration of Generation Cost Uncertainties

In this case, the linear multiparametric optimization problem (31–35) integrates in the cost vector C^T the parameters Δ_k to model generation cost uncertainties:

$$\min \mathrm{Z} = \sum \mathrm{c_k}(\Delta_k).Pg_k + G.\sum PNS_k \tag{31}$$

$$\text{s.t.} \sum Pg_k + \sum PNS_k = \sum Pl_k^{ctr} \tag{32}$$

$$Pg_k^{\min} \leq Pg_k \leq Pg_k^{\max} \tag{33}$$

$$PNS_k \leq Pl_k^{ctr} \tag{34}$$

$$P_b^{\min} \leq \sum a_{bk}.\left(Pg_k + PNS_k - Pl_k^{ctr}\right) \leq P_b^{\max} \tag{35}$$

To identify nonredundant constraints leading to the definition of the critical regions in the uncertainty space, we will now use the optimality condition (21) in terms of the parameters Δ_k. In this case, the critical regions are identified performing a primal pivoting over the initial optimal and feasible basis as well as over all new optimal and feasible basis identified along the search procedure.

When solving problems (31–35), we should recall that each variable (generations, branch flows, and PNS) is constant for all generation cost combinations inside each identified critical region. Therefore, the values of these output variables will change only if there is a basis change, which means moving from one critical region to another one.

Accordingly, for each critical region, the partial membership function of any variable is built considering the nonredundant inequalities defining that region and the maximum membership degree of that output variable. This is simply done by solving the linear system formed by the inequalities that define each critical region to check if at least one point of a given cut level belongs to the critical region under analysis. Once all these pairs are obtained, they are aggregated using the fuzzy union operator to obtain the final membership function of the output variable under analysis. Therefore, when addressing generation cost uncertainties, the possible behavior of the output variables is represented by pairing each of them, including the value of the output variable and the corresponding membership degree. This result corresponds, in some way, to the dual of the type of results obtained for load uncertainties. This should not be an entire surprise since when addressing load uncertainties we are multiparameterizing the right hand side vector, while for generation cost uncertainties we are running a cost function multiparameterization study. From a conceptual point of view, this means that these two problems correspond to a primal and dual version of the same problem.

5.4 Simultaneous Integration of Cost and Load Uncertainties

The linear multiparametric optimization problem (36–40) includes parameters to model generation cost uncertainties, Δ_k, and parameters related with load uncertainties, λ_k, so that we can address, simultaneously, the impact of these uncertainties in the output variables:

$$\min \mathrm{Z} = \sum \mathrm{c_k}(\Delta_k).Pg_k + G.\sum PNS_k \tag{36}$$

$$\text{s.t.} \sum Pg_k + \sum PNS_k = \sum Pl_k^{ctr} + \sum \lambda_k \tag{37}$$

$$Pg_k^{\min} \leq Pg_k \leq Pg_k^{\max} \tag{38}$$

$$PNS_k \leq Pl_k^{ctr} + \lambda_k \tag{39}$$

$$P_b^{\min} \leq \sum a_{bk}.\left(Pg_k + PNS_k - \left(Pl_k^{ctr} + \lambda_k\right)\right) \leq P_b^{\max} \tag{40}$$

The presence of the parameters Δ_k and λ_k will eventually turn the optimal and feasible basis identified for the initial DC-OPF problem as unfeasible or not optimal in some regions of the uncertainty space. Thus a basis change means that we identify critical regions defined by sets of nonredundant constraints related both with the feasibility (20) and optimality (21) conditions. These constraints are linear functions of the uncertainty parameters λ_k and Δ_k.

Once all critical regions are identified, the algorithm proceeds with the construction of the membership functions of the final results. To do this, we once again recognize that in each critical region, the behavior of each variable is expressed by a linear expression, and so for each of them and for each critical region, we solve optimization problems as (27–30) to get the widest possible behavior of that variable in that region. In this case, however, the number of constraints is larger because we are now considering the nonredundant inequalities coming from the application of both (20) and (21).

5.5 Integration of Active Losses

Active losses in branch b from node i to node j are calculated in an exact way by expression (41), which depends on voltage phases and magnitudes (θ_i, θ_j, V_i, and V_j) and on the branch conductance g_{ij}. Considering that voltage magnitudes are approximated to 1.0 pu in the DC model, we obtain the approximated expression (42) (Saraiva 1999).

$$Loss_{ij} = g_{ij}.\left(V_i^2 + V_j^2 - 2.V_i.V_j.\cos\theta_{ij}\right) \tag{41}$$

$$Loss_{ij} = 2.g_{ij}.(1 - \cos\theta_{ij}) \tag{42}$$

A simple way of integrating this estimate of active losses in the crisp formulation (5–9) is detailed in the following iterative process:

1. Perform the deterministic DC OPF study formulated by (5–9)
2. Compute the nodal voltage phases, according to the DC model
3. Compute an estimate of the active losses in each branch of the system

4. Add half of the active losses estimated for each branch to the load connected to the corresponding extreme buses
5. Perform a new deterministic DC OPF study to update the generation strategy
6. Compute the nodal voltage phases according with the DC model
7. Finish if the difference between every voltage phase in two successive iterations is smaller than a specified value. Otherwise return to step 3

To take into account the effect of the active losses on the results of the algorithms presented in Sects. 5.2, 5.3, and 5.4, for each extreme point of the identified critical regions we must run the algorithm just detailed. As it will be shown in Sect. 6, in general the impact of active losses is reduced, and so this procedure typically originates small deviations regarding the initial results.

5.6 Computation of Nodal Marginal Prices

Marginal pricing is broadly recognized as the core approach to the economic evaluation of generation and transmission services. However, marginal pricing has several drawbacks. In the first place, tariff schemes based on short-term marginal prices may lead to perverse effects since, for example, more frequent transmission congestion could enlarge the nodal marginal price dispersion and so increase the marginal remuneration or congestion rent. Second, pure short-term nodal prices do not take into account transmission investment costs, and so it would not be possible to recover these costs. This under recovery problem is well known in the literature and it is termed as the revenue reconciliation problem (Leão and Saraiva 2003). Third, since marginal prices depend on several factors, they are typically very volatile. Therefore, following Saraiva (1999), their computation should internalize several aspects as follows:

- Uncertainties clearly affect load values especially in market environment. Therefore, load uncertainties and their consequences on nodal marginal prices are certainly a major issue that should be investigated.
- Uncertainties also affect generation costs. Changes on generation costs will originate changes in the dispatch policy and thus in nodal marginal prices.
- Nodal marginal prices depend on the system components that are available in instant t. Therefore, the consequence of the nonideal nature of the system components, that is, their reliability on nodal marginal prices should be investigated.

The nodal marginal price volatility makes it difficult to predict the marginal remuneration that could be recovered by the transmission provider. In this sense, several works were developed to integrate uncertainties in the nodal marginal price computation. In general, these methods treat load uncertainties using probabilistic models (Baughman Martin and Lee Walter 1992; Rivier and Pérez-Arriaga 1993) or fuzzy approaches (Certo and Saraiva 2001; Jesus and Leão 2004; Leão and Saraiva 2003; Saraiva 1999).

The short-term nodal marginal price in bus k can be defined as the impact on the cost function of a short-term operation planning problem regarding a unit variation of the load in node k. Let us consider the crisp DC OPF model (5–9), including the branch loss estimate as described in Sect. 5.5. According to Saraiva (1999), the nodal marginal price in node k is given by (43):

$$\rho_k = \gamma + \gamma.\frac{\partial Losses}{\partial P_{Lk}} + \sigma_k - \sum_{all\ branches} \eta_b.\frac{\partial P_b}{\partial P_{Lk}} \tag{43}$$

In this expression,

- γ represents the dual variable of the generation/load balance (6).
- The second term represents the impact on the cost function, Z, from varying branch losses in the whole network due a variation of the load in bus k.
- The third term represents the contribution from constrains (8) that are eventually on their limits. σ_k is the dual variable of the constraint in node k.
- The fourth term represents the contribution to the cost function, Z, from each branch flow constraint that is on its limit. In this expression, η_b represents the dual variable of the corresponding constraint and the derivative of the branch flow in branch b, P_b, regarding the load in bus k, P_{Lk}, is the symmetric of the corresponding sensitivity coefficient.

The algorithm developed to compute the nodal marginal prices membership functions comprises three distinct stages. In the first stage, the algorithm determines the maximum and minimum possible values of all variables in each cut level. Once this initial stage is completed, the algorithm can evolve to include the impact of the active transmission losses for each identified operating point in each cut level or this impact can be neglected. At last, the algorithm determines the nodal marginal prices using (43). To build the membership function of nodal marginal prices and similarly to what was described in Sects. 5.2, 5.3, and 5.4, one also obtains partial membership functions of nodal marginal prices. When they are all known, the final membership function is obtained applying the fuzzy union operator to all of them.

5.7 Final Remarks

Regarding the expected results, it should be mentioned that three distinct situations can occur:

- When modeling only load uncertainties, each variable in each critical region is represented by a linear expression. Therefore, generations, branch flows, and PNS are described by linear (at least by segments) membership functions. In a different way, since nodal marginal prices are related with the dual variables of this optimization problem, their membership functions are described by pairs of price/membership value.

- When modeling only generation cost uncertainties, each variable is constant inside each critical region. This means that their corresponding membership functions are described by pairs of power/membership value. Since the nodal marginal prices are related with the dual variables of the original problem, in this case the membership functions of the nodal marginal prices are described by linear, at least by segments, membership functions.
- Finally, when considering both load and generation cost uncertainties, primal and dual variables of the original problem are represented by linear expressions of the parameters used to model uncertainties. As a result, generation, branch flows, PNS, and nodal marginal prices are described by linear (at least by segments) membership functions.

6 Case Study

6.1 Data

The algorithms described so far were used to build the membership functions of generations, branch flows, PNS, and nodal marginal prices considering a case study based on the IEEE 24 bus/38 branch test system. The original data for this system is given in Task Force of Application of Probabilistic Methods Subcommittee (1979). Regarding the data in this reference, the load was increased to 4,060.05 MW. Table 1 presents the installed capacities and the central values of the generation costs and Table 2 indicates the central values of the loads. The total installed capacity is 5,226 MW according to the data in Table 1. Branch data can be obtained from Task Force of Application of Probabilistic Methods Subcommittee (1979) considering

Table 1 Installed system capacity

Bus/Gen	Capacity (MW)	Cost (€ /MWh)	Bus/Gen	Capacity (MW)	Cost (€/MWh)
1/1	40.0	30.0	16/1	310.0	55.0
1/2	40.0	32.0	19/1	800.0	87.0
1/3	152.0	40.0	21/1	800.0	80.0
1/4	152.0	43.0	22/1	100.0	15.0
2/1	40.0	36.0	22/2	100.0	17.0
2/2	40.0	38.0	22/3	100.0	19.0
2/3	152.0	41.0	22/4	100.0	15.0
2/4	152.0	42.0	22/5	100.0	17.0
7/1	150.0	45.0	22/6	100.0	25.0
7/2	200.0	43.0	23/1	200.0	50.0
13/1	250.0	61.0	23/2	50.0	49.0
13/2	394.0	62.0	23/3	310.0	47.0
13/3	394.0	67.0	–	–	–

Table 2 Load central values

Bus	Load (MW)	Bus	Load (MW)	Bus	Load (MW)
1	220.48	9	385.82	17	0.00
2	270.80	10	216.49	18	226.76
3	3.94	11	40.00	19	265.53
4	32.67	12	10.00	20	103.92
5	105.94	13	162.45	21	50.00
6	187.65	14	262.88	22	10.00
7	218.77	15	650.36	23	0.00
8	398.09	16	225.50	24	12.00

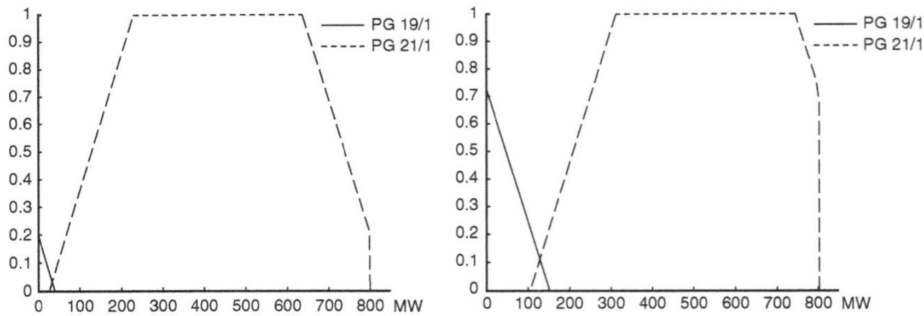

Fig. 5 Membership functions of generators 19/1 and 21/1 not considering the effect of the transmission losses (at the left) and considering this effect (at the right)

that the transformers have a capacity of 400 MW, the capacity of the branches 1 to 6 and 8 to 13 was set at 175 MW and the capacity of the remaining branches was set at 500 MW.

6.2 Results Considering Only Load Uncertainties

To test the algorithm presented in Sect. 5.2, we considered trapezoidal fuzzy numbers to model loads. These numbers have at the 0.0 level the uncertainty range from $\pm 10\%$ and at the 1.0 level from $\pm 5\%$ of its central value. Figure 5 presents the membership functions of generators 19/1 and 21/1 considering and not considering the effect of transmission losses.

As expected, the load variation determines changes on the generation of generator 21/1 as it is the marginal one. When this generator reaches its maximum capacity of 800 MW, generator 19/1 becomes the marginal one. Another important point illustrated by these functions is related with the impact of active losses. In this case, these generators have larger generation values for the same level of uncertainty due to the compensation of losses.

Figure 6 presents the nodal marginal prices at node 1 for the two situations presented previously. As referred in Sect. 5.7, since nodal marginal prices are related with the dual variables of the original problem, their membership functions are

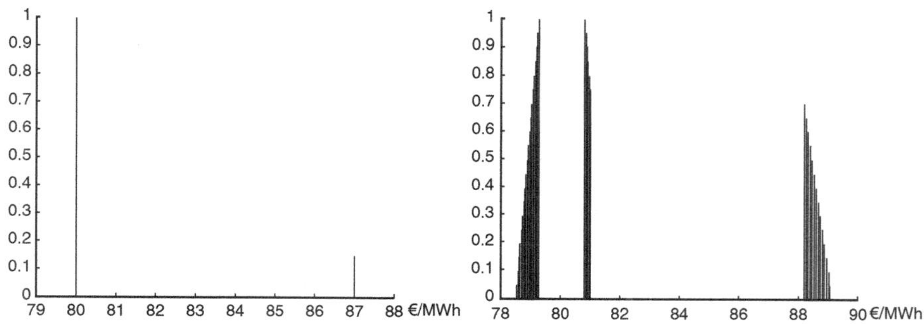

Fig. 6 Membership functions of the nodal marginal price in node 1 not considering the effect of transmission losses (at the left) and considering this effect (at the right)

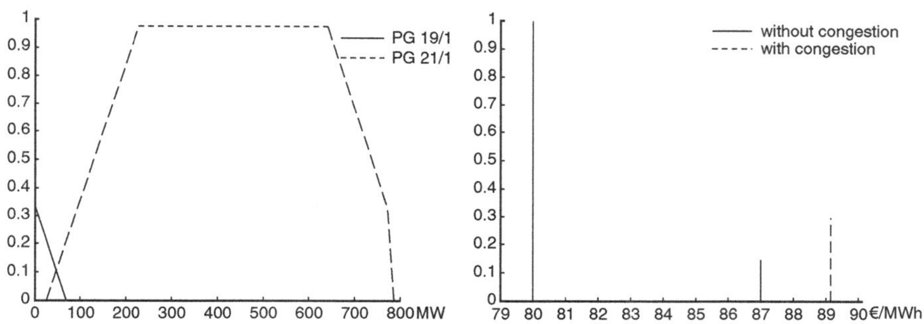

Fig. 7 Membership functions of generators 19/1 and 21/1 (at the left) and of the nodal marginal price in node 15 (at the right)

described by pairs of price/membership values. From Fig. 6 it is possible to see that, in the absence of the transmission losses, branch congestions, or PNS effect, the marginal price in node 1 equals the generation cost of the marginal generator. When the effect of active losses is considered, the nodal marginal prices increase or decrease depending on the impact of load variations in active losses. For instance, when generator 19/1 is the marginal one, an increase of the load in node 1 increases active losses and so the marginal price in node 1 also increases.

To check the impact of congested branches, the flow limit of branches 15–21 was reduced to 350 MW. As a consequence, for some combination of loads these branches get congested. Figure 7 presents the membership functions of generators 19/1 and 21/1 and of the marginal price in node 15 in this case.

Analysis of these results indicates that the congestion on branches 15–21 implies a change of the marginal prices in most of the nodes, but more particular in the ones that are closer to the congested branches. In case of node 15 in Fig. 7, the marginal price increase means that a generation increase or a load reduction in this node contributes to alleviate the congestion. The membership function of generator 21/1 in Fig. 7 reveals that now this generator does not reach its maximum limit of

800 MW, different from what was indicated in Fig. 5. This is due to the congestion of branches 15–21.

6.3 Results Considering Only Generation Cost Uncertainties

In this simulation, the trapezoidal fuzzy numbers (44–49) were used to model the cost of generators 1/1, 2/1, 7/1, 19/1, 22/2, and 23/2. In this case, Fig. 8 presents the membership functions of generators 19/1 and 21/1:

$$\mathrm{CP}_{\mathrm{G1/1}} = (26.0, 27.5, 32.5, 34.0) \tag{44}$$

$$\mathrm{CP}_{\mathrm{G2/1}} = (33.0, 34.5, 37.5, 39.0) \tag{45}$$

$$\mathrm{CP}_{\mathrm{G7/1}} = (42.0, 43.5, 46.5, 48.0) \tag{46}$$

$$\mathrm{CP}_{\mathrm{G19/1}} = (74.0, 82.0, 92.0, 100.0) \tag{47}$$

$$\mathrm{CP}_{\mathrm{G22/2}} = (14.0, 15.5, 18.5, 20.0) \tag{48}$$

$$CP_{G23/2} = (46.0, 47.5, 50.5, 52.0) \tag{49}$$

As it was mentioned in Sects. 5.3 and 5.7, in this case the membership functions of generators are represented by pairs of power/membership values. The functions in Fig. 8 reflect two different generation strategies according to the specified cost uncertainties. In fact, when the cost of generator 19/1 is smaller than 80 €/MWh (which corresponds to the cost of generator 21/1), this generator will generate 434.05 MW. For larger costs this generator will be at 0 MW and generator 21/1 will generate 434.05 MW. Figure 8 also indicates that when active losses are included, there is, in general, an increase of generation values. This increase depends on the adopted generation strategy because of their different impacts on active losses.

Figure 9 presents the nodal marginal price of node 1 in these two situations. As mentioned in Sect. 5.7, the membership functions of the nodal prices are linear, at

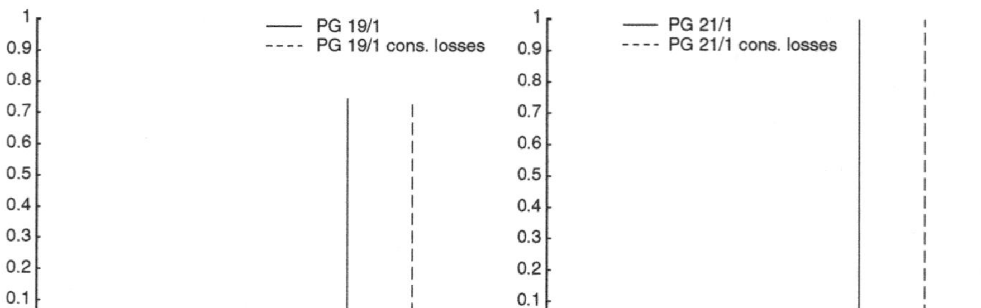

Fig. 8 Membership functions of generators 19/1 and 21/1 considering and not considering the effect of transmission losses

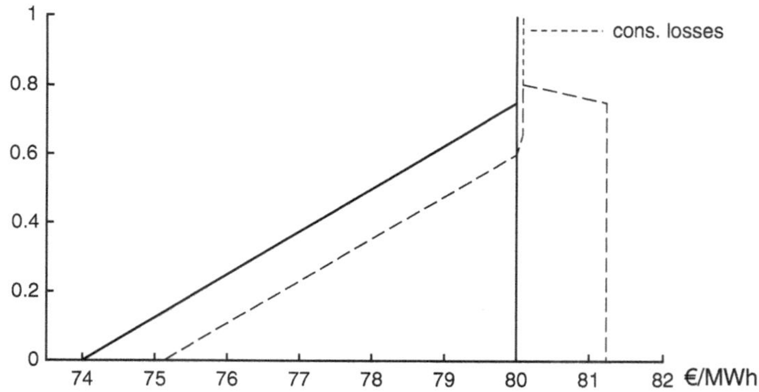

Fig. 9 Membership function of the nodal marginal price at node 1 considering and not considering the transmission losses effect

least by segments for cost uncertainties. Considering the effect of active losses, there is an increase of the nodal price in node 1. These results are in line with the ones in Fig. 6 as we observed that a load increase in node 1 contributed to increase active losses.

6.4 *Results Considering Load and Generation Cost Uncertainties*

In this simulation we considered the load uncertainties used in Sect. 6.2 and the generation cost uncertainties specified in Sect. 6.3. Figure 10 presents the membership functions of generators 13/3, 19/1, and 21/1. Simultaneously modeling load and generation cost uncertainties turns the problem and the results more complex since they reflect characteristics of both cases analyzed in Sects. 6.2 and 6.3. In fact, the different generation strategies and generation values become function not only of load uncertainties but also of the generation cost uncertainties. As a result, for some combinations of the specified uncertainties, there is congestion on the branches 8–9, 16–19, and 11–13. The congestion on branches 16–19 and 11–13 prevents generators 19/1 and 13/3 from having larger outputs. If active losses are considered as in the right side of Fig. 10, the generation values are larger for some uncertainty levels.

Figure 11 presents the membership functions of the nodal marginal prices in nodes 15 and 21. This figure indicates that these two prices are very similar. This was expected as the nodal marginal prices incorporate the dual variable of the generation/load balance equation and the impact of increasing the load regarding congestion and active losses. For nodes 15 or 21, this impact is very similar, explaining that these two membership functions are similar.

In line with what was mentioned in Sect. 6.2, Fig. 12 presents the results obtained when the limits of branches 15–21 are reduced to 350 MW. Comparing these results with the ones in Fig. 7, we can see that in this case the effect of the congestion

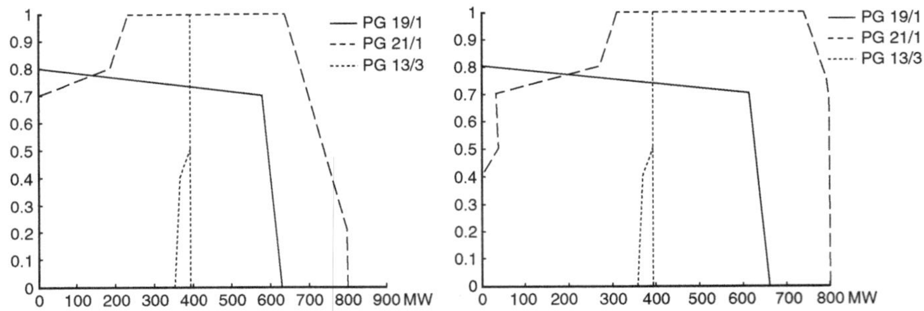

Fig. 10 Membership functions of generators 13/3, 19/1 and 21/1 when they are not considered the transmission losses effect (at the left) and when they are considered (at the right)

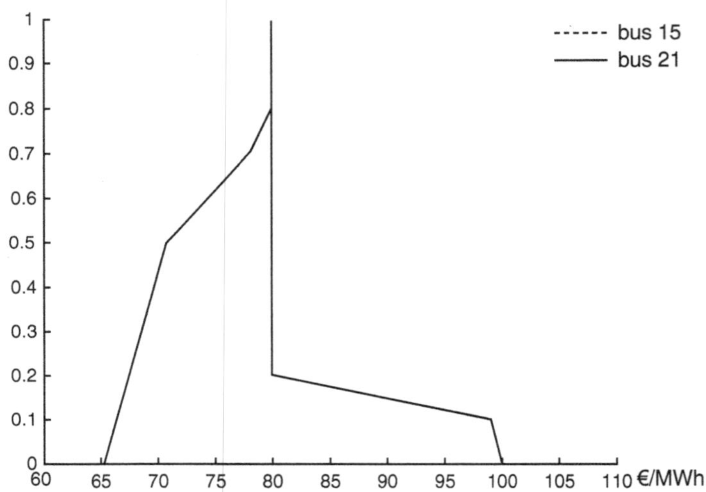

Fig. 11 Membership functions of nodal marginal prices on buses 15 and 21

on branches 15–21 is not so clear because the maximum generation of generator 19/1 is no longer determined by congested branches but, instead, by the specified generation cost uncertainties.

Finally, Fig. 13 presents the membership functions of the nodal marginal prices on nodes 15 and 21 when the limits of branches 15–21 are set to 350 MW. As expected, the membership functions of these prices are no longer similar, as they were on Fig. 11. The congestion of branches 15–21 determine a maximum price of 80 €/MWh for node 21 and a value larger than 105 €/MWh for node 15. This situation is in line with what was mentioned in Sect. 6.2, because the nodal marginal price increase in bus 15 means that a generation increase or a load reduction in this node contributes to alleviate congestion in branches 15–21.

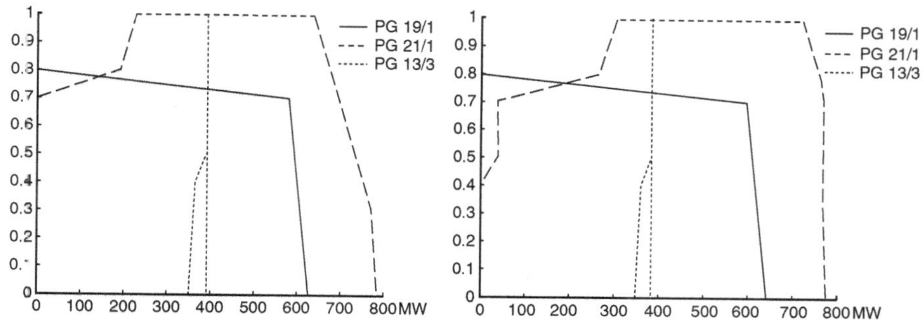

Fig. 12 Membership functions of generators 13/3, 19/1 and 21/1 when they are not considered the transmission losses effect (at left) and when they are considered (at right)

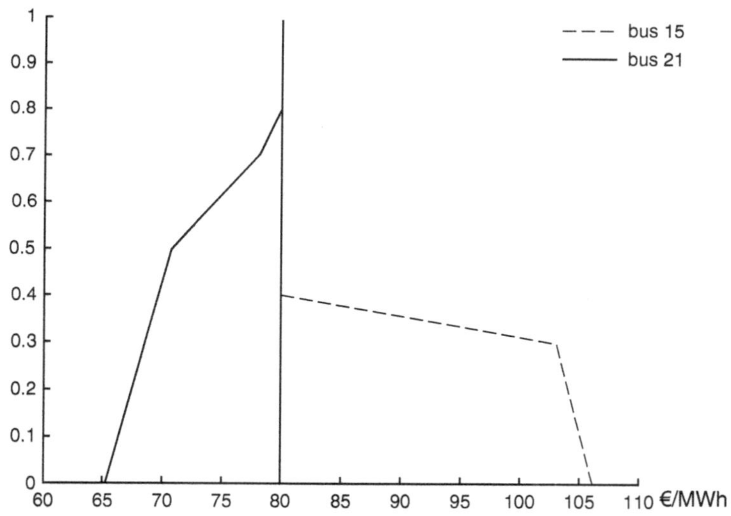

Fig. 13 Membership functions of nodal marginal prices on buses 15 and 21

7 Conclusions

In this paper we presented the most relevant concepts of the original Fuzzy DC Optimal Power Flow developed back in 1990s and also the New Fuzzy DC Optimal Power Flow model. As it was shown, this new approach can be used to model not only load uncertainties but also generation cost uncertainties or both, simultaneously. Since it uses linear multiparametric optimization techniques, it is possible to obtain more accurate results than with the original model. Additionally, the algorithm developed to integrate an estimate of active transmission losses on the results and also the approach used to compute and build the nodal marginal price membership functions were described.

This model can be very useful for a variety of agents acting in the electricity sector, given the current volatility of several inputs to this type of studies. Generation cost uncertainties, namely fuel costs, and the demand level uncertainties, related with the wide spread economic recession, are just two elements that should be internalized. Apart from their volatility, it should be mentioned that for several of these changes there is no or little history meaning that the derivation of probabilistic functions to model these recent events can be questioned. As a result, fuzzy set-based models can gain a new area of application. These models can also be used in a profitable way to get more insight on the possible system behavior by generation companies, transmission providers, or regulatory agencies to help them taking more sounded decisions at different levels, such as expansion planning, operation planning, or regulatory and tariff aspects.

Acknowledgements The first author thank Fundação para a Ciência e Tecnologia, FCT, which funded this research work through the PhD grant no. SFRH/BD/34314/2006.

References

Allan RN, Al-Shakarchi MRG (1976) Probabilistic A.C. load flow. Proc IEE 123:531–536

Allan RN, Al-Sharkarchi M (1977) Probabilistic techniques in A.C. load-flow analysis. Proc IEE 124:154–160

Allan RN, Borkowska B, Grigg CH (1974) Probabilistic analysis of power flows. Proc IEE 121:1551–1556

Allan RN, Grigg CH, Newey DA, Simmons RF (1976) Probabilistic power-flow techniques extended and applied to operation decision making. Proc IEE 123:1317–1324

Allan RN, Leite da Silva AM (1981) Probabilistic load flow using multilinearisations. Proc IEE 128:280–287

Allan RN, Leite da Silva AM, Burchett RC (1981) Evaluation methods and accuracy in probabilistic load flow solutions. IEEE Trans PAS PAS-100:2539–2546

Baughman Martin L, Lee Walter W (1992) A Monte Carlo model for calculating spot market prices of electricity. IEEE Trans Power Syst 7(2):584–590

Borkowska B (1974) Probabilistic load flow. IEEE Trans PAS PAS-93 12:752–759

Certo J, Saraiva JT (2001) Evaluation of target prices for transmission congestion contracts using a Monte Carlo accelerated approach. IEEE Porto Power Tech Conf 1:6

Chaturvedi A, Prasad K, Ranjan R (2006) Use of interval arithmetic to incorporte the uncertainty of load demand for radial distribution system analysis. IEEE Trans Power Deliv 21(2):1019–1021

Dimitrivski A, Tomsovic K (2004) Boundary load flow solutions. IEEE Trans Power Syst 19(1):348–355

El-Hawary ME, Mbamalu GAN (1989) Stochastic optimal load flow using a combined Quasi-Newton and conjugated gradient technique. Electric Power Energy Syst 11(2):85–93

Gal T (1979) Postoptimal analysis, parametric programming and related topics. McGraw-Hill, New York

Gomes BA, Saraiva JT (2007) Calculation of nodal marginal prices considering load and generation price uncertainties. Proc IEEE Lausanne Power Tech 849–854

Jesus PM, Leão TP (2004) Impact of uncertainty and elastic response of demand in short term marginal prices. 8th International Conference on Probabilistic Methods Applied to Power Systems 32–37

Karakatsanis TS, Hatziargyriou ND (1994) Probabilistic constrained load flow based on sensitivity analysis. IEEE Trans Power Syst 9:1853–1860

Leão MT, Saraiva JT (2003) Solving the revenue reconciliation problem of distribution network providers using long-term marginal prices. IEEE Trans Power Syst 18(1):339–345

Leite da Silva AM, Arienti VL (1990) Probabilistic load flow by a multilinear simulation algorithm. IEE Proc 137(Pt. C):256–262

Leite da Silva AM, Allan RN, Soares SM, Arienti VL (1985) Probabilistic load flow considering network outages. IEE Proc 132(Pt. C):139–145

Chun Lien (2005) Probabilistic load flow computation using point estimate method. IEEE Trans Power Syst 20(4):1843–1851

Madrigal M, Ponnambalam K, Quintana VH (1998) Probabilistic optimal power flow. Proc IEEE Canadian Conf Electrical Comput Eng 385–388

Miranda V, Matos MA, Saraiva JT (1990) Fuzzy load flow – new algorithms incorporating uncertain generation and load representation. Proc 10th Power Sys Comput Conf 621–625

Miranda V, Saraiva JT (1992) Fuzzy modelling of power system optimal load flow. IEEE Trans Power Syst 7:843–849

Muela E, Schweickardt G, Garcés F (2007) Fuzzy possibilistic model for medium-term power generation planning with environmental criteria. Energy Policy 35:5643–5655

Rivier M, Pérez-Arriaga IJ (1993) Computation and decomposition of spot prices for transmission pricing. 11th Power Systems Computation Conference

Saraiva JT (1999) Evaluation of the impact of load uncertainties in spot prices using fuzzy set models. 13th Power Systems Computation Conference

Saraiva JT, Miranda V (1993) Impacts in power system modelling from including fuzzy concepts in models. Proc Athens Power Tech 1:417–422

Saraiva JT, Miranda V (1996) Identification of Hedging policies in generation/transmission systems. Proc 12th Power Syst Comput Conf 2:779–785

Saraiva JT, Miranda V, Pinto LMVG (1994) Impact on some planning decisions from a fuzzy modelling of power systems. IEEE Trans Power Syst 9:819–825

Saraiva JT, Miranda V, Pinto LMVG (1996) Generation/Transmission power system reliability evaluation by Monte-Carlo simulation assuming a fuzzy load description. IEEE Trans Power Syst 11:690–695

Soderholm P (2008) The political economy of international green certificates market. Energy Policy 36:2051–2062

Task Force of Application of Probabilistic Methods Subcommittee (1979) IEEE reliability test system. IEEE Trans PAS PAS-98 2047–2054

Verbic G, Canizares CA (2006) Probabilistic optimal power flow in electricity markets based on a two-point estimate method. IEEE Trans Power Syst 21(4):1883–1893

Wang Z, Alvarado FL (1992) Interval arithmetic in power flow analysis. Trans Power Syst 7(3):1341–1349

Zadeh L (1965) Fuzzy sets. Inf Control 8:338–353

Zhang P, Lee ST (2004) Probabilistic load flow computation using the method of combined cumulants and Gram–Charlier expansion. IEEE Trans Power Syst 19(1):676–682

OBDD-Based Load Shedding Algorithm for Power Systems

Qianchuan Zhao, Xiao Li, and Da-Zhong Zheng

Abstract Load shedding has been extensively studied for years. It has been used as an important measure for emergency control. This paper shows that the problem is NP-hard and introduce a way to obtain load shedding strategies based on ordered binary decision diagram (OBDD). The advantages of our method include that priority relationships among different loads are explicitly characterized and all solutions that violate static constrains including power balance, priority, and real power flow safety are excluded. This will make search for load shedding schemes that satisfy transient stability much more efficient.

Keywords Load shedding · NP-hardness · OBDD · Partial order · Power balance · Power flow

1 Introduction

Load shedding has been studied for years for different purposes (e.g., Medicherla et al. (1979); Concordia et al. (1995); Feng et al. (1998); Fernandes et al. (2008); Faranda et al. (2007); Voumvoulakis and Hatziargyriou (2008)) and many progresses have been made. For example, in Medicherla et al. (1979), load shedding was used as a tool to avoid line overloads; in Feng et al. (1998), load shedding was considered to avoid voltage collapse. It is encouraging to notice that load shedding is a potential tool to stop large-scale system blackout by preventing cascading failure. It has been shown that a dropping (shedding) of about 0.4% of the total network load for 30 min would have prevented the cascading effects of the black out in August 1996 in the western North American grid Amin (2001). Although there are many progresses, further study of ways to find good load shedding strategies is still needed. In general, as an important emergency control measure, load shedding

Q. Zhao (✉)
Center for Intelligent and Networked Systems (CFINS), Department of Automation and TNList Lab, Tsinghua University, Beijing 100084, China
e-mail: zhaoqc@tsinghua.edu.cn

S. Rebennack et al. (eds.), *Handbook of Power Systems I*, Energy Systems,
DOI 10.1007/978-3-642-02493-1_10, © Springer-Verlag Berlin Heidelberg 2010

is challenging since there are many complicated constraints and objective functions. These constraints include, for example, load shedding only happens for interruptible loads (Faranda et al. 2007), and some loads may be more important than others and are more desirable to be kept and there are loads that are out of the control region (Fernandes et al. 2008), only a small fraction of the total load power is desired to be shed.

In this paper, we first formulate a load shedding problem as an emergency control problem satisfying three key constrains: power balance, priority among loads and no line overloading. We show that the problem is NP-hard and introduce a way to obtain load shedding strategies based on the ordered binary decision diagram (OBDD) technique (Bryant 1986, 1992), which has been used in the study of controlled islanding operation (Zhao et al. 2003; Sun et al. 2003; Sun 2004). The advantages of our method include that priority relationships among different loads are explicitly characterized and all solutions that violate static constrains including power balance, priority and real power flow safety are excluded. This will make search for load shedding schemes that satisfy additional constraints such as transient stability much more efficient.

2 Literature Review

As pointed out in Introduction, load shedding problem has been widely studied.

Load shedding principles has been reviewed in Concordia et al. (1995) as a backup protection for the system in cases that might occur, which were not covered by the power system design process. It is important that load shedding strategies are designed on the basis of mature understanding of the characteristics of the system involved, including system topology and dynamic characteristics of its generation and its load. A poorly designed load shedding program may be ineffective, or worse may exacerbate stresses on the transmission network leading to its cascading disruption. Frequency threshold, step size and number of steps, time delay and priorities and distributions are all important aspects that should be considered when designing load shedding strategies.

As reviewed in Amraee et al. (2007), the load-shedding schemes proposed so far can be classified into three categories. In the first group, the amount of load to be shed is fixed a priori. This scheme is similar to the under-frequency load-shedding scheme. Here, the minimum amount of load to be shed is determined using time simulation analysis, incorporating dynamic aspects of the instability phenomenon. Obviously, dynamic simulation is time-consuming and is suitable for special cases such as transient voltage-instability analysis. In addition, it is more difficult to incorporate a time simulation study into an optimization model. The second group tries to determine a minimum load for shedding by estimating dynamic load parameters. In this approach, results are very sensitive to dynamic load model parameters. Finally, in the third group, minimum load shedding is determined using optimal power-flow equations based on a static model of the power system. The dynamics associated with voltage stability are often slow, and hence static approaches may represent a good approximation. The basic idea behind this approach is to identify a feasible solution to the power-flow equations.

From computation method point of view, as pointed out in Concordia et al. (1995), under emergency condition, the ability to decide load shedding strategies in short time is very important and time consuming optimization methods fail to do this will unlikely be useful in practice. This observation is also behind many existing literature.

In Concordia et al. (1995), to generate efficient strategies, heuristic designs for under-frequency load shedding schedules were given based on experience.

Shandilya et al. (1993) presents a method based on the concept of local optimization trying to find a new secure operating point with minimum control actions, that is, rescheduling of generators and load shedding in the vicinity of the overloaded line, with little deviation from the preadjustment state. The problem is formulated as a nonlinear optimization problem. To make the problem tractable, the problem is approximated by assuming that only DC power flow constraints are active and only a small number of buses away from the terminal buses of the overloaded line for local optimization are considered.

Feng et al. (1998) presents an approach to determine the minimum load shedding to restore the system solvability for cases when no equilibrium point exists due to a severe contingency (such as tripping of a heavily loaded transmission line or outage of a large generating unit). Through parameterizing a given control strategy, the continuation method is applied to find the unique equilibrium point associated with the control strategy on the system post-contingency boundary. Then invariant subspace parametric sensitivity (ISPS) is used to determine the most effective control strategy so that a practical minimum load shedding can be derived.

In summary, although it is widely believed that the load shedding problem is difficult, there is no formal justification analysis on it. Since direct search based on simulation is time consuming, existing optimization methods usually introduce local sensitivity analysis to reduce the computation load.

Our contribution is to show that the load shedding problem is a NP-hard problem. This explains why the problem stand out for a long time and remain not fully solved. Different from existing optimization-based methods, we propose a new framework to attack the load shedding problem based on the OBDD computing technique.

3 Preliminary

3.1 Boolean Expressions and Their OBDDs

A Boolean expression can be used to describe a set equivalent to it (i.e., satisfying set), and evaluating the intersection of several sets is equivalent to AND operation upon several equivalent Boolean expressions. Depending on the associative law of

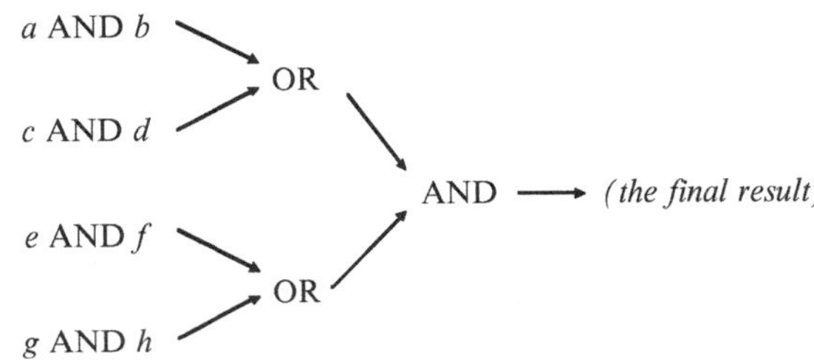

Fig. 1 Illustration of Boolean expression evaluation

Boolean expression, a complex Boolean expression can be split to many parts for separate evaluation before evaluating the final value of the whole expression. For example, there is a Boolean expression

$$(a \otimes b \oplus c \otimes d) \otimes (e \otimes f \oplus g \otimes h).$$

It can be evaluated as shown in Fig. 1. In this case, a AND b, c AND d, e AND f, and g AND h can be evaluated simultaneously. And the two OR operations can also be done simultaneously.

OBDD (Bryant 1986, 1992) is an efficient data structure to maintain complex Boolean expression, which makes it convenient to store and transfer Boolean expression. The key idea behind the OBDD algorithms is that the construction of an OBDD representing the complete solution set can usually be efficiently carried out by the combination of OBDDs representing part of the constraints. This feature benefits the inter-node transmission of Boolean expressions for intermediate evaluating results of OBDD-based parallel algorithm in a computer cluster. The BuDDy package (Jørn Lind-Nielsen's BuDDy Package online) is a powerful open source OBDD implementation, which will be used in this paper.

3.2 Signed Integers and Their OBDD Vectors

A decimal integer can be written as boolean vector according to its binary form. For example,

$$(35)_{10} = (100011)_2 \Rightarrow \begin{pmatrix} 1 \\ 0 \\ 0 \\ 0 \\ 1 \\ 1 \end{pmatrix} \tag{1}$$

Generally, an integer X can be written as

$$(X)_{10} = (b_{n-1}b_{n-2}...b_2b_1b_0)_2 \Rightarrow \begin{pmatrix} b_{n-1} \\ b_{n-2} \\ \cdots \\ b_2 \\ b_1 \\ b_0 \end{pmatrix} \tag{2}$$

Here n must be large enough for the binary form of X to contain all necessary bits to avoid overflow. If $b_{n-1}, b_{n-2}, \ldots, b_2, b_1, b_0$ are not constant values but Boolean expressions stored as OBDDs, the vector on the right of the above is an OBDD vector.

In modern computer, a negative integer $-X$ usually is expressed in its complement code $\overline{X-1}$. Here the overline means Boolean NOT operation upon every bit of the binary form for the integer. For example,

$$(-29)_{10} \Rightarrow (\overline{29-1})_{10} = (\overline{28})_{10} = (\overline{011100})_2 = (100011)_2 \Rightarrow \begin{pmatrix} 1 \\ 0 \\ 0 \\ 0 \\ 1 \\ 1 \end{pmatrix} \tag{3}$$

An 2-bit binary integer can be used to contain signed decimal integers as

Signed Decimal	−2	−1	0	+1
Binary	10	11	00	01

Similarly, 3-bit binary integer can be used to contain signed decimal integers as

Signed Decimal	−4	−3	−2	−1	0	+1	+2	+3
Binary	100	101	110	111	000	001	010	011

Generally, an n-bit binary integer can be used to contain signed decimal integers ranging from $-(2^{n-1})$ to $+(2^{n-1}-1)$. On the other hand, for unsigned decimal integers, the range is from 0 to $(2^n - 1)$. And the highest bit (the most significant bit) of the binary integer can be used to judge whether the integer is positive ($0 \Rightarrow +$ or 0) or negative($1 \Rightarrow -$).

Note that, in (1) and (3), different integers (the unsigned integer 35 and the signed integer -29) share the same binary form and OBDD vector. This means that a single binary integer or its OBDD vector can be differently interpreted for signed integer and unsigned integer.

$$\begin{pmatrix}1\\0\\0\\0\\1\\1\end{pmatrix} \Rightarrow (100011)_2 \begin{array}{c}\overset{\text{signed}}{\Longrightarrow}\\ \underset{\text{unsigned}}{\Longrightarrow}\end{array} \begin{array}{r}-29\\35\end{array}$$

One more example, 3-bit binary integer can be differently interpreted as

Binary	100	101	110	111	000	001	010	011
Signed Decimal	−4	−3	−2	−1	0	+1	+2	+3
Unsigned Decimal	4	5	6	7	0	1	2	3

The advantage of complement code is that there is no difference of addition and subtraction operations between positive and negative integers. During the course of addition or substraction operations, binary overflow is ignored. For examples,

$$\begin{aligned}
&(+1)_{10} + (+2)_{10} \Rightarrow (001)_2 + (010)_2 = (011)_2 \Rightarrow (+3)_{10}\\
&(-1)_{10} + (-2)_{10} \Rightarrow (111)_2 + (110)_2 = (\underline{1}101)_2 \Rightarrow (101)_2 \Rightarrow (-3)_{10}\\
&(-2)_{10} + (+3)_{10} \Rightarrow (110)_2 + (011)_2 = (\underline{1}001)_2 \Rightarrow (001)_2 \Rightarrow (+1)_{10}\\
&(-4)_{10} + (+2)_{10} \Rightarrow (100)_2 + (010)_2 = (110)_2 \Rightarrow (-2)_{10}
\end{aligned}$$

Here, the underlined symbol "$\underline{1}$" indicates ignored overflow. This feature benefits the sum operation to get power total of a power network.

The Buddy package (Jørn Lind-Nielsen's BuDDy Package online) provides only OBDD vector comparison operators (functions) for unsigned integers:

C/C++ function	C++ operator	Meaning (Unsigned comparison)
`bvec_gte(a, b)`	`a>=b`	greater than or equal to
`bvec_gle(a, b)`	`a<=b`	less than or equal to
`bvec_gth(a, b)`	`a>b`	greater than
`bvec_glh(a, b)`	`a<b`	less than

To handle regular computation for power network, comparison operations for signed integer need to be imported. Based on the rule of complement code described above and the implementation of comparison operations for unsigned integer, we can get the result of comparison operations for signed integer. First, we can judge the signs of the two integers by its highest bit (the most significant bit, usually abbreviated as "MSB"), then different measures can be chosen for every combination of their signs. For example, the comparison "$a \geq b$," greater than or equal to, results:

MSB of a	Sign of a	MSB of b	Sign of b	Result($a \geq b$)	Explanation
`true`	−	`true`	−	`bvec_gte(a, b)`	Both negative, more operation is required
`true`	−	`false`	+ or 0	`false`	Negative always less than non-negative
`false`	+ or 0	`true`	−	`true`	Non-negative always greater than negative
`false`	+ or 0	`false`	+ or 0	`bvec_gte(a, b)`	Both non-negative, more operation is required

Here, the unsigned comparison operators (functions) of (Jørn Lind-Nielsen's BuDDy Package online) are reused when a and b share the same sign.

3.3 *Key Constraints*

To make sure power system security after emergency control measures such as load shedding are taken, there are several key constraints that must be satisfied. They are listed below. It should be made clear that they are *necessary* instead of *sufficient* conditions for system to survive.

- *Power Balance Constraint (PBC)*: In network generation, load must be balanced up to a tolerance level, which is usually no more than small fraction (e.g., 3–5%) of the total power generation
- *Priority Constraint (PRC)*: In network, some loads are more important than others. And these loads should be ensured prior to other loads
- *Power Flow Constraint (PFC)*: All transmission lines, transistors, capacitors in a power network can only burden a special power limit. Typically, we consider only the transmission line capacity limits

4 The Load Shedding Problem(LSP)

Before we present that formulation of our load shedding problem, we would like to specify the information needed to define the three constraints listed in Sect. 3.3.

Denote L as the set of load buses whose load could be cut off. It is a subset of $\{1, \ldots, n\}$. We denote $\overline{L} = \{1, \ldots, n\} - L$ as the set of all other buses whose load and generation will definitely remain in the system. Buses with pure generation will be classified into $\overline{L}$ since here we consider only the search for load shedding schemes and will not consider turning off generators.

The priority among load buses can be naturally specified by the decision maker as a set of priority rules R. Each priority rule is given as a pair (i, j) of load buses in L. The meaning of a priority rule is that the load on bus i should always be kept if the load on bus j is kept. In this case, we say that bus i has higher priority than bus j, that is, bus i is more important than bus j. It should be clear that the priority rules should be consistent, that is if both (i, j) and (j, i) are set as priority rules, then load on bus i and load on bus j are equally important and will both remain in the system or will both be cut off in a valid load shedding scheme.

We follow the idea of Zhao et al. (2003) to model the power flow constraint where for each transmission line l_{ij} connecting bus i and bus j, we set a predefined limit PSL_{ij}, we require that the real power flow f_{ij} on the line l_{ij} is less than $\alpha\mathrm{PSL}_{ij}$ after the load shedding scheme is given. Here $\alpha = 0.9$ is the safety coefficient. The task to calculate real power flow can be done quickly (Wood and Wollenberg 2003) by many software tools such as MATPOWER (MATPOWER Package by Zimmerman et al. online). For example, we will use the function runpf() in MATPOWER (implementing the default full AC Newton's method (Tinney and Hart 1967)) to compute f_{ij} as its real part in our numerical examples.

With previous description of the key constraints, our load shedding problem now can be stated formally as follows.

Load shedding problem (LSP): Given a power system with n buses. Let P_i be the total power of bus i and d be the power in-balance tolerance. Assume that a set of buses $L \subset \{1, \ldots, n\}$ is given as the set of load buses whose load could be cut off. For a set of given priority rules R among load buses, is there a subset C of L such that R is satisfied,

$$|\sum_{i \in \overline{C}} P_i| \leq d \tag{4}$$

holds and the power flow in the resulting system after load shedding is safe, that is,

$$|f_{ij}| \leq \alpha \mathrm{PSL}_{ij} \tag{5}$$

for all branches l_{ij}? Here $\overline{C} = \{1, \ldots, n\} - C$ is the set of buses remaining in the system after load shedding, f_{ij} and PSL_{ij} are the power flow and predefined safety limit of the branch l_{ij}, respectively, $\alpha (= 0.9)$ is the safety coefficient.

Remark 1. In the more general setting where only part of the total power of bus i is interruptible, (4) needs to be generalized to

$$|\sum_{i \in \overline{C}} P_i^{\mathrm{int}} + \sum_{j=1}^{n} P_j^{\mathrm{fix}}| \leq d, \tag{6}$$

where P_i^{int} and P_i^{fix} are interruptible amount of power and fixed (non-interruptible) amount of power on bus i, respectively.

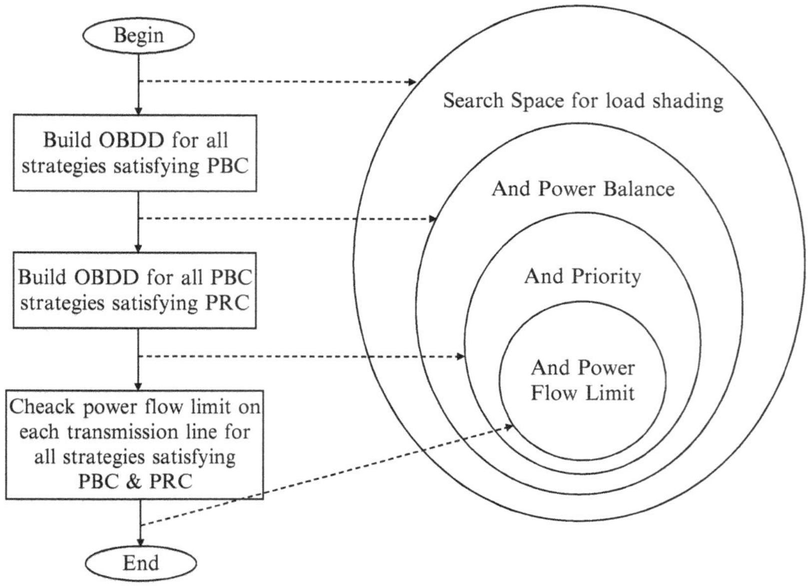

Fig. 2 Flow chart of algorithm and the reduction of search space

5 Solution of LSP Based on OBDD

As will be proved in Sect. 6, the load shedding problem LSP is NP-hard. This implies that brute force search would not work when the problem scale is large. To solve this problem, we employ OBDD technique, which has been widely used in the verification of large scale industry applications.[1] More specifically, by converting our key constraints PBC and PRC into Boolean expressions, OBDDs can be built for them and combined to provide a final OBDD, which represent all load shedding schemes that satisfy both constrains. Sampling technique then could be used to further generate promising solutions with real power flow checking. Our method and its steps are illustrated in Fig. 2.

The generation of the constraints PBC and PRC will be introduced in Sects. 5.1 and 5.2, respectively. Once the two OBDDs are built, we can build the OBDD corresponding to the load shedding schemes satisfying PBC and PRC simultaneously by simply carrying out an AND operation.

A prototype of our algorithm can be evaluated on the website `http://obdd.cfins.au.tsinghua.edu.cn/`

[1] It should be pointed out that the power of OBDD in solving large scale problems comes from the fact that problem knowledge is efficiently utilized, but in their worst cases, for NP-hard problems, theoretically saying, OBDD may still suffer long computation time.

5.1 Boolean Expression for the Power Balance Constraint (PBC)

We will use a Boolean variable S_i to denote the ON/OFF switch status of power on bus i in the power system. So, for a bus i in L, S_i could be either 0 or 1 depending on whether the load on it will be cut off or not. Let P_{sum} be the total power of the system. We have

$$P_{\text{sum}} = \sum_{i=1}^{n} P_i S_i. \tag{7}$$

For a bus j in $\overline{L}$, we set $S_j \equiv 1$.[2] It is easy to see that (7) can be written in the form of

$$P_{\text{sum}} = \sum_{i \in L} P_i S_i + \sum_{j \in \overline{L}}^{n} P_i. \tag{8}$$

We use the Boolean vector $(S_1, \ldots, S_n)$ to denote a load shedding scheme, which has to be decided by the decision maker: if a load shedding scheme requires to cut off loads on bus set $C \subset L$, then the corresponding Boolean variables S_i are such that $S_i = 0$ for $i \in C$ and $S_j = 1$ for $j \in \overline{C} = \{1, \ldots, n\} - C$. For a load shedding scheme, the total power expression (7) can then be further reduced to the form

$$P_{\text{sum}} = \sum_{i \in \overline{C}} P_i. \tag{9}$$

To convert the inequality requirement of power balance constraint (PBC) (4) for load shedding, we have the following observation. In a modern computer, an arithmetic number is often expressed in binary form, the same to the OBDD-based algorithm. We use the example in Fig. 3 as an example. The power of bus 4 is +20 MW, and its binary form is 010100, which can be written as a Boolean vector as below.

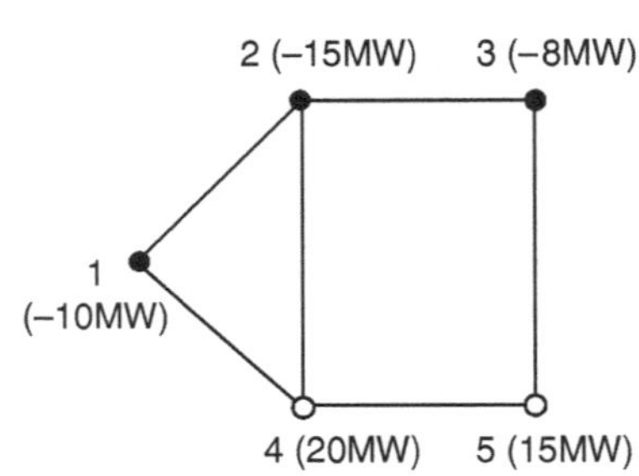

Fig. 3 A 5 bus example

[2] If in practice, generator shedding is also possible, then this assumption might need to be changed accordingly.

$$P_4 = (20)_{10}(MW) = (010100)_2 \Rightarrow \begin{pmatrix} 0 \\ 1 \\ 0 \\ 1 \\ 0 \\ 0 \end{pmatrix}$$

In general, we treat real numbers as scaled integer values and the precision of the vector P_i for power of a bus i is chosen long enough to contain the largest intermediate value during the full evaluating course.

For the example in Fig. 3, the total power can be expressed as (negative power value is represented as complement code: $\overline{x-1}$)

$$\begin{aligned} P_{\text{sum}} &= P_1S_1 + P_2S_2 + P_3S_3 + P_4S_4 + P_5S_5 \\ &= 20S_1 + (-10)S_2 + (-15)S_3 + (-8)S_4 + 15S_5 \\ &= \begin{pmatrix} 0 \\ 1 \\ 0 \\ 1 \\ 0 \\ 0 \end{pmatrix} S_1 + \begin{pmatrix} 1 \\ 1 \\ 0 \\ 1 \\ 1 \\ 0 \end{pmatrix} S_2 + \begin{pmatrix} 1 \\ 1 \\ 0 \\ 0 \\ 0 \\ 1 \end{pmatrix} S_3 + \begin{pmatrix} 1 \\ 1 \\ 1 \\ 0 \\ 0 \\ 0 \end{pmatrix} S_4 + \begin{pmatrix} 0 \\ 0 \\ 1 \\ 1 \\ 1 \\ 1 \end{pmatrix} S_5 \\ &= \begin{pmatrix} 0 \\ S_1 \\ 0 \\ S_1 \\ 0 \\ 0 \end{pmatrix} + \begin{pmatrix} S_2 \\ S_2 \\ 0 \\ S_2 \\ S_2 \\ 0 \end{pmatrix} + \begin{pmatrix} S_3 \\ S_3 \\ 0 \\ 0 \\ 0 \\ S_3 \end{pmatrix} + \begin{pmatrix} S_4 \\ S_4 \\ S_4 \\ 0 \\ 0 \\ 0 \end{pmatrix} + \begin{pmatrix} 0 \\ 0 \\ S_5 \\ S_5 \\ S_5 \\ S_5 \end{pmatrix} \end{aligned} \tag{10}$$

If we assume only buses 1, 2, 3 are considered to be able to cut off, then $L = \{1, 2, 3\}$, $\overline{L} = \{4, 5\}$. So $S_4 = S_5 = 1$. The total power under a load shedding scheme $(S_1, S_2, S_3, 1, 1)$ can be further written as

$$P_{\text{sum}} = \begin{pmatrix} 0 \\ S_1 \\ 0 \\ S_1 \\ 0 \\ 0 \end{pmatrix} + \begin{pmatrix} S_2 \\ S_2 \\ 0 \\ S_2 \\ S_2 \\ 0 \end{pmatrix} + \begin{pmatrix} S_3 \\ S_3 \\ 0 \\ 0 \\ 0 \\ S_3 \end{pmatrix} + \begin{pmatrix} 1 \\ 1 \\ 1 \\ 0 \\ 0 \\ 0 \end{pmatrix} + \begin{pmatrix} 0 \\ 0 \\ 1 \\ 1 \\ 1 \\ 1 \end{pmatrix}. \tag{11}$$

Based on the expression (8), we can decide the power balance condition as

$$F_{PBC} = (|P_{\text{sum}}| \leq d) = true, \tag{12}$$

where d is the tolerated limit in difference between power generation and load in a system that could lead to stable transient. In the above expressions, arithmetic addition and comparison operations can be converted into combinations of Boolean operations AND, OR, and NOT, as all modern commercial CPU's do. (Documentation of Jørn Lind-Nielsen's BuDDy Package (online) also describes these Boolean operation combinations for arithmetic operations.) So, the arithmetic–Boolean mixed expression can be converted into a pure Boolean expression. With this pure Boolean expression for the power balance condition, we can build an OBDD encoding all load shedding schemes that achieve power balance.

5.2 *Boolean Expression for the Priority Constraint (PRC)*

The priority rules can be described directly as a set of inequalities over the Boolean variables S_i representing load switch status on bus i. For example, if we require that only the load on bus 1 has higher priority than the loads on all other buses, that is, we need to keep the load on bus 1 when cut off loads on some other buses are sufficient to achieve power balance, we can require

$$S_1 \geq S_i, \quad i = 2, \ldots, n$$

or equivalently (with the operator IMPLICATION)

$$S_1 \leftarrow S_2 \quad S_1 \leftarrow S_3 \quad \ldots \quad S_1 \leftarrow S_n$$

In general, the given set R of priority rules can be captured by a partial order (Suppes 1999) among $S_1, \ldots, S_n$. Equivalently, we can use Hasse diagram (Suppes (1999)) (a directed acyclic graph) $D = (V, E)$ of the partial order to concisely describe the set of priority rules. In the diagram, $V = 1, \ldots, n$ are the vertex set of D and E is the edge set of D. An edge $(i, j) \in E$ if and only if we require that bus i has higher priority than bus j when we do load shedding, or equivalently, the bus switch status S_i and S_j satisfy the constraint that $S_i \geq S_j$ (or $S_i \leftarrow S_j$). Overall, the priority constraint can be described as an equation:

$$F_{PRC} = \left(\prod_{(i,j) \in E} S_i \leftarrow S_j \right) = \text{true}. \quad (13)$$

6 NP-Hardness of the Problem

The power balanced load shedding problem is NP-hard (Garey and Johnson 1979). To establish this result, we need to cite the following well-known NP-complete problem.

The 0–1 KNAPSACK Problem: Given integers $c_j \geq 0$, $j = 1, \ldots, n$ and $K \geq 0$, is there a subset S of $\{1, \ldots, n\}$ such that $\sum_{j \in S} c_j = K$?

Lemma 1. *The 0–1 KNAPSACK Problem is NP-complete.*

Proof. The proof given in Theorem 15.8 of Papadimitriou (1982) is valid for our version of 0–1 KNAPSACK Problem, although the statement of 0–1 KNAPSACK Problem in Papadimitriou (1982) requires only c_j, $j = 1, \ldots, n$ and K to be integers.

Theorem 1. *The load shedding problem **LSP** is NP-hard.*

Proof. We shall transform load shedding problem to the 0-1 KNAPSACK Problem. Given any instance $c_1, \ldots, c_n$, K of 0-1 KNAPSACK, we construct a power system as follows. The system has n load buses (bus 1 to bus n) and 1 generation bus (bus $n + 1$). The total power on the load buses are $P_i = -c_i$, $i = 1, \ldots, n$. The total power on the generation bus is $P_{n+1} = K$. Let the in-balance tolerance be $d = 1/2$. We assume that all buses for the constructed power system are connected through transmission lines with sufficiently large power flow limit and furthermore there is no priority among the n load buses. So, only the power balance condition (PBC) given by (12) is needed to consider when we solve the load shedding problem for this system. We claim that the instance of 0–1 KNAPSACK has a solution if and only if LSP for the constructed power system has a solution and thus can establish the NP-hardness of LSP based on Lemma 1.

If: When there is a load shedding scheme S_i, $i = 1 \ldots, n + 1$ satisfying (12) with the generation bus always on, that is, $S_{n+1} = 1$, we know that

$$|\sum_{i=1}^{n+1} P_i S_i| = |\sum_{i=1}^{n} (-c_i) S_i + K| \leq d = 1/2. \tag{14}$$

By defining $S = \{i | S_i = 1, i = 1, \ldots, n\}$ we can see from (14) that

$$\sum_{j \in S}^{n} (c_j) = K$$

by noting $|\sum_{i=1}^{n} (-c_i) S_i + K|$ is an integer and (14) implies $|\sum_{i=1}^{n} (-c_i) S_i + K| = 0$.

Only if: When the instance of 0–1 KNAPSACK has a solution, say a subset S of $\{1, \ldots, n\}$ is such that $\sum_{j \in S} c_j = K$, by defining $S_i = 0$ if $i \notin S$ and $S_i = 1$ otherwise we can introduce a load shedding scheme. We can verify for this scheme that

$$|\sum_{i=1}^{n+1} P_i S_i| = |\sum_{i=1}^{n} (-c_i) S_i + K| = 0 \leq d = 1/2.$$

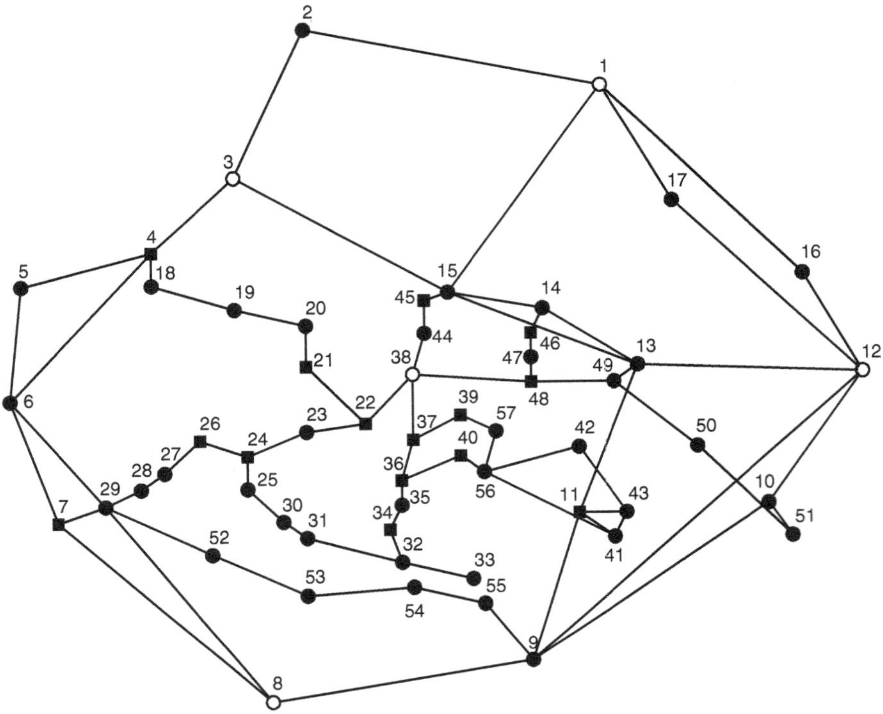

Fig. 4 The IEEE 57-bus test case

This implies that the load shedding scheme satisfies the power balance constraint. Recall that we assume that the power flow limit and priority among load buses are not a problem in this system. So the load shedding scheme is a solution to LSP. The proof is completed.

7 Case Study

To test the method proposed in this paper, we study a IEEE 57-bus system (Power Systems Test Case Archive online). The topology of the system is shown in Fig. 4. The total generation is about 300 MW less than the total load. For simplicity, we assume that the power injections from generators are fixed, and total amount of load on load buses can be shed but with two different ways of setting detailed priority.

(1) If we set the tolerance for power in-balance as 20 MW and set the priority among all load buses as $S_2 \le S_3 \le S_5 \le S_6 \le S_9 \le S_{10} \le S_{12} \le S_{13} \le S_{14} \le S_{15} \le S_{16} \le S_{17} \le S_{18} \le S_{19} \le S_{20} \le S_{23} \le S_{25} \le S_{27} \le S_{28} \le S_{29} \le S_{30} \le S_{31} \le S_{32} \le S_{33} \le S_{35} \le S_{38} \le S_{41} \le S_{42} \le S_{43} \le S_{44} \le S_{47} \le S_{49} \le S_{50} \le S_{51} \le S_{52} \le S_{53} \le S_{54} \le S_{55} \le S_{56} \le S_{57}$, we

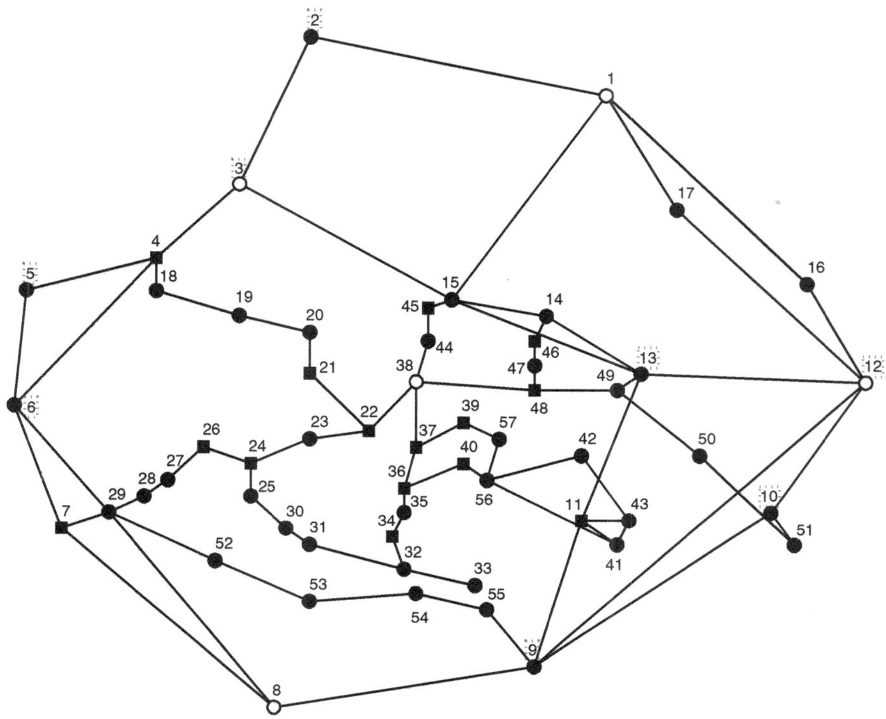

Fig. 5 A load shedding scheme for the IEEE 57-bus test case

can find the OBDD containing all the load shedding schemes that satisfy the power balance condition and priority requirement. Note that not all buses should be considered in load shedding procedure, for example, buses containing generators. We require the buses listed below always to be connected to the system: buses 1, 4, 7, 8, 11, 21, 22, 24, 26, 34, 36, 37, 39, 40, 45, 46, and 48.

Totally there are 32 solutions. The computing platform is Intel Xeon E5335 (2.00 GHz, Quad-Core, 8MB L2 Cache). The time to obtain OBDD for PBC is 0.117 s. The time to obtain OBDD considering further PRC is negligible. Figure 5 shows one of the solution in which loads need to be cut off are shown in RED dashed boxes. The load shedding happens on buses 2, 3, 5, 6, 9, 10, 12, and 13. The power generation is less than the total load by 18.9 MW. From the power flow (injection) calculated by MATPOWER (runpf()) list in Table 1, we can see that the load shedding scheme could pass the power flow check if the power safety limit of all lines PSL_{ij} are set as uniformly as 200 MW and the safety coefficient is set as $\alpha = 0.9$. Of course, further examination such as transient stability analysis is needed if one want to apply this specific load shedding scheme.

(2) If we set the tolerance for power in-balance as 5 MW and the priority among only some of the buses as $S_{28} \le S_{27} \le S_{25} \le S_{23} \le S_{20} \le S_{19} \le S_{18} \le S_{17} \le S_{16} \le S_{15} \le S_{14} \le S_{13} \le S_{12} \le S_{10} \le S_9 \le S_6 \le S_5 \le S_3 \le S_2$. As a result

Table 1 Power flow for load shedding scheme in Fig. 5

No.	Bus_{in}	Bus_{out}	P_{in}(MW)	No.	Bus_{in}	Bus_{out}	P_{in}(MW)	No.	Bus_{in}	Bus_{out}	P_{in}(MW)
1	1	2	14.12	2	2	3	13.22	3	3	4	−23.12
4	4	5	−21.95	5	4	6	−32.95	6	6	7	−14.27
7	6	8	−41.75	8	8	9	175.46	9	9	10	38.74
10	9	11	51.17	11	9	12	23.51	12	9	13	39.49
13	13	14	29.85	14	13	15	0.33	15	1	15	34.60
16	1	16	24.32	17	1	17	38.45	18	3	15	36.11
19	4	18	13.89	20	4	18	17.78	21	5	6	−22.33
22	7	8	−81.21	23	10	12	1.95	24	11	13	25.08
25	12	13	1.87	26	12	16	19.02	27	12	17	3.91
28	14	15	−23.27	29	18	19	4.47	30	19	20	1.06
31	21	20	1.24	32	21	22	−1.24	33	22	23	3.57
34	23	24	−2.73	35	24	25	7.35	36	24	25	7.06
37	24	26	−17.20	38	26	27	−17.20	39	27	28	−27.05
40	28	29	−32.12	41	7	29	66.90	42	25	30	8.11
43	30	31	4.39	44	31	32	−1.50	45	32	33	3.81
46	34	32	6.92	47	34	35	−6.92	48	35	36	−12.96
49	36	37	−15.13	50	37	38	−18.06	51	37	39	2.82
52	36	40	2.08	53	22	38	−4.81	54	11	41	10.30
55	41	42	10.17	56	41	43	−13.07	57	38	44	−12.23
58	15	45	24.63	59	14	46	42.48	60	46	47	42.48
61	47	48	12.25	62	48	49	−4.90	63	49	50	3.33
64	50	51	−17.72	65	10	51	36.15	66	13	49	34.57
67	29	52	17.34	68	52	53	12.01	69	53	54	−8.11
70	54	55	−12.37	71	11	43	15.07	72	44	45	−24.27
73	40	56	2.07	74	56	41	−6.63	75	56	42	−2.81
76	39	57	2.82	77	57	56	−3.88	78	38	49	−8.13
79	38	48	−16.88	80	9	55	19.49				

of relaxing priority requirement among buses, we have much more solutions even with more strict balance tolerance (totally 512,728 solutions) compared to the first case where we set priority among all load buses. On the same computing platform, the time to obtain OBDD for PBC is 0.398 s. The time to obtain OBDD considering further PRC is negligible. Figure 6 shows one of the solution in which loads need to be cut off are shown in RED dashed boxes. The load shedding happens on buses 10, 12, 13, 14, 15, 16, 17, 18, 19, 20, 23, 25, 27, 28, 29, 30, 31, 32, 33, 35, 38, and 43. The power generation is less than the total load by 1.3 MW. The power flow (injection) calculated by MATPOWER (runpf()) is shown in Table 2. We can see the load shedding scheme could not pass the power flow check if the power safety limit of all lines PSL_{ij} are set as uniformly as 200 MW and the safety coefficient is set as $\alpha = 0.9$ (noticing branch 8 from bus 8 to bus 9) but will pass if PSL_{ij} are increased to 210 MW.

We have also done more tests on other examples. The rough estimation of the problem size that can be solved in short time is about 60 decision variables. Of course, since the problems we are studying contain real numbers, the complexity

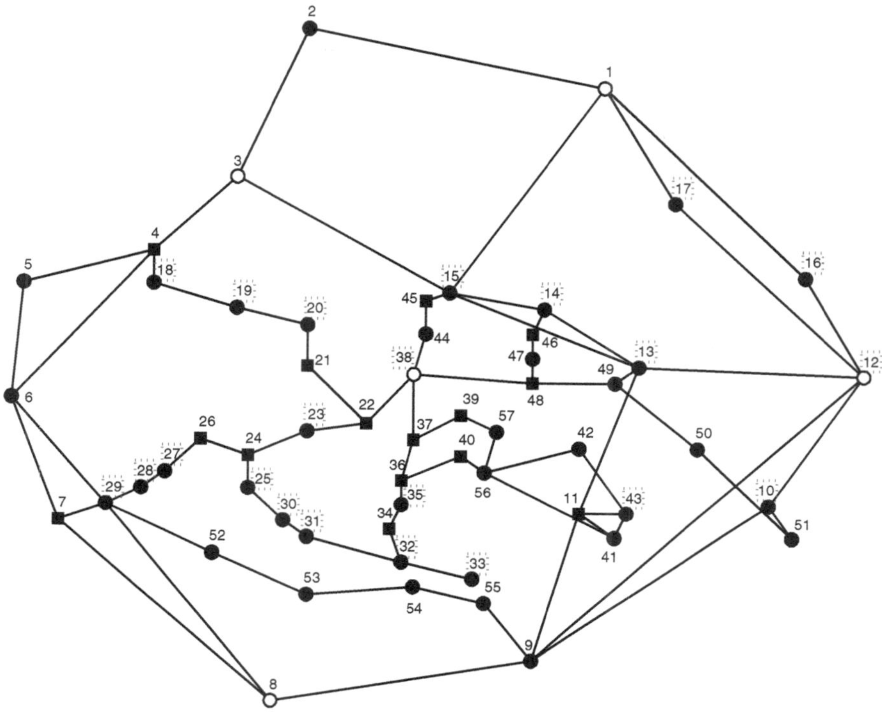

Fig. 6 A load shedding scheme for the IEEE 57-bus test case

of the problem should also depend on these parameters. For large scale problems, some aggregation may be needed to apply our method.

8 Conclusions

Load shedding is a useful tool for emergency control and other purposes. This paper shows that the problem is a NP-hard problem. This explains why the problem remains not fully solved. Because of this reason, we formulate and propose to solve the load shedding problem in a new way based on the OBDD technique. We hope the method could provide some insights to the understanding of the challenges and the efficient way to consider difficult constraints when making load shedding decisions.

With some additional work, the method proposed in this paper may be extended to combine with the splitting surface searching algorithms developed in Zhao et al. (2003), Sun et al. (2003), Sun (2004). It may also be extended to formulate generator shedding (Jin et al. 2007) or generation rescheduling to reduce power injection. These are possible future work.

Table 2 Power flow for load shedding scheme in Fig. 6

No.	Bus_{in}	Bus_{out}	P_{in}(MW)	No.	Bus_{in}	Bus_{out}	P_{in}(MW)	No.	Bus_{in}	Bus_{out}	P_{in}(MW)
1	1	2	25.52	2	2	3	21.64	3	3	4	4.74
4	4	5	3.83	5	4	6	−1.42	6	6	7	−36.62
7	6	8	−49.02	**8**	**8**	**9**	**181.96**	9	9	10	14.66
10	9	11	16.52	11	9	12	2.81	12	9	13	8.15
13	13	14	7.28	14	13	15	−11.25	15	1	15	35.22
16	1	16	12.72	17	1	17	12.74	18	3	15	15.62
19	4	18	1.01	20	4	18	1.30	21	5	6	−9.19
22	7	8	−67.52	23	10	12	−13.55	24	11	13	−1.72
25	12	13	14.25	26	12	16	−12.60	27	12	17	−12.62
28	14	15	−23.64	29	18	19	2.31	30	19	20	2.29
31	21	20	−2.28	32	21	22	2.28	33	22	23	−7.63
34	23	24	−7.64	35	24	25	0.64	36	24	25	0.61
37	24	26	−8.97	38	26	27	−8.97	39	27	28	−9.15
40	28	29	−9.21	41	7	29	30.62	42	25	30	1.25
43	30	31	1.24	44	31	32	1.24	45	32	33	−0.00
46	34	32	−1.22	47	34	35	1.22	48	35	36	1.22
49	36	37	−3.76	50	37	38	−8.64	51	37	39	4.87
52	36	40	4.98	53	22	38	9.91	54	11	41	7.82
55	41	42	7.80	56	41	43	−10.33	57	38	44	−2.92
58	15	45	15.09	59	14	46	30.91	60	46	47	30.91
61	47	48	0.94	62	48	49	2.61	63	49	50	11.19
64	50	51	−9.95	65	10	51	28.08	66	13	49	24.27
67	29	52	21.36	68	52	53	15.83	69	53	54	−4.37
70	54	55	−8.54	71	11	43	10.33	72	44	45	−14.94
73	40	56	4.96	74	56	41	−3.94	75	56	42	−0.55
76	39	57	4.86	77	57	56	−1.84	78	38	49	2.35
79	38	48	1.72	80	9	55	15.54				

Acknowledgements This work was supported by NSFC Grant Nos. (60574067, 60736027, 60721003). This is an invited paper submitted to the "Power Systems Handbook" organized by Dr. Mario V. Pereira and Prof. Dr. Panos M. Pardalos. The manuscript is intended to be considered as a contribution in "Computing Technologies" (D)– Computing Technologies in Energy systems.

References

Amin M (2001) Toward self-healing energy infrastructure systems. IEEE Comput Appl Power 14(1):20–28

Amraee T, Ranjbar AM, Mozafari B, Sadati N (2007) An enhanced under-voltage load-shedding scheme to provide voltage stability. Elec Power Syst Res 77:1038–1046

Bryant RE (1986) Graph-based algorithms for Boolean function manipulation. IEEE Trans Comput 35:677–691

Bryant RE (1992) Symbolic boolean manipulation with ordered binary decision diagrams. ACM Comput Surv 24(3):293–318

Concordia C, Fink LH, Poullikkas G (1995) Load shedding on an isolated system. IEEE Trans Power Apparat Syst 10(3):1467–1472

Faranda R, Pievatolo A, Tironi E (2007) Load shedding: A new proposal. IEEE Trans Power Syst 22(4):2086–2093

Feng Z, Ajjarapu V, Maratukulam DJ (1998) A practical minimum load shedding strategy to mitigate voltage collapes. IEEE Trans Power Syst 13(4):1285–1291

Fernandes TSP, Lenzi JR, Mikilita MA (2008) Load shedding strategies using optimal load flow with relaxation of restrictions. IEEE Trans Power Syst 23(2):712–718

Garey MR, Johnson DS (1979) Computers and intractability: A guide to the theory of NP-completeness. San Francisco, CA, Freeman

Jin M, Sidhu TS, Sun K (2007) A new system splitting scheme based on the unified stability control framework. IEEE Trans Power Syst 22(1): 433–441

Medicherla TKP, Billinton R, Sachdev MS (1979) Generation rescheduling and load shedding to alleviate line overloads-analysis. IEEE Trans Power Apparat Syst 98(6):1876–1884

Papadimitriou CH (1982) Combinatorial optimization: algorithms and complexity, Englewood Cliffs, Prentice-Hall, NJ

Shandilya A, Gupta H, Sharma J (1993) Method for generation rescheduling and load shedding to alleviate line overloads using local optimisation. IEE Proc C 140(5):337–342

Sun K, Zheng D, Lu Q (2003) Splitting strategies for islanding operation of large-scale power systems using OBDD-based methods. IEEE Trans Power Syst 18(2):912–923

Sun K (2004) An OBDD-based three-phase method for searching for splitting strategies of large-scale power networks against blackouts. Ph.D dissertation, Tsinghua University, Beijing

Suppes P (1999) Introduction to logic. Dover Publications Inc., Mineola, NY

Tinney WF, Hart CE (1967) Power flow solution by Newton's method. IEEE Trans Power Apparat Syst PAS-86(11):1449–1460

Voumvoulakis EM, Hatziargyriou ND (2008) Decision trees-aided self-organized maps for corrective dynamic security. IEEE Trans Power Syst 23(2):622–630

Wood AJ, Wollenberg BF (2003) Power generation, operation and control. 2nd edn, Wiley, NY

Zhao Q, Sun K, Zheng D, Lu Q, Ma J (2003) A study of system splitting strategies for island operation of power system: A two-phase method based on OBDDs. IEEE Trans Power Syst 18(4):1556–1565

Jørn Lind-Nielsen's BuDDy Package. [Online] `http://buddy.sourceforge.net/`

MATPOWER Package by Zimmerman RD, Murillo-Sánchez CE, Gan D [Online] `http://www.pserc.cornell.edu/matpower/`

Power Systems Test Case Archive. [Online] `http://www.ee.washington.edu/research/pstca/`

Solution to Short-term Unit Commitment Problem

Md. Sayeed Salam

Abstract The Lagrangian relaxation approach to solve the unit commitment problem for a large system comprising both thermal and hydro generating units is presented. Commitment states of thermal units are obtained by solving thermal subproblems of Lagrangian dual problem. To get the output levels of hydro units, the hydrothermal scheduling is performed with a thermal unit commitment schedule obtained by solving thermal subproblems. Extensive constraints are considered. Nonlinear functions are used for thermal generation cost, water discharge rate and sulfur oxide emission. A general transmission loss formula is utilized for incorporating transmission loss. The variable metric method is used for updating the Lagrangian multipliers during maximization of the dual function. The Lagrangian multipliers are adjusted by the linear interpolation method during searching for a feasible suboptimal solution near the dual optimal point. A refinement algorithm is used to fine tune the schedule. A unit commitment expert system is employed for checking the feasibility of the solution and for handling constraints, which are difficult or impractical to be implemented in commitment algorithm. Results of the implementation on a utility are shown.

Keywords Expert system · Hydrothermal scheduling · Lagrangian relaxation · Unit commitment

List of Symbols

M, H	Number of thermal and hydro units, respectively
i, h	Index of thermal and hydro units, respectively
T	Number of periods for dividing the scheduling time horizon

M.S. Salam
BRAC University, Dhaka, Bangladesh
e-mail: sayeed@bracuniversity.ac.bd

S. Rebennack et al. (eds.), *Handbook of Power Systems I*, Energy Systems,
DOI 10.1007/978-3-642-02493-1_11,

t	Time index
P	MW power output of a generating unit
$\overline{P}, \underline{P}$	Maximum and minimum MW power of a generating unit, respectively
r	Spinning reserve contribution
$\overline{r}$	Maximum spinning reserve contribution of a generating unit
x	State variable, e.g., $x = \begin{cases} 2, & \text{unit has been in service for 2 successive periods} \\ -3, & \text{unit has been shutdown for 3 successive periods} \end{cases}$
U	Commitment state, when $U = \begin{cases} 1, & \text{unit is on} \\ 0, & \text{unit is off} \end{cases}$
D_t, R_t	Demand and MW spinning reserve contribution in period t, respectively
$Ploss_t$	Transmission loss in period t
$Tu\min_i, Td\min_i$	Minimum up and down time of ith thermal unit, respectively
Δ_i	Ramp rate limit of the ith thermal unit
f_h	Water inflow rate of the hth hydro unit
$qtot_h$	Prespecified volume of water available for the hth hydro unit
Si_h, Sf_h	Initial and final volume of water of the reservoir of the hth hydro unit, respectively
$C_i(.), S_i(.)$	Production cost and start-up cost functions of the ith thermal unit, respectively
$E_i(.)$	Emission function of the ith thermal unit
$q_h(.)$	Water flow rate function for the hth hydro unit
$b0, b1, b2$	Coefficients for production cost function
$e0, e1, e2$	Coefficients for emission function
$a0, a1, a2$	Coefficients for water flow rate function
$B0, B1, B$	Transmission loss coefficients
λ, μ	Vector of Lagrangian multipliers associated with power balance and spinning reserve constraints, respectively
λ_t, μ_t	tth component of λ, μ respectively
$\theta_{s,k}$	Phase angle difference between voltage phasors of bus s and bus k
Rs,k	The real element at row s and column k of network bus impedance matrix
Vk	Voltage at bus k
γ_h	Lagrangian multiplier for the hth hydro unit

1 Introduction

The unit commitment problem determines the combination of available generating units and scheduling their respective outputs to satisfy the forecasted demand with the minimum total production cost under the operating constraints enforced by the system for a specified period that usually varies from 24 h to 1 week. Attempts to develop rigid unit operating schedules for more than 1 week in advance are extremely curtailed due to uncertainty in hourly load forecasts at lead times greater than 1 week. The operating constraints reduce freedom in the choice of starting up and shutting down generating units. The constraints to be satisfied are usually the status restriction of individual generating units, minimum up time, minimum down time, capacity limits, generation limit for the first and last hour, limited ramp rate, group constraint, power balance constraint, spinning reserve constraint, hydro constraint, etc.

The high dimensionality and combinatorial nature of the unit commitment problem curtail attempts to develop any rigorous mathematical optimization method capable of solving the whole problem for any real-size system. Nevertheless, in the literature, many methods using some sort of approximation and simplification have been proposed. The available approaches for solving unit commitment problem can usually be classified into heuristic methods and mathematical programming methods (Kuloor et al. 1992).

Heuristic methods are non-rigorous computer-aided empirical methods, which make the unit commitment decisions according to a precalculated priority list and incorporate all the operating constraints heuristically. The metaheuristic methods (Liyong et al. 2006; Pappala and Erlich 2008; Saber et al. 2007) are iterative techniques that are able to search for local optimal solutions and a global optimal solution. In the metaheuristic methods, the techniques frequently applied are simulated annealing, tabu search, genetic algorithm, greedy randomized adaptive search procedure, evolutionary programming, and particle swarm optimization.

Heuristic methods are flexible and allow for the consideration of practical operating constraints. The main shortcoming of heuristic methods is that they cannot guarantee the optimal solutions or even furnish an estimate of the magnitude of their sub-optimality. This aspect becomes rather significant in large-scale power systems, as a small percentage, for example, 0.5%, in the costs of unit commitment schedules represents a substantial financial annual saving. The main difficulty of metaheuristic approaches is their sensitivity to the choice of parameters and they also generate infeasible solutions. Therefore, it is advantageous to employ more rigorous methods compared with heuristics methods to generate more economical solutions as the size of a system grows.

The mathematical programming approaches are dynamic programming, Lagrangian relaxation, Benders decomposition, and mixed integer programming (Kuloor et al. 1992). Benders decomposition is the least promising and is reflected by the lack of enough published works reporting its success. In the literature, dynamic programming and Lagrangian relaxation have been used extensively to develop industry-grade unit commitment programs. Their major advantage seems to be the requirement of reasonable computation time (Salam 2004). Recent advances in mixed integer programming (MIP) make MIP approach rigorous, robust, flexible, and efficient. The MIP formulation allows for more sophisticated modeling and flexibility of complex resources and constraints, yielding low cost reliable commitment solutions in less computation time (Carrion and Arroyo 2006; Hur et al. 2007).

In dynamic programming, it is relatively easy to add constraints that affect operations at an hour (such as power balance constraints) since these constraints mainly affect the economic dispatch and solution method. However, the dynamic programming suffers from the curse of dimensionality. Hence, it is required to limit the commitments considered at any hour through some simplification techniques such as truncation and fixed priority ordering. This simplification, particularly for large scale systems, can lead to suboptimal schedules.

Lagrangian relaxation method has the advantage of being easily modified to model characteristics of specific utilities. It is more advantageous due to its flexibility in dealing with different types of constraints. It is relatively easy to add unit constraints. Lagrangian relaxation is flexible to incorporating additional coupling constraints that have not been considered so far. The only requirement is that constraints must be additively separable in units. Such constraints could be area reserve constraint, area interchange constraint, etc. To incorporate such constraint into the framework of Lagrangian relaxation, a Lagrangian multiplier is defined for each constraint for each time period and the constraints are adjoined into the objective function of the relaxed problem. The Lagrangian relaxation method is also more flexible than dynamic programming because no priority ordering is imposed. The amount of computation varies linearly with the number of units. Hence, it is computationally much more attractive for large systems.

One weakness of the Lagrangian relaxation method is that the dual optimal solution seldom satisfies the once relaxed coupling constraints. Another weakness is the sensitivity problem that may cause unnecessary commitments of some units. Therefore only a near optimal feasible solution can be expected. However, the degree of suboptimality decreases as the number of units increases.

In Lagrangian relaxation approach, the commitment schedule may be so sensitive to the variations of the Lagrange multipliers that a slight modification of the multipliers may change the status of several units. This sensitivity problem is more serious for systems having several groups of identical units. Even though fuel costs of identical units can be slightly modified to make small differences among cost characteristics, this sensitivity problem still exists. In other words, unnecessary commitment of some units may be possible in the solution given by this method. To overcome this difficulty, a refinement process may be developed for Lagrangian relaxation approach. This refinement process inspects some candidate units whose shutdown may result in additional reduction of the operating cost (Tong and Shahidehpour 1990).

In this paper, the Lagrangian relaxation approach (Salam et al. 1998) to solve the unit commitment problem for a large system comprising both thermal and hydro generating units is presented. Commitment states of thermal units are obtained by solving thermal subproblems of Lagrangian dual problem using dynamic programming without discretizing generation levels. To get the output levels of hydro units, the hydrothermal scheduling is performed with a thermal unit commitment schedule obtained from solutions of thermal subproblems. Extensive constraints such as status restriction of individual generating units, that is, must run, must out, base load, cycling, and peaking, power balance, spinning reserve, minimum up/down time, capacity limits, ramp rate, limited generation for the first and last hour, sulfur oxide emission, and hydro constraints are taken into account. Nonlinear functions are used for thermal generation cost, water discharge rate, and sulfur oxide emission. A general transmission loss formula (Elgerd 1971) whose expression has a similar quadratic form to the B matrix loss formulation is utilized for incorporating transmission loss. A refinement algorithm is developed and used to fine tune the schedule. A unit commitment expert system (Salam et al. 1991) is employed as a preprocessor as well as a postprocessor to the unit commitment program to check and alter commitment results by adjusting the input data if necessary. The expert system checks the feasibility of the solution and handles constraints that are difficult or impractical to be implemented in commitment algorithm such as cycling of gas turbine and steam turbine units, group constraints, etc. Results of the implementation on a utility are shown.

2 Generating Units

In most utilities, two major types of generating units, that is, thermal and hydro, are generally available. Their characteristics are described below:

2.1 Thermal Units

The input to the thermal units is usually expressed either in terms of heat energy requirement or in terms of total cost per hour and the output in terms of electrical power. There are two types of thermal units, steam turbine and gas turbine units.

2.1.1 Steam Turbine Unit

This unit consists of a boiler that generates steam to drive a turbine generator set. Because of the operational limitations of boiler and turbine, steam turbine unit is characterized by the presence of capacity limits, generation for the first and last hour of operation (assuming 1 h start-up and shut-down sequence), ramp rate limit, minimum up and down times, and time-dependent start up cost. Each of these is described below:

Capacity limits: When a unit is online, it can only be operated within a certain range. Otherwise, the output becomes zero.

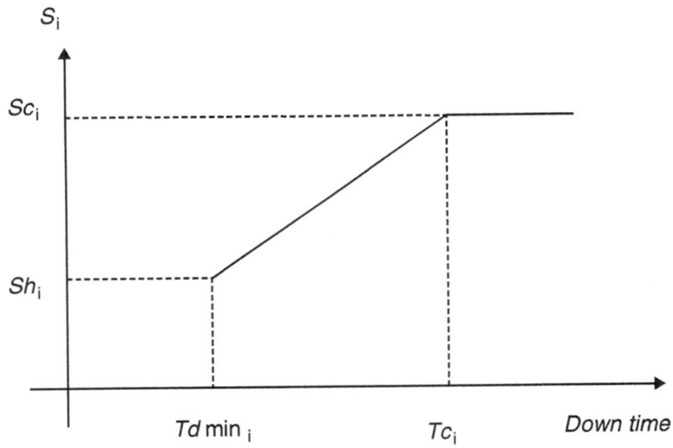

Fig. 1 Start up cost

Generation for the first and last hour: It may be necessary to keep the output at minimum level for the first hour online and for the last hour before it is shut down.

Ramp rate: Ramp rate is the change of output power of a unit between two successive time steps. This ramp rate always has a maximum limit.

Minimum up time: If a unit is started, it must remain in running mode for at least a certain period, called minimum up time, before it is shut down again.

Minimum down time: If a unit is shut down, it must remain down for at least a certain period, called minimum down time, before it is started up again.

Start up cost: If a unit is not running continuously, a cost called start up cost is incurred on it every time it is started up. Several methods of representing start up costs have been proposed. Figure 1 shows a very simple representation (Guan et al. 1992). The time varying start up cost is a linear function of time since last shut down as shown in the figure. It is given by

$$S_i(x_{i,t}, U_{i,t}) = Sh_i + (Sc_i - Sh_i)(|x_{i,t}| - Tdmin_i)/(Tc_i - Tdmin_i)$$

. The start up cost remains constant after the cold start up time.

2.1.2 Gas Turbine Unit

This unit consists of a gas turbine that drives a generator. A gas turbine unit has no minimum up and down time. It has high ramp rate, high running cost, and negligible start up cost.

2.2 Hydro Units

This unit consists of a water turbine that drives a generator. The input to the hydro units is usually expressed in terms of volume of water per unit time and the output in terms of electrical power. Hydro units have capacity limits, limited water resources, and high ramp rate.

3 Operating Constraints

The unit commitment schedule needs to satisfy a number of constraints. These constraints reduce the freedom in the choice of starting up and shutting down generating units. Unit commitment generally considers system and unit constraints. They are described in the following sections.

3.1 System Constraints

These are the constraints imposed by the system. They are as follows:
Demand: Unit commitment must meet the forecasted demand.
Spinning reserve requirement: The required spinning reserve can be specified as a percentage of the largest unit online. Thus the loss of generation of the largest unit online can be recovered within a specified period, for example, 10 min.
Scheduling period: A scheduling period of 24 h or 1 week is usually considered.

3.2 Unit Constraints

The unit may impose a number of constraints that guide the selection of starting up and shutting down generating units. Unit commitment usually considers the following unit constraints.

Initial conditions: These include the number of hours that each unit has been running (or shutdown) and the dispatch of each unit before the unit commitment begins.

Unit's minimum and maximum generation limits: These limits outline the region of the dispatch of a unit.

Generation levels for the first and last hour: Should consider generation levels for these 2 h as described in Sect. 2.1.1.

Limited ramp rates: Should consider ramp rate limits as described in Sect. 2.1.1.

Minimum up and down time: Should consider minimum up and down time constraints that are discussed in Sect. 2.1.1.

Must run units: These units are always online due to high efficiency or operational reliability or high initial capital cost such as nuclear units.

Must out units: These units are out of service because of maintenance and forced outage and are unavailable for commitment.

Base load units: These units are online and have their generation specified for certain time period. Nuclear or very large steam units are usually used as base load units.

Cycling units: These units are cycled, that is, switched on and off satisfying the operating constraints.

Peaking units: These are the units that may be online during peak load periods. Gas turbine units are usually used as peaking units.

Group constraint: It limits the summation of dispatches of a group of units due to the limitation of transmission line capacity, etc.

Start up cost: It is a function of down time of the unit as described in Sect. 2.1.1.

Hydro constraints: It limits the amount of water to be used over a scheduling period.

4 Objective Function and Constraints of Unit Commitment Problem

The objective of unit commitment is to minimize the system's total operating cost. This includes fuel cost and start-up cost and is given as

$$\sum_{t=1}^{T}\left\{\sum_{i=1}^{M}[C_i(P_{i,t}) + S_i(x_{i,t}, U_{i,t})]\right\}. \tag{1}$$

The total emission from the system is

$$\sum_{t=1}^{T}\{\sum_{i=1}^{M}[E_i(P_{i,t})U_{i,t}]\}. \tag{2}$$

The main objective of a unit commitment problem considering emissions is to minimize the functions in (1) and (2). The problem may be formulated by treating (2) as a duality constraint rather than as a dual objective (Kuloor et al. 1992). The cost objective (1) is then augmented by (2) using Lagrangian multiplier ω as follows:

$$\min\sum_{t=1}^{T}\{\sum_{i=1}^{M}[C_i(P_{i,t}) + S_i(x_{i,t}, U_{i,t}) + \omega_i E_i(P_{i,t})U_{i,t}]\}. \tag{3}$$

Varying ω_i, called emission weighting factor, leads to different solutions for $P_{i,t}$ and hence different values to (2). In this manner, we can vary ω_i to mitigate the impact of the emission of the corresponding unit even though it is not possible to eliminate emission entirely.

The constraints to be considered are as follows:

1. Power balance constraints

$$\sum_{i=1}^{M} P_{i,t} + \sum_{h=1}^{H} P_{h,t} = D_t + Ploss_t. \tag{4}$$

2. Spinning reserve constraints

$$\sum_{i=1}^{M} [r_{i,t} + P_{i,t}] + \sum_{h=1}^{H} \overline{P_h} \geq R_t + D_t + Ploss_t \tag{5}$$
$$r_{i,t} + P_{i,t} \leq \overline{P_i}$$
$$r_{i,t} \leq \overline{r_i}.$$

3. Capacity limits of generating units

$$\underline{P_i} \leq P_{i,t} \leq \overline{P_i}$$
$$0 \leq P_{h,t} \leq \overline{P_h}. \tag{6}$$

4. Ramp rate constraints

$$(P_{i,t+1} - \Delta_i) \leq P_{i,t} \leq (P_{i,t+1} + \Delta_i). \tag{7}$$

5. Minimum generation for the first and last hour

$$P_{i,t} = \underline{P_i} \qquad for\ t = 1\ and\ t = T. \tag{8}$$

6. Minimum up time constraints

$$x_{i,t} \geq Tumin_i. \tag{9}$$

7. Minimum down time constraints

$$x_{i,t} \geq Tdmin_i. \tag{10}$$

8. Hydro constraints

$$\sum_{t=1}^{T} q_h(P_{h,t}) = qtot_h, \tag{11}$$

where

$$qtot_h = Si_h - Sf_h + f_h * T. \tag{12}$$

C_i, E_i, and q_h are considered as quadratic functions of power output, and S_i is a function of down time. The transmission loss is represented using a general transmission loss formula (Elgerd 1971) whose expression has a similar quadratic form to the B matrix loss formulation as

$$Ploss_t = \sum_{s=1}^{M+H} \sum_{k=1}^{M+H} B_{s,k} P_{s,t} P_{k,t} + \sum_{s=1}^{M+H} B1_s P_{s,t} + B0, \tag{13}$$

where

$$B_{s,k} = \frac{R_{s,k}}{|V_s||V_k|} Sin\theta_{s,k}.$$

The unknown variables are the commitment states of thermal units and the output powers of thermal and hydro units.

5 Lagrangian Relaxation Approach

In the Lagrangian relaxation approach, unit commitment problem is formulated in terms of (1) a cost function, that is, the sum of terms each involving a single unit, (2) a set of constraints involving a single unit, and (3) a set of coupling constraints, one for each hour in the study period, involving all units. An approximate solution to this problem can be obtained by adjoining the coupling constraints onto the cost using Lagrangian multipliers to form a Lagrange function and a dual function. The Lagrange function is then decoupled into small subproblems, which are solved separately with the remaining constraints. Meanwhile, the dual function is maximized with respect to the Lagrangian multipliers, usually by a series of iterations.

5.1 Lagrangian Dual Problem

The unit commitment problem defined by (3)–(13) is known as the primal problem, and the unknown variables are denoted as the primal variables. In the primal problem, only the power balance constraints (4) and the spinning reserve constraints (5) are coupling constraints that link the operation of the generating units. These coupling constraints are relaxed by adjoining them onto the cost function using two sets of Lagrangian multipliers λ and μ with the components of μ(i.e., μ_t) confined to be positive. If we use the transmission loss represented by (13) in (4) and (5), we find difficulty in separating the Lagrange function into thermal and hydro subproblem because of the off-diagonal loss coefficients. We can overcome this problem by carrying out the following modifications.

The transmission loss:

$$
\begin{aligned}
Ploss_t &= \sum_{s=1}^{M+H} \sum_{k=1}^{M+H} B_{s,k} P_{s,t} P_{k,t} + \sum_{s=1}^{M+H} B1_s P_{s,t} + B0 \\
&= \sum_{s=1}^{M+H} P_{s,t} \sum_{k=1}^{M+H} B_{s,k} P_{k,t} + \sum_{s=1}^{M+H} B1_s P_{s,t} + B0 \\
&= \sum_{s=1}^{M+H} P_{s,t} \left[\sum_{k=1}^{M} B_{s,k} P_{k,t} + \sum_{k=M+1}^{M+H} B_{s,k} P_{k,t} \right] + \sum_{s=1}^{M+H} B1_s P_{s,t} + B0 \\
&= \sum_{s=1}^{M} P_{s,t} \left[\sum_{k=1}^{M} B_{s,k} P_{k,t} + \sum_{k=M+1}^{M+H} B_{s,k} P_{k,t} \right] + \sum_{s=1}^{M} B1_s P_{s,t} \\
&\quad + \sum_{s=M+1}^{M+H} P_{s,t} \left[\sum_{k=1}^{M} B_{s,k} P_{k,t} + \sum_{k=M+1}^{M+H} B_{s,k} P_{k,t} \right] + \sum_{s=M+1}^{M+H} B1_s P_{s,t} + B0.
\end{aligned}
$$

The cross coupling terms involved are for $B_{s,k}$, where $s \neq k$, s and k are either a hydro and a thermal unit, respectively, or a thermal and a hydro unit, respectively. The term $B_{s,k}$ is dependent on $R_{s,k}$. Since $R_{s,k}$ is much less than $R_{s,s}$ and $R_{k,k}$, which is its diagonal elements for either its row or column of the bus impedance matrix, the cross coupling terms can be neglected to make the dual problem separable. The coefficients involved are two identical matrices of rank $M \times H$. The resulting equation for the transmission loss then becomes

$$
\sum_{s=1}^{M} P_{s,t}^2 B_{s,s} + \sum_{s=1}^{M} P_{s,t} B1_s + \sum_{s=M+1}^{M+H} P_{s,t}^2 B_{s,s} + \sum_{s=M+1}^{M+H} P_{s,t} B1_s + B0.
$$

This results in the following Lagrangian dual problem:

$$
\max F(\lambda, \mu)\, with\ all\ \mu_t \geq 0, \tag{14}
$$

where

$$
\begin{aligned}
F(\lambda, \mu) = \min_{P_{i,t}, U_{i,t}, P_{h,t}} \sum_{t=1}^{T} \Bigg\{ & \sum_{i=1}^{M} [C_i(P_{i,t}) + S_i(x_{i,t}, U_{i,t}) + \omega_i E_i(P_{i,t}) U_{i,t}] \\
& -\lambda_t \left(\sum_{i=1}^{M} \left[P_{i,t} - P_{i,t}^2 B_{i,i} - P_{i,t} B1_i \right] + \sum_{h=1}^{H} \left[P_{h,t} - P_{h,t}^2 B_{h,h} - P_{h,t} B1_h \right] \right)
\end{aligned}
$$

$$-\mu_t\left(\sum_{i=1}^{M}\left[r_{i,t}+P_{i,t}-(r_{i,t}+P_{i,t})^2B_{i,i}-(r_{i,t}+P_{i,t})B1_i\right]\right.$$
$$\left.\left.+\sum_{h=1}^{H}\left[\overline{P_h}-\overline{P_h}^2B_{h,h}-\overline{P_h}B1_h\right]\right)+\lambda_t(D_t+B0)+\mu_t(D_t+R_t+B0)\right\} \tag{15}$$

subject to (6)–(12).

The inequalities, $\mu_t \geq 0$, bound the domain of the dual function $F(\lambda, \mu)$ such that it cannot exceed the objective function of the primal problem. Hence, the maximal of $F(\lambda, \mu)$ is the closest bound that can be provided by this function for the primal optimal. Although $B_{s,k}$, $s \neq k$ coefficients are neglected in the formulation of the dual problem, the same is not done in the primal problem, that is, during checking the violation of the power balance and the spinning reserve constraints.

5.2 Solution of the Dual Problem

The solution process to the dual problem is an iterative that has the following two major steps:

1. $F(\lambda, \mu)$ is determined by minimizing the right-hand side of (15) for a fixed setting of Lagrangian multipliers. This proposes a solution for the primal variables.
2. The convergence test is performed. If all the convergence criteria described below are fulfilled, the dual optimal is found. Otherwise, the multipliers are updated and step (1) is performed.

The following two convergence criteria are used:

(a) The relative difference between the dual function value at the end of $(m+1)$th iteration and at the end of mth iteration is less than or equal to a tolerance value ε_d.
(b) The relative difference between the square norm of the Lagrangian multiplier vector at the end of $(m+1)$th iteration and at the end of mth iteration is less than or equal to a tolerance value ε_L.

Since the criterion for the proper choice of λ and μ is the maximization of the dual function, methods such as subgradient and variable metric methods for updating the Lagrangian multipliers may be used to achieve the goal. As the application of the variable metric method has shown a faster convergence rate in this updating process (Aoki et al. 1987), this method is used in this paper.

If the constant terms are removed, (15) can be separated into two sets of subproblems, where each subproblem deals with only one generating unit. The classification of the subproblems is based on the types of generating units, as follows:

(a) Thermal subproblem:

$$\min_{P_{i,t}, U_{i,t}} L_i,$$

where

$$L_i \equiv \sum_{t=1}^{T}\{[C_i(P_{i,t}) + S_i(x_{i,t}, U_{i,t}) + \omega_i E_i(P_{i,t})U_{i,t}]$$
$$-\lambda_t[P_{i,t} - P_{i,t}^2 B_{i,i} - P_{i,t}B1_i] - \mu_t[r_{i,t} + P_{i,t} - (r_{i,t} + P_{i,t})^2 B_{i,i}$$
$$-(r_{i,t} + P_{i,t})B1_i]\} \tag{16}$$

subject to (6)–(10).
(b) Hydro subproblems:

$$\min_{P_{h,t}} \sum_{t=1}^{T}(-\lambda_t[P_{h,t} - P_{h,t}^2 B_{h,h} - P_{h,t}B1_h] - \mu_t[\overline{P_h} - \overline{P_h}^2 B_{h,h} - \overline{P_h}B1_h]) \tag{17}$$

subject to (6), (11), and (12).

5.3 *Solving Thermal Subproblems*

Thermal subproblems may be solved independently using dynamic programming without discretizing generation levels (Guan et al. 1992). The formulation may be modified to take into account nonlinear functions for thermal generation cost and variable transmission losses. Two types of thermal units, steam turbine and gas turbine, are considered. Some of the thermal units even have ramp rate constraints. This variation of thermal units calls for various solution methods as described below.

5.3.1 Steam Turbine Unit Without Ramp Rate Constraints

The solution method for a thermal (steam turbine unit) subproblem without ramp rate constraints is presented first. For the objective function in (16) with a fixed setting of λ and μ, the non-start-up cost is defined as

$$f_i(P_{i,t}, x_{i,t}) \equiv C_i(P_{i,t}) + \omega_i E_i(P_{i,t}) - \lambda_t[P_{i,t} - P_{i,t}^2 B_{i,i} - P_{i,t}B1_i]$$
$$-\mu_t[r_{i,t} + P_{i,t} - (r_{i,t} + P_{i,t})^2 B_{i,i} - (r_{i,t} + P_{i,t})B1_i]. \tag{18}$$

Equation (16) can be rewritten as

$$L_i = \sum_{t=1}^{T}[f_i(P_{i,t}, x_{i,t}) + S_i(x_{i,t}, U_{i,t})]. \tag{19}$$

We know that L_i is step-wise additive, there are no dynamics on the generation levels, and the start-up cost $S_i(x_{i,t}, U_{i,t})$ is independent of generation $P_{i,t}$. Hence, the optimal generation level $P^*_{i,t}$ at time t for an up state ($x_{i,t} > 0$) can be obtained by minimizing $f_i(P_{i,t}, x_{i,t})$ subject to the first and last hour generation constraint, (8). That is,

$$P^*_{i,t} = \arg\min_{P_{i,t}} f_i(P_{i,t}, x_{i,t}) \tag{20}$$

provided (8) is not active. Otherwise, $P^*_{i,t} = \underline{P_i}$. Substituting quadratic functions for cost and emission,

$$\begin{aligned} C_i(P_{i,t}) &= b0_i + b1_i P_{i,t} + b2_i P^2_{i,t} \\ E_i(P_{i,t}) &= e0_i + e1_i P_{i,t} + e2_i P^2_{i,t}, \end{aligned}$$

into (18), we obtain

$$\begin{aligned} f_i(P_{i,t}, x_{i,t}) = b0_i + b1_i P_{i,t} + b2_i P^2_{i,t} + \omega_i(e0_i + e1_i P_{i,t} + e2_i P^2_{i,t}) \\ -\lambda_t[P_{i,t} - P^2_{i,t} B_{i,i} - P_{i,t} B1_i] \\ -\mu_t[\overline{P_i} - \overline{P_i}^2 B_{i,i} - \overline{P_i} B1_i]. \end{aligned}$$

Using the condition for optimum $\partial f_i / \partial P_{i,t} = 0$ and constraining $P_{i,t}$ between the minimum and maximum generation levels, we obtain the solution to (20) as

$$b1_i + 2b2_i P_{i,t} + \omega_i(e1_i + 2e2_i P_{i,t}) - \lambda_t[1 - 2P_{i,t} B_{i,i} - B1_i] = 0$$

or

$$\begin{aligned} P_{i,t} &= \{\lambda_t(1 - B1_i) - b1_i - \omega_i e1_i\}/\{2(b2_i + \omega_i e2_i + \lambda_t B_{i,i})\} \\ P^*_{i,t} &= \min\{\max\{P_{i,t}, \underline{P_i}\}, \overline{P_i}\}. \end{aligned} \tag{21}$$

For each unit, the spinning reserve contribution $r_{i,t}$ is given by

$$r_{i,t} = \overline{P_i} - P^*_{i,t}.$$

If $r_{i,t} > \overline{r_i}$, then after fixing $r_{i,t} = \overline{r_i}$ (18) becomes

$$\begin{aligned} f_i(P_{i,t}, x_{i,t}) = b0_i + b1_i P_{i,t} + b2_i P^2_{i,t} + \omega_i(e0_i + e1_i P_{i,t} + e2_i P^2_{i,t}) \\ -\lambda_t[P_{i,t} - P^2_{i,t} B_{i,i} - P_{i,t} B1_i] \\ -\mu_t[\overline{r_i} + P_{i,t} - (\overline{r_i} + P_{i,t})^2 B_{i,i} - (\overline{r_i} + P_{i,t}) B1_i]. \end{aligned}$$

Using the condition for optimum as before, we get

$$\begin{aligned} b1_i + 2b2_i P_{i,t} + \omega_i(e1_i + 2e2_i P_{i,t}) - \lambda_t[1 - 2P_{i,t} B_{i,i} - B1_i] \\ -\mu_t[1 - 2(\overline{r_i} + P_{i,t}) B_{i,i} - B1_i] = 0 \end{aligned}$$

or

$$P_{i,t} = \frac{\lambda_t(1 - B1_i) + \mu_t(1 - 2\overline{r_i}B_{i,i} - B1_i) - b1_i - \omega_i e1_i}{2(b2_i + \omega_i e2_i + \lambda_t B_{i,i} + \mu_t B_{i,i})}$$

$$P^*_{i,t} = \min\{\max\{P_{i,t}, \underline{P}_i\}, \overline{P_i}\},$$

provided (8) is not active. Otherwise, $P^*_{i,t} = \underline{P}_i$.

The time varying start-up cost is a linear function of down time as shown in Fig. 1. The start-up cost remains constant after the cold start up time. The number of down states required to describe the different start-up costs at a particular hour is therefore equal to the cold start-up time Tc_i. Again, a unit can be kept on and shut down after it is up for the minimum up time. Hence the required number of up states is the minimum up time plus one, where the extra one is needed to consider last hour generation. Using the above analysis for down and up states, the state transition diagram can be depicted as in Fig. 2. In the figure, each node indicates a state and each edge with an arrow represents a possible state transition. The non-start-up and start-up costs are associated with nodes (states) and edges (state transition), respectively. All the nodes corresponding to up states at hour t have the same generation level

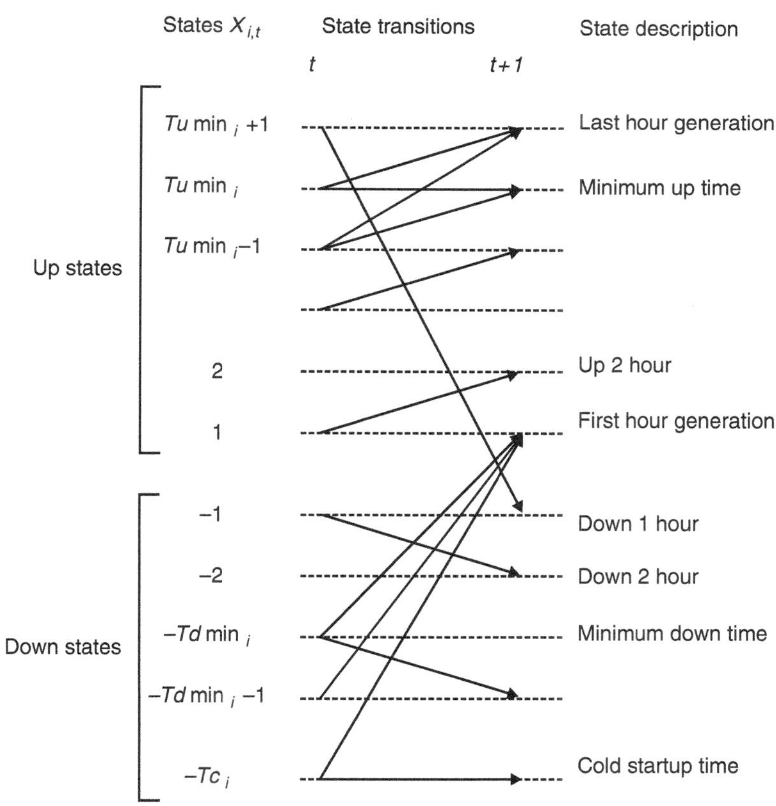

Fig. 2 State transition diagram for steam turbine unit

and therefore the same non-start-up cost with the possible exceptions of those for the first and last hour generations. The first and last hour generations for units with constraint (8) must satisfy the constraint. The generation level and cost are zero for all the down state nodes. Following this state transition diagram, the optimal commitment and generation of unit i can be obtained by using dynamic programming that requires a few states and well-structured state transitions at each hour.

5.3.2 Steam Turbine Unit with Ramp Rate Constraints

If we consider the ramp rate constraint (7) for a steam turbine unit, then the generation levels at two consecutive hours are coupled. Hence by using additional sets of Lagrangian multipliers $v1_i$ and $v2_i$ to relax the ramp up and ramp down constraints, respectively, the optimal generation at hour t is obtained. The cost function in (16) then becomes

$$L_i^{/}(\lambda, \mu, v1_i, v2_i) \equiv L_i(\lambda, \mu) + \sum \{v1_{i,t}[P_{i,t+1} - \Delta_i - P_{i,t}] + v2_{i,t}[P_{i,t} - P_{i,t+1} - \Delta_i]\}. \tag{22}$$

By rearranging terms in (22) according to hours, subproblem can be rewritten as

$$\min_{\substack{U_{i,t}\\P_{i,t}}} L_i^{/}(\lambda, \mu, v1_i, v2_i),$$

with

$$L_i^{/}(\lambda, \mu, v1_i, v2_i) \equiv \sum [h_i(P_{i,t}, x_{i,t}, v1_i, v2_i) + S_i(x_{i,t}, U_{i,t}) - \Delta_i(v1_{i,t} + v2_{i,t})] \tag{23}$$

subject to (6) and (8–10).

In (23),

$$h_i(P_{i,t}, x_{i,t}, v1_i, v2_i) \equiv f_i(P_{i,t}, x_{i,t}) + [v2_{i,t} - v1_{i,t} + v1_{i,t-1} - v2_{i,t-1}]P_{i,t} \quad t = 2, 3, \ldots, T-1 \tag{24}$$

$$h_i(P_{i,1}, x_{i,1}, v1_i, v2_i) \equiv f_i(P_{i,1}, x_{i,1}) + [v2_{i,1} - v1_{i,1}]P_{i,1} \tag{25}$$

$$h_i(P_{i,T}, x_{i,T}, v1_i, v2_i) \equiv f_i(P_{i,T}, x_{i,T}) + [v1_{i,T-1} - v2_{i,T-1}]P_{i,T}, \tag{26}$$

where $v1_i$ and $v2_i$ are slack vectors like λ and μ.

The optimal generation for each hour can be obtained following (20)

$$P_{i,t}^{*} = \arg\min_{P_{i,t}} h_i(P_{i,t}, x_{i,t}, v1_i, v2_i), \tag{27}$$

provided (8) is not active. Otherwise, $P_{i,t}^{*} = \underline{P}_i$. Dynamic programming can then be applied to get the optimal commitment and generation of unit i based on the state transition diagram shown in Fig. 2.

Finding Optimal Generation at Hour 1

Using the condition for optimum $\partial h_i/\partial P_{i,1} = 0$ and constraining $P_{i,1}$between the minimum and maximum generation levels, we obtain the solution to (27) as

$$b1_i + 2b2_i P_{i,1} + \omega_i(e1_i + 2e2_i P_{i,1}) - \lambda_1[1 - 2P_{i,1}B_{i,i} - B1_i] + v2_{i,1} - v1_{i,1} = 0$$

or

$$\begin{aligned} P_{i,1} &= \{\lambda_1(1 - B1_i) - b1_i - \omega_i e1_i - v2_{i,1} + v1_{i,1}\}/\{2(b2_i + \omega_i e2_i + \lambda_1 B_{i,i})\} \\ P^*_{i,1} &= \min\{\max\{P_{i,1}, \underline{P_i}\}, \overline{P_i}\}. \end{aligned}$$

Now, $r_{i,1} = \overline{P_i} - P^*_{i,1}$. If $r_{i,1} > \overline{r_i}$, then after fixing $r_{i,1} = \overline{r_i}$, (25) becomes

$$\begin{aligned} h_i(P_{i,1}, x_{i,1}, v1_i, v2_i) = {} & b0_i + b1_i P_{i,1} + b2_i P^2_{i,1} + \omega_i(e0_i + e1_i P_{i,1} + e2_i P^2_{i,1}) \\ & -\lambda_1[P_{i,1} - P^2_{i,1}B_{i,i} - P_{i,1}B1_i] - \mu_1[\overline{r_i} + P_{i,1} \\ & -(\overline{r_i} + P_{i,1})^2 B_{i,i} - (\overline{r_i} + P_{i,1})B1_i] + [v2_{i,1} - v1_{i,1}]P_{i,1}. \end{aligned}$$

Using the condition for optimum as before, we get

$$\begin{aligned} & b1_i + 2b2_i P_{i,1} + \omega_i(e1_i + 2e2_i P_{i,1}) - \lambda_1[1 - 2P_{i,1}B_{i,i} - B1_i] \\ & -\mu_1[1 - 2(\overline{r_i} + P_{i,1})B_{i,i} - B1_i] + v2_{i,1} - v1_{i,1} = 0 \end{aligned}$$

or

$$P_{i,1} = \frac{\lambda_1(1 - B1_i) + \mu_1(1 - 2\overline{r_i}B_{i,i} - B1_i) - b1_i - \omega_i e1_i - v2_{i,1} + v1_{i,1}}{2(b2_i + \omega_i e2_i + \lambda_1 B_{i,i} + \mu_1 B_{i,i})}$$

$$P^*_{i,1} = \min\{\max\{P_{i,1}, \underline{P_i}\}, \overline{P_i}\},$$

provided (8) is not active. Otherwise, $P^*_{i,1} = \underline{P_i}$.

Finding Optimal Generation at Hour T

Using the condition for optimum $\partial h_i/\partial P_{i,T} = 0$ and constraining $P_{i,T}$ between the minimum and maximum generation levels, we obtain the solution to (27) as

$$P_{i,T} = \{\lambda_T(1 - B1_i) - b1_i - \omega_i e1_i - v1_{i,T-1} + v2_{i,T-1}\}/\{2(b2_i + \omega_i e2_i + \lambda_T B_{i,i})\}$$

$$P^*_{i,T} = \min\{\max\{P_{i,T}, \underline{P_i}\}, \overline{P_i}\}.$$

Now, $r_{i,T} = \overline{P_i} - P^*_{i,T}$. If $r_{i,T} > \overline{r_i}$, then after fixing $r_{i,T} = \overline{r_i}$ in (26) and using the condition for optimum as before, we get

$$P_{i,T} = \frac{\lambda_T(1 - B1_i) + \mu_T(1 - 2\overline{r_i}B_{i,i} - B1_i) - b1_i - \omega_i e1_i - v1_{i,T-1} + v2_{i,T-1}}{2(b2_i + \omega_i e2_i + \lambda_T B_{i,i} + \mu_T B_{i,i})}$$

$$P^*_{i,T} = \min\{\max\{P_{i,T}, \underline{P_i}\}, \overline{P_i}\},$$

provided (8) is not active. Otherwise, $P^*_{i,T} = \underline{P_i}$.

Finding Optimal Generation at Hour t (other than 1 and T)

Using the condition for optimum $\partial h_i/\partial P_{i,t} = 0$ and constraining $P_{i,t}$ between the minimum and maximum generation levels, we obtain the solution to (27) as

$$\begin{aligned} P_{i,t} &= \{\lambda_t(1 - B1_i) - b1_i - \omega_i e1_i - v2_{i,t} + v1_{i,t} - v1_{i,t-1} + v2_{i,t-1}\}/ \\ &\quad \{2(b2_i + \omega_i e2_i + \lambda_t B_{i,i})\} \\ P^*_{i,t} &= \min\{\max\{P_{i,t}, \underline{P_i}\}, \overline{P_i}\}. \end{aligned}$$

Now, $r_{i,t} = \overline{P_i} - P^*_{i,t}$. If $r_{i,t} > \overline{r_i}$, then after fixing $r_{i,t} = \overline{r_i}$ in (24) and using the condition for optimum as before, we get

$$P_{i,t} = \frac{\lambda_t(1 - B1_i) + \mu_t(1 - 2\overline{r_i}B_{i,i} - B1_i) - b1_i - \omega_i e1_i - v2_{i,t} + v1_{i,t} - v1_{i,t-1} + v2_{i,t-1}}{2(b2_i + \omega_i e2_i + \lambda_t B_{i,i} + \mu_t B_{i,i})}$$

$$P^*_{i,t} = \min\{\max\{P_{i,t}, \underline{P_i}\}, \overline{P_i}\},$$

provided (8) is not active. Otherwise, $P^*_{i,t} = \underline{P_i}$.

Let $L'_i(\lambda, \mu, v1_i, v2_i)$ be the optimal Lagrangian for (23). The multipliers $v1_i$ and $v2_i$ are updated at an intermediate level by a variable metric method to maximize the Lagrangian, that is,

$$\begin{matrix} L'^*_i(\lambda, \mu, v1_i, v2_i) \\ v1_i \\ v2_i \end{matrix} \tag{28}$$

and the subproblem is solved at the low level as if there were no ramp rate constraints. The variable metric method to update the multipliers λ, μ, $v1_i$, and $v2_i$ is presented in Sect. 6.1.

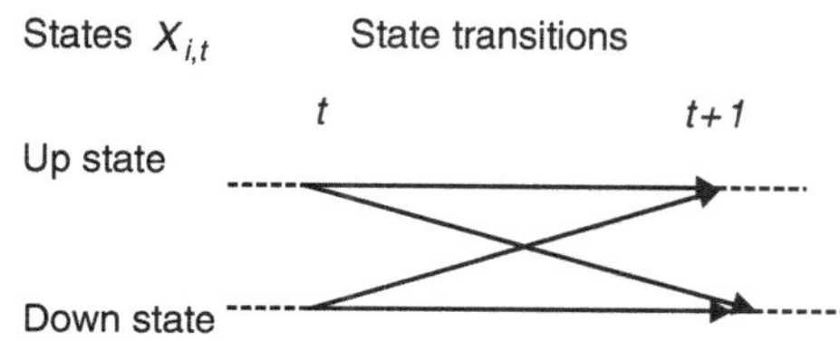

Fig. 3 State transition diagram for gas turbine unit

5.3.3 Gas Turbine Unit

A gas turbine unit does not have ramp rate and minimum up/down time constraints. It has negligible start-up cost. Hence its cost function calculation is the same as described above but neglecting the start-up cost. Since minimum up/down time is not considered here, only two states (up and down) are needed in the state transition diagram as shown in Fig. 3.
Hence, (19) can be rewritten as

$$L_i = \sum_{t=1}^{T} f_i(P_{i,t}, x_{i,t}),$$

and the optimal generation level $P_{i,t}^*$ at time t for an up state ($x_{i,t} > 0$) can therefore be obtained by minimizing $f_i(P_{i,t}, x_{i,t})$. That is,

$$P_{i,t}^* = \arg\min_{P_{i,t}} f_i(P_{i,t}, x_{i,t}).$$

Now, rewriting (21), we get

$$P_{i,t} = \{\lambda_t(1 - B1_i) - b1_i - \omega_i e1_i\}/\{2(b2_i + \omega_i e2_i + \lambda_t B_{i,i})\}$$
$$P_{i,t}^* = \min\{\max\{P_{i,t}, \underline{P_i}\}, \overline{P_i}\}.$$

Now, $r_{i,t} = \overline{P_i} - P_{i,t}^*$. If $r_{i,t} > \overline{r_i}$, then we get

$$P_{i,t} = \frac{\lambda_t(1 - B1_i) + \mu_t(1 - 2\overline{r_i} B_{i,i} - B1_i) - b1_i - \omega_i e1_i}{2(b2_i + \omega_i e2_i + \lambda_t B_{i,i} + \mu_t B_{i,i})}$$

$$P_{i,t}^* = \min\{\max\{P_{i,t}, \underline{P_i}\}, \overline{P_i}\}.$$

The generation level and cost are zero for all the down state nodes. Based on the state transition diagram shown in Fig. 3, the optimal commitment and generation of unit i can be obtained by using dynamic programming with two states and well structured state transitions at each hour.

5.4 Solving Hydro Subproblems

In the hydro subproblem defined by (17), the unknown variables are the output levels of a hydro unit. To solve for the output levels of hydro units, any hydrothermal scheduling solution approach, such as nonlinear network flows or nonlinear programming techniques, may be used with a thermal unit commitment schedule obtained by solving thermal subproblems. An efficient hydrothermal scheduling algorithm that is capable of handling nonlinear functions for water discharge characteristics, thermal cost, and transmission loss constraints is described in Sect. 7.

6 Solution Methodology

To solve hydrothermal scheduling, we need only the commitment states of thermal units, not those of hydro units. In the Lagrangian relaxation approach, commitment states of thermal units can be obtained by solving thermal subproblems only. Hence the unit commitment solution approach follows the flow presented in Fig. 4.

In the initial iteration, the hydro units' generations are set to zero, and thermal subproblems are solved. During maximization of the dual function, Lagrangian multipliers are updated using the hydro units' generations and the solution of the thermal subproblems. The variable metric method as shown in Sect. 6.1 is used for updating the multipliers. The solution of hydrothermal scheduling is used to reset the values of the hydro units' generations. In the subsequent iteration, thermal subproblems and hydrothermal scheduling are solved as shown in the figure. If the power balance and/or the spinning reserve constraints are not satisfied, a suboptimal feasible solution is searched where the Lagrangian multipliers are adjusted by the linear interpolation method (Tong and Shahidehpour 1990) described in Sect. 6.2.

The refinement algorithm described in Sect. 6.3 is used. The expert system recommends modifying specific input data for the program if the result is found operationally unacceptable. The operator resets the input data and repeats the whole cycle until an operationally feasible and/or preferable solution is found.

6.1 Variable Metric Method for Dual Optimization

Let $\delta_{\ell,t}$ and $\xi_{\ell,t}$ be the subgradients for λ_t and μ_t, respectively, where the subscript ℓ indicates the iteration counts. Then $\delta_{\ell,t}$ and $\xi_{\ell,t}$ are defined as follows:

$$\delta_{\ell,t} = D_t + Ploss_t - \left(\sum_{i=1}^{M} P_{i,t} + \sum_{h=1}^{H} P_{h,t} \right) \tag{29}$$

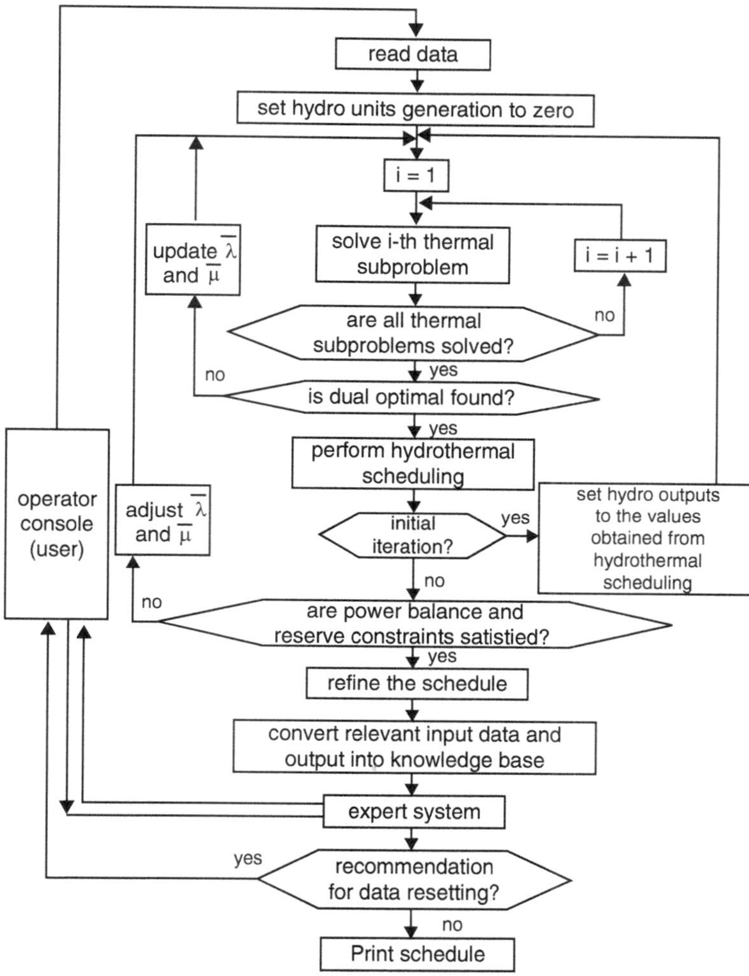

Fig. 4 Unit commitment algorithm

$$\xi_{\ell,t} = R_t + D_t + Ploss_t - \left(\sum_{i=1}^{M} [r_{i,t} + P_{i,t}] + \sum_{h=1}^{H} \overline{P_h} \right). \tag{30}$$

By using $\delta_{\ell,t}$ and $\xi_{\ell,t}$ defined by (29) and (30), the Lagrangian multipliers λ and μ are updated to maximize the dual function by the following variable metric updating rule (Aoki et al. 1987):

$$\begin{bmatrix} \lambda_{t+1} \\ \mu_{t+1} \end{bmatrix} = \max \left(0, \begin{bmatrix} \lambda_t \\ \mu_t \end{bmatrix} + \omega_\ell H_\ell \begin{bmatrix} \delta_t \\ \xi_t \end{bmatrix} \right) \tag{31}$$

$$\omega_\ell = \beta_\ell / \left| H_\ell \begin{bmatrix} \delta_t \\ \xi_t \end{bmatrix} \right| \tag{32}$$

$$\beta_\ell = 1/(a + b * \ell),\ a > 0,\ b > 0 \tag{33}$$

$$H_{\ell+1} = H_\ell + \gamma (H_\ell \eta_\ell)(H_\ell \eta_\ell)^T / \eta_\ell^T H_\ell \eta_\ell \tag{34}$$

$$H_0 = E,\ 0 < \gamma < 1 \tag{35}$$

$$\eta_\ell = \begin{bmatrix} \delta_\ell \\ \xi_\ell \end{bmatrix} - \begin{bmatrix} \delta_{\ell-1} \\ \xi_{\ell-1} \end{bmatrix}, \tag{36}$$

where δ_ℓ and ξ_ℓ are vectors composed of $\delta_{\ell,t}$ and $\xi_{\ell,t}$ respectively. "T" and "E" represent the transpose operation and the identity matrix.

Note that β_ℓ in (33) represents the step size satisfying the following two conditions:

1. Converges to zero, that is, $\lim_{\ell \to \alpha} \beta_\ell = 0$
2. Does not converge to a point excepting the solution, that is, $\sum_\ell \beta_\ell \to \alpha$

The multipliers $v1_i$ and $v2_i$ for relaxing the ramp rate constraint are updated in the same way as in (29)–(36) with λ and μ replaced by $v1_i$ and $v2_i$. The iteration index and the subgradients, however, are different. For each high level iteration with fixed λ and μ,$v1_i$ and $v2_i$ are updated at the intermediate level until the dual cost function in (28) cannot be improved. The subgradients of $L_i^{/*}$ for $v1_i$ and $v2_i$ are

$$g_{v1,t} = P_{i,t+1} - \Delta_i - P_{i,t}$$

and

$$g_{v2,t} = P_{i,t} - [P_{i,t+1} + \Delta_i].$$

6.2 Linear Interpolation Method for Suboptimal Feasible Solution

Because of no-convexity of the primal problem, the combined solution of the subproblems that corresponds to the dual optimal point seldom satisfies the relaxed constraints, that is, the power balance and the reserve constraints. However, the dual optimal is a sharp lower bound of the primal problem. A feasible suboptimal solution is therefore searched near the dual optimal point. The 2T Lagrangian multipliers λ and μ cause the interactions of the decoupled subproblems in (14) through (17). These multipliers resemble the suggested market prices of energy and spinning reserve, respectively. After λ and μ are fixed, the solutions of the subproblems yield the optimal generation schedules of the units based on the suggested prices. The relationship between the unit commitment and the Lagrangian multipliers is the basis of the search for a suboptimal feasible solution near the dual optimal point. The values of λ and μ are updated repeatedly according to the violation of the relaxed constraints, and the subproblems are solved after every updating process. This iterative process is continued until all those relaxed constraints are satisfied.

To show the updating process of the searching algorithm, it is assumed that λ and μ have been updated m times and have already taken the values λ_m and μ_m, respectively. If the power balance and the reserve constraints are not yet satisfied in the period t, the tth components of the two Lagrangian vectors, λ_t and μ_t, will be changed by adding $\delta\lambda_{t,m}$ and $\delta\mu_{t,m}$ to them, respectively. The following expressions are used for determining $\delta\lambda_{t,m}$ and $\delta\mu_{t,m}$:

$$\delta\lambda_{t,m} = \frac{(D_t - G_{t,m}) * \lambda_{t,m}}{D_t}$$

$$\delta\mu_{t,m} = \frac{(D_t + R_t - H_{t,m}) * \mu_{t,m}}{R_t},$$

where

$$G_{t,m} = \left[\sum_{i=1}^{M} P_{i,t} + \sum_{h=1}^{H} P_{h,t} - Ploss_t\right]_{\lambda=\lambda_m;\mu=\mu_m}$$

$$H_{t,m} = \left[\sum_{i=1}^{M} [r_{i,t} + P_{i,t}] + \sum_{h=1}^{H} \overline{P_h} - Ploss_t\right]_{\lambda=\lambda_m;\mu=\mu_m}.$$

Each thermal unit is considered independently, and hydrothermal scheduling is not needed during the optimization of the dual function. However, during the search for a feasible suboptimal solution, the whole system is no longer decoupled and the hydrothermal scheduling is performed to determine the total generation in every hour.

6.3 Development of the Refinement Algorithm

Unnecessary commitment of some units may be possible in the solution given by the Lagrangian relaxation. To overcome this problem, a refinement process is introduced at the final stage of the algorithm. This refinement process examines some candidate units whose shutdown may contribute to further operating cost reduction. The basic building block of this refinement algorithm directly follows the approach in Tong and Shahidehpour (1990). For every period t, candidate units are searched. The candidate units are inefficient units whose shutdowns do not violate the minimum up/down time constraints. The reason for choosing the inefficient units as candidates is that the cost reduction may be achieved by shifting the loadings of these units to more efficient units. The selection process of the candidate units for a period t is as follows:

1. Choose those committed thermal units whose outputs are found to be equal to their corresponding minimum MW capacities.
2. Check the minimum up and down time constraints of the units chosen in step (1). Discard those units whose shutdown will violate the constraints. The same units may be selected as candidates in different periods. So in performing this step, the selection of candidate units in other periods should be considered too.

The selection process does not consider the hydro units because commitment variables are not associated with them. Since it is necessary to check whether it is more economical to shut down a particular candidate unit in a specific period, the total costs of operation for the following two cases should be examined:

(a) The output of the unit in that period is zero.
(b) The output of the unit in that period is assigned to be equal to its minimum capacity.

To substitute the operation of candidate units selected in each period, fictitious units are used. If a unit is chosen as a candidate for several different periods, corresponding number of fictitious units are needed to represent it in those periods. The fictitious unit used for substituting the operation of a thermal unit that is chosen as a candidate unit in a period t is designed to have the characteristics of operating from zero to the minimum output of the candidate unit. Hence, the examination of the operating limits of the fictitious unit will be used for evaluating the economic merits of the two possible operational modes of the candidate unit.

Hydrothermal scheduling is performed to determine the power generation of various units in different periods. However, additional steps are needed to determine the commitment states and the actual loading of candidate units based on the assigned output loading of fictitious units given in the dispatch solution. The following three cases will be observed in the solution:

1. The output of the fictitious unit i_t is zero. Then, the ith thermal unit in period t can be shut down.
2. The output of the fictitious unit i_t is equal to $\underline{P}_i$. The ith thermal unit should be kept on in period t and its output is equal to $\underline{P}_i$.
3. The output of the fictitious unit i_t is between zero and $\underline{P}_i$.

If the outputs of all fictitious units satisfy case (1) or (2), then no further refinement is possible. The schedule and the loading of units generated by the solution can then be used for operating the system. However, if the outputs of some of the candidate units satisfy case (3), the following steps are performed:

(a) Shut down all these candidate units
(b) Perform hydrothermal scheduling
(c) If turning all of them off violates the reserve constraints, then the units with lower incremental costs will be discarded from the candidate units until the violated constraints are satisfied and go to step (a) again. Otherwise stop.

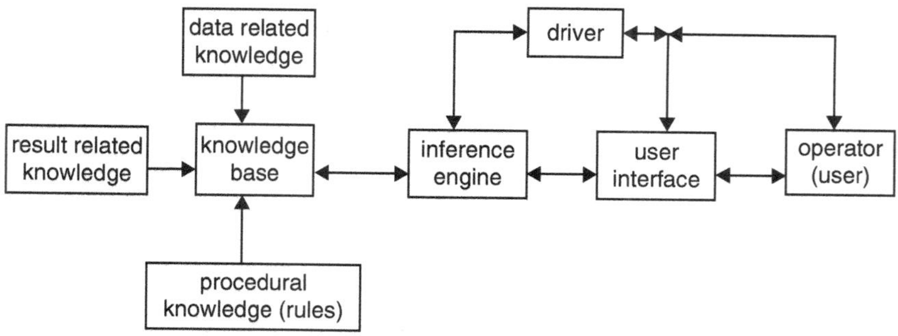

Fig. 5 Structure of the expert system

In the refinement algorithm (Tong and Shahidehpour 1990), linear programming technique is applied to find the dispatch of units, and the cost function of the fictitious unit is considered linear. They introduced a linear relaxation, which utilizes fictitious units to substitute the operation of candidate units. However, in the refinement algorithm developed in this work, the hydrothermal scheduling algorithm described in Sect. 7 is used and the nonlinear cost function is used for fictitious unit.

6.4 Unit Commitment Expert System

The structure of the expert system is shown in Fig. 5. The knowledge base consists of data-related knowledge, commitment result-related knowledge, and procedural knowledge. The data- and commitment result-related knowledge are generated by the unit commitment program. The procedural knowledge consists of the rules that direct the use of knowledge for yielding specific recommendation. The rules are developed by combining the knowledge of unit commitment experts and experienced power system operators. The inference engine uses backward chaining to find a proper recommendation. The user interface is used for asking the user to obtain information that is unavailable in the data and results of the unit commitment program. It is also used for showing the recommendation of a consultation session and reasoning behind the recommendation. The driver co-ordinates the functions of the inference engine and user interface (Salam et al. 1991).

The expert system checks for the unit commitment and loading problems. Under the unit commitment problem, the following subproblems are considered:

Gas turbine unit's cycling
Combined commitment of gas and steam turbine units
Steam turbine unit's cycling
Commitment for adequate voltage control
Commitment at a particular plant

On the other hand, the following subproblems are considered in the unit loading problem:

Gas turbine unit's loading
Largest loaded unit's loading
Loadings of units for group constraints

The user has to select one subproblem. This starts the execution of the unit commitment expert system and generates a recommendation for that subproblem.

7 Hydrothermal Scheduling

Optimal scheduling of power plant generation is the determination of the generation for every generating unit such that the total system generation cost is minimum while satisfying the system constraints. However, with insignificant marginal cost of hydro electric power, the problem of minimizing the operational cost of a hydrothermal system essentially reduces to minimizing the fuel cost for thermal units constrained by the generating limits, available water, and the energy balance condition in a given period of time [10]. Many approaches have been suggested to solve the hydrothermal scheduling problem. The proposed approaches include dynamic programming, functional analysis technique, method of local variations, principle of progressive optimality, general mathematical programming techniques, and evolutionary algorithm (Rashid and Nor 1991; Somasundaram et al. 2006). Dynamic programming has the ability to handle all the constraints enforced by the hydro subsystem. The computational requirements are, however, considerable with this technique for a realistic system size. The incremental dynamic programming technique keeps the computational requirements in a reasonable range. The method of local variations algorithm shows better performance than the incremental dynamic programming-based algorithm. Again the application of progressive optimality algorithm performs better than the method of local variations.

Generally all those methods have slow convergence characteristics. Investigations on the use of Newton–Raphson method have been carried out. Formulation of the scheduling problem in Newton–Raphson method for solving a set of nonlinear equations produces a large matrix expression. The drawbacks of Newton's method are the computation of the inverse of a large matrix, the ill-conditioning of the Jacobian matrix, and the divergence caused by starting values. Powell's hybrid method is used to avoid the divergence problem encountered by Newton–Raphson method. A method using LU factorization of the matrix in Newton's method formulation shows superiority over the hybrid Powell's method; however, the size of the matrix still remains very large requiring substantial computations.

The coordination equations may be linearized so that the Lagrangian of the water availability constraint is determined separately from the unit generations (Rashid and Nor 1991). This water availability constraint Lagrangian multiplier determines the Lagrangian multiplier for the power balance constraint and hence leads to the computation of the generation of thermal and hydro units. The algorithm requires small computing resources. It has global-like convergence property so that even if the starting values are far from the solution, convergence is still achieved rapidly. The formulation (Rashid and Nor 1991) may be modified to cope with emission constraint as described below.

Mathematically, the hydrothermal scheduling problem can be expressed as follows:

$$\sum_{t=1}^{T}\sum_{i=1}^{M} C_i(P_{i,t}) \tag{37}$$

subject to the energy balance equation

$$EB_t = \sum_{i=1}^{M} P_{i,t} + \sum_{h=1}^{H} P_{h,t} - D_t - Ploss_t = 0$$

and to the water availability constraint

$$W_h = \sum_{t=1}^{T} q_h(P_{h,t}) = qtot_h,$$

with

$$\underline{P_i} \le P_{i,t} \le \overline{P_i}$$
$$0 \le P_{h,t} \le \overline{P_h}.$$

The total emission from the system is

$$\sum_{t=1}^{T}\sum_{i=1}^{M} E_i(P_{i,t}). \tag{38}$$

The cost objective function in (37) is augmented by constraint (38) using Lagrangian multiplier ω, called the emission weighting factor, as follows:

$$\min_{P_{i,t}} z = \sum_{t=1}^{T}\sum_{i=1}^{M}\{C_i(P_{i,t}) + \omega_i E_i(P_{i,t})\}.$$

Representing cost and sulfur oxide emission of thermal unit as quadratic functions of thermal generation and water discharge rate of hydro unit as quadratic function of hydro generation, we get

$$C_i(P_{i,t}) = b0_i + b1_i P_{i,t} + b2_i P_{i,t}^2$$
$$E_i(P_{i,t}) = e0_i + e1_i P_{i,t} + e2_i P_{i,t}^2$$
$$q_h(P_{h,t}) = a0_h + a1_h P_{h,t} + a2_h P_{h,t}^2.$$

The transmission loss is represented by the following expression:

$$Ploss_t = \sum_{s=1}^{M+H} \sum_{j=1}^{M+H} B_{s,j} P_{s,} P_{j,t} + \sum_{s=1}^{M+H} B1 P_{s,t} + B0.$$

The augmented Lagrangian function L is

$$L(P_{i,t}, P_{h,t}, \lambda, \gamma) = z - \sum_{t=1}^{T} \lambda_t EB_t - \sum_{h=1}^{H} \gamma_h (W_h - qtot_h).$$

The set of coordination equations for a minimum cost operating condition is then given by

$$\begin{aligned}
\frac{\partial L}{\partial P_{i,t}} &= \frac{d}{dP_{i,t}}[C_i(P_{i,t}) + \omega_i E_i(P_{i,t})] - \lambda_t \frac{\partial EB_t}{\partial P_{i,t}} = 0 \\
\frac{\partial L}{\partial P_{h,t}} &= -\lambda_t \frac{\partial EB_t}{\partial P_{h,t}} - \gamma_h \frac{dW_h}{dP_{h,t}} = 0 \\
\frac{\partial L}{\partial \lambda_t} &= -EB_t = 0 \\
\frac{\partial L}{\partial \gamma_h} &= -W_h + qtot_h = 0.
\end{aligned}$$

Substituting the above system parameters into the minimum condition yields the following coordination equations:

$$\begin{aligned}
b1_i + 2b2_i P_{i,t} + \omega_i e1_i + 2\omega_i e2_i P_{i,t} &= \lambda_t (1 - K_{i,t}) \\
\gamma_h (a1_h + 2a2_h P_{h,t}) &= \lambda_t (1 - K_{h,t}) \\
\sum_{i=1}^{M} P_{i,t} + \sum_{h=1}^{H} P_{h,t} &= D_t + Ploss_t \\
\sum_{t=1}^{T} (a0_h + a1_h P_{h,t} + a2_h P_{h,t}^2) &= qtot_h,
\end{aligned}$$

where

$$\begin{aligned}
K_{i,t} &= \frac{\partial Ploss_t}{\partial P_{i,t}} = 2 \sum_{j=1}^{M+H} B_{i,j} P_{j,t} + B1_i \\
K_{h,t} &= \frac{\partial Ploss_t}{\partial P_{h,t}} = 2 \sum_{j=1}^{M+H} B_{h,j} P_{j,t} + B1_h.
\end{aligned}$$

These are a set of nonlinear equations of unknown variables $P_{i,t}$ (steam), $P_{h,t}$ (hydro), λ_t, γ_h, and they can only be solved iteratively. Let $P_{i,t}^{old}$, $P_{h,t}^{old}$, γ_h^{old}, λ_t^{old} be the approximate solutions at the previous iterative stage. The next iterates, that is,

$$\begin{aligned} P_{i,t}^{new} &= P_{i,t}^{old} + \delta P_{i,t}, P_{h,t}^{new} = P_{h,t}^{old} + \delta P_{h,t} \\ \gamma_h^{new} &= \gamma_h^{old} + \delta \gamma_h, \lambda_t^{new} = \lambda_t^{old} + \delta \lambda_t, \end{aligned}$$

generated by Newton's method are given as follows:

$$\begin{aligned} &2(b2_i + \omega_i e2_i)\delta P_{i,t} + 2\lambda_t^{old} \sum_{j=1}^{M+H} B_{i,j}\delta P_{j,t} - \lambda_t^{new}(1 - K_{i,t}^{old}) \\ &= -2(b2_i + \omega_i e2_i) P_{i,t}^{old} - b1_i - \omega_i e1_i \end{aligned} \quad (39)$$

$$\begin{aligned} &2a2_h \gamma_h^{old} \delta P_{h,t} + 2\lambda_t^{old} \sum_{j=1}^{M+H} B_{h,j}\delta P_{j,t} - \lambda_t^{new}(1 - K_{h,t}^{old}) \\ &+(2a2_h P_{h,t}^{old} + a1_h)\lambda_t^{new} = 0 \end{aligned} \quad (40)$$

$$\begin{aligned} &\sum_{i=1}^{M}(1 - K_{i,t}^{old})\delta P_{i,t} + \sum_{h=1}^{H}(1 - K_{h,t}^{old})\delta P_{h,t} \\ &\quad = Ploss_t^{old} + D_t - \sum_{i=1}^{M} P_{i,t}^{old} - \sum_{h=1}^{H} P_{h,t}^{old} \end{aligned} \quad (41)$$

$$\begin{aligned} \sum_{t=1}^{T}(2a2_h P_{h,t}^{old} + a1_h)\delta P_{h,t} = qtot_h - \sum_{t=1}^{T}[a2_h(P_{h,t}^{old})^2 \\ + a1_h P_{h,t}^{old} + a0_h]. \end{aligned} \quad (42)$$

Consider (39) and (40). To simplify calculations, the equations are diagonalized by neglecting all the terms with $B_{s,j}, s \neq j$ coefficients in the terms $\sum_{j=1}^{M+H} B_{i,j}\delta P_{j,t}$, $\sum_{j=1}^{M+H} B_{h,j}\delta P_{j,t}$. This yields the equations as follows:

$$\begin{aligned} &[2(b2_i + \omega_i e2_i) + 2\lambda_t^{old} B_{i,i}]\delta P_{i,t} - \lambda_t^{new}(1 - K_{i,t}) \\ &\quad = -2(b2_i + \omega_i e2_i) P_{i,t}^{old} - b1_i - \omega_i e1_i \end{aligned}$$

or

$$\delta P_{i,t} = \frac{\lambda_t^{new}(1 - K_{i,t}^{old}) - [2(b2_i + \omega_i e2_i)P_{i,t}^{old} + b1_i + \omega_i e1_i]}{2(b2_i + \omega_i e2_i) + 2\lambda_t^{old} B_{i,i}} \tag{43}$$

and $(2\gamma_h^{old} a2_h + 2\lambda_t^{old} B_{h,h})\delta P_{h,t} - \lambda_t^{new}(1 - K_{h,t}^{old})$

$$+(2a2_h P_{h,t}^{old} + a1_h)\gamma_h^{new} = 0$$

or

$$\delta P_{h,t} = \frac{\lambda_t^{new}(1 - K_{h,t}^{old}) - (2a2_h P_{h,t}^{old} + a1_h)\gamma_h^{new}}{2\gamma_h^{old} a2_h + 2\lambda_t^{old} B_{h,h}}. \tag{44}$$

Next inserting these expressions for $\delta P_{i,t}$ and $\delta P_{h,t}$ into (41) and (42), we get

$$A_t \lambda_t^{new} - \sum_{h=1}^{H} \alpha_{h,t} \gamma_h^{new} = C_t \tag{45}$$

$$\sum_{t=1}^{T} \alpha_{h,t} \lambda_t^{new} - \beta_h \gamma_h^{new} = \delta_h, \tag{46}$$

where

$$A_t = \sum_{i=1}^{M} \frac{(1 - K_{i,t}^{old})^2}{2(b2_i + \omega_i e2_i) + 2\lambda_t^{old} B_{i,i}} + \sum_{h=1}^{H} \frac{(1 - K_{h,t}^{old})^2}{2\gamma_h^{old} a2_h + 2\lambda_t^{old} B_{h,h}}$$

$$\alpha_{h,t} = \frac{(1 - K_{h,t}^{old})(2a2_h P_{h,t}^{old} + a1_h)}{2\gamma_h^{old} a2_h + 2\lambda_t^{old} B_{h,h}}$$

$$\beta_h = \sum_{t=1}^{T} \frac{(2a2_h P_{h,t}^{old} + a1_h)^2}{2\gamma_h^{old} a2_h + 2\lambda_t^{old} B_{h,h}}$$

$$C_t = Ploss_t^{old} + D_t - \sum_{i=1}^{M} P_{i,t}^{old} - \sum_{h=1}^{H} P_{h,t}^{old} + \sum_{i=1}^{M} \frac{(1 - K_{i,t}^{old})[2(b2_i + \omega_i e2_i)P_{i,t}^{old} + b1_i + \omega_i e1_i]}{2(b2_i + \omega_i e2_i) + 2\lambda_t^{old} B_{i,i}}$$

$$\delta_h = qtot_h - \sum_{t=1}^{T} [a2_h (P_{h,t}^{old})^2 + a1_h P_{h,t}^{old} + a0_h].$$

Finally, eliminating λ_t^{new} from (45) and (46), we get

$$\sum_{t=1}^{T}\sum_{j=1}^{H}\frac{\alpha_{h,t}\alpha_{j,t}\gamma_j^{new}}{A_t} - \beta_h\gamma_h^{new} = \delta_h - \sum_{t=1}^{T}\frac{\alpha_{h,t}C_t}{A_t}.$$

This is a set of H equations for γ_h^{new}. Having obtained γ_h, we solve for λ_t from (45). Then we can get the equations for $\delta P_{i,t}$ and $\delta P_{h,t}$ from (43) and (44).

8 Numerical Results

The unit commitment program implementing the algorithm (excluding the expert system) shown in Fig. 4 is written in the C language. The expert system is developed in Prolog. The test system is a practical utility system consisting of 32 thermal and 12 hydro generating units, of which seven are gas turbine units. The total thermal capacity of the system is 3,640 MW and the total hydro capacity is 848 MW. The characteristics of generating units are given in Tables 1 and 2. Numerical results presented here are based on three data sets: Case 1, Wednesday; Case 2, Saturday; and Case 3, Sunday. The daily demand curves are shown in Fig. 6. The scheduling horizon is 24 h in all cases.

The results are monitored at two stages. At the first stage, the expert system is not utilized and the environmental constraint is not enforced (i.e., $\omega = 0$ for all units). The results of this stage in terms of total production cost and CPU time requirements are shown in Table 3. The results show that the schedules are obtained in reasonable time in all cases. The schedules for cases 2 and 3 are shown in Tables 4 and 5, respectively.

At the second stage, the complete algorithm, which is iterative in nature, is applied. The complete solution process for Case 3 is given below: Step 1: Using the unit commitment solution method excluding the expert system and the environmental constraint ($\omega = 0$ for all units), a schedule with cost of operation $1,641,302 is obtained. The schedule is shown in Table 6. The total emission from this commitment is 648.95 tons. The emissions from each unit are shown in Table 7. Emissions from units 1, 2, 9, and 10 are found to be quite high. The CPU time requirement for obtaining the schedule is 42 s.

Note that the schedule and cost of this step are different from those of Case 3 of the first stage. This is because thermal units 1, 2, and 9 are used as base load units and thermal units 11, 12, 17, 18, 24, and 25 are used as must run units in this step. This status restriction has not been imposed in any case of first stage.

Step 2: To reduce emission from the above mentioned four units, unit commitment program is run again using $\omega = 200$ for these units. A schedule with operating cost $1,654,434 is obtained. The deviation of this schedule from that in Step 1 is as follows:

Thermal unit 5 is not committed at hour 10 and is committed during hours 16–23
Thermal units 6 and 8 are committed during hours 11–23

Table 1 Characteristics of thermal units

Unit	Plant	Type[a]	Minimum up time h	Minimum down time h	Maximum generation MW	Minimum generation MW	Ramp rate MW/h
1	1	ST	6	6	300	100	174
2	1	ST	6	6	300	100	174
3	2	ST	4	4	145	90	180
4	2	ST	4	4	145	90	180
5	2	ST	4	4	145	90	180
6	2	ST	4	4	145	90	180
7	2	ST	4	4	145	90	180
8	2	ST	4	4	145	90	180
9	1	ST	6	6	300	100	60
10	1	ST	6	6	300	100	60
11	3	ST	6	6	120	70	90
12	3	ST	6	6	120	70	90
13	4	ST	3	3	60	30	60
14	4	ST	3	3	60	30	60
15	4	ST	3	3	60	30	60
16	4	ST	3	3	60	30	60
17	4	ST	3	3	120	70	60
18	4	ST	3	3	120	70	60
19	4	ST	3	3	120	70	60
20	5	ST	6	6	30	12.5	30
21	5	ST	6	6	30	12.5	30
22	5	ST	6	6	30	12.5	30
23	5	ST	6	6	120	65	60
24	5	ST	6	6	120	65	60
25	5	ST	6	6	120	65	60
26	6	GT	0	0	90	30	108
27	6	GT	0	0	90	30	108
28	3	GT	0	0	20	5	21.6
29	4	GT	0	0	20	5	21.6
30	5	GT	0	0	20	5	21.6
31	7	GT	0	0	20	5	21.6
32	1	GT	0	0	20	5	21.6

[a] *ST* steam turbine unit, *GT* gas turbine unit

Thermal unit 7 is committed during hours 13–23
Thermal unit 10 is not committed at all
Hydro units 1–4 have nonzero generation during hours 7–10
Hydro units 5–8 have nonzero generation at hours 7 and 8
Hydro units 9–12 have nonzero generation at hour 3 and during hours 5–12.

The total emission from this commitment is 608.71 tons. The CPU time requirement for obtaining the schedule is 40 s. The increase of $13,132, that is, 0.8% in total cost in this step over the previous step, is due to the inclusion of the emission constraint. But this causes a reduction of 40.24 tons, that is, 6.2% in total emission.

Table 2 Characteristics of hydro units

Unit	Plant	Maximum generation MW	Minimum generation MW
1	A	100	0
2	A	100	0
3	A	100	0
4	A	100	0
5	B	87	0
6	B	87	0
7	B	87	0
8	B	87	0
9	C	25	0
10	C	25	0
11	C	25	0
12	C	25	0

Table 3 Cost and CPU time

Data set	Cost, $	CPU, s
Case 1	2,113,407	52
Case 2	1,983,504	51
Case 3	1,582,802	53

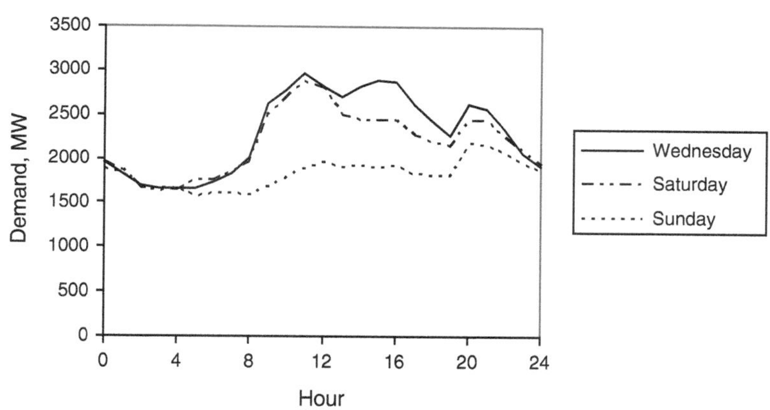

Fig. 6 Daily demand curves

The result is then analyzed using the unit commitment expert system. The expert system recommends keeping thermal unit 3 "must run" at hour 13. It also gives a reasoning shown in Fig. 7, stating violation of rules for the steam turbine unit's cycling.

Step 3: The information to keep thermal unit 3 "must run" at hour 13 is added to the data. The unit commitment program is rerun. The schedule obtained corresponds to an operating cost of $1,654,693. The schedule is shown in Table 8. The total emission from this commitment is 608.37 tons. The CPU time requirement for obtaining the schedule is 40 s. The must run constraint inclusion causes the increase of $259

Table 4 Schedule of Case 2

Hour	Thermal units[a]				Hydro units[b]
	000000000	1111111111	2222222222	333	000000001111
	123456789	0123456789	0123456789	012	123456789012
1	110111101	1000000000	0000000000	000	111111110000
2	110111101	1000000000	0000000000	000	111111110000
3	110111101	1000000000	0000000000	000	111111110000
4	110111101	1000000000	0000000000	000	111111110000
5	110111101	1000000000	0000000000	000	000011110000
6	110111101	1000000000	0000000000	000	000011110000
7	111111111	1000000000	0000000000	000	111111110000
8	111111111	1000000000	0000000000	000	000011110000
9	111111111	1000000001	0000000000	000	111111111111
10	11111111	1000000001	0000000000	000	111111111111
11	111111111	1001111001	0000000000	000	111111111111
12	111111111	1001111001	0000000000	000	111111111111
13	111111111	1001111001	0000000000	000	111111111111
14	111111111	1001111001	0000000000	000	111111111111
15	111111111	1000000000	0000000000	000	111111111111
16	111111111	1000000000	0000000000	000	111111111111
17	111111111	1000000000	0000000000	000	111111111111
18	111111111	1000000000	0000000000	000	111111110000
19	111111111	1000000000	0000000000	000	111111110000
20	111111111	1000000000	0000000000	000	111111111111
21	111111111	1000000000	0000000000	000	111111111111
22	111111111	1000000000	0000000000	000	111111111111
23	111111111	1000000000	0000000000	000	111111110000
24	111111111	1000000000	0000000000	000	000011110000

[a] 1-unit is on; 0-unit is off

[b] 1-nonzero generation; 0-zero generation

Table 5 Schedule of Case 3

Hour	Thermal units[a]				Hydro units[b]
	000000000	1111111111	2222222222	333	000000001111
	123456789	0123456789	0123456789	012	123456789012
1	110111101	0000000000	0000000000	000	111111111111
2	110111101	0000000000	0000000000	000	111111111111
3	110111101	1000000000	0000000000	000	111111110000
4	110111101	1000000000	0000000000	000	111111110000
5	110111101	1000000000	0000000000	000	111111110000
6	110111101	1000000000	0000000000	000	111111110000
7	110111101	1000000000	0000000000	000	111111110000
8	110111101	1000000000	0000000000	000	111111110000
9	110111101	1000000000	0000000000	000	111111111111
10	110111101	1000000000	0000000000	000	111111111111

(Continued)

Table 5 (Continued)

Hour	Thermal units[a]				Hydro units[b]
11	110111101	1000000000	0000000000	000	111111111111
	000000000	1111111111	2222222222	333	000000001111
	123456789	0123456789	0123456789	012	123456789012
12	111111111	1000000000	0000000000	000	111111111111
13	111111011	1000000000	0000000000	000	111111111111
14	111111011	1000000000	0000000000	000	111111111111
15	111111011	1000000000	0000000000	000	111111111111
16	111111011	1000000000	0000000000	000	111111111111
17	111111011	1000000000	0000000000	000	111111111111
18	111111011	1000000000	0000000000	000	111111111111
19	111111011	1000000000	0000000000	000	111111111111
20	111111011	1000000000	0000000000	000	111111111111
21	111111011	1000000000	0000000000	000	111111111111
22	111111011	1000000000	0000000000	000	111111111111
23	111111011	1000000000	0000000000	000	111111111111
24	111111011	1000000000	0000000000	000	111111111111

[a] 1-unit is on; 0-unit is off

[b] 1-nonzero generation; 0-zero generation

Keep thermal unit 3 must run at hour 13
was derived by rule 36 as follows

unit 3 is not switched on again
was derived by rule 31 as follows

steam turbine unit 3 at plant 2 is switched off at 13th hour
as extracted from the schedule

and

steam turbine unit 3 at plant 2 is never switched on

and

steam turbine unit 7 at plant 2 is switched on at 13th hour
as extracted from the schedule

and

unit 3 and unit 7 are not the same

and

13th hour is within 24 hours of 13th hour
as found from calculation

and

unit 3 is not on scheduled outage at any hour
as given by you

Fig. 7 Reasoning behind a recommendation

Table 6 Schedule of Step 1

Hour	Thermal units[a]				Hydro units[b]
	000000000	1111111111	2222222222	333	000000001111
	123456789	0123456789	0123456789	012	123456789012
1	111100001	0110000110	0000110000	000	111111111111
2	111100001	0110000110	0000110000	000	111111111111
3	111100001	0110000110	0000110000	000	111111110000
4	111100001	0110000110	0000110000	000	111111111111
5	111100001	0110000110	0000110000	000	111111110000
6	111100001	0110000110	0000110000	000	111111110000
7	111100001	1110000110	0000110000	000	000000000000
8	111100001	1110000110	0000110000	000	000000000000
9	111100001	1110000110	0000110000	000	000011110000
10	111110001	1110000110	0000110000	000	000011110000
11	111110001	1110000110	0000110000	000	111111110000
12	111110001	1110000110	0000110000	000	111111110000
13	110010001	1110000110	0000110000	000	111111111111
14	110010001	1110000110	0000110000	000	111111111111
15	110010001	1110000110	0000110000	000	111111111111
16	110000001	1110000110	0000110000	000	111111111111
17	110000001	1110000110	0000110000	000	111111111111
18	110000001	1110000110	0000110000	000	111111111111
19	110000001	1110000110	0000110000	000	111111111111
20	110000001	1110000110	0000110000	000	111111111111
21	110000001	1110000110	0000110000	000	111111111111
22	110000001	1110000110	0000110000	000	111111111111
23	110000001	1110000110	0000110000	000	111111111111
24	110000001	1110000110	0000110000	000	111111111111

[a] 1-unit is on; 0-unit is off
[b] 1-nonzero generation; 0-zero generation

Table 7 Emissions from thermal units

Unit	Emission (tons)	Unit	Emission (tons)	Unit	Emission (tons)
1	122.85	12	22.03	23	0.00
2	128.23	13	0.00	24	20.43
3	15.40	14	0.00	25	20.45
4	15.40	15	0.00	26	0.00
5	7.33	16	0.00	27	0.00
6	0.00	17	21.97	28	0.00
7	0.00	18	22.00	29	0.00
8	0.00	19	0.00	30	0.00
9	132.40	20	0.00	31	0.00
10	99.04	21	0.00	32	0.00
11	21.41	22	0.00		

Table 8 Schedule of Step 3

Hour	Thermal units[a]				Hydro units[b]
	000000000	1111111111	2222222222	333	000000001111
	123456789	0123456789	0123456789	012	123456789012
1	111100001	0110000110	0000110000	000	111111111111
2	111100001	0110000110	0000110000	000	111111111111
3	111100001	0110000110	0000110000	000	111111111111
4	111100001	0110000110	0000110000	000	111111111111
5	111100001	0110000110	0000110000	000	111111111111
6	111100001	0110000110	0000110000	000	111111111111
7	111100001	0110000110	0000110000	000	111111111111
8	111100101	0110000110	0000110000	000	111111110000
9	111100101	0110000110	0000110000	000	111111111111
10	111100101	0110000110	0000110000	000	111111111111
11	111100101	0110000110	0000110000	000	111111111111
12	111111111	0110000110	0000110000	000	111111111111
13	111011111	0110000110	0000110000	000	111111111111
14	111011011	0110000110	0000110000	000	111111111111
15	111011011	0110000110	0000110000	000	111111111111
16	111011011	0110000110	0000110000	000	111111111111
17	111011011	0110000110	0000110000	000	111111111111
18	111011011	0110000110	0000110000	000	111111111111
19	111011011	0110000110	0000110000	000	111111111111
20	111011011	0110000110	0000110000	000	111111111111
21	111011011	0110000110	0000110000	000	111111111111
22	111011011	0110000110	0000110000	000	111111111111
23	111011011	0110000110	0000110000	000	111111111111
24	110000001	0110000110	0000110000	000	111111111111

[a] 1-unit is on; 0-unit is off

[b] 1-nonzero generation; 0-zero generation

in operating cost in this step over the previous step. The expert system analyzes the results and finds no problem in unit commitment and unit loading. Thus it suggests that the schedule obtained in this step is operationally feasible.

The expert system has led to obtaining an operationally feasible solution by adjustment of the input data for the unit commitment program. Each consultation with the expert system about a subproblem takes between 5 and 15 s.

9 Conclusions

The Lagrangian relaxation approach for solving the unit commitment problem for a large system is presented. Extensive constraints are considered. Nonlinear functions are used for thermal generation cost, water discharge rate, and sulfur oxide emission. Transmission loss is included using a general transmission loss formula.

The hydro subproblems have not been solved independently. To get the values of unknown variables, that is, the output levels of hydro units of hydro subproblems, the hydrothermal scheduling is performed using an efficient algorithm with a thermal unit commitment schedule obtained by solving thermal subproblems using dynamic programming without discretizing generation levels. A refinement algorithm is used to fine tune the schedule. Constraints that are difficult or impractical to be implemented in commitment algorithm are handled by the unit commitment expert system. The expert system also checks the feasibility of the solution. Numerical results show that feasible solutions are obtainable within reasonable time.

References

Aoki A, Satoh T, Itoh M, Ichimori T, Masegi K (1987) Unit commitment in a large scale power system including fuel constrained thermal and pumped storage hydro. IEEE Trans Power Syst 2(4):1077–1084

Carrion M, Arroyo JM (2006) A computationally efficient mixed-integer linear formulation for the thermal unit commitment problem. IEEE Trans Power Syst 21(3):1371–1378

Elgerd OI (1971) Electric energy systems theory: an introduction. McGraw-Hill, New York, pp. 294–296

Guan X, Luh PB, Yan H, Amalfi JA (1992) An optimization-based method for unit commitment. Electr Power Energ Syst 14(1):9–17

Hur D, Jeong HS, Lee HJ (2007) A performance review of Lagrangian relaxation method for unit commitment in Korean electricity market. Power Tech, 2007 IEEE Lausanne, pp. 2184–2188

Kuloor S, Hope GS, Malik OP (1992) Environmentally constrained unit commitment. IEE Proc Gener Transm Distrib 139(2):122–128

Liyong S, Yan Z, Chuanwen J (2006) A matrix real-coded genetic algorithm to the unit commitment problem. Electr Power Syst Res 76(9–10):716–728

Padhy NP (2004) Unit commitment – a bibliographical survey. IEEE Trans Power Syst 19(2): 1196–1205

Pappala VS, Erlich I (2008) A new approach for solving the unit commitment problem by adaptive particle swarm optimization. IEEE Power and Energy Society General Meeting – Conversion and Delivery of Electrical Energy in the 21st Century, 20–24 July, pp. 1–6

Rashid AHA, Nor KM (1991) An efficient method for optimal scheduling of fixed head hydro and thermal plants. IEEE Trans Power Syst 6(2):632–636

Saber AY, Senjyu T, Yona A, Funabashi T (2007) Unit commitment computation by fuzzy adaptive particle swarm optimization. IET Gener Transm Distrib 1(3):456–465

Salam MS (2004) Comparison of Lagrangian relaxation and truncated dynamic programming methods for solving hydrothermal coordination problems. Int. Conf. on Intelligent Sensing and Information Processing, Chennai, India, 4–7 January, pp. 265–270

Salam MS, Hamdan AR, Nor KM (1991) Integrating an expert system into a thermal unit commitment algorithm. IEE Proc Gener Transm Distrib 138(6):553–559

Salam MS, Nor KM, Hamdan AR (1998) Hydrothermal scheduling based Lagrangian relaxation approach to hydrothermal coordination. IEEE Trans Power Syst 13(1):226–235

Somasundaram P, Lakshmiramanan R, Kuppusamy K (2006) New approach with evolutionary programming algorithm to emission constrained economic dispatch. Power Energ Syst 26(3):291–295

Tong SK, Shahidehpour SM (1990) An innovative approach to generation scheduling in large-scale hydro-thermal power systems with fuel constrained units. IEEE Trans Power Syst 5(2):665–673

A Systems Approach for the Optimal Retrofitting of Utility Networks Under Demand and Market Uncertainties

O. Adarijo-Akindele, A. Yang, F. Cecelja, and A.C. Kokossis

Abstract This paper presents a systematic optimization approach to the retrofitting of utility systems whose operation faces uncertainties in the steam demand and the fuel and power prices. The optimization determines retrofit configurations to minimize an (expected) annualised total cost, using a stochastic programming approach deployed at two levels. The upper level optimizes structural modifications, while the second level optimizes the operation of the network. Uncertainties, modelled by distribution functions, link the two stages as the lower layer produces statistical information used, in aggregate form, by the upper level. The approach uses a case study to demonstrate its potential and value, producing evidence that uncertainties are important to consider early and that the two-level optimization effectively screens networks capable to afford unexpected changes in the parameters.

Keywords Modelling · Retrofit design · Stochastic programming · Uncertainty · Utility system

1 Introduction

Utility systems are an integral part of many industrial sites. Their operation varies from the generation of steams needed on site to more complex co-generation that provides the heat and power needed for the site and external customers. A utility system features several degrees of freedom in its structural design and operation. The former includes the number and values of steam levels, the number and capacities of individual units such as boilers and turbines, and the connectivity of these units. The latter involves operational variables such as types of fuels and their consumptions and the load of individual units. The existence of these degrees of freedom renders opportunities for the improvement of the performance of the utility system

A. Yang (✉)
Department of Chemical and Process Engineering, University of Surrey, Guildford GU2 7XH, UK
e-mail: a.yang@surrey.ac.uk

S. Rebennack et al. (eds.), *Handbook of Power Systems I*, Energy Systems,
DOI 10.1007/978-3-642-02493-1_12,

by means of optimization at different stages of its lifecycle, for example the grass-root design where the utility system is designed from scratch as a new system or the retrofit design where the utility system is partially redesigned for improvement (or retrofitting) on the basis of its existing form.

In the last two decades, a number of approaches have been reported for optimization of utility systems. Papoulias and Grossmann (1983) proposed an MILP approach for the synthesis of flexible utility systems capable of coping with a multi-period pattern of utility demand. Hui and Natori (1996) presented a mixed-integer formulation for multi-period synthesis and operation planning for utility systems and discussed the industrial relevance. Mavromatis and Kokossis (1998a) developed an approach to targeting and steam level optimization based on an advantageous turbine hardware model. Bruno et al. (1998) reported a rigorous MINLP model for both synthesis and operational optimization of utility plants. Iyer and Grossmann (1998) proposed a multi-period MILP approach for the synthesis and planning of utility systems under multiple periods. Varbanov et al. (2004) addressed the operational optimization of utility systems by solving iteratively a MILP model to avoid the complication of directly solving the original MINLP model. Shang and Kokossis (2004, 2005) developed systematic approaches to the optimization of stream levels (based on a transhipment model) and to the synthesis and design of utility systems (based on an effective combination of mathematical optimization and thermodynamic analysis). More recently, Smith and co-workers (Aguilar et al. 2007a,b) presented a comprehensive modelling framework and its applications in grassroots design, retrofit and operational optimization of utility systems.

The existing publications highlight the importance of variations in utility systems optimization. A common factor across the studies has been the variation in energy demand over different operational periods (Papoulias and Grossmann 1983; Hui and Natori 1996; Maia and Qassim 1997; Iyer and Grossmann 1998; Shang and Kokossis 2005). The consideration, further extended in more recent research (Aguilar et al. 2007a), is coping with both internal factors such as energy demand as well as external factors such as the variations of the price of power and fuels. A common feature has been to characterise variations or uncertainties associated with a certain factor (e.g. steam demand, fuel price, etc.) by considering a few discrete quantitative levels of this factor (e.g. several representative values of steam demand). The optimal design subsequently takes into account all these levels typically by assigning a specific weighting factor to each of them in the course of making the design decision.

This work considers a different approach to the characterisation of uncertainties. Instead of specifying a few definite levels of a certain factor, the work assumes that this factor follows a particular statistical distribution proposed based on the historical data or other knowledge about the plant operation or the market. It is argued that this approach is more suitable for characterising uncertainties associated with a factor in a longer term (e.g. the life time of a utility system) and is more applicable when factors with a greater degree of uncertainties (e.g. fuel prices) are involved. Under these circumstances, using a few predefined levels may be inadequate for rendering a realistic representation of the uncertainties. This approach to uncertainty representation supports stochastic programming which, as a framework to carry out optimization under uncertainties (Sahinidis 2004), has been applied to a number of areas, including the optimal design of chemical processes. In this work, a stochastic programming approach is employed to determine retrofit designs.

In the remainder of the paper, a problem description is first given in Sect. 2. Section 3 presents the general stochastic programming approach, its customization, as well as the implementation as adopted in this work. A case study on the retrofit design of a specific utility system is subsequently reported in Sect. 4, followed by conclusions and discussions on the future work in Sect. 5.

2 Problem Description

Consider a utility system generating power through several pressures and using a number of turbines and boilers to meet the demand from different plants. The retrofit design of such a utility system, that is the design for improving this existing system as opposed to developing a completely new system, is to adjust the system at a minimum cost to meet certain steam and power demand. This is equivalent to an optimization problem:

$$\min_{R} \; c = f(R, S, U), \tag{1}$$

where c is the total annualised cost, which comprises the annualised retrofit investment cost and the annualised operating cost; c is a function f of the retrofit arrangement R, configurations of the existing utility system S, as well as the uncertainties of the internal and external factors that affect the operation of the system U. The retrofit arrangement refers to the type of equipments (boilers and turbines), including their sizes and locations. The investment on new equipments accounts for the retrofitting cost that augments the objective function. The operating cost includes the cost of fuels and the (net) imported electricity. In principle, the operating cost depends on the configuration of the retrofitted utility system, the work load of the system, and the fuel and power prices, which are often subject to uncertainty.

The operating conditions or factors in a utility system that may be subject to uncertainty include the following:

- Fuel price (external factor)
- Power tariffs (external factor)
- Steam demand (internal factor)
- Power demand (internal factor)

A review of the history of energy markets shows that there is a constant fluctuation in the prices of energy products, with the general trend being upwards. Based on the data provided by the Energy Information Administration (EIA 2008), the prices of fuel and power follow the same trend and vary at a similar rate. As a result, the paper makes use of a single price indicator for the uncertainties of the prices of both fuel and power. Among other alternatives, the price indicator, defined as the ratio of a future price to the current price, is set to follow a particular statistical distribution that reflects the predicted spread across the lifetime of the retrofitted utility system.

In addition to the uncertainties associated with external factors, the nature of the plant operation served by a utility system is such that there exist high and low steam demand periods. Demand fluctuations might be due to throughput variations and production rates. Similar to external uncertainties, fluctuations of internal demands are part of a retrofit optimization problem represented by means of statistical distributions. Uncertainties associated with external and internal factors affect only operating costs. Because of the presence of uncertainties, the evaluation of the operating cost is not deterministic but, in relation to the objective function in (1), represented in the form of a statistical variable.

The next section explains the steps and the formulation of the approach; the application of this approach is demonstrated with an industrial case study in Sect. 4.

3 The Stochastic Programming Based Approach

The section explains the stochastic programming framework and the customization required to address the optimal retrofit design presented earlier. A particular realization of this approach is then described, which is the tool actually utilized in the case study.

3.1 The Stochastic Programming Framework

According to Sahinidis (2004), stochastic programming handles optimization under uncertainties by addressing the following problem:

$$\min_{x} \; c = f(x) + E_{\omega \in \Omega}\left[Q(x, \omega)\right] \tag{2}$$

$$s.t. \;\; g(x) \leq 0, \tag{3}$$

with

$$Q(x, \omega) = \min_{y} \; F(\omega, x, y) \tag{4}$$

$$s.t. \;\; G(\omega, x, y) \leq 0, \tag{5}$$

where ω is a random variable from a probability space with Ω as the sample space, E refers to the expectation of the value of the function Q, which is a random variable as well due to the presence of the random variable ω in the arguments of this function. This problem formulation essentially implies performing optimization at two different stages. The first stage involves variables that need to be determined before the introduction of uncertainties; these variables are denoted by x in (2).

The variables of the second stage, denoted by y in (4), are those that form an *operational-level* decision following the first stage plan and the realisation of the uncertainties (i.e. the occurrence of particular values of uncertain factors). Denoted by (4), the second stage optimization addresses such *operational-level* decisions. Apparently, the first stage optimization requires the result of the second stage optimization. Accordingly, the overall objective is to determine the first stage variables to minimize the sum of the first stage costs (denoted by the first term in (2)) and the (expected) second stage costs (denoted by the second term in (2)).

Applying this stochastic programming paradigm, the retrofit design can be reformulated. First, the objective function in (1) is decomposed into two parts, namely the *capital cost*, corresponding to $f(x)$, and the *operating cost* corresponding to $E_{\omega\in\Omega}\left[\Omega(x,\omega)\right]$ – both in (2). The coordination between (2) and (1) are explained below:

- x is equivalent to R in (1), representing variables that characterise a retrofit arrangement
- ω represents a random sample from the uncertainty sample space originally denoted by U in (1) and is specified by an arbitrary combination of plausible values of uncertain external and internal factors considered in the retrofit design
- Q stands for the minimum operating cost given x and ω; this minimum cost results from applying the optimal operating policy denoted by y in (4) as the solution of the second stage optimization (operational optimization)

This optimal operating policy is usually in terms of (i) the amount of fuel(s) consumed by the boilers and possibly some gas turbines as well as (ii) the distribution of steams among different steam turbines and let-down stations. Note that the configuration of the existing utility system (denoted as S in (1)) does not explicitly appear in this new problem formulation (i.e. (2–5)), but will be taken as a known parameter and utilised when the operating cost (denoted by F in (4)) is evaluated. Regarding the constraints for each of the two stages, (3) usually refers to constraints that enforce bounds to the retrofit arrangement variables, whilst (5) refers to the mathematical model of the utility system, which will be discussed later in this section.

The stochastic programming paradigm leads to solutions by means of a two-level optimization approach, namely *structural optimization* and *operational optimization*. Figure 1 illustrates the case. For a particular retrofit arrangement, the figure illustrates that the expected minimum (operating) cost is the result of averaging the minimum operating costs computed for a set of representative samples drawn from the sample space of the uncertain factors.

3.2 Realisation of the Proposed Approach

There are four key elements addressed in a realisation of the proposed approach: the *modelling of the utility systems*, the *sampling of the space of uncertain factors*, the *operational optimization* and the *structural optimization*. Several authors in the past

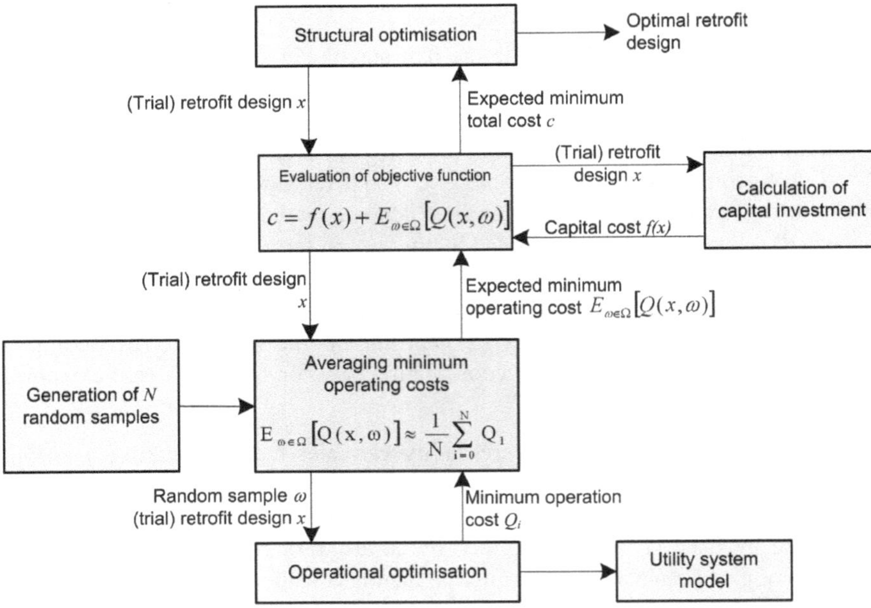

Fig. 1 Stochastic programming based approach to optimal retrofit of utility systems

have addressed the modelling of utility systems. In this work, the models of boilers, turbines, let-down stations and steam headers and the handling of mass and energy balances follow the formulation of Varbanov et al. (2004). The calculation of the isentropic enthalpy change across steam turbines, as required by the turbine model, follows the work by Mavromatis and Kokossis (1998a). Regarding the sampling of the space of uncertain factors, the work adopts the Latin Hypercube Sampling (LHS) (Iman et al. 1981) as a well-tested and efficient technique.

The operational optimization adopts the algorithm proposed by Varbanov et al. (2004). The operational optimization of a utility system with known configuration is in principle a non-linear problem with the non-linearity introduced by the energy balance element in the model to compute the isentropic enthalpy changes. The algorithm adopted here solves this problem by iteratively solving a linear programming problem assuming a fixed value for each isentropic enthalpy change ΔH_{is} involved in the model. At the end of each iteration, the operating policy resulting from the linear programming is used to compute a set of new values for ΔH_{is} according to the rigorous process model. The new isentropic enthalpy values then become the input of the next iteration. This process is repeated until there is no significant change in the values of ΔH_{is}. The optimal operating policy obtained at the end of this iterative process is regarded as the solution of the operational optimization.

Regarding structural optimization, it is important to realise from previous discussions that the evaluation of its objective function involves a number of runs of the operational optimization. The number of runs equals the size of the set of samples representing the space of uncertain factors. For instance, for the analysis of ten samples, ten operational optimization runs will be required within the structural optimization. Although a relatively efficient algorithm has been selected for the operational optimization as explained earlier, this need of repeating the operational optimization makes the evaluation of the structural optimization objective function remain an inevitably time-consuming task. Furthermore, the derivatives of this objective function are not directly available. To cope with an objective function with such characteristics, CONDOR (Berghen and Bersini 2004) is adopted in this work. CONDOR implements a trust-region-based optimization algorithm with the objective function approximated through local quadratic interpolation; the search for the optimal solution at each optimization step is then based on this approximation rather than the original objective function. This approach does not require derivatives of the original objective function and also has the potential of reducing the number of its evaluations.

In this work, a computational framework realising the proposed approach has been developed in MATLAB, which makes use of the MATLAB version of the CONDOR optimiser.

4 Case Study

A case study has been carried out to demonstrate the proposed approach. In this section, the utility system as the target of the retrofit design is first depicted. The specification of the retrofit problem and the solution choices are then given, with the results presented in the subsequent section.

4.1 The Targeted Utility System

A retrofit design was assumed to be carried out for a particular utility system as shown in Fig. 2 (excluding T7–T12), which is one of the scenarios presented by Varbanov et al. (2004). The utility system includes the following components:

- Two steam boilers (B1, B2) producing HP steam
- One gas turbine (GT) with heat recovery steam generator (HRSG) producing high pressure (HP) steam
- Two extraction turbines (T1, T2) between HP and mid-pressure (MP)/low pressure (LP) mains
- Two back pressure turbines (T3, T4) between HP and MP mains
- Two driver turbines (DRV1, DRV2) between MP and LP mains
- Two condensing turbines (T5, T6) between LP and condensing mains
- Three let down stations between the various pressured steam stages
- Seven pressured mains and one sub-atmospheric condensing mains

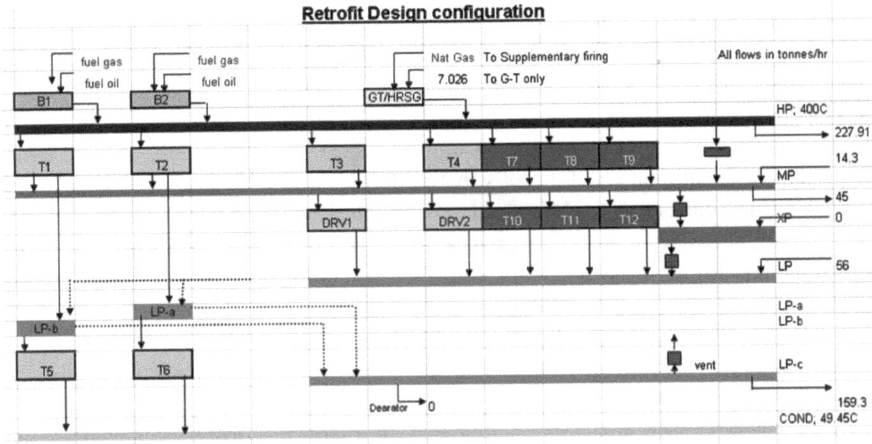

Fig. 2 Case study utility system: existing and retrofit configuration

The maximum amount of power that can be imported and exported is 50 and 10 MW, respectively. The system has three fuel options, namely fuel oil, fuel gas and natural gas. The system, before retrofitting, utilises only fuel gas for the boilers and only natural gas for the gas turbine/HRSG unit.

4.2 The Retrofit Problem Specification and Solution Choices

It was assumed that the retrofit arrangement will introduce six (6) new back pressure turbines to the initial configuration; these include three turbines to be located between the HP and MP mains (numbered T7, T8 and T9) and three turbines between the MP and LP mains (numbered T10, T11 and T12). The gas turbine remains the same and its operation is kept unchanged. Furthermore, introduction of new boilers is not considered in this retrofit design. Thus, this optimal retrofit design becomes a problem of determining the optimal sizes of the new turbines, which yield the minimum total annualised cost. The size of each new turbine, in terms of maximum steam flow rate t/h, is restricted to be within [30, 100].

The capital cost part of the objective function (cost of the new turbines) is computed according to a correlation between the turbine work at full load and annualised investment cost given by Bruno et al. (1998). The operating cost in this case study includes the cost for purchasing a varied combination of fuels (fuel oil, fuel gas and natural gas) and the net cost of importing electricity. The default unit prices of fuels and power specified in Varbanov et al. (2004) were adopted with 103.41, 70.82 and 159.95 \$/tonne for fuel gas, fuel oil and natural gas and 0.045 and 0.06 \$/kwh for imported and exported power, respectively. The fuel combinations have been organised such that there are two fuel choices for the boilers (fuel gas, fuel oil) but only

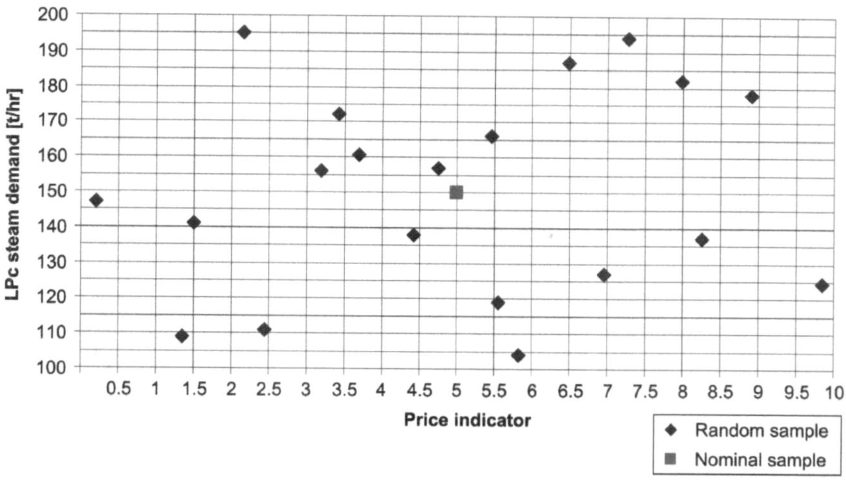

Fig. 3 Random uncertainty combinations in full spectrum and nominal uncertainty (not cross-referenced)

one for supplementary firing in the HRSG unit (natural gas). This is in keeping with the set up presented by Varbanov et al. (2004).

Regarding the uncertainties, both the prices of fuel and power and the demand of LPc steam were considered. According to the treatment outlined earlier, the prices are represented by means of a single price indicator, which was assumed to follow a uniform distribution within the range of 1–10. The demand of LPc steam was also assumed to follow a uniform distribution, within the range of 100–200 t/h, representing low to high demand. The number of samples N generated within the space of uncertain factors was determined by following an empirical rule: $N = 10 \cdot K$, where K is the number of uncertain factors considered and is equal to 2 in this particular case (price indicator and LPc steam demand) (Fig. 3).

The parameters required for modelling equipment in the existing utility system are set according to Varbanov et al. (2004). The modelling parameters of all the new turbines, as required for correlating the work produced by a turbine with its steam flow rate, are assumed to be the same as those of turbine T3 in the existing system.

5 Results and Discussion

In this section, the result of utilizing the proposed approach is first presented. Afterwards, two other different designs are discussed, in which their results are compared with that of the proposed approach.

5.1 Optimal Retrofit Design Resulting from the Proposed Approach

The result of the case study as described in the previous section is summarised in Tables 1 and 2.

Table 1 shows the optimal sizes of the new turbines to be introduced to the utility system. In Table 2, the price indicator and the LPc steam demand for each of the 20 random samples generated by Latin Hypercube sampling are listed. The table also shows the minimum operating cost of the retrofitted plant corresponding to the optimal design and the specific values of the two factors subject to uncertainty as indicated by each sample. The average minimum operating cost, the capital cost, as well as the total annualised cost are given at the bottom of this table. This retrofit design is referred to as *Optimal Design* in the sequel.

5.2 Comparison with Other Designs

To investigate how the way of treating uncertainties will influence the result of retrofit design, two further retrofit design studies were conducted.

The first one, referred to as *Alternative Design A* in the sequel, is the retrofit optimization of the existing system with the current values of the factors upon the retrofit design despite the fact that these factors will be subject to uncertainties when the retrofitted system operates in the future. That means, this design did not take into account the future fluctuations of the price indicator and the LPc steam demand. The resulting retrofit design was then used to estimate the expected cost if the utility system was to operate under the uncertainties as specified earlier in Sect. 4. This was carried out by performing calculations similar to those resulting in Table 2, except that the new turbine sizes used here were the result of the optimization without considering uncertainties. The expected cost estimated this way is shown in Fig. 4, with comparison to the cost corresponding to the *Optimal Design* resulting from the proposed approach, that is one that properly addresses the uncertainties. It is evident that the cost originally reported by this design, via the optimal value of the objective function without considering uncertainties, is seriously underestimated. Furthermore, when the cost of this design is re-estimated by taking into account uncertainties following the procedure explained above, it appears to be

Table 1 Retrofit turbines optimal sizes in terms of the maximum steam flow rates

Turbine	Optimal size (tone/h)
T7	30.564
T8	39.933
T9	32.455
T10	37.607
T11	47.344
T12	47.908

Table 2 Uncertainty samples for retrofit optimization and their corresponding minimum operating cost

	Price indicator	LPc steam demand (t/h)	Minimum operating cost $\times 10^8$ (\$/yr)
Sample w1	9.866167	124.030534	2.7705
Sample w2	5.927877	165.475938	1.6646
Sample w3	6.988069	187.153971	1.9623
Sample w4	2.957452	111.84105	0.8305
Sample w5	6.407655	103.827935	1.7993
Sample w6	5.251297	152.256089	1.4746
Sample w7	2.619967	195.342083	0.7421
Sample w8	1.810545	108.324013	0.5084
Sample w9	7.46866	126.722345	2.0973
Sample w10	6.069987	118.040479	1.7045
Sample w11	8.480852	181.863426	2.3815
Sample w12	3.695556	156.504599	1.0378
Sample w13	4.861064	133.01408	1.365
Sample w14	1.110849	147.931856	0.3119
Sample w15	1.992758	141.723935	0.5596
Sample w16	4.186019	160.812075	1.1755
Sample w17	7.819875	193.14196	2.2101
Sample w18	9.384511	177.873734	2.6353
Sample w19	3.870639	172.026683	1.0869
Sample w20	8.770925	137.171397	2.463
		Average min. operating cost	*1.539*
		Capital cost	*0.005*
		Total cost for retrofit	*1.544*

\$899,000/*annum* higher than the cost of the *Optimal Design* resulting from the proposed approach.

Another design performed for comparison purposes, referred to as *Alternative Design B* in the sequel, was based on the average values of the uncertain factors, assuming the same uncertainty factors and ranges as specified in Sect. 4. A procedure similar to *Alternative Design A* was performed, with results as shown in Fig. 5. It can be seen that the cost originally reported by *Alternative Design B*, via the optimal value of the objective function without considering uncertainties, is no much different from the expected cost of a retrofitted system, which implements this design and is operated under the uncertainties assumed in the main case study. However, this design still leads to an expected cost (when exposing the design to uncertainties), which is \$160,000/annum higher than that of the *Optimal Design* resulting from the proposed approach.

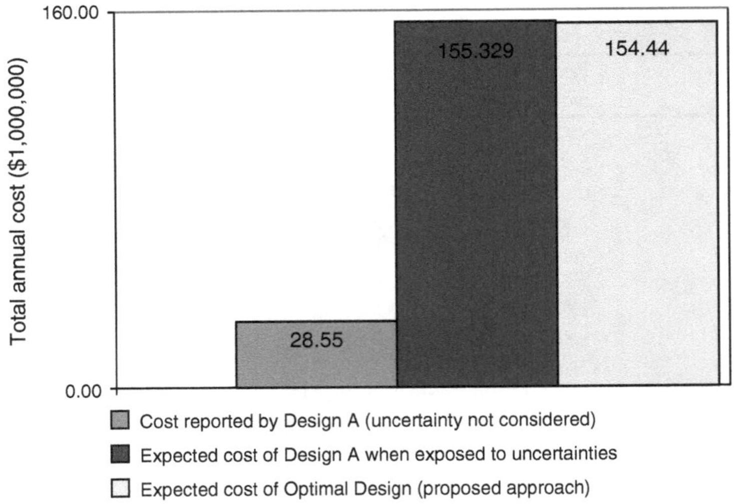

Fig. 4 Results of design A in comparison with that of the main case study

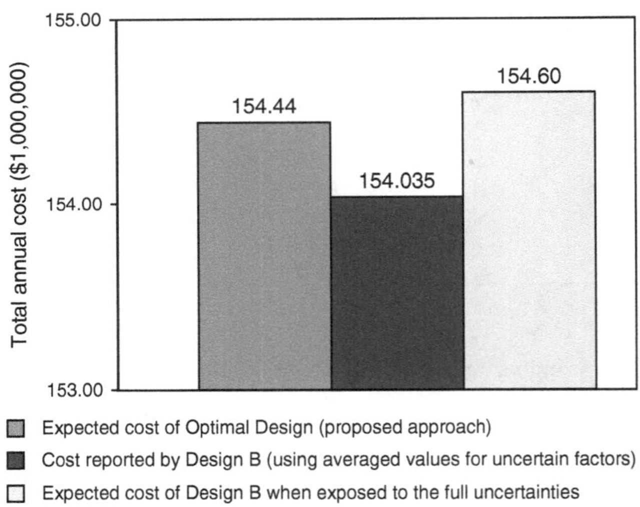

Fig. 5 Results of alternative design B in comparison with that of the main case study

6 Conclusions and Future Work

A systematic approach to the optimal retrofit of utility systems operated under uncertainties has been developed. It is suggested to use statistical distributions as a general means to represent the uncertainties associated with any external factors such as prices of fuels and power and any internal factors such as demand of steams.

The problem of optimal retrofit design can thus be handled following a two-stage stochastic programming framework, which involves both the structural optimization and the operational optimization of the utility system under consideration. This approach has been applied to a concrete case study. It has revealed that the retrofit design following this approach has an expected total annualised cost lower than those resulting from designs where the uncertainties are not handled properly.

The developed approach has been exposed to limited tests so far. Further work could be done to observe the effect of other uncertainty factors like the site power demand. Furthermore, the sensitivity of the resulting design to the different assumptions on the statistic distribution of an uncertain factor remains to be studied. Besides, the number of retrofit turbines was fixed in the case study reported here. This could be changed and thus introduce an additional degree of freedom in the optimization. Finally, the current realization of the proposed approach makes use of a local optimizer; an implementation that utilizes a global optimizer may bring better results in some applications.

References

Aguilar O, Perry SJ, Kim J-K, Smith R (2007a) Design and optimization of flexible utility systems subject to variable conditions. Part 1: Modelling framework. Chem Eng Res Des 85(A8):1136–1148

Aguilar O, Perry SJ, Kim J-K, Smith R (2007b) Design and optimization of flexible utility systems subject to variable conditions. Part 2: Methodology and applications. Chem Eng Res Des 85(A8):1149–1168

Berghen F, Bersini H (2004) CONDOR, an new parallel, constrained extension of Powell's UOBYQA algorithm. Experimental results and comparison with the DFO algorithm. Technical report, IRIDIA, Universite Libre de Bruxelles, Belgium. http://iridia.ulb.ac.be/~fvandenb/CONDORInBrief/CONDORInBrief.html

Bruno JC, Fernandez F, Castells F, Grossmann IE (1998) A rigorous MINLP model for the optimal synthesis and operation of utility plants. Chem Eng Res Des 76:246–258

Energy Information Administration (2008) Selected National Average Natural Gas Prices, 2004–2009. Available online at http://www.eia.doe.gov/pub/oil_gas/natural_gas/data_publications/natural_gas_monthly/current/pdf/table_03.pdf. Accessed March 2010

Hui CW, Natori Y (1996) An industrial application using mixed integer-programming technique: a multi-period utility system model. Comput Chem Eng 20:s1577–s1582

Iman RL, Helton JC, Campbell JE (1981) An approach to sensitivity analysis of computer models: Part I–introduction, input variable selection and preliminary variable assessment. J Qual Technol 13(3):174–183

Iyer RR, Grossmann IE (1998) Synthesis and operational planning of utility systems for multiperiod operation. Comput Chem Eng 22:979–993

Maia LOA, Qassim RY (1997) Synthesis of utility systems with variable demands using simulated annealing. Comput Chem Eng 21:947–950

Mavromatis SP, Kokossis AC (1998a) Conceptual optimisation of utility networks for operational variations-1: targets and level optimisation. Chem Eng Sci 53:1585–1608

Mavromatis SP, Kokossis AC (1998b) Conceptual optimisation of utility networks for operational variations-2: network development and optimisation. Chem Eng Sci 53:1609–1630

Papalexandri KP, Pistikopoulos EN, Kalitventzeff B (1998) Modelling and optimization aspects in energy management and plant operation with variable energy demands-application to industrial problems. Comput Chem Eng 22:1319–1333

Papoulias SA, Grossmann IE (1983) A structural optimization approach in process synthesis-I utility systems. Comput Chem Eng 7:695–706

Sahinidis NV (2004) Optimization under uncertainty: state-of-the-art and opportunities. Comput Chem Eng 28:971–983

Shang Z, Kokossis AC (2004) A transhipment model for the optimization of steam levels of total site utilitysy stem for multiperiod operation. Comput Chem Eng 28:1673–1688

Shang Z, Kokossis A (2005) A systematic approach to the synthesis and design of flexible site utility systems. Chem Eng Sci 60:4431–4451

Varbanov PS, Doyle S, Smith R (2004) Modelling and optimization of utility systems. Chem Eng Res Des 82(A5):561–578

Co-Optimization of Energy and Ancillary Service Markets

E. Grant Read

Abstract Many electricity markets now co-optimize production and pricing of ancillary services such as contingency reserve and regulation with that of energy. This approach has proved successful in reducing ancillary service costs and in providing consistent pricing incentives for potential providers of both energy and ancillary services Here we discuss the basic concepts involved, the optimization formulations employed to clear such co-optimized markets, and some of the practical issues that arise.

Keywords Ancillary services · Contingency reserve · Co-optimization · Electricity markets · Frequency control · Pricing · Regulation

1 Introduction

In many parts of the world, traditional electricity sector arrangements have now been replaced by markets into which competing generators sell energy and from which competing wholesale customers buy energy. Physically, though, such markets still operate in the context of a transmission system controlled in a more or less traditional fashion by a system operator (SO). This leaves the status of what have traditionally been called "ancillary services" open to debate. Such services may include frequency regulation, fast response to contingency events, reactive support, and provision of longer term "backup" capacity, generation facilities capable of performing a "black start" in a collapsed power system. Although more radical proposals have been suggested, it is generally agreed that these services should be purchased by the SO and deployed in a fairly traditional centralized fashion. Although not all services can be procured via competitive market processes, this chapter focuses on

E. Grant Read (✉)
Energy Modelling Research Group, University of Canterbury, Private Bag 4800, Christchurch 8140, New Zealand
e-mail: grant.read@canterbury.ac.nz

S. Rebennack et al. (eds.), *Handbook of Power Systems I*, Energy Systems,
DOI 10.1007/978-3-642-02493-1_13, © Springer-Verlag Berlin Heidelberg 2010

contingency reserve and regulation services, which we will collectively refer to as "reserve."[1]

If reserve could always be supplied efficiently by "marginal" or "supra-marginal" plant, that is, from capacity deemed too expensive to generate electricity at the time, the decision as to which plant should provide reserve would become secondary to that of generation dispatch. But the need for quick response limits the reserve MW available from each unit, and means that reserve must typically be supplied by plant that is already operating, but at less than its full economic output. Thus the SO must decide, in every trading period, which participants will be on reserve or regulation duty rather than producing energy. Compensation is also an issue, since plant on reserve duty will suffer an "opportunity cost" loss, from operating at less than full economic output, or efficiency.

The trade-off between optimal energy and ancillary service dispatch is quite complex, and changes dynamically and unpredictably, for all kinds of reasons. So a sequential approach, in which ancillary service dispatch is determined either before or after energy dispatch, may be expected to consistently yield sub-optimal outcomes. Accordingly, an increasing number of markets, starting with New Zealand in 1996, have adopted "co-optimization" formulations to clear both energy and reserve markets simultaneously and efficiently.[2] Here we focus solely on this co-optimization approach, as it applies to the procurement and deployment of regulation and contingency response ancillary services within trading intervals.

Once "contingency response" requirements extend beyond a single trading interval, this ancillary service starts to merge with the concept of providing "backup capacity" and eventually with the concept of providing "MW capacity" generally. Many markets do include payments for some kind of MW capacity, but provision of MW capacity is not generally described as an ancillary service, and market arrangements relating to this will not be considered here.[3] Thus the term "energy/reserve co-optimization", as used here, refers primarily to co-optimization of energy with intra-interval ancillary services, not with longer term capacity provision. Some comment is made, though, on co-optimization involving ancillary service obligations extending beyond a trading interval.

We focus particularly on markets in New Zealand, Australia, and Singapore, which each illustrate different aspects of the general approach. Initial experience with the New Zealand market is described by Read et al. (1998) while, from direct experience of the other two markets cited, we can say that industry studies indicate

---

[1] In many markets the term "reserve" refers to any MW capacity held in excess of expected peak requirements, but some of the markets discussed here use "reserve" and particularly "spinning reserve" to refer to quick acting contingency response services.

[2] Ring et al. (1993) first published the co-optimization formulation, albeit with a simplified representation of the energy market. Read and Ring (1995) describe the formulation in the context of a full AC nodal pricing model.

[3] In general, we favor defining "ancillary services" as being all those services required to maintain stable power system operation at a finer time scale, or locational scale, than the blocks traded in the energy market.

substantial savings, of the order of 30–50% of ancillary service cost, when a co-optimized ancillary service market was introduced after experience with an initial non-co-optimized market. We should comment, though, that the importance of this kind of quick-acting ancillary service depends greatly on the size of the power system involved. The Singapore and New Zealand power systems, in particular, are much smaller than the kind of integrated system that can be developed on a continental scale, with the New Zealand system being further sub-divided into two AC sub-systems, one on each main island, joined by an HVDC link.

Within such a (sub-) system, loss of a 250 MW unit may be regarded as a significant contingency, and loss of the HVDC link even more so. There have actually been times when loads have been so low, and inward transfers so high, that New Zealand's South Island has had more MW on contingency response duty, to cover the possible loss of the link, than on generation duty.[4] In such systems, adequate provision of ancillary services can become nearly as critical as provision of energy, and quite costly. With only a small set of units to choose from, capacity can also be so tight that pre-determining which units are to be placed on reserve duty can sometimes make it difficult to achieve feasible, let alone optimal, dispatch. Thus simultaneous co-optimization is a much higher priority than it would be elsewhere, and of similar, and sometimes greater, importance than "nodal pricing," for example. Particularly since publication of FERC's "Standard Market Design" in the US, though, the advantages of co-optimization have also become accepted in much larger systems. But our purpose here is not to document the practical or economic benefits of the co-optimization approach, but to describe how it is implemented. As will be seen, the basic approach is to modify the linear programming (LP) formulation used to clear the energy market. After clarifying some basic concepts, we describe a basic formulation for this purpose and several variations on it.

2 Basic Concepts

Frequency is a property of each AC (sub-) system, so that any disturbance (almost) instantly affects frequency throughout that (sub-) system, but not in other (sub)-systems that may be interconnected via HVDC links. "Contingency reserve" and "regulation" are both required to maintain system frequency and are jointly referred to as "frequency control ancillary services" (FCAS) in the Australian market, but they differ in ways that create specific issues for the formulation.

Contingency reserve is held to cover the sudden loss of a generation unit, or import link, because any such loss will lead to frequency falling rapidly toward unacceptable levels. In the New Zealand system, for example, contingency reserve

[4] This is typical of many systems around the world, particularly on islands, and in developing countries. A similar formulation is used in the Philippines, for example, with many more island sub-systems than New Zealand, and correspondingly greater localized ancillary service requirements.

must be able to respond so as to arrest that fall within 6 s. To respond that quickly, units cannot wait to receive instructions from the SO, and each unit must respond directly to the frequency drop it observes. The traditional way of providing contingency response is through "partially loaded spinning reserve" (PLSR), that is, from units that are already spinning at system frequency, but not at full load. Such units are able to quickly ramp up their output until the frequency is restored to normal, thus implicitly replacing the energy input from the generator/link. But that means each unit is responding blindly, not knowing where the contingency has occurred, or coordinating its response with other units. As will be seen, this has important implications for the formulation.

Next, once the fall in frequency has been arrested, it must be restored to normal levels, within 60 s for the New Zealand system. This response is coordinated in a similar manner, but not necessarily from the same units.[5] The response characteristics of different unit types differs, so that some units are better at arresting the initial frequency fall in the 6 s time frame, while others are better at restoring frequency in the 60 s time frame. In the limit, a single contingency response service could be defined, requiring supply of 6 and 60 s reserve in some fixed proportion. But participation of many reserve sources would then be significantly restricted, because they are more capable of responding in one time frame than the other. Efficiency requires that two separate services be defined, and these are actually defined from a zero base in both time frames. Thus a supplier of 6 s contingency response need not sustain that response through into the 60 s time frame, and vice versa. Accordingly, the formulation allows units to provide both types of reserve independently of the other, and without any jointly limiting constraints.

Many systems now employ mechanisms other than "spinning reserve" for this purpose. In a hydro system, contingency response can be supplied by "tail water depressed" hydro units, kept spinning at system frequency, but with no water flow, thus enabling them to start generating as soon as the input gates are opened. Some markets, including New Zealand, also allow "interruptible load" to compete with generator response as a means of restoring frequency. Both technologies provide a sudden response, after some delay, rather than a gradual response as frequency drops, and are thus better suited to restoring frequency than to arresting frequency drop.[6] So their impact on system frequency will differ from that of traditional spinning reserve on a per MW basis.

From a primal optimization perspective, it may seem obvious that constraints should be added to reflect these characteristics, and perhaps limit reliance on particular reserve sources. But each such restriction creates a different "product," which only a few suppliers can compete to provide. To create a competitive market, we must reduce a variety of response patterns, reflecting the characteristics of differing

---

[5] Many markets also trade longer term contingency reserves, intended to restore reserve margins after stable operation has been restored. The same principles apply, except that response from such services will be centrally coordinated, unlike the quick response services discussed here.

[6] Interruptible load response may be uncertain, too, since the SO cannot generally monitor loads in real time.

technologies, down to a simple categorization into discrete requirement classes. What is really being traded, then, is not "spinning reserve," or "interruptible load," but a degree of impact on frequency, in a particular time frame. But we can apply "efficiency factors" to scale the contributions of, and prices for, different types of supplier providing reserve in the same class.[7]

By way of contrast, the "regulation" or "frequency keeping" service is not used to respond to sudden contingency events, but to constant small fluctuations, both positive and negative, in the supply/demand balance, due to load variation, breakdown of small generation units, and variations in output from wind generators, for example. For technical reasons, having several units all attempting to perform this task in an uncoordinated fashion risks creating instability. So some systems, such as that in New Zealand, have traditionally relied upon having a single "frequency keeping" generator, typically a large hydro station, designed for and dedicated to that task. Market processes may be employed to determine which generator will play that role in any particular period, but a competitive real time market for regulation duty can not really be developed under those circumstances. There are clear advantages in spreading this duty across multiple units, though, and many systems do this using "automated generation control" (AGC) to send signals coordinating responses from all the units involved. We will see that this complicates the formulation, but also introduces possibilities for more precise optimization.

For both contingency reserve and regulation, what is being optimized by the market is not the way in which those services respond within a trading interval, and nor is that what participants are paid for. What is optimized, and paid for, is the assignment of reserve/regulation duty at the beginning of the trading interval.[8] Thus generators on contingency reserve duty are paid for being on standby, whether or not they are called on to respond, and market prices must be at least high enough to compensate them for withdrawing capacity from the energy market for that purpose. The "opportunity cost" of doing this typically represents a major component of the price.[9] Participants are expected to account for all other costs, including the cost of

---

[7] The Singapore formulation introduced these, and allows them to be piece-wise linear, reflecting the fact that there may be concern about the system placing too much reliance on any one type of reserve supply. In reality, piece-wise linearisation has not been applied, and an alternative construct has been introduced, to constrain the amount of interruptible load accepted in particular regions of the network. Either way, once constraints limiting the contribution, or effectiveness, of particular reserve supply groups bind, the marginal value of further supply from that group reduces, possibly to zero, and the price payable should really drop accordingly. Effectively, these supplier groups are not really competing in the same market any more. There is no ideal solution to this market design problem, only compromise, but that is not our concern here.

[8] This may be contrasted with theoretical proposals that contingency response be induced, coordinated, and rewarded by a very, very short-term market, with prices peaking to extremely high levels for a few seconds or minutes following a contingency event.

[9] An alternative paradigm would pay participants constrained on/off compensation when placed on ancillary service duty. This may seem simple, and even "fair", in the short run. But it is actually easier to determine one price, for all providers, from the shadow price of the relevant LP constraint. This also gives appropriate long run incentives for provision of ancillary service capabilities

operating in a more flexible generation mode, and the expected nett cost of actually responding, via a per MW "fee"[10] in their offers.[11]

3 Formulation

3.1 Energy Market Formulation

The co-optimization formulation may be regarded as an extension of the energy market formulation, and energy market formulations can involve a great deal of detail to deal with the locational aspects of the market, including definition of "nodal injections," treatment of line flows, limits and losses, and implementation of security constraints etc. that may be required for various purposes. But this detail is largely irrelevant to the basic exposition and will be ignored here.[12] Later sections describe extensions introduced to deal with real life complications that have proved important in particular markets, but the simplified formulation described here is generic.

The markets discussed here all use an LP-based market clearing engine (MCE) to optimize end-of-period (EOP) dispatch targets, given an observed beginning-of-period (BOP) dispatch pattern. If we ignore the possibility of demand side bids, and co-optimization, the basic structure of an electricity market-clearing formulation, for a single dispatch period, is as follows:

- An objective function defined entirely by the "energy" offers of participants
- A set of simple bounds on generation in each offer tranche or "block"
- A set of simple ramp rate constraints
- An equality specifying a zero nett energy balance for each node, or zone, for which a distinct price will be determined by the market
- A set of equality constraints defining the way in which power will flow through the transmission system, at the level of detail to which it is modeled[13]
- A set of inequality constraints defining the acceptable limits on such power flows

---

on different plant types, just as paying all energy providers the same energy price gives correct incentives for provision of generation capability from different plant types.

[10] We will refer to this as an offer "price," although we will see that participants are not actually prepared to supply at this "price" until it is adjusted for opportunity costs.

[11] Since the extra generation occurs over a relatively small period and generators will be paid for it at prevailing energy prices, this nett cost is typically quite low, for generation. It may be the dominant consideration for interruptible load, though.

[12] The Australian formulation is not publicly available, but detailed formulations for Singapore may be found at http://www.emcsg.com/n916,12.html and for New Zealand at http://www.ctricitycommission.govt.nz/pdfs/opdev/servprovinfo/servprovpdfs/SPD-v4–3.pdf.

[13] All the models used in real markets, and discussed here, employ a DC approximation to the optimal (active) power flow equations. But Read and Ring (1995) and Wu et al. (2004) both discuss the formulation and interpretation of a co-optimization model in a full (active and reactive) AC power flow formulation.

To this formulation, co-optimization adds the following elements, each of which is discussed further below:

- Inequality constraints defining reserve requirements for one or more reserve classes (e.g., 6 and 60 s reserve, or regulation)
- Offers to supply reserve in each such class from participants in various reserve supplier groups
- Joint capacity constraints to ensure that participants are not dispatched to supply more energy and reserve, in aggregate, than they can physically supply

3.2 Defining Requirements

A MW reserve requirement could be defined by the SO, outside the market optimization, and a simple lower bound is generally specified so as to ensure that some minimum reserve level is always supplied. But, even though the SO might adjust the requirement, exogenously, to match observed conditions, this formulation does not fully exploit the potential for co-optimization if reserve requirements can be reduced by reducing the critical contingency size, as is often the case for smaller power systems. So this minimum requirement constraint is often supplemented by constraints to ensure that reserve is sufficient to cover the most serious "credible" contingency implied by the dispatch determined, endogenously, by the MCE. If the standard is failure of the largest unit, this can be assured by including constraints forcing the requirement to be greater than that needed to cover failure of each one of the units considered large enough to create a problem.[14]

It might be thought that the reserve required to cover a failure of a unit would be just its MW generation at the time. But the provision of reserve is a mutual undertaking, and so even the units defining the maximum contingency requirement may supply reserve, while themselves relying on reserve from all other units. We could form constraints that exclude each unit from covering its own failure, but that would create as many different reserve services as there are critical units. Rearranging those constraints, though, we can define a single system reserve requirement to which all units may contribute, in which the loss of a unit implies loss of its reserve contribution, as well as its energy.

---

[14] Similar constraints can be formulated for key transmission links and for multi-unit contingencies. This is simple, in principle, but can become complex if failure of one link element can be partially covered by "fail-over" to another. The New Zealand formulation includes a complex integer module to determine the optimal HVDC link configuration and corresponding reserve requirements. The Singapore formulation includes "secondary" contingencies due to plant tripping off when frequency drops, thus exacerbating the primary contingency. If such plant exists, the RHS of (1) must be modified to reflect this, by adding the generation/reserve contribution from that plant on top of each primary contingency modeled. It may also be important to design a cost recovery mechanism so that such plant faces the incremental costs they impose on the system.

That is, if we let req^c be the system's requirement for reserve class c, gen_i the generation from unit i, and res_i^c its contribution to meeting reserve requirement req^c, then for each reserve class traded, c, we must have[15]

$$\sum_j res_j^c \geq req^c \geq gen_i + res_i^c - SYSRESP^c \quad \forall \text{ large units, } i \tag{1}$$

3.3 Defining Participant Offers

On the supply side, the direct costs, to the market, are defined by participant offers, and the simplest possible form for these is a set of offer blocks, just as for energy. The energy offers are still summed to form *Energyobjective*, as in a non-co-optimized market. But, if we let the $offer_{jb}$ be the price at which block b is offered by participant j, with $RESLIM_{jb}$ being the bounds on these offer blocks, we now get a summation constraint and an additional term in the objective, as follows:

$$0 \leq res_{jb}^c \leq RESLIM_{jb}^c \quad \forall \text{ blocks } b, \text{ of unit } j\text{'s offer for class } c \tag{2}$$

$$res_j^c = \sum_b res_{jb}^c \quad \forall \quad \text{units } j, \text{ and classes } c \tag{3}$$

$$Objective = Energy\,objective + \sum_j \sum_b \sum_c res_{jb}^c * offer_{jb}^c \tag{4}$$

The New Zealand market, where this co-optimization concept was first introduced, uses a different offer form, designed to model the fact that much of the spinning reserve is provided by hydro stations containing several relatively small units, offered to the market at the aggregate station level.[16] Such units operate best at around 80% loading, producing an optimal balance between generation and reserve provision, and the optimal station operating strategy is to progressively load up more units, each operating at the same balance point. If market conditions make it more desirable to produce more energy and less reserve, or vice versa, all units simultaneously shift away from that balance point. This can be represented using a "radial" offer form, in which the offer blocks become segments radiating from the origin, rather than flat bands as in Fig. 1. Read et al. show how that offer form can also be manipulated to provide quite a general representation of the energy/reserve

[15] Here the constant *SYSRESP* includes a variety of factors, such as machine inertia and involuntary governor response, which tend to limit frequency fall, even with no specific contingency response. It is typically updated to reflect the system conditions observed in each trading interval. In Singapore, though, it is actually calculated from the dispatch variables, and hence co-optimized within the LP.

[16] Alvey et al. (1998) describe the whole formulation while Read et al. (1998) explain the co-optimization aspect in greater detail, using a slightly different offer form.

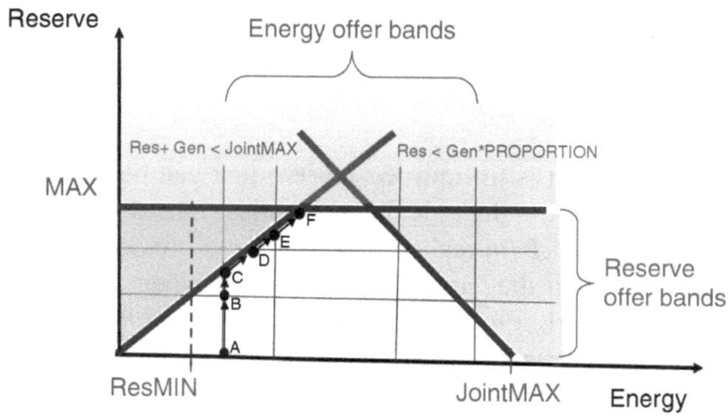

Fig. 1 Energy/reserve offer space

capabilities of any unit, hydro or thermal, including hydro plant operating in "tail water depressed" mode.[17]

3.4 *Defining Performance Limits*[18]

There is no reason why a unit cannot provide a mix of energy and reserve. And there is no reason why 6 and 60 s contingency reserve cannot be supplied by the same MW of capacity simply by maintaining the 6 s MW response through into the 60 s time frame. But we cannot have energy and reserve provided by the same MW capacity. A unit that is already operating at its maximum cannot provide any reserve, while a unit operating below its maximum capacity can only ramp up to its maximum, *JOINTMAX*[19]. So we get

$$gen_j + res_j^c \leq JOINTMAX_j^c \qquad \forall j, c \tag{5}$$

Such a constraint is shown in Fig. 1, along with the energy offer bands and the reserve offer bands described above, and the aggregate upper bound they imply.[20] The figure also shows another constraint in the form of a ray rising from the origin,

[17] That is, already synchronized to the system and ready to provide reserve response, but spinning in air and providing no energy.

[18] Our discussion relates to what may be termed "raise response," required to deal with under-frequency due to the sudden failure of a major generator or import link. The Australian market also trades "lower response" to prevent over-frequency due to the sudden failure of a major load or export link. The discussion still applies, except that some aspects are reversed. Thus plant is never "constrained off," but may be "constrained on," to provide lower reserve.

[19] This may differ by reserve class, depending on short-term overload capacity.

[20] This upper bound will be determined, for each time frame, by the maximum ramp achievable by the unit in that time frame, but this is accounted for in forming the offer.

often given a name such as "reserve proportion constraint":

$$res_j^c \leq PROPORTION_j^c * gen_j \; \forall j \qquad (6)$$

The role of this constraint is to limit the reserve that can be supplied at low generation levels, but it is a less than ideal compromise, forced by the desire to make the MCE formulation an LP. In reality, most units can provide no reserve at all, in these short time frames, if they are not generating or when operating below some minimum level (*ResMIN* in Fig. 1). Thus we would like to formulate this problem with an integer on/off variable. Such formulations are employed in markets that centralize unit commitment decisions, as is common in North America, but imply that participants can be "constrained on" to operate at a loss and may seek compensation. So other markets, including Australia and New Zealand, have sought to preserve the principle of "self-commitment" by using an LP formulation, but relying on participants to structure offers (using negatively priced offer blocks) to avoid being dispatched below minimum running levels, and to withdraw reserve offers when they are unable to provide reserve.

Many markets also allow the simple "capability envelope" illustrated in Fig. 1 to be modified in various ways, either via the offer form or in standing data. Thus the joint capacity constraint may be shifted to the right if a unit has temporary overload capacity that can be used for reserve purposes, but not for sustained generation. Provided that convexity is maintained, constraints may also be added to "shave off' parts of the feasible region, towards the top or bottom end of its operating range, where a unit cannot ramp as fast as it can in the middle. None of this changes the basic formulation, though, or its economic interpretation.

A slightly different formulation is required for regulation, though, because regulation can be only supplied by generators operating in a generation range where AGC can be employed. This is often narrower than the range over which a participant is willing and able to operate the unit manually. Further, since the unit must stay within its AGC range throughout the time when it delivers its regulation service, its base-point dispatch must lie deep enough within that range that it will not move out of the range when supplying the maximum "swing" implied by its regulation dispatch. If we think of regulation as a single symmetrical service, then a unit providing ± 10 MW of regulation cannot be dispatched, for energy purposes, closer than 10 MW to either its upper or lower AGC limit. This implies constraints sloping up at 45° from those upper and lower limits, as in Fig. 2, where the true feasible region may be thought of as including the convex region shown, but also extending further along the generation axis, in both directions.[21]

In principle, this creates an integer formulation, but an LP formulation was retained in Australia. When a unit offers regulation, the feasible region is restricted to be just the convex region shown in Fig. 2. It is the participant's responsibility

[21] In reality, the Australian market trades raise and lower regulation services separately, so that the upper limit applies only to the raise service and the lower only to the lower service. But this does not greatly alter the situation discussed here.

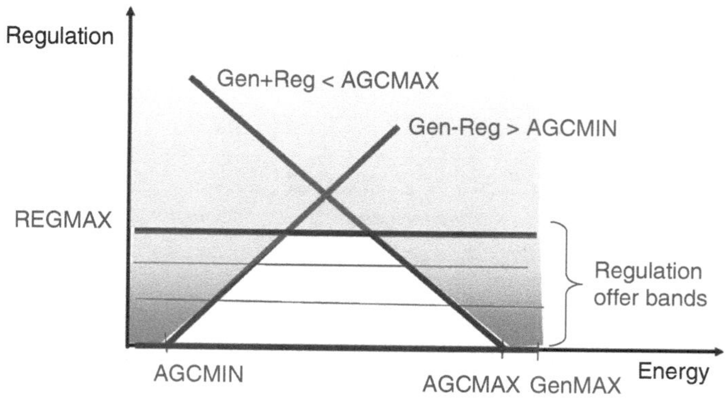

Fig. 2 Feasible offer region for symmetric regulation service

to withdraw its regulation offers if it wishes to de-commit, or operate outside this range, and to recognize when it might be profitable to move back into the AGC range and offer regulation. Although an integer formulation has been implemented in Singapore, to date, it is only employed after an initial LP run indicates that units may be "trapped" inside their AGC range.

The feasible region shown above, defined in terms of the regulation/energy offer, is not the only constraint applied, though. Since a unit on both regulation and contingency reserve duty will not be able to provide its full potential contingency response at a time when it had already been dispatched to provide its full regulation response, regulation and contingency reserve cannot be provided from the same MW capacity. So, letting reg_j be the regulation contribution from unit j, constraint (5) above must be replaced or supplemented by

$$gen_j + res_j^c + reg_j \leq JOINTMAX_j^c \qquad \forall j, c \tag{7}$$

There may also be joint ramping constraints. Since (raise) contingency reserve response requires units to ramp up rapidly, we need to consider whether their ability to do so will be affected by the rate at which they may already be ramping, for either energy or regulation purposes. The maximum ramp that can be achieved over a few seconds may not be sustainable over a whole trading interval, and the technological mechanisms involved in providing these two kinds of response can differ significantly, so that they hardly interact at all.[22] When time scales are very different, as in the original New Zealand market (6/60 s out of a dispatch interval of 1,800 s), the interaction can be ignored. Joint ramping constraints are needed when time-scales

[22] For example, conventional thermal units may provide very short-term contingency response capability by throttling back the flow of steam during normal generation, thus incurring costs, which may be reflected in their reserve offers, and then suddenly releasing the pressure during a contingency. By way of contrast, longer-term response can be provided only by increasing the rate at which fuel is supplied to the boilers, a process that involves significant time lags.

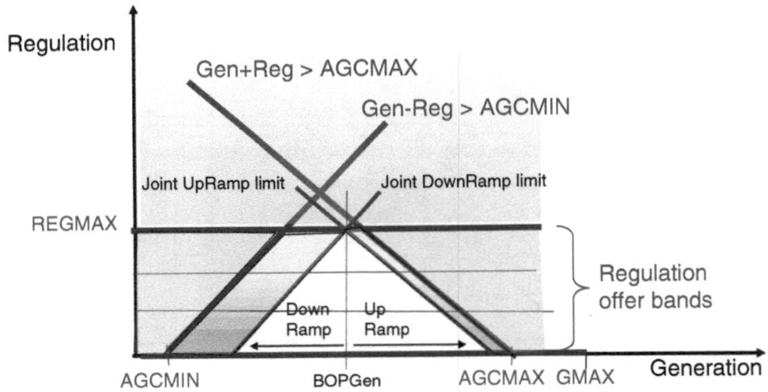

Fig. 3 Joint ramping constraints for a symmetric regulation service

are similar, though, as in Australia, with 5 min contingency response traded in a market with a 5 min dispatch interval.[23]

These constraints further restrict the feasible region, so that the availability of reserve/regulation response reduces as the EOP energy dispatch target shifts further from its observed BOP level, *BOPGen*. Figure 3 illustrates the effect for a symmetrical regulation service. If a joint *MAXRamp* is expressed in MW/second, say, and, D, R, and C, respectively, represent the response time frames, in seconds, required for energy dispatch (i.e. the dispatch interval), regulation, and one class of contingency reserve, the simple ramp limits in the non-co-optimized formulation become joint ramp rate constraints of the general form[24]

$$(gen_j - BOPGen_j)/D + Reg_j/R + \mathrm{Res}_j^c/C \leq MAXRamp \qquad \forall j, c \qquad (8)$$

Finally, constraints may be introduced to further limit reserve provision in ways that depend on a unit's observed starting generation level. The Singapore formulation includes constraints whose goal is really just to limit reserve dispatch as far as possible, in a region where units are probably unable to provide much reserve at all, particularly when units are ramping up. Of itself, the reserve proportion constraint actually implies that units ramping up can supply more, not less reserve. A suitable constraint, (6), referred to earlier maintaining a convex feasible region and consistency with other constraints, would have the same form as the *UpRamp* limit shown in Fig. 3 and intersect the generation axis at the same point *(BOPGen + MAXRAMP,*

[23] These constraints were first introduced in Singapore, where they apply only to the longest contingency reserve class, which is 10 min response, traded in a market with a 30 min dispatch interval. Similar constraints have since been retro-fitted to the Australian formulation, though.

[24] This is for raise response. In the contingency reserve case, a unit ramping down for energy purposes might actually have an increased ability to provide raise response, by simply slowing or reversing its ramp down rate. Technologically, this may not be so easy, though, and these formulations generally err on the side of caution by dis-allowing such a constraint relaxation.

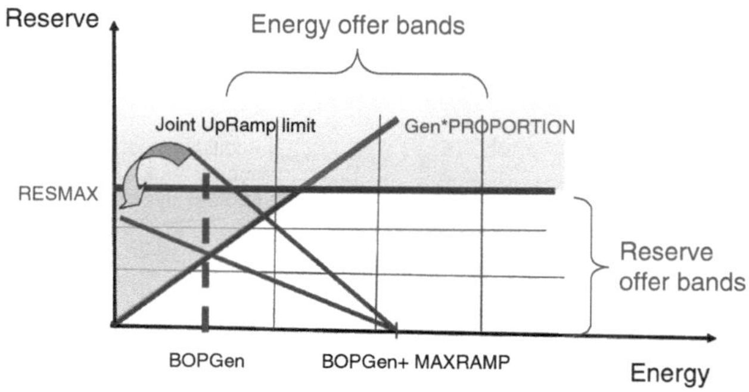

Fig. 4 Tightening the joint ramping constraint at low generation levels

0). But it would be rotated counter-clockwise so as to also pass through the point (*BOPGen, BOPGen * PROPORTION*), as in Fig. 4.[25] Such a constraint effectively over-rides the joint ramping limit, but only in the low generation region where the reserve proportion constraint applies.

4 Economic Interpretation

Read et al. (1998) give a numerical example of this kind of formulation, albeit using the more complex offer form used in the New Zealand market. They also discuss the economic interpretation of that example, in terms of the implied supply curve for reserve, and the profitability of suppliers. From a primal perspective, co-optimization will obviously produce better solutions than sequential optimization, the critical issue being whether the extra complexity is justified. As noted earlier, our own experience suggests that it is amply justified in the smaller markets discussed here, while the increasing adoption of this kind of formulation suggests it is worthwhile in larger markets too.

In our experience, though, understanding the dual is just as important as understanding the primal. Thus we focus on that aspect here because it is often misunderstood. Commercially, participants want to understand why they have been charged or paid what they have, and analysts seek to explain prices in terms of the underlying offers. Practically, this means that the primal formulation must often be crafted with a careful eye to the dual implications, in which constraints and variables are sometimes created to ensure that the model reports specific shadow prices required for market purposes.

In all these formulations, the price of each class of reserve is taken to be the shadow price on the reserve requirement constraint (1). Unlike the energy balance constraint, this is an inequality, and so reserve prices can never be negative. This price applies to all cleared offers, which are all thus paid the same, per MW, for

[25] But note that this is not actually the constraint form in the current Singapore formulation.

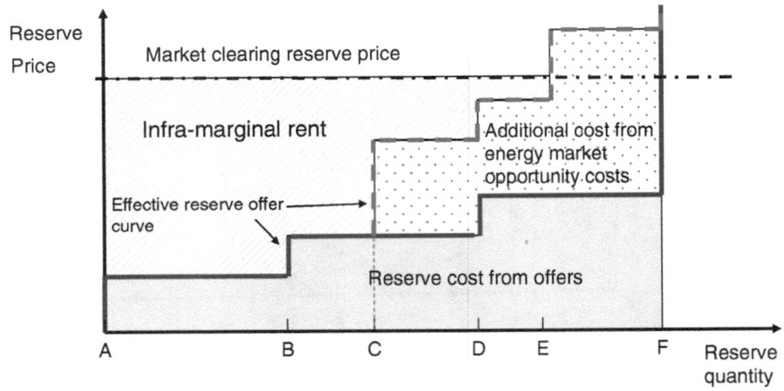

Fig. 5 Effective reserve offer curve

delivering the same service, just as in the energy market. For the marginal supply block, this price should just cover the cost of supply, as determined by the LP from the offers, and we are relying on competition to discipline those offers to be a reasonable reflection of costs.[26] For infra-marginal supply blocks, there will be a "rental" element, as in Fig. 5, providing an appropriate return on the capital investment ultimately required to provide reserve in this more efficient way.[27]

But identification of "the marginal supplier" is often not straightforward. In non-co-optimized markets, ignoring losses and network constraints, the "supply curve" for energy consists of a simple stack of energy offers. Thus, unless there are identically priced offers, a unique marginal supplier can be found by working up that stack, accepting offers until the demand is met. Nodal energy markets capture more real-world complexity, but there is usually still only one marginal supplier, if no network constraints bind.[28] It can be shown that each binding network constraint then brings one more marginal supplier into play, with its output being constantly balanced against that of the other marginal suppliers in such a way that flows over each constrained network line are kept just at the constraint level. So, if N network constraints bind, there will be N + 1 marginal suppliers, with their energy offers all mutually determining the pattern of nodal prices, via solution of

[26] "Market power" can be an issue, but it is often much less of an issue in the reserve market than in the energy market. Since generators pay for this service and since interruptible load now supplies a significant portion of it, generators are, on average, nett buyers from this market, with incentives to push prices down, not up.

[27] Technically this rent is the shadow price on the reserve (tranche) capacity bound. It represents the difference between the MCP for reserve and its effective cost, as determined by the participant's joint energy/reserve offer. It rewards suppliers for making capacity more flexible.

[28] There can be more than one, if their offers are effectively tied, after accounting for marginal losses, but then they will all have the same effective marginal cost, and any one of them can be taken as "the" marginal" supplier, when computing prices.

a set of simultaneous pricing equations corresponding to a corner point of the dual problem.[29]

Co-optimization complicates this picture by introducing one further binding constraint, and creating one more marginal supplier, for each reserve class. So, with two reserve classes, there will now be N + 3 marginal suppliers. In simple cases there may be a unique marginal supplier for energy and one more unique marginal supplier for each reserve class. In general, though, any or all these marginal suppliers may be simultaneously marginal for both energy and reserve, and the corresponding reserve and energy price patterns will be jointly determined by solution of a set of simultaneous equations involving both their energy and reserve offers. Thus the market reserve price is not determined solely by the "reserve offer price" of a specific marginal reserve supplier, and nor is the market energy price determined solely by the "energy offer price" of a specific marginal energy supplier. To understand this, it may be helpful to consider what the effective offer curves actually are for energy and reserve.

Suppose a participant who had been optimally dispatched to operate at point A in Fig. 1, where it produces only energy and no reserve, was asked to provide some reserve to the market. Parametric programming on the reserve output quantity would trace out the dispatch path shown, from A to F. It first moves up through its first and into its second reserve offer blocks, and those offer prices form the first two steps of its effective reserve offer curve, as in Fig. 5. At point C, though, it cannot provide more reserve without also increasing generation. This implies an "opportunity cost," because the reason it was dispatched at point A, at the top of an energy offer block, was presumably because the market price of energy is less than the price at which it is prepared to produce, in its next energy offer block.[30] This opportunity cost is a legitimate part of the cost of providing reserve, and this creates a new step in the effective reserve offer curve. Similarly for the energy market opportunity costs, and reserve offer costs, of every other reserve or energy block traversed, along the path shown, until no more reserve is available, at point F. Such situations, where generators are "constrained on" to supply reserve, are relatively rare, but it is very common for generators to be "constrained off" to supply reserve, creating an analogous effective offer curve at the other end of their operating range. The effective energy offer curve can be formed similarly by movement along a horizontal path from any dispatch point, say E, until supplying more generation ultimately incurs an opportunity cost when reserve must be withdrawn from the market.

Finally, since frequency is a property of the whole system, all participants will benefit from maintaining it, and no participant can maintain frequency at their own location by simply purchasing a quantity of this ancillary service for its own use. So,

---

[29] Of course these prices are automatically provided by the standard LP sensitivity analysis.

[30] Some authors seem to think it necessary to explicitly compute these opportunity costs. This can be useful when interpreting results, and might be necessary if the reserve market was to be cleared after the energy prices had been finalized, as in some of the models discussed by Gan and Litvinov (2003). But adding them into the objective function, as in Tan and Kirschen (2006)), would double count these costs and produce sub-optimal outcomes.

unlike the energy market, where participants both buy and sell, the SO purchases on behalf of the whole system. The allocation of the costs incurred by such purchases is another topic. Mathematically, the requirement for the contingency reserve service is set, in the formulation, by the unit(s) or link(s) whose failure would constitute the single largest contingency, and it might be thought reasonable to assign all costs to those units/links. But that would create perverse incentives for all units to avoid being the largest, thus effectively removing useful generation capacity from the system. Nor is it realistic, because the formulation is really only a proxy for a very complex stochastic formulation in which the optimal requirement would be influenced by the size and probability of all unit breakdowns. Details lie beyond the scope of this chapter but, in practice, costs are recovered from participants using pricing formulae which attempt to approximate the optimal stochastic price structure by giving incentives, broadly, to reduce the size and probability of contingency events across the spectrum.

5 Multi-zone Formulations

The generic formulation above relates to markets trading various types of "reserve" in a single zone. In New Zealand, contingency reserve cannot be traded across the HVDC link because it does not respond rapidly in contingency situations, and is itself often the critical contingency. So the market trades reserve in each AC subsystem independently.[31] But the original Australian market formulation allowed for multiple contingency zones, within a single AC system, with limited transfer capability between them.[32]

The objective function does not change, but (1) is modified so that a zonal reserve requirement, req^z, is defined for each zone, z, by the critical contingency for that zone. Part of that requirement may need to be covered from purchases in that zone, but the rest may be imported subject to a set of contingency power flows within the MCE, ensuring that the pattern of reserve purchases allowed a feasible flow pattern to meet each contingency. Since a single contingency will elicit a different response in each time frame, the model must now include a complete set of generation and flow variables, zonal energy balances, and power flow equations, for each zonal contingency, z, and reserve class, c, as well as for the base case. If $resp_j^{zc}$ is the response from unit j to contingency z in the time frame for reserve class c and gen_j^{zc} is the resultant generation level in that time frame, we require

$$resp_j^{zc} = gen_j^{zc} - gen_j \quad \forall c, z, j \tag{9}$$

[31] They do not quite form separate markets, though, because co-optimization links the price of reserve in each island to the energy price there, and those energy prices are themselves linked.

[32] This formulation is highly stylized. This aspect of the formulation has now been disabled, while the original is not publicly available, and not the same as that in Ma et al. (1999).

$$\sum_{j \in Y} resp_j^{zc} \leq \sum_{j \in Y} res_j^c \quad \forall c, z, Y \tag{10}$$

If we also define a new load variable for zone z in contingency z, by adding req^z to the normal load there, the inter-zonal flow pattern implied by this load/generation pattern must then be feasible for each contingency z and class c. No intra-zonal transmission limits are applied, though, and inter-zonal transmission limits may be considerably more relaxed than the base case limits. This reflects the fact that limits are normally set conservatively so as to allow for the possibility of short-lived contingency flows of this nature. The reason for this simplification, though, is not just to reduce computational effort, but also to simplify the market by allowing all reserve, and response, provided in a zone to be treated interchangeably, as a single commodity.[33] This means that we can state (10) in this aggregated zonal form, and the model can report the price to be paid for reserve purchased in each zone as the shadow price on this constraint.[34]

There is no necessary connection between pricing zones for energy and reserve purposes, but more complex constraint structures imply more complex pricing structures. In the limit, a multi-contingency model such as that of Chen et al. (2005) could produce nodal prices for each contingency class. That may not be a viable market design option for both computational and complexity reasons, but the New England ISO formulation described by Zheng and Litvinov (2008) does allow for nested reserve zones. Constraint priorities are set by penalty factors, which may ultimately set prices if constraints must be violated.[35] Chattopadhyay et al. (2003) describe a multi-contingency formulation, applied to contingencies impacting on voltage stability. Since those contingencies are assumed to be proportional to base-case loads at particular locations, the formulation implies impacts on base-case energy prices for those zones, even though the contingencies are assigned a zero probability of actually occurring, in the objective function.

On the other hand, Chen et al. describe a formulation that assigns explicit probabilities to contingency events and seeks to minimize expected costs. Formulation-wise, the dispatches for each contingency scenario must all still be feasible, irrespective of the probabilities assigned. The probabilities only appear in the objective function and can affect optimal dispatch and prices. Siriariyaporn and Robinson (2008) envisage both the probability and the cost of potential outages being reflected directly in the objective function. But the practical significance of this complexity seems moot, given the quite small probabilities typically involved.

Another issue is that all these formulations assume that participants will know where a contingency has occurred in sufficient time to respond appropriately, as determined by the power flow for that contingency in the MCE. This is unrealistic

---

[33] Limits may also be placed on offers from units subject to more severe localized constraints.

[34] The shadow price on the zonal version of (1) would be the price for meeting that zone's requirement. But this is irrelevant, because costs are recovered according to a different formula.

[35] Otherwise prices are set entirely by opportunity costs, since participants cannot specify any fee component.

for time frames so short that generators must respond blindly to frequency drop without knowing where the contingency has occurred. Thus the proposed purchase pattern may not actually be implementable, because a response that would be appropriate for a local contingency, say, could overload lines if the contingency is in another zone. The problem may be avoided, at some expense, by operating interconnectors conservatively so that they can accommodate large flow swings during contingency events. Ideally, though, constraints should be added to the formulation, linking the regional response across contingencies so that reserve providers are modeled as responding in proportion to the frequency drop caused by each contingency, irrespective of its location.

An intermediate situation arises with respect to trading services coordinated by AGC, since the AGC algorithm will determine the degree of locational differentiation possible. Thus Read (2008) presents a formulation currently under consideration in New Zealand to allow trading of a symmetric regulation service across an HVDC link.[36] In real time, the AGC system is assumed to calculate the response required in each AC island zone and the corresponding transfer swing, given the observed frequency deviation in each zone. This aggregate island swing is then apportioned according to participation factors set, at the beginning of each dispatch interval, in proportion to regulation MW cleared by the MCE, at the start of each dispatch interval. The basic formulation meets an aggregate national requirement, from either zone, subject to HVDC swing capacity limits set by technical factors and by the freeboard between its base energy market dispatch and the upper/lower transfer limits in that flow direction. Thus one zone relies, in part, on regulation services provided in the other. At the limits, this "trading" leaves no spare swing capacity for "sharing" of regulation duties when zonal load fluctuations offset one another. More balanced solutions do allow such sharing, which provides operational gains, and arguably could reduce national purchase requirements, too. If so, the national requirement constraint can be replaced by two constraint segments, reflecting increasing regulation requirements as trading increases, in either direction.

6 Multi-period Formulations

Finally, the formulations here are relatively straightforward because they are expressed solely in terms of optimization within a single dispatch interval, without considering the possibility that contingency response might be provided, in part, by re-dispatch at the start of the next dispatch interval. The possibility of ad hoc re-dispatch offers a further opportunity to provide an enhanced response, particularly in markets with longer dispatch intervals. So the system actually has further

[36] See http://www.electricitycommission.govt.nz/pdfs/opdev/comqual/consultationpdfs/freq-reg/AppendixD.pdf

mechanisms, not recognized by the MCE formulation, for dealing with contingencies, and for restoring satisfactory margins within the assumed contingency response time frame. By ignoring this possibility, these formulations all err on the side of caution. They provide enough reserve capacity to cope with contingencies occurring under worst case assumptions that re-dispatch will not be possible until after the contingency has been dealt with.

This does not invalidate the formulations or the prices they produce. If the standard is set so as to be able to cope with the worst case scenario, then these formulations do find the least cost dispatch to meet that standard, and prices to match. The fact that the system may actually be more secure at some points in the dispatch cycle than at others is just a factor that needs to be taken into account in setting the standard. Conversely, there is no reason why the single-period formulations discussed above cannot be extended to become a multi-period formulation, with or without unit commitment, in markets where such formulations are employed for energy dispatch purposes. Such formulations imply more complex price structures, with prices in one period being partially set by offers in other periods, and/or create the potential for price/dispatch inconsistencies. But those issues are not restricted to the ancillary service aspects of multi-period market formulations.

Similar observations apply, too, to standards that require "reserve" to be carried that can only respond in time frames longer than the dispatch interval. There is no reason why such services cannot be co-optimized, just as for the intra-interval reserve services discussed. Longer time frames allow for more ramping, thus allowing more MW capacity to compete, and also making it easier to direct contingency-specific responses. So, with respect to feasibility, ignoring contingent re-dispatch options just means that co-optimization may err on the conservative side and supply greater security than strictly necessary.

There is an optimality issue, though, because the ultimate cost of a proposed energy/reserve dispatch pattern will depend, in part, on how the re-dispatch process might play out after a contingency. In principle, this makes a stochastic formulation attractive. Chen et al. assign probabilities to contingencies in the objective function, thus implying that prices calculated for those states have some impact on base-case prices, but do not present a multi-period model. Ehsani et al. (2009) present a multi-period, integer, co-optimization formulation in which a constraint is imposed representing the composite risk of generator and transmission failures.[37] But this constraint is set in a loop exogenous to the co-optimization, and there is no endogenous modeling of dispatch during contingency conditions. Doorman and Nygreen (2002) discuss a hypothetical model in which participants submit an entire formulation for inter-temporal unit commitment and co-optimization as an offer. But no model in the literature seems to include endogenous optimization of the management of contingencies over several dispatch intervals after the contingency is assumed to occur. And the real advantages of doing so are not clear.

---

[37] The pricing implications of such a constraint form are not explored, because generators do not make reserve offers, and are only paid their energy market opportunity costs.

In markets where contingency responses are assumed to play out over a period longer than the dispatch interval, there certainly will be a non-zero probability that new prices will be determined during a contingency.[38] But prices are re-calculated for each dispatch interval anyway; participants are always uncertain as to what they will be, and all participants are affected, whether they are on reserve duty or not. Thus it may be argued that a stochastic formulation is required for energy markets, too. In reality, stochasticity is a feature of all markets, in any sector, and market participants must always factor this into their decision-making. In this case, while many electricity markets provide information to participants by optimizing for several independent alternative market scenarios, none, so far as we are aware, employ a proper stochastic optimization over the kind of time frame relevant to ancillary services. Rather, participants receive explicit payments for being on standby duty and are assumed to factor the relatively low probability of contingencies, and of contingent re-dispatch, into their offers.

7 Conclusions

LP-based co-optimization has proved successful in maintaining integrated coordination of energy and ancillary services in a market environment, thus creating what is essentially a multi-commodity market. It was implemented first in smaller electricity systems, where ancillary service provision is most critical, but is now commonly implemented in larger markets, too.

References

Alvey T, Goodwin D, Xingwang M, Streiffert D, Sun D (1998) A security-constrained bid-clearing system for the New Zealand wholesale electricity market. IEEE Trans Power Syst 13:340–346

Chattopadhyay D, Chakrabarti BB, Read EG (2003) A spot pricing mechanism for voltage stability. Int J Electr Power Energy Syst 25:725–734

Chen J, Thorpe JS, Thomas RJ, Mount TD (2005) Location-based scheduling and pricing for energy and reserves: a responsive reserve market proposal. Decision Support Syst 40(3-4): 563–577

Ehsani A, Ranjbar AM, Fotuhi-Firuzabad M (2009) A proposed model for co-optimization of energy and reserve in competitive electricity markets. Appl Math Model 33:92–109

Doorman GL, Nygreen B (2002) An integrated model for market pricing of energy and ancillary services. Electric Power Syst Res 61(3): 169–177

Gan D, Litvinov E (2003) Energy and reserve market designs with explicit consideration to lost opportunity costs. IEEE Trans Power Syst 18:53–59

Ma X, Sun D, Cheung K (1999) Energy and reserve dispatch in a multi-zone electricity market. IEEE Trans Power Syst 14:913–991

[38] This is true in a market where contingencies are dealt with in a shorter time frame, too, because a contingency can occur just before prices are re-calculated for the next dispatch interval.

Read EG (2008) An expanded co-optimisation formulation for New Zealand's electricity market. Presented to OR Society of New Zealand Conference, Wellington. Available from http://www.mang.canterbury.ac.nz/research/emrg/conferencepapers.shtml
Read EG, Drayton-Bright G, Ring BJ (1998) An integrated energy/reserve market for New Zealand. In: Zaccours G (ed) Deregulation of electric utilities, pp. 297–319. Kluwer, Boston
Read EG, Ring BJ (1995) Dispatch based pricing: theory and application. In: Turner A (ed) Dispatch based pricing for the New Zealand power system. Trans Power New Zealand, Wellington
Ring BJ, Read EG, Drayton GR (1993) Optimal pricing for reserve electricity generation capacity. Proceedings of the OR society of New Zealand. pp. 84–91
Siriariyaporn V, Robinson M (2008) Co-optimization of energy and operating reserve in real-time electricity markets. Proceedings DRPT, Nanjing. pp. 577–582
Tan YT, Kirschen DS (2006) Co-optimization of energy and reserve in electricity markets with demand-side participation in reserve services. Proceedings of the IEEE Power Systems Conference and Exposition. Atlanta, GA. pp. 1182–1189
Wu T, Rothleder M, Alaywan Z, Papalexopoulos AD (2004) Pricing energy and ancillary services in integrated market systems by an optimal power flow. IEEE Trans Power Syst 19(1):339–347
Zheng T, Litvinov E (2008) Contingency-based zonal reserve modeling and pricing in a co-optimized energy and reserve market. IEEE Trans Power Syst 23:77–86

Part II
Expansion Planning

Investment Decisions Under Uncertainty Using Stochastic Dynamic Programming: A Case Study of Wind Power

Klaus Vogstad and Trine Krogh Kristoffersen

Abstract The present paper adopts a real options approach to value wind power investments under uncertainty. Flexibility arises from the possibility to defer the construction of a wind farm until more information is available, the alternative to abandon the investment, and the options to select the scale of the project and up-scale the project. Taking into account uncertainties in future electricity prices, subsides, and investment costs, the problem is solved by dynamic stochastic programming. The motivation rests on a real business case of the major Norwegian power producer Agder Energi and experience from the Nordic power market at Nord Pool.

Keywords Investment · Real options · Renewable energy · Stochastic dynamic programming · Wind power

1 Introduction

As a result of sustainability enhancements, including reduction of emissions, security of supply, and adequacy of resources, the deployment of renewable energy has become increasingly important. Utilities contribute to supply-side adequacy of the system by investments in renewable energy technologies. The profitability and the timing of such investments is therefore crucial.

In general, renewable energy technologies are characterized by higher investment costs and lower marginal costs than conventional power plants. To assess the value of investments, it must be taken into account that, in contrast to conventional supply that can be largely controlled, renewable energy generation may be only partly predictable or the value of renewable energy technologies can be affected by uncertainties in investment costs, market prices, subsidies, etc.

K. Vogstad (✉)
Agder Energi Produksjon and NTNU Dept of Industrial Economics
e-mail: klausvogstad@gmail.com

S. Rebennack et al. (eds.), *Handbook of Power Systems I*, Energy Systems,
DOI 10.1007/978-3-642-02493-1_14,

Mostly, renewable energy investments are flexible in the sense that they can be deferred until uncertainty is, possibly only partly, resolved. Flexibility can also stem from options to break a project into an exploration phase, a permit phase, an investment phase, etc. to up-scale or down-scale generation, to switch generation mode, for example, between standby, sleep, and active, to expand or contract operations in related industries, for example, heat and transport, or to shut down, restart, or abandon the investment.

Recently, investment theory has moved towards the application of principles from financial option valuation for appraisal of investments in real assets. Assuming an investment is irreversible, can be deferred, and involves uncertainty, the opportunity to invest is considered a real option. As opposed to static net present value calculations, real options theory provides a dynamic framework for timing investments and takes into account the value of flexibility, including the possibility of waiting for uncertainty to resolve etc.

Utilizing the similarities to financial option pricing (Hull 2003), real options can be valued in a risk neutral fashion, constructing a replicating portfolio and applying arbitrage arguments (Dixit and Pindyck 1994). However, in valuing real options, the appropriate assets may be illiquid. At the expense of estimating a risk-adjusted discount rate, stochastic dynamic programming requires no information beyond that of the given investment (Bertsekas 1995). At the same time, the method is suitable for including the complexities such as uncertainty unfolding according to an advanced stochastic process (Oksendal 1995).

Considering the Norwegian power producer Agder Energi and the Nordic power market at Nord Pool, the aim of the present paper is to assess the profitability and the timing of investments in wind power projects under uncertainty. With wind sites being scarce, it is relevant to value exclusive property rights, referred to as real options, taking the following into account:

- Assuming a competitive market and a price-taking investor, uncertainties arise from future investment costs, electricity prices, and subsidies.
- Flexibility to defer an investment (call option), select the scale of a project, up-scale a project (switching option), or abandon an investment (put option).

It is assumed that exploration is completed, that the wind sites at which to valuate exclusive property rights have been identified, that permits have been granted from the land owners and authorities, and that wind measurements at the sites are finished. Although wind may be only partly predictable, expected wind resources have been assessed using the measurements and related long-term data. Given a wind site, this paper aims at valuing the real option on the project.

Similar investment studies include Fleten et al. (2007), who compare real options valuations with net present value assessments for investments in renewable power generation and wind power in particular. Assuming long-term electricity prices follow a geometric Brownian motion whose parameters are estimated from forward contracts, investments are subjected to risk neutral valuation, using the risk-free rate for discounting. For other sources of uncertainties that may have a significant impact on investment values, forward markets may not exists, making the approach of the present paper relevant. In contrast, the methodology of Botterud (2007) is very similar to that of the present paper. The authors apply stochastic dynamic programming to evaluate power plant investments considering the Nordic electricity market under different market designs. Accordingly, it is examined how capacity payments influence the timing and the scaling of investments. However, although the present paper considers multiple sources of uncertainty, only long-term uncertainty in load growth is taken into account.

2 Real Options On Wind Power

The aim is to determine the profitability and the timing of a wind power investment. Consider a finite time horizon of T years discretized into yearly time periods and denoted by $\{1,\dots,T\}$. The value of the investment is affected by uncertainties in yearly investment costs, market prices, and subsidies, represented by the random variables $c_t : \Omega \to \mathbb{R}, p_t : \Omega \to \mathbb{R}, s_t : \Omega \to \mathbb{R}, t = 1,\dots,T$ defined on some probability space $(\Omega, \mathcal{F}, \mathbb{P})$. Using the real options approach, the value of the investment takes into account the flexibility to invest at any time period, that is, an option to defer the investment, and options to scale, to up-scale, and to abandon the investment. Flexibility is represented by the decision and state variables $x_1,\dots,x_T$. For the current study, it is assumed that $x_t \in \{1,\dots,8\}$ where the states are listed in Table 1 and transitions between states are shown in Fig. 1. However, the problem easily generalizes to allow for other states and transitions.

The value of the investment is given by the maximum discounted expected future profits resulting from investment costs, IC_t, and yearly revenues, YR_t. Denoting by $\mathbb{E}$ the expectation with respect to all the random variables, this computes as

$$\max_{x_1,\dots,x_T} \mathbb{E}\Big[\sum_{t=1}^{T-1}(1+r^{int})^{-t}(YR_t(x_t,p_t,s_t) - IC_t(x_t,c_t)) \\ +(1+r^{int})^{-T}YR(x_T,p_T,s_T)\Big], \qquad (1)$$

where r^{int} denotes a risk-adjusted discount rate.

Table 1 Investment states

Index	State
1	Wait
2	Abandon
3,4	Invest, low capacity
5	Operate, low capacity
6,7	Invest, high capacity
8	Operate, high capacity

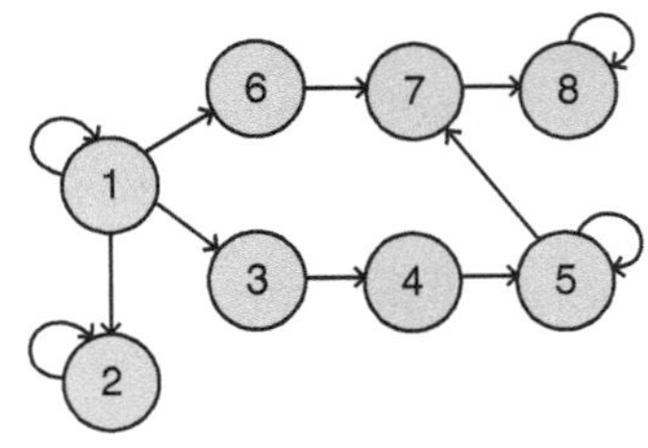

Fig. 1 Investment state transitions

Table 2 Investment costs. Example for the Bjerkreim wind farm

State	1,2	3,4	5	6,7	8
Site cost (MEUR)	0	203	0	280	0
Turbine price (MEUR/MW)	0	1.25	0	1.25	0
Capacity (MW)	0	125.0	125.0	172.5	172.5

Yearly revenues are received in operation states only. During year t, load l is sold at the market price, using the average yearly price p_t, and receives a subsidy s_t. At the same time, it incurs an operation cost oc_t. To account for diurnal and seasonal variations, yearly revenues are adjusted with the multipliers α^{day} and α^{season}. Hence,

$$YR_t(x_t, p_t, s_t) = \begin{cases} l(x_t)(p_t + s_t - oc_t)\alpha^{day}\alpha^{season} & \text{, if } x_t \in \{5, 8\} \\ 0 & \text{, otherwise.} \end{cases}$$

Subsidies are given for a maximum of T' years. However, to avoid an enlargement of the state space, subsidies are adjusted such as to correspond to a fixed yearly annuity over the time horizon T considered for investment valuation, that is,

$$s_t \Big(\sum_{t'=1}^{T'} (1 + r^{int})^{-t'} \Big) \Big(\sum_{t''=1}^{T} (1 + r^{int})^{-t''} \Big)^{-1}.$$

Investment states incur costs composed of a site and a turbine component. Letting $sc(x_t)$ denote the cost of property rights, which depend on the scale of the investment, c_t denote the cost of turbine capacity, which depend on the development in wind turbine prices and $L(x_t)$ denote the scale of capacity, investment costs amount to

$$IC_t(x_t, c_t) = \begin{cases} (sc(x_t) + c_t L(x_t))/2 & \text{, if } x_t \in \{3, 4, 6, 7\} \\ 0 & \text{, otherwise,} \end{cases}$$

where costs are divided between the years of construction, the construction time of the current study being 2 years. As an example, the components of yearly investment costs are listed in Table 2 for the Bjerkreim wind farm, which is part of the wind power portfolio of Agder Energi. Because of the option to defer the investment, the remainder of the time horizon may be less than the lifetime of the investment. To avoid end effects from a finite time horizon, investment costs are adjusted according to a fixed yearly annuity to be paid for the remainder of the time horizon such that

$$IC_t(x_t, c_t)\Big(\sum_{t'=1}^{T-t+1}(1+r^{int})^{-t'}\Big)\Big(\sum_{t''=1}^{LT}(1+r^{int})^{-t''}\Big)^{-1},$$

where LT denotes the lifetime of the investment.

For valuation of wind power investments, tax on income must be considered. Taxable income is the sum of yearly revenues subtracted tax deductions due to depreciations over the lifetime of the investment. Calculations show that losses to be carried forward are usually zero or very small so that the investment always induces tax. Ignoring loss carry-forwards, discounted tax deductions due to depreciation of the investment can be accounted for in the investment costs without loss of generality. Let r^{dep} denote the depreciation rate and r^{tax} the tax rate. The taxable income in year t for an investment made in year t' is therefore

$$YR_t(x_t, p_t, s_t) - IC_{t'}(x_{t'}, c_{t'})(1 - r^{dep})^{t-t'-1} r^{dep}.$$

The tax effect can be divided into revenue tax to be paid in year t,

$$r^{tax} YR_t(x_t, p_t, s_t),$$

and tax deductions to be subtracted from the investment cost of year t', and the total effect for the remainder of the time horizon being

$$IC_{t'}(x_{t'}, c_{t'}) \sum_{t=t'}^{LT+t'-1} (1+r^{int})^{-1}(1-r^{dep})^{t-t'-1} r^{dep}.$$

For ease of notation, let $\xi_t = (c_t, p_t, s_t)$ be the stochastic vector composed of the uncertainties at time period t. Owing to separability of (1), the optimality principle by Bellman (1957) applies. Thus, given a stage t, an investment decision x_t and a state of uncertainty ξ_t, the value function $\Pi_t(x_t, \xi_t)$ accumulates current and expected future profits of stages $t, \ldots, T$ such that the value of the investment $\Pi_1(x_1, \xi_1)$ can be obtained from the following stochastic dynamic programming recursion. The conditional expectation of ξ_{t+1} given ξ_t is denoted by $\mathbb{E}_{\xi_{t+1}|\xi_t}$. For stage t,

$$\Pi_t(x_t, \xi_t) = YR_t(x_t, \xi_t) - IC_t(x_t, \xi_t) + (1+r^{int})^{-1} \max_{x_{t+1}} \{\mathbb{E}_{\xi_{t+1}|\xi_t}[\Pi_{t+1}(x_{t+1}, \xi_{t+1})] \mid x_{t+1} \in X_t(x_t)\}, \quad t = 1, \ldots, T-1 \qquad (2)$$

and for stage T,

$$\Pi_T(x_T, \xi_T) = YR_T(x_T, \xi_T), \tag{3}$$

assuming no investments occur in the last stage. If the value of the investment $\Pi_1(x_1, \xi_1)$ is positive it is profitable to invest in the first stage.

3 Uncertainty

Uncertainties in wind resources, investment costs, electricity prices, and subsidies have a significant impact on the net present value of a wind power project. However, since the uncertainty of wind resources do not change over time, without loss of generality, average load can be used for real option valuations.

Uncertainty in investment costs, electricity prices, and subsidies is represented by a three-dimensional discrete-time stochastic process $\{\xi_t\}_{t=1}^T$ on the probability space $(\Omega, \mathcal{F}, \mathbb{P})$.

The three stochastic processes of costs, prices, and subsidies are known to be autocorrelated. However, to facilitate the application of stochastic dynamic programming, including a limited dimension of the state space, the stochastic processes are assumed to form three independent Markov chains. The independence between subsidies and prices and subsidies and investment costs can be justified in Norwegian case although independence may not generally apply and an improved description of uncertainty would include correlations.

The estimation of Markov chains is usually based on historical data. For some factors, however, there is insufficient historical data available to estimate the long-term models used for investment valuation. Alternatives for estimating Markov chains are therefore the following:

- The use of data from forward markets
- The use of fundamental models for simulation of future scenarios
- The use of expert-based scenarios provided by market analysts or external reports

where a scenario refers to a possible path of realizations. The use of forward data assumes the existence of a market for the factor of interest. This is the case for electricity prices but generally not for investment costs and subsidies. Still, the forward market for electricity prices does not have a sufficiently long time horizon for investment valuation of wind power projects. The development of fundamental models is feasible although very few models are capable of simulating over long time horizons and at the same time take uncertainty into account. For such reasons, power producers commonly use expert-based scenarios.

3.0.1 Market Prices

Historical data is often available for estimating price models. Making use of this, the current power producer Agder Energi maintains a time series model based on mean reversion properties of the prices. The model is suitable for generating future scenarios that facilitate the estimation of a Markov chain.

With a slight abuse of notation, consider therefore the one-dimensional discrete-time Markov chain $\{\xi_t\}_{t=1}^T$ with finite state space $\{1,\ldots,S\}$. Given a number of future price scenarios, the state space is constructed by dividing prices into disjoint intervals. To use stochastic dynamic programming, transition probabilities are estimated using maximum likelihood estimation. The transition probability between states i to j at time t is given by $\pi_t^{ij} = \mathbb{P}(\xi_{t+1} = \xi_{t+1}^j \mid \xi_t^i)$ and is estimated by the observed relative frequency of transitions between the states

$$\hat{\pi}^{ij} = \sum_{t=1}^{T}(\hat{\xi}_{t-1}^i, \hat{\xi}_t^j) / \sum_{t=1}^{T}\sum_{k=1}^{S}(\hat{\xi}_{t-1}^i, \hat{\xi}_t^k),$$

where $(\hat{\xi}_{t-1}^i, \hat{\xi}_t^j)$ denotes a transition between states i to j at time t. To achieve a distribution with similar statistical properties as the time series model, transition probabilities may occasionally be modified. In the same spirit, the mean reversion of the Markov chain has been tested against that of the time series model. For further reference of the entire estimation procedure, see Mo et al. (2001).

3.0.2 Investment Costs and Subsidies

Market analysts often provide subjective scenarios for future investment costs and subsidies on the basis of some fundamental drivers. For the one-dimensional processes of investment costs and subsidies, such scenarios are given by the realizations $\{\xi_t^s\}_{t=1}^T, s = 1,\ldots,S$ and the probabilities $\pi_t^s = \mathbb{P}(\xi_t = \xi_t^s), s = 1,\ldots,S$, where S denotes the number of scenarios.

This probabilistic information is used to estimate one-dimensional discrete-time Markov chains $\{\xi_t\}_{t=1}^T$ with finite state space $\{1,\ldots,S\}$. Denoting by $\hat{\pi}_t^{ij}$ the estimate for the transition probability $\mathbb{P}(\xi_t = \xi_t^j \mid \xi_t^i)$ between states i to j at time t, transition probabilities have to satisfy the linear equations

$$\sum_{k=1}^{S}\hat{\pi}_t^{ik} = 1, \quad \sum_{k=1}^{S}\pi_t^k\hat{\pi}_t^{ki} = \pi_{t+1}^i, \quad \hat{\pi}_t^{ij} \geq 0, \quad i,j = 1,\ldots,S, t = 1,\ldots,T.$$

Using the equations as constraints in an optimization problem, the conditional probability distribution is determined as the one that minimizes some objective function criteria. Experiments have been made with different criteria.

The estimation of transition probabilities has been implemented as an interactive procedure, making market analysts able to provide additional information and make subjective adjustments.

4 Value of Wind Power Investments

The stochastic dynamic programming recursion (2) and (3) have been implemented in the modeling language Mosel and, using the solver Xpress version 1.18.05, run on a ProLiant DL380 G4 server.

To illustrate the real options valuation of wind power investments we consider the Bjerkreim wind farm, which is part of the wind power portfolio of Agder Energi. Data consists of average number of full load hours measured at the relevant sight, estimated operation costs and investment costs for property rights, as well as approximate interest and discount rates. The interest rate was set according to the company's internal estimates for rates of returns on wind power projects, see other studies,[1] whereas the discount rate is taken to be the official in Norway. Investment costs for turbine capacity, electricity prices, and subsidies were described in the previous sections. The main results are listed in Table 3, where the first column holds investment values in terms of (maximum) discounted expected future profits. The second column shows the probability of profitability after investment has been undertaken. As seen, replacing the NPV criteria with ROV results in a higher expected profitability at reduced risk. The reason is that NPV does not take into account the flexibility of the investment. In particular, the options to defer the investment, select the scale of the project, up-scaling the project, or abandoning the investment in the future may generate higher future profits. Including such options ROV suggests delaying investments until more information is revealed through path-dependencies captured in the multiperiod Markov models representing stochastic price-, subsidy and investment cost processes.

We further explore the differences in using NPV and ROV assessments for making investment decisions. In reality, such decisions would be made dynamically as information is updated and it is relevant to evaluate the performance of the decision rules that arise from the NPV and ROV approaches (cf. Botterud (2003)). The decision rule of ROV is determined by the stochastic programming recursion, while that of NPV is to invest if the expected net present value becomes positive and if so in the investment of highest value. To evaluate the performance of the decision rules, we have simulated updated information. This is done by means of recursive sampling via inversion of the conditional cumulative distribution functions. For a given time period, conditional on the previous time period, we have sampled from a uniform distribution to obtain a random value of the cumulative distribution function whose corresponding quantile was used for calculating profits. The distributions of

Table 3 Investment value. Example for the Bjerkreim wind farm

	Value [MEUR]	P(NPV>0)
NPV	−2.9	43%
ROV	11	80%

[1] www.enova.no

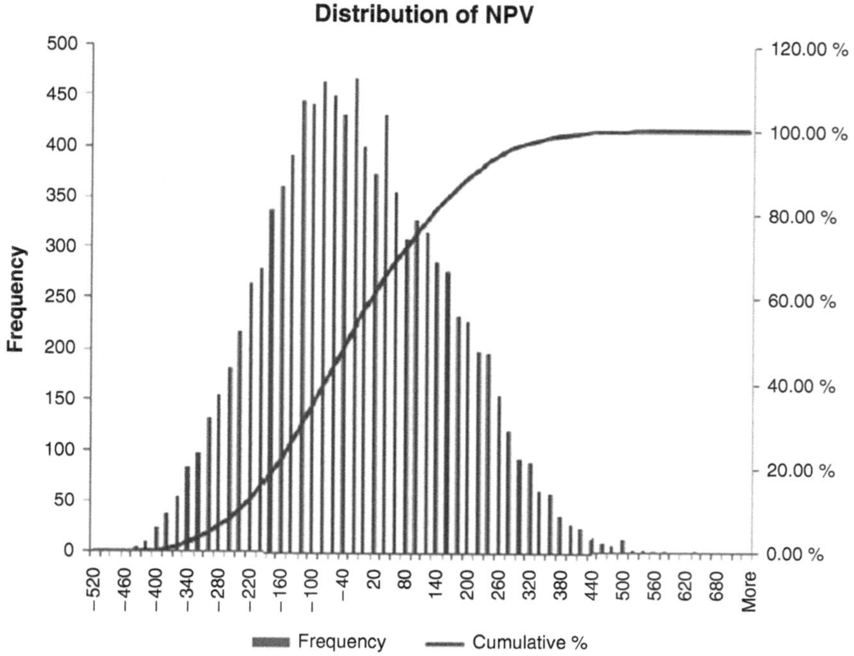

Fig. 2 Monte Carlo simulations for NPV

discounted profits are shown in Figs. 2 and 3. It can be seen that the ROV simulations have a higher expected value than the NPV simulations, which is explained by the value of flexibility. Applying NPV, the investor is equally exposed to positive and negative changes in investment costs, electricity prices, and subsidies. In contrast, using ROV the investor is more likely to obtain a positive investment value being able not only to defer but also to time the investment, scale, up-scale, and abandon the investment if changes in investment costs are positive or changes in electricity prices and subsidies are negative. For the same reason, the ROV simulations show lower down-side risk than the NPV simulations. Hence, compared to net present value calculations, the power producer is subject to higher profits and lower risks by applying the real options valuation as a basis for making investment decisions.

5 Conclusion

The present paper values wind power investments under uncertainty, considering a real business case on exclusive property rights to wind sites. To reflect reality, the valuation takes into account uncertainty in both future electricity prices, subsides, and investment costs, and the possibility to defer the construction of a wind farm until more information is available, the alternative to abandon the investment and

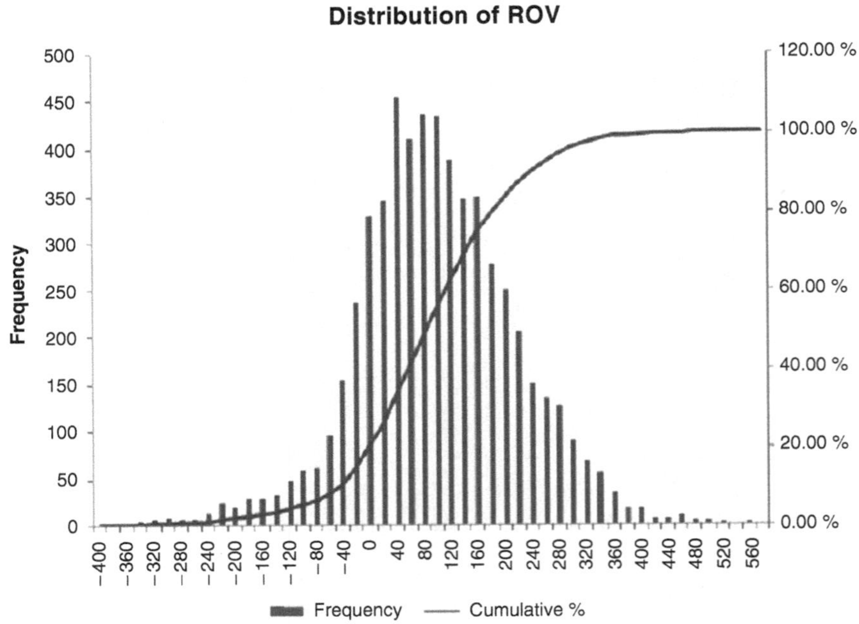

Fig. 3 Monte Carlo simulations for ROV

the options to select the scale of the project and up-scale the project. Stochastic dynamic programming at the same time facilitates the inclusion of advanced stochastic processes and provides the dynamic framework for timing the investment. Further research will be made in the direction of including correlated stochastic processes.

Acknowledgements The authors thank Dalane Vind and Agder Energi for providing the case study Bjerkreim. We would also Silje K. Usterud, Mari B. Jonassen, and Cecilie Å. Nyrønning for their important contribution to the modeling as part of their summer jobs at Agder Energi and project theses at NTNU Department of Industrial Economics.

References

Bellman R (1957) Dynamic programming. Princeton University Press, NJ

Bertsekas DP (1995) Dynamic programming and optimal control. Athena Scientific, MA

Botterud A (2003) Long-term planning in restructured power systems. Dynamic modelling of investments in new power generation under uncertainty. PhD Thesis, The Norwegian University of Science and Technology

Botterud A (2007) A stochastic dynamic model for optimal timing of investments in new generation capacity in restructured power systems. Electr Power Energ Syst 29:167–174

Dixit AK, Pindyck RJ (1994) Investment under uncertainty. Princeton University Press, NJ

Fleten S-E, Maribu KM, Wagensteen I (2007) Optimal investment startegies in decentralized renewable power genration under uncertainty. Energy 32:803–815

Hull J (2003) Options, futures and other derivatives. University of Toronto, Prentice-Hall, NJ
Mo B, Gjelsvik A, Grundt A, Kåsen K (2001) Optimization and hydropower operation in a liberalized market with focus on price modelling. IEEE Porto Power Tech Conference, Porto, Portugal, 2001
Oksendal BK (1995) Stochastic differential equations: an introduction with applications. Springer, Berlin

The Integration of Social Concerns into Electricity Power Planning: A Combined Delphi and AHP Approach

P. Ferreira, M. Araújo, and M.E.J. O'Kelly

Abstract The increasing acceptance of the principle of sustainable development has been a major driving force towards new approaches to energy planning. This is a complex process involving multiple and conflicting objectives, in which many agents were able to influence decisions. The integration of environmental, social and economic issues in decision making, although fundamental, is not an easy task, and tradeoffs must be made. The increasing importance of social aspects adds additional complexity to the traditional models that must now deal with variables recognizably difficult to measure in a quantitative scale. This study explores the issue of the social impact, as a fundamental aspect of the electricity planning process, aiming to give a measurable interpretation of the expected social impact of future electricity scenarios. A structured methodology, based on a combination of the Analytic Hierarchy Process and Delphi process, is proposed. The methodology is applied for the social evaluation of future electricity scenarios in Portugal, resulting in the elicitation and assignment of average social impact values for these scenarios. The proposed tool offers guidance to decision makers and presents a clear path to explicitly recognise and integrate the social preferences into electricity planning models.

Keywords Analytic Hierarchy Process (AHP) · Delphi · Electricity power planning · Social impact · Social sustainability

1 Introduction

Sustainable long-range electricity planning involves tradeoffs between multiple goals. Rationally, the multiple attributes of each competing and acceptable electricity generation technology or portfolio, in terms of the attainment of goals, must be

P. Ferreira (✉)
Department of Production and Systems, University of Minho, Azurem, 4800–058 Guimarães Portugal
e-mail: paulaf@dps.uminho.pt

S. Rebennack et al. (eds.), *Handbook of Power Systems I*, Energy Systems,
DOI 10.1007/978-3-642-02493-1_15, © Springer-Verlag Berlin Heidelberg 2010

assessed. Integrated resource planning should seek to identify the mix of resources that can best meet the future electricity needs of consumers, the economy, the environment and the society.

Environmental impacts of electricity generation activities become increasingly critical. The need to control atmospheric emissions of greenhouse and other gases and substances requires the full evaluation of the environmental characteristics of each electricity generation technology and the inclusion of environmental objectives in the electricity planning process. Cost minimisation is also perceived as fundamental to ensure the competitiveness of the economy. Attaining this objective involves the careful selection of technologies and efficient management of the system operation and power reserve requirements. Long-term electricity planning frequently relies on complex optimisation models drawn to minimise cost, restricted by technical and environmental conditions.

The complexity of the problem increased further due to the desirable inclusion of a number of unquantifiable or subjectively valued objectives, thus making energy planning decisions prone to some degree of controversy. For example, renewable energy projects, and in particular wind power plants, frequently have to face local opposition, and the spread of these renewable technologies may be slowed down by low social acceptability. The electricity planning process needs to rely on a formal approach for the assessment of the overall social outcome of each particular generation mix. This work addresses this matter and deals with the complexity of the social issues surrounding electricity planning, providing guidance on how to integrate the social dimension into the development of sustainable electricity plans for the future.

The structure of this paper is presented subsequently. Following this introduction, Sect. 2 aims to bring a theoretical background to the study. The close relationship between sustainable development and energy is described. The integration of sustainable development concerns into energy planning is examined, addressing in particular the social dimension of the problem. In Sect. 3 a structured methodology is proposed to assess the social sustainability of different electricity generation technologies, aiming to incorporate this information into the overall electricity planning process. Section 4 deals with the application of the proposed methodology to the case of sustainable electricity planning in Portugal. The main conclusions are summarised in Sect. 5, pointing out also directions for further research.

2 Energy and Sustainable Development

Energy use and availability are central issues in sustainable development. Energy is essential for economic development and for improving society's living standards. However, political decisions regarding the use of sustainable energy must take into account social and environmental concerns. Until recently, sustainable development was perceived as an essential environmental issue, concerning the integration of environmental concerns into economic decision-making (Lehtonen 2004). For example, for the particular case of the role of renewable energy sources (RES) to sustainable development, Del Río and Burguillo (2008) support the view that much emphasis is being put on the environmental benefits, while socioeconomic impacts have not received a comparable attention.

The three dimensions of sustainable development are intrinsically linked. As the G8 Renewable Energy Task Force (2001) [p. 14] recognises: "Economies can only grow if they are not threatened by environmental catastrophe or social unrest. Environmental quality can only be protected if basic economic needs are fulfilled and individuals take responsibility for public goods. Finally, social development rests on economic growth as well as a healthy environment." The economic, social and environmental perspectives are all included in the key elements of a sustainable energy system listed by Jefferson (2006): sufficient growth of energy supplies to meet human needs, energy efficiency and conservation measures, addressing public health and safety issues and protection of the biosphere. Thus, the sustainable development and sustainable energy planning are based on the same three dimensions, viz., economic, environmental and social.

The increasing acceptance of the principle of sustainable development has been a major driving force towards new approaches to energy planning. Achieving the goal of sustainable development implies recognising and including the social and environmental impacts of the energy sector in the decision making process. Under conditions of sustainable energy planning, the profitability of energy companies and the financial viability of energy projects become highly dependent on non-financial factors. The simultaneous assessment of economic, strategic, social, environmental and technical aspects is fundamental for making professionally correct investment decisions in any sector.[1] However, it is particularly important for the energy sector, traditionally associated with large-scale projects with strong and conflicting social impacts: on the one hand, these projects are absolutely indispensable for the social welfare of the population, but on the other hand, they are frequently associated with environmental problems and have to deal with social opposition.

The evaluation of technologies and future scenarios needs to expand beyond financial cost alone and the appraisal process must be based on a framework recognising full social cost. Proper selection of energy technologies for the future represents a valuable contribution to meeting sustainable energy development targets. Hepbasli (2008) states that energy resources and their utilisation is intimately related to sustainable development and a sustainable energy system must fulfil the requirements of being cost-efficient, reliable and environmental friendly. Also Dincer and Rosen (2005) state that sustainable development requires a sustainable supply of clean and affordable energy resources that do not cause negative societal impacts. Similarly, Lund (2007) supports the view that sustainable development involves energy savings on the demand side, efficiency improvements in the energy production and replacement of fossil fuels by various sources of renewable energy.

[1] The inclusion of non-financial aspects in project evaluation is debated at some length in a previous work from the authors (Ferreira et al. 2004).

2.1 Sustainable Energy Planning

Electricity power planning is, using the definition of Hobbs (1995) [p. 1], “the selection of power generation and energy efficiency resources to meet customer demands for electricity over a multi-decade time horizon”. This author presents three reasons for the increased complexity of the energy planning process: the increasing number of options; the great uncertainty in load growth, fuel markets, technological development and government regulation; and finally, the inclusion of new objectives other than cost. In fact, the changes in the electricity sector along with the need for sustainable development required traditional electricity planning to expand beyond pure financial analysis and even beyond direct environmental impact analysis. The increasing use of RES in electricity systems adds additional considerations to the traditional planning models, in particular the need to take into account (a) their frequent priority access to the grid system, (b) the impacts that technologies of variable output such as wind energy can have on the overall operation of the electricity system and (c) the public attitude towards these technologies. In addition, the central electricity planning process based on a single decision maker is no longer acceptable, and the importance of examining tradeoffs amongst objectives is now well recognised. Considering the three dimensions of sustainable development has gradually increased the importance of the social aspect in the decision process.

The energy planner has now the task of designing electricity strategies for the future, with the view of enhancing the financial performance of the sector while simultaneously addressing environmental and social concerns. Thus, the planners must deal not only with variables that may be quantified and simulated but also with the social impact assessment. As Bruckner et al. (2005) note, this is an ever changing field depending on aspects like policy issues, advances in computer sciences and developments in economics, engineering and sociology.

The electricity planning process has been addressed by a large number of authors, proposing different approaches and methods to solve these problems. Most of these approaches include diverse multicriteria tools, expressing each criterion in its own units or involving some kind of cost benefit analysis, in which environmental criteria are expressed in economic terms. The process frequently requires the planner to work with quantitative and qualitative information. However, continuous models focus mainly on the cost and economic dimensions of the problem. Some of the less quantifiable issues associated with the social impacts of electricity generating activities have been covered by multicriteria methods, using well recognised methods like the ones from the outranking family such as the Analytic Hierarchy Process (AHP), or by the economic valuation of externalities like the ExternE study (European Commission 2003).

The literature, for long, has been debating the planning methods available and providing some examples of the application. A detailed analysis of the subject may be found in studies such as Loring (2007) or Pohekar and Ramachandran (2004), where the authors review a large number of publications on the use of multicriteria decision making for energy planning. Also Hobbs and Meier (2003), [p. 123] present what they call a “representative sample” of multicriteria decision making applications to energy planning and policy problems. Huang et al. (1995) present a comprehensive literature review on decision analysis on energy and environmental modelling, including studies published from 1960 to 1994. Greening and Bernow (2004) collect some examples describing the application of several multicriteria methods to energy and environmental issues. Diakoulaki et al. (2005) analysed a large number of publications addressing the use of multicriteria methods to energy related decisions and Jebaraj and Iniyan (2006) review several energy models, including planning and optimisation models, among others.[2]

2.2 *Importance of the Social Dimension*

The thinking about social sustainability is not yet as advanced as for the other two pillars (World Bank 2003). However, the Brundtland Report (World Commission on Environment and Development 1987) made clear the need to expand the sustainable development concept beyond ecological concerns and fully recognise and integrate the social dimensions of sustainability, reflecting the need to ensure equitable social progress and overall social welfare. Recent studies involving energy indicators for sustainable development already reveal increasing concern with the social dimension of the concept, especially for developing countries, as in Pereira et al. (2007) or Vera and Langlois (2007).

The question of public acceptance is now generally viewed as a fundamental aspect to be included in the social dimension of the sustainable energy planning process, frequently addressed by participatory methods. Energy planning often involves many decision makers and can affect numerous and heterogeneous stakeholders, with different value systems and different concerns (Greening and Bernow 2004). Because of the great variety of ethical positions, the perception of the stakeholders involved may differ significantly.

The consultation of relevant experts and competent authorities is an essential element in the decision process, and multicriteria applications frequently involve a large and interdisciplinary group of stakeholders (Diakoulaki et al. 2005). The World Commission on Dams (2001) underlines the need to implement participatory decision-making for improving the outcome of dams and water development projects and points, gaining public acceptance as a strategic priority of the projects. Del Río and Burguillo (2008) stressed the importance of the participatory approach which takes into account the opinions and interests of all stakeholders. The authors argued that the assessment of a project's sustainability should focus not only on the impact of the proposal, but also on how this impact is perceived by the local population, how the benefits are distributed among the different players and how this perception and distribution affects the acceptance of the project. In conclusion, the acceptance or rejection of a renewable energy project by the local population can

[2] A review of recent papers proposing different approaches to energy planning, with predominant emphasis on the particular sector of electricity, may be found in Ferreira (2008), Chap. 3.

make its implementation and its contribution to local sustainability either a success or a failure. Loring (2007) analysed the factors affecting wind energy projects' success and concluded that projects with high levels of participatory planning are more likely to be publicly accepted and successful. Also Wolsink (2007) drew attention to the need to take into consideration public attitude on wind implementation decisions, not only at a general level but also at the local project level, and stressed the importance of including the public in the decision making process.

The creation of clear energy strategies merging cost effectiveness with environmental and social issues is the main challenge for energy planners. Cost oriented approaches, where the monetary assessment is the only basis for the decision making, are no longer an option, and information on the ecological and social impacts of the possible energy plans need to be combined with traditional economic monetary indicators. The existence of different perspectives and values must also be acknowledged and fully incorporated in the planning process, avoiding centralised decisions based on restricted judgements. The evolution of the market conditions and the increasing concerns with sustainable development have brought about profound changes in the approach to the energy decision process and to the priority assigned to each objective during the energy planning process. Sustainable energy planning should now be seen as a multidisciplinary process, where the economic, environmental and social impacts must be taken into consideration, at local and global levels, and where the participatory approaches can bring considerable benefits.

As highlighted by Söderholm and Sundqvist (2003), many impacts of the power generation sector involve moral concerns, and economic valuation provides an insufficient basis for social choice. The social dimension of sustainable development is much more elusive and recognizably difficult to analyse quantitatively. Thus, the social analysis cannot be addressed with the same analytic toolbox as the environment and economic ones (Lehtonen 2004). This study explores the issue of the social impact as a fundamental aspect of the electricity planning process, with strong implications for the policy decision making and for the effective realisation of the drawn plans. A structured methodology is presented, establishing a possible way of quantifying the expected overall social impact of future electricity generation scenarios.

3 Methodology

The core elements of the proposed methodology are the Delphi survey and the AHP analysis. By subdividing the problem into its constituent parts (Analytic Hierarchy), the problem is simplified and allows information on each separate issue to be examined. The relative strength or priority of each objective can be established (Delphi process) and the results synthesised to derive a single overall priority for all activities (Hemphill et al. 2002).

The combination of the AHP and Delphi has been used in different fields, with the aim of quantifying the value judgment obtained in a group decision-making

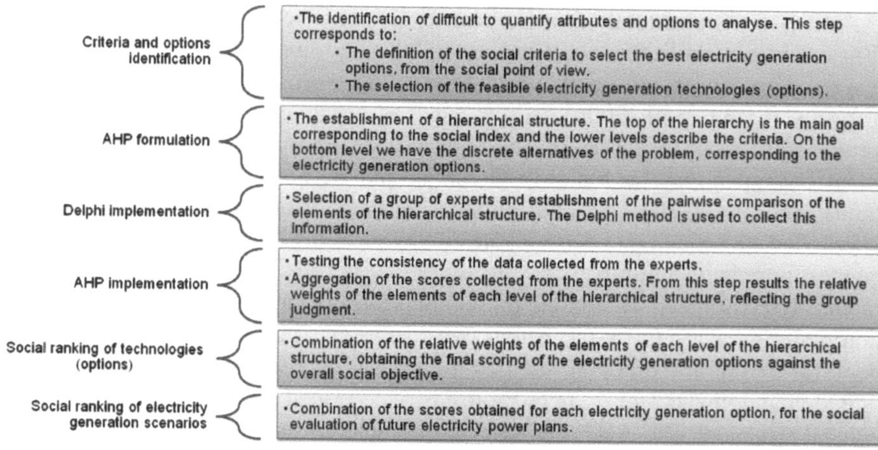

Fig. 1 Proposed methodology for integrating social concerns into an electricity power planning

process. Some recent examples include works from Hemphill et al. (2002), on the weighting of the key attributes of sustainable urban regeneration, or Zhong-Wu et al. (2007), for the appraisal of the eco-environmental quality of an ecosystem. Liang et al. (2006) presented a power generation expansion model that uses AHP and Delphi methodologies, and more recently Torres Sibille et al. (2009a,b) applied these methods on the definition of an indicator for the quantification of the objective aesthetic impact of solar power plants and wind farms.

The proposed methodology for establishing a possible way of allocating weights to the major social impacts and resulting in a final social impact index for future electricity power plans or scenarios is shown in Fig. 1.

3.1 Suitability of the AHP Approach

The analytical hierarchy process was developed by Saaty (1980) and is based on the formulation of the decision problem in a hierarchical structure. Typically, the overall objective or goal of the decision-making process is represented at the top level, criterion or attribute elements affecting the decision at the intermediate level and the decision options at the lower level (Nigim et al. 2004). The user chooses weights by comparing attributes two at a time, assessing the ratios for their importance. These ratios are used not only to compute the weights of individual attributes, but also to measure the consistency of the user's assessments (Hobbs and Meier 2003). The method incorporates the researcher's subjective judgment aided, if needs be, by expert opinion during the analysis and by expressing the complex system in a hierarchical structure. Thus, AHP assists the decision-making process to be systemic, numerical and computable.

AHP is a popular method in problem evaluation (see, e.g., Hobbs and Meier (2003), Limmeechokchaia and Chawana (2007) or Liang et al. (2006)). It is recognized as a robust and flexible tool for dealing with complex decision-making problems (Liang et al. 2006) and its use has been largely explored in the literature, with many examples in the energy decision-making field. An extensive list of examples may be found in Greening and Bernow (2004) or Pohekar and Ramachandran (2004). The latter authors presented a literature review on multi-criteria decision-making on sustainable energy planning, and observed that AHP is the most popular technique.

The AHP is especially suitable for complex systems where multiple options and multiple criteria are to be taken into consideration. The computation of a social index for a complex problem, such as the electricity generation options, involves individual judgments and it can be more easily described and analysed using a hierarchical structure. The AHP was selected because of its simplicity and ability to deal with qualitative/subjective data. The method is well suited for group decision making (Lai et al. 2002) and its integration with the Delphi method is also well documented. Using the hierarchical structure, the experts compare the electricity generation options against different criteria. It is possible to recognize conflicts among experts for each element of the hierarchy, how it affects the final ranking and also the consistency of the judgments. The qualitative scale used simplifies the judgement but at the same time allows for the mathematical treatment of the results. The final outcome is the global ranking of the options.

3.2 *Suitability of the Delphi Approach*

The main objective of the Delphi technique is to describe a variety of alternatives and to provide a constructive forum in which consensus may occur (Rayens and Hahn 2000). The three basic conditions of the process are anonymity of the respondents, statistical treatment of the responses and controlled feedback in subsequent rounds.

The anonymity of the answers gives group members the freedom to express their opinion, avoiding possible negative influences due to previous assumed positions, status of the participating experts and reluctance on assuming positions different from the general opinion or from a dominant group. The statistical treatment of the responses allows the assembly of collective information. This phase feature ensures that all the opinions are accounted for the final answer and that these opinions may be communicated to the panel without revealing individual judgments. The controlled feedback ensures that the panel individuals have access to the responses of the whole group as well as their own response for reconsideration. The basic sequence of the Delphi method may be resumed as an interactive questionnaire that is passed around several times in a group of experts, keeping the anonymity of the individual responses.

As Wright and Giovinazzo (2000) stated, the Delphi is not a statistically representative study but a process of collecting opinions from a group of experts, who from their knowledge and exchange of information may achieve comprehensive opinions on the proposed questions. This issue is also pointed out by Okoli and Pawlowsk (2004), who underlined that the questions that a Delphi study investigates are those of high uncertainty and speculation. Thus a general population or sample might not be sufficiently knowledgeable to answer the questions accurately. Also Alberts (2007) demonstrated the advantages of experts' consultation. In his recent study on how to address wind turbine noise and potential wildlife impacts, the author showed that the participants with insufficient experience were unable to participate effectively in the decision-making process, demonstrating that it can be more productive to seek input from technical experts than to seek consensus from all stakeholders.

For this particular research, the questions addressed are complex and highly subjective. Using a panel of experts with previous knowledge and interest in the matter in question seems to be the most productive way to collect opinions. Also, the structured questionnaire ensures a proper collection of information, in a way that may easily be incorporated in the AHP analysis.

4 Implementation of the Proposed Methodology

The proposed methodology was applied for the evaluation of future electricity plans for Portugal. The process started with the identification of the components of the hierarchal structure, namely (1) the electricity generation options that should be included in the analysis and (2) the relevant criteria to consider. Following this, a group of experts was invited to participate in the process and the combination of Delphi and AHP methodologies was used to characterize and systematize the experts' preferences.

4.1 Selection of Options (Electricity Generation Technologies)

Portugal is strongly dependent on external energy sources, in particular oil. In 2007, 83% of the primary energy came from imports, and oil represented about 54% of the primary energy consumed. The main national resources come from renewable energy sources (RES), especially the hydro sector for electricity production. The electricity and heat production activities accounted for 39% of the total primary energy consumption, and about 70% of electricity consumed in Portugal came from imported fuels and from electricity imports from Spain. Electricity production was then the largest consumer of primary energy and the largest consumer of imported energy resources. As the main domestic resource for electricity production

Table 1 Distribution of installed power and electricity production in mainland Portugal, 2008. Source: (REN 2008)

	Installed power (MW)	Electricity production (GWh)[a]
Thermal power plants	5,820 (39%)	23,797 (57%)
Large hydro power plants	4,578 (31%)	6,436 (15%)
Special regime producers	4,518 (30%)	11,551 (28%)
Total	14,916 (100%)	41,784 (100%)

[a]Injected in the public grid

is hydro power, the system is highly dependent on fuel importations and on rainfall conditions.[3]

In 2008, the total electricity consumption reached 53,587 GWh (DGGE 2008). Predictions for the next years indicate that the electricity consumption will continue rising at an average annual rate of 4.3% between 2009 and 2019. At present, the Portuguese electricity generating system is basically a mixed hydrothermal system. The total installed power reached about 14,916 MW in 2008, distributed between thermal power plants (coal, fuel oil, natural gas and gas oil), hydro power plants and Special Regime Producers (SRP)[4], as presented in Table 1.

Based on REN (2008) forecasts, an increase of about 85% of the total installed electricity generation capacity between 2008 and 2019 may be expected. According to these forecasts, there will be a reduction of both thermal and large hydro power quotas and a large increase of the SRP quota, in relative weight. All the energy sources will grow in absolute terms, with the exception of oil power due to the dismantling of the power plants presently consuming it. Thermal power is expected to increase exclusively due to the growth of the natural gas power groups up to 2015. After that, REN (2008) scenarios assume mainly new investments in coal and a few investments in natural gas. The growth of the SRP will be mainly driven by the increase of the renewable energy sector, in particular wind. According to these scenarios, the wind sector will achieve about 29% of the total installed power by 2019.

The Portuguese strategy for the electricity sector represents a clear effort for the promotion of endogenous resources, reduction of external energy dependency and diversification of supply. The combined growth of coal, natural gas and wind power seem to be fundamental for the accomplishment of these goals. As so, these three technologies have crucial importance for future electricity scenarios and its social sustainability should be evaluated.

4.2 Selection of Criteria

The criteria should be able to represent the main social (or non quantifiable) features of the system. From existing literature addressing the social impact of the electricity

[3] Own calculations based on DGGE online information (www.dgge.pt, April 2009).

[4] Includes the small hydro generation, the production from other non-hydro renewable sources and the cogeneration.

generation technologies and from discussions with experts in the energy field, the criteria considered relevant were defined.

The public perception of wind power is addressed by several authors for a number of countries or regions. Some examples of research studies on this field include Ek (2005), Wolsink (2007), Manwell et al. (2002) or Bergmann et al. (2006) among many others. Most of the studies identified as positive aspects the renewable characteristic of wind power and the avoided emissions. On the other hand, in most of the publications there is a predominant emphasis on the negative visual impact on the landscape. Other identified negative impacts include the impacts on wildlife, the noise pollution, the unreliability of wind energy supply and the possible financial cost, with particular emphasis on the first two aspects.

Studies addressing the coal and gas power plants' impacts deal mainly with the cost and environmental emissions (see Rafaj and Kypreos 2007 or Söderholm and Sundqvist 2003). The environmental impact usually focuses on the damages caused to health and on the impact on climate change. In the ExternE project (European Commission 1995a,b), the external effects from coal and natural gas power plants were mainly associated with their pollutant emissions and their impact on public and occupational health, agriculture and forests. The noise problem was also pointed out, including operational and traffic impact.

Based on the literature and on the non-structured interviews conducted with Portuguese experts from the academic field, energy consultants, members of environmental associations, environmental public organisms' staff and researchers, a set of non-quantitative criteria were chosen to illustrate the proposed process for the social evaluation of the electricity generation technologies:

1. *Noise impact*. This impact is often referred on the literature as an important criterion to take into account in the valuation of wind and thermal power plant projects. Noise levels can be measured quantitatively, but the public's perception of the noise impact is highly subjective. The interviews also revealed that this is a critical issue for the Portuguese population and, that most complaints, when existing, are due to the noise impact of the energy projects.
2. *Impact on birds and wildlife*. The Portuguese experts revealed concerns about this impact, in particular in relation to wind power projects. It is also stressed in most international studies and included in the list of potential disadvantages.
3. *Visual impact*. According to the interviews, this aspect seems to be still of minor importance in Portugal. However, with the expected increase of wind turbines, people may become more aware of its presence and the aesthetical concerns may become more important. For this reason and also because this is the strongest impact reported in international literature, it was decided to include it in the analysis.
4. *The social acceptance*. The experts' interviews indicate that public opposition is not a fundamental criterion to take into account during the energy planning process. However, Wolsink (2007), for example emphasised the need to take into consideration public attitude on wind implementation decisions, not only at a general level but also at the local project level and stressed the importance of including the public in decision-making process. Also Cavallaro and

Ciraolo (2005) support that social acceptability is extremely important since it may heavily influence the amount of time needed to complete the energy project. The public acceptance of a project may not be sufficient to ensure its viability, but represents a clear contribution to its success. This last criterion aims to synthesise the experts' perception of the general social acceptance of the electricity generation alternatives.

As the questionnaire will involve pairwise comparisons, it was decided to limit the number of criteria included, avoiding a long and complex process that might reduce the experts' willingness to participate.

4.3 Hierarchical Structure Formulation

The problem was subdivided into a hierarchy, in which the main objective is placed in the top vertex, the criteria placed in the intermediate level and the options placed on the bottom level. Combining the electricity generation option and the previously identified criteria, the hierarchical structure of this particular problem may be represented as in Fig. 2.

Based on this hierarchy tree, the process should follow to the pairwise comparison for the evaluation of the criteria against the overall social ranking objective and for the estimation of the relative performance of each option on each of the criteria, evaluated in a numerical scale. For this particular research, the aim was to address the negative social impact of each generating technology.[5] For comparison a scale based on Saaty (1980) proposal was used, detailed in Table 2.

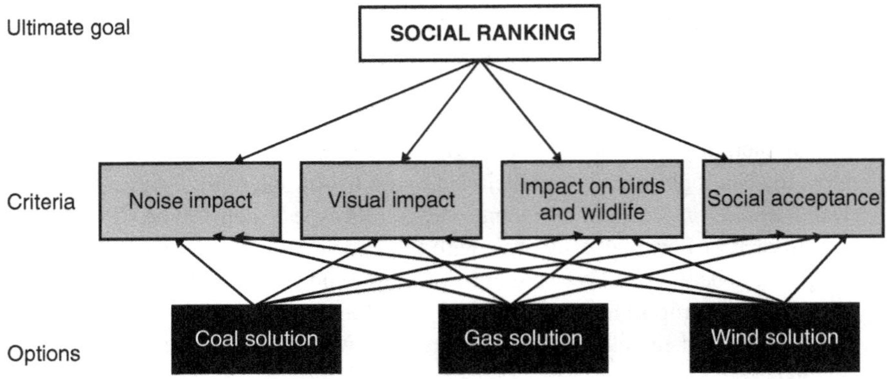

Fig. 2 AHP model for the prioritisation of electricity generation options

[5] A particular technology assigned a higher score is considered "worst" from the social point of view than a technology assigned with lower score.

Table 2 Scale preferences used in the pairwise comparison process

Range	Category	Score
Superior	Absolutely superior	9
	Very strongly superior	7
	Strongly superior	5
	Moderately superior	3
Equal	Equal	1
	Absolutely inferior	1/9
Inferior	Very strongly inferior	1/7
	Strongly inferior	1/5
	Moderately inferior	1/3

Table 3 Pairwise comparison of the alternatives with respect to the noise impact

	Coal	Gas	Wind
Coal	1	1	1/3
Gas	1	1	1/5
Wind	3	5	1

Table 4 Vector of weights of the alternatives with respect to the noise impact

	Noise impact
Gas	0.156
Coal	0.185
Wind	0.659
CR	0.0280

To illustrate the kind of results obtained, Table 3 presents a pairwise comparison matrix drawn from the information provided from one of the experts for the evaluation of the three possible generation technologies against noise impact.

The matrix above (Table 3) shows that for this expert the noise impact of coal solution is equal to the noise impact of gas solution. The noise impact of the wind solution is strongly superior to the noise impact of the gas solution and moderately superior to the coal solution.[6]

The pairwise comparisons of each expert were used as input for the AHP analysis using the scale presented in Table 2. The consistency of each comparison matrix was tested and the relative weights of the elements on each level were computed for each expert. For the given example, the vector of weights may be computed along with the consistency ratio, as presented in Table 4.

For this particular expert as far as noise is concerned, gas is the most desirable solution, followed by coal generation plants with wind generation being the least

[6] Wind technology is then considered "strongly worst to society" than gas technology, from the noise impact point of view. The same way, wind technology is considered "moderately worst to society" than coal technology, from the noise impact point of view.

desirable. Since the consistency ratio (CR) is below 10%, then the judgements are considered consistent (Hon et al. 2005; Kablan 2004; Zhong-Wu et al. 2007).

4.4 Delphi Implementation

The focus of the Delphi process was on the comparison of three electricity generation technologies (wind, coal and natural gas) in what concerns their major impacts from the social point of view.

The experts were selected from Portuguese universities. With the support of the internet, university staff involved in energy projects or lecturing subjects on this area were identified. Additional experts came from contacts made in the course of the research. Twelve experts who would be appropriate to include in the pilot group were identified. Although all the experts came from the same professional field, they have different opinions and hold a variety of positions for and against each one of the options analysed. The results obtained by using the questionnaire were to be used in the AHP analysis. As so, the questionnaire was written using a pairwise comparison structure and Saaty scale of response, as seen in Table 2. The questionnaire included pairwise comparisons of both the options and the social criteria.

The Delphi implementation involved two iterations and lasted for less than 3 months. Nine of the 12 experts concluded the process, although it was necessary to encourage their involvement through electronic and telephone reminders. The obtained responses were statistically examined using Excel Analysis Toolpack, concerning the frequency distribution and the interquartil range as a measure of consensus. Figure 3 summarises the Delphi process followed to assess experts' opinions on the social impact of the electricity generation technologies in Portugal.

The experts were asked to give their individual view on the pairwise comparison of criteria and options. For the social acceptance criterion, the experts were expected to give their response based on their experience and on what they perceive is the view of the population. In general, the results of the Delphi analysis revealed lack of consensus among experts in some questions. This was not an unexpected outcome due to the subjectivity of the analysed issues and the different awareness and individual perception of the experts. Other studies, such as Shackley and McLachlan (2006), suggest also that there is unlikely to be a wide-ranging consensus amongst energy stakeholders on the desirability of specific future forms of energy generation. However, the results seem to be stable with only few response changes between the first and second round.

4.5 Determination of Weights for the Electricity Generation Options

This phase of the research combines the information obtained from the Delphi process with AHP, to convert pairwise comparison of the elements of the hierarchical

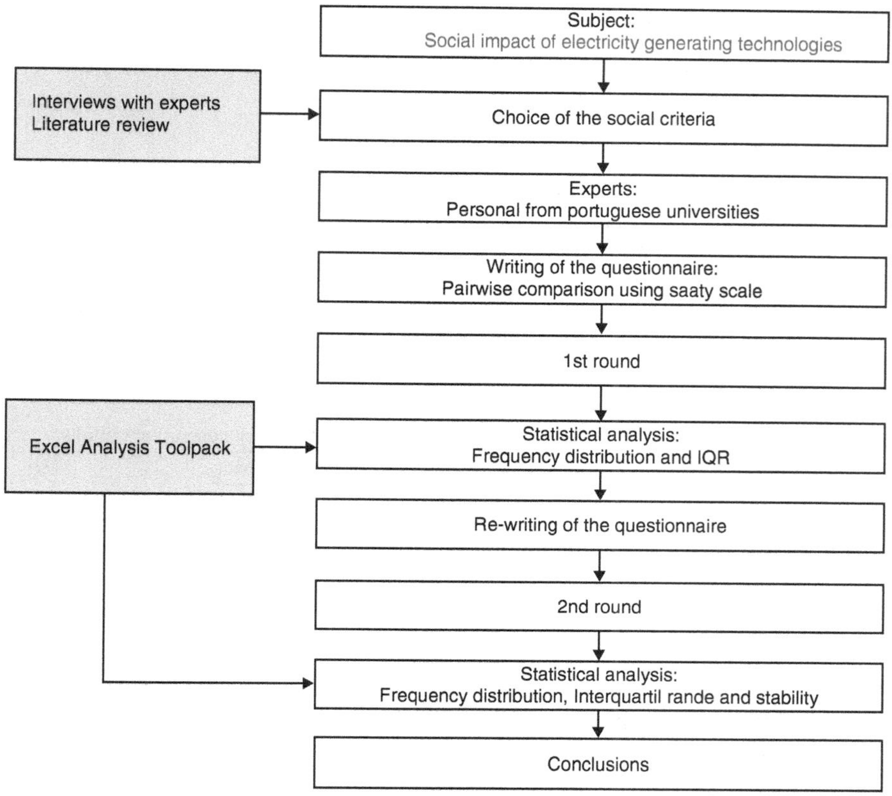

Fig. 3 Delphi process for the social evaluation of electricity generation options

structure in an overall social index, allowing for the ranking of the alternatives. The pairwise comparisons of each expert were used as input for the computation using the scale presented in Table 2. The consistency of each comparison matrix was tested and the relative weights of the elements on each level were computed for each expert.

The group view was represented by the aggregation of each individual's resulting priorities. As the consistency is low for some of the resulting matrixes, only the relative scores of individual matrixes passing the consistency test were included in the aggregation process.

Tables 5 and 6 give the aggregated comparison matrix for the criteria and for the alternatives under each criterion using the geometric mean for the aggregation of the experts' opinions into the final judgement.

The priority vector ranking of criteria with respect to the general goal indicates that social acceptance ranked first followed by impact on birds and wildlife. All these criteria reflect negative aspects for society. For the sake of the consistency of the analysis, social acceptance criterion was computed as the reciprocal corresponding to "social rejection".

Table 5 Aggregated normalised comparison matrix for the criteria

Criteria	Priority ranking
Impact on birds	0.293
Noise impact	0.227
Social acceptance	0.346
Visual impact	0.134

Table 6 Aggregated normalised comparison matrix for the alternatives under each criterion

Solution	Impact on birds	Noise impact	Social acceptance	Visual impact
Coal	0.298	0.230	0.721	0.411
Gas	0.239	0.197	0.203	0.261
Wind	0.463	0.573	0.076	0.328

Table 7 Aggregated score for the overall social impact of the electricity generation options

Solution	Social impact
Coal	0.444
Wind	0.220
Gas	0.336

According to the group assessment, the wind solution ranked first with respect to the impact on birds and wildlife and to the noise impact. It means that, of the three solutions, wind is the one with strongest negative impacts on birds and wildlife and on noise level. For the other two criteria (visual impact and social acceptance) coal ranked first, meaning that, of the three solutions, coal is the one with strongest negative impacts on visual perception and social acceptance.

Combining the relative weights of the elements at each level of the hierarchical structure, the final scoring of the electricity generation options against the overall social objective is obtained. Table 7 synthesises the overall normalised priorities for the three solutions.

According to the results of the group judgment, the coal solution presents the highest social impact followed by the wind solution. The gas solution seems to be the one ranking better from the global social point of view. The high weight of the social acceptance criterion combined with the low social acceptance of coal comparatively to gas or wind solutions led to a score translating high negative social impact for the coal solution. The gas solution ranked in last for all but social acceptance criteria, resulting in a low overall social impact for this option.

The results clearly demonstrate the importance of the social acceptance criterion to the final ranking. In fact, as may be inferred from Table 6, the differences between the alternatives are particularly remarkable for this criterion. A sensitivity analysis of the results was conducted by withdrawing each criterion from the aggregation procedure. The obtained solutions indicate that the final ranking is exactly the same in sequence when the impact on birds and wildlife, or the noise impact, or the visual impact are excluded from the analysis, although the aggregated score for the overall social impact of each electricity generation option changes. The only exception is the social acceptance. If this criterion was excluded from the analysis and all the others remained unchanged, wind power would be the solution with the highest overall social impact, followed by coal and then gas solution.

According to this simple sensitivity exercise, it seems that future work should focus on the social acceptance criterion, both because of the high weight assigned to it and also because of the high differences detected among the social acceptance of each alternative. Further work can decompose and express this criterion in a number of sub-criteria, allowing for a deeper analysis of the results and contributing also to guide the experts in the Delphi process.

4.6 Social Impact of Future Electricity Generation Scenarios

The use of AHP and Delphi techniques for the social evaluation of electricity generation technologies was proposed and demonstrated in the last sections. The results obtained can be described by a weight vector characterizing the overall social impact of each one of the technologies considered: coal, gas and wind. However, as seen in Sect. 4.1, future scenarios for the Portuguese electricity system are expected to be based on a mix of different technologies, with coal, gas and wind power playing a key role. To obtain a final ranking of these possible scenarios, the overall social scores of the three alternatives need to be combined by means of a mathematical algorithm. The aim is to get a final index for each possible plan, combining more than one of the available electricity generation technologies. The weights were aggregated using an additive function.

This additive function assumes that the weights assigned to each electricity generation option are constants and satisfy the additive independence, that is these weights do not depend on the relative levels of each option. This additive value model offers a simple way of evaluating multiattribute alternatives. This simplicity makes it widely used in energy planning and policy, as described by Hobbs and Meier (2003).

Equation (1) presents the computation of the average social index (ASI) for each possible electricity plan, depending on the installed power of each electricity generation option and on the weights derived from the AHP:

$$\mathrm{ASI} = \frac{\sum\limits_{t} \mathrm{P_{coal}} \mathrm{W_{coal}} + \sum\limits_{t} \mathrm{P_{gas}} \mathrm{W_{gas}} + \sum\limits_{t} \mathrm{P_{wind}} \mathrm{W_{wind}}}{\sum\limits_{t} (\mathrm{P_{coal}} + \mathrm{P_{gas}} + \mathrm{P_{wind}})} \quad (1)$$

where $\mathrm{W_{coal}}$, $\mathrm{W_{gas}}$ and $\mathrm{W_{wind}}$, represent the overall normalised weights for the coal, gas and wind solutions, described in Table 7, and $\mathrm{P_{coal}}$, $\mathrm{P_{gas}}$ and $\mathrm{P_{wind}}$, represent the installed power of coal, gas and wind power plants in each scenario.

To illustrate the process, a set of possible plans for 2017 were drawn from Ferreira (2008). All these plans ensure that the average Kyoto protocol limits imposed to Portugal would not be overcome and are consistent with the objective

Table 8 Possible electricity plans obtained from Ferreira (2008)

		Plan 1	Plan 2	Plan 3	Plan 4
Total installed power (MW)	Coal (new)	–	2,400	–	600
	Coal (existing)[a]	1,820	1,820	1,820	1,820
	Natural gas (new)	5,110	1,860	5,110	3,720
	Natural gas (existing)[a]	2,916[b]	2,916[b]	2,916[b]	2,916[b]
	Wind (new)	3,225	6,514	3,225	6,500
	Wind (existing)[a]	1,515	1,515	1,515	1,515
	Large hydro	5,805	5,805	5,805	5,805
	NWSRP	3,245	3,245	3,245	3,245
	Total	23,636	26,075	23,636	26,121
Share of RES (%)		39	45	39	46
External dependency (%)		65	58	65	57
Cost (€/MWh)		33.627	34.961	34.365	36.950
CO_2 (ton/MWh)		0.379	0.379	0.332	0.332
ASI		0.292	0.341	0.292	0.316

[a]Existing at the end of 2006

of reaching 39% share of electricity produced from RES, as required by Directive 2001/77/EC. Table 8 describes these plans,[7] detailing the expected total installed power, the share of RES, average cost, average CO_2 emissions and ASI.

This analysis allows the decision maker to recognise the differences between the possible electricity generation alternatives and foresee their estimated impacts. The final selection of an electricity strategy for the future depends on the priority that the decision maker chooses to assign to each one of the objectives considered. For example, with respect to the plans described in Table 8, the results reveal that it will be possible to achieve average CO_2 emission of 0.379 ton/MWh at a minimal cost of 33.6 €/MWh, investing mainly on new natural gas power plants (Plan 1). As natural gas is a socially well accepted solution, the social impact of this strategy should be low,[8] but the external dependency of the electricity generation sector will remain high. If the decision maker is willing to increase cost by about 4% (Plan 2), it will be possible to keep CO_2 emissions at the same level and the external dependency of the electricity production sector may be reduced by 7%. Also, a more balanced mix between coal and natural gas may be achieved, resulting in considerable advantages from the security of supply point of view. However, as this strategy requires less natural gas power plants and additional investments in new coal and wind power plants, this solution presents a higher ASI reflecting a worst social impact outcome.

[7] For additional information on the characterization of the electricity plans and design see Ferreira (2008).

[8] In what concerns the four social criteria analysed: impact on birds and wild life, visual impact, noise level and social acceptance.

5 Conclusions

From the work presented, a structured decision making process for the electricity power planning problem can be outlined, combining AHP and Delphi analysis for the social evaluation of future electricity scenarios. The proposed methodology highlights the importance of the social dimension of sustainable planning and recognises that energy decisions should be guided by a context that reflects economic, environmental and social concerns. This distinct and comprehensive referential framework is a useful instrument to distinguish and evaluate different energy strategies and plans, thereby contributing to facilitating explicit discussion and informed decision making.

The combination of social, environmental and economic evaluations will benefit the energy plan formulation, ensuring the robustness of the process and leading to a defensible choice aimed at reducing conflict. However, it is clear that the integration of the social criteria issues on the evaluation of future electricity plans, although being fundamental, is not an easy and consensual task. The suggested tool can be developed as a guideline for providing a structured process to accomplish this task. It can be used for the identification of the relevant social impacts of electricity generation options, for the evaluation of the overall social impact of each electricity generation option and for the assessment of the relative importance of these impacts for the society, giving a measurable interpretation of the expected social impact of future electricity scenarios.

The application of the proposed methodology or the evaluation of future electricity plans in Portugal revealed that the broad diversity of interests and values of the decision makers make consensus difficult to achieve in the energy planning process. It should be highlighted that the difficulty on reaching consensus and consistent results is not a completely unexpected result. Regardless of these difficulties, the proposed tool offers a clear path to explicitly recognise and integrate the social dimension into the electricity planning process, resorting to a structured participatory approach.

According to AHP results, the rank of gas solution was the first in the order of priority, most probably because it represents a compromise solution. It is seen as an electricity generation solution with low environmental problems, which increases its social acceptance over coal. It has also reduced impacts on wildlife, especially when compared to wind power, and lower visual and noise impact than both other alternatives. Coal ranks in last, mainly due to the reduced social acceptance of this alternative. The impact on birds and wildlife and the noise impact are the most severe effects reported for wind power comparatively to both coal and gas solutions. However, the social perception of each technology can be highly volatile and influenced by public groups or opinion makers. The new clean coal technologies and the prices development may easily change this general opinion. Likewise, the spreading of wind power plants may demonstrate that the social impacts of this technology are more or less important than the level assumed by these experts.

Despite the aforementioned various advantages of the proposed approach for the integration of social concerns into electricity power planning, there are a few scalability issues that may increase the computational complexities and the number of judgments required. This issue may be particularly relevant when the method is to be used with a large number of criteria or options or when the decision making may involve multiple stages and several decision makers. Therefore, good judgement must prevail to select a proper but limited number of criteria that would make both the Delphi process and the AHP analysis executable. The same holds for the number of experts and decision makers involved in the analysis, although the process could be conducted locally or regionally using the same analytical framework in different stages and, therefore, handling more information in a still efficient way.

The application of the model was presented through a pilot experiment. It is the authors' conviction that this research could still be extended and should be able to accommodate additional data without compromising the resources needed and the efficiency of the process. Two main points must be considered on further research: (1) The increase of the number of experts for enriching the information obtained. This would avoid the influence of each individual on the consensus or stability decision of the group. Further insights' of the subject would be obtained and additional results could be derived from different statistical tools. (2) The inclusion of other social criteria and even the inclusion of quantitative aspects like cost, external dependency and CO_2 emissions. Although these last aspects may be measured by quantitative scales, the proposed methodology can give an important contribution on the elicitation of the relative importance of these impacts for the society.

References

Alberts D (2007) Stakeholders on subject matter experts, who should be consulted. Energy Policy 35(4):2236–2346

Bergmann A, Hanley N, Wright R (2006) Valuing the attributes of renewable energy investments. Energy Policy 34(9):1004–1014

Bruckner T, Morrison R, Wittmann T (2005) Public policy modelling of distributed energy technologies: strategies, attributes, and challenges. Ecol Econ 54(2–3):328–345

Cavallaro F, Ciraolo L (2005) A multicriteria approach to evaluate wind energy plants on an Italian island. Energy Policy 33(2):235–244

Del Río P, Burguillo M (2008) Assessing the impact of renewable energy deployment on local sustainability: Towards a theoretical framework. Renew Sust Energ Rev 12:1325–1344

DGGE (2008) Renováveis. Estatísticas rápidas. December 2008 (in Portuguese). www.dgge.pt

Diakoulaki D, Antunes C, Martins A (2005) MCDA and energy planning. In: Figueira J, Greco S, Erghott M (eds.) Multiple criteria decision analysis – State of the Art survey, int. series in operations research and management science, vol. 78. Springer, pp. 859–897

Dincer I, Rosen M (2005) Thermodynamic aspects of renewables and sustainable development. Renew Sust Energ Rev 9:169–189

Ek K (2005) Public and private attitudes towards "green" electricity: the case of Swedish wind power. Energ Policy 33(13):1677–1689

European Commission (1995a) ExternE. Externalities of Energy, vol. 3. Coal and Lignite, EUR 16522, 1995

European Commission (1995b) ExternE. Externalities of Energy, vol. 4. Oil and Gas, EUR 16523, 1995

European Commission (2003) External Costs. Research results on socio-environmental damages due to electricity and transport, Directorate-General for Research

Ferreira P (2008) Electricity power planning in Portugal: the role of wind energy. PhD Dissertation, University of Minho, Portugal

Ferreira P, Araújo M, O'Kelly M (2004) Including non-financial aspects in project evaluation – a survey. 15th Mini-EURO Conference Managing Uncertainty in Decision Support Models, Coimbra, Portugal, 22–24 Sept 2004

G8 Renewable Energy Task Force (2001) Final Report. July, 2001. (http://www.worldenergy.org/wec-geis/focus/renew/g8.asp)

Greening L, Bernow S (2004) Design of coordinated energy and environmental policies: use of multi-criteria decision making. Energy Policy 32(6):721–735

Hemphill L, McGreal S, Berry J (2002) An aggregating weighting system for evaluating sustainable urban regeneration. J Prop Res 19(4):553–575

Hepbasli A (2008) A key review on exergetic analysis and assessment of renewable energy resources for a sustainable future. Renew Sust Energ Rev 12(3):593–661

Hobbs B (1995) Optimisation methods for electric utility resource planning. Eur J Oper Res 83(1):1–20

Hobbs B, Meier P (2003) Energy decisions and the environment: A guide to the use of multicriteria methods, 2nd edn. Kluwer

Hon C, Hou J, Tang L (2005) The application of Delphi method for setting up performance evaluation structure and criteria weights on warehousing of third party logistics. Proceedings of the 35th International Conference on Computers and Industrial Engineering, Istanbul, Turkey, June 2005

Huang J, Poh K. Ang B (1995) Decision analysis in energy and environmental modelling. Energy 20(9):843–855

Jebaraj S, Iniyan S (2006) A review of energy models. Renew Sust Energ Rev 10(4):281–311

Jefferson M (2006) Sustainable energy development: performance and prospects. Renew Energ 31(5):571–582

Kablan M (2004) Decision support for energy conservation promotion: an analytic hierarchy process approach. Energ Policy 32(10):1151–1158

Lai V, Wong B, Cheung W (2002) Group decision making in multiple criteria environment: A case using AHP in software selection. Eur J Oper Res 137(1):134–144

Lehtonen M (2004) The environmental–social interface of sustainable development: capabilities, social capital, institutions. Ecol Econ 49(2):199–214

Liang Z, Yang K, Yuan S, Zhang H, Zhang Z (2006) Decision support for choice optimal power generation projects: Fuzzy comprehensive evaluation model based on the electricity market. Energ Policy 34(17):3359–3364

Limmeechokchaia B, Chawana S (2007) Sustainable energy development strategies in the rural Thailand: The case of the improved cooking stove and the small biogas digester. Renew Sust Energ Rev 11(5):818–837

Loring J (2007) Wind energy planning in England, Wales and Denmark: Factors influencing project success. Energ Policy 35(4):2648–2660

Lund H (2007) Renewable energy strategies for sustainable development. Energy 32(6):912–919

Manwell J, McGowan J, Rogers A (2002) Wind energy explained: Theory, design and application. Willey, England

Nigim K, Munier N, Green J (2004) Pre-feasibility MCDM tools to aid communities in prioritizing local viable renewable energy sources. Renew Energ 29(11):1775–1791

Okoli C, Pawlowsk S (2004) The Delphi method as a research tool: an example, design considerations and applications. Inform Manag 42(1):15–29

Pereira A, Soares J, Oliveira R, Queiroz R (2007) Energy in Brazil: Toward sustainable development? Energ Policy 36(1):73–83

Pohekar S, Ramachandran M (2004) Application of multi-criteria decision making to sustainable energy planning – A review. Renew Sust Energ Rev 8(4):365–381

Rafaj P, Kypreos S (2007) Internalisation of external cost in the power generation sector: Analysis with Global Multi-regional MARKAL model. Energ Policy 35(2):828–843

Rayens M, Hahn E (2000) Building consensus using the policy Delphi method. Policy Polit Nurs Pract 1(4):308–315

REN (2008) Plano de desenvolvimento e investimento da rede de transporte 2009–2014 (2019) (In Portuguese) (www.ren.pt)

REN (2008) Dados técnicos electricidade 2008. Valores provisórios. (in Portuguese)

Saaty T (1980) The analytic hierarchy process. McGraw Hill, New York

Shackley S, McLachlan C (2006) Trade-offs in assessing different energy futures: a regional multi-criteria assessment of the role of carbon dioxide capture and storage. Environ Sci Policy 9: 376–391

Söderholm P, Sundqvist T (2003) Pricing environmental externalities in the power sector: ethical limits and implications for social choice. Ecol Econ 46(3):333–350

Torres Sibille A, Cloquell-Ballester V, Cloquell-Ballester V, Darton R (2009a) Aesthetic impact assessment of solar power plants: An objective and a subjective approach. Renew Sust Energ Rev 13:986–999

Torres Sibille A, Cloquell-Ballester V, Cloquell-Ballester V, Darton R (2009b) Development and validation of a multicriteria indicator for the assessment of objective aesthetic impact of wind farms. Renew Sust Energ Rev 13(1):40–66

Vera I, Langlois L (2007) Energy indicators for sustainable development. Energy 32(6):875–882

Wolsink M (2007) Wind power implementation: The nature of public attitudes: Equity and fairness instead of 'backyard motive. Renew Sust Energ Re 11(6):1188–1207

World Bank (2003) World development report 2003. Sustainable development in a dynamic world – transforming institutions, growth, and quality of life

World Commission on Dams (2001) Dams and development: A new framework for decision-making. Overview of the report by the World Commission on Dams. December 2001. (www.poptel.org.uk/iied/docs/drylands/dry_ip108eng.pdf)

World Commission on Environment and Development (1987) Our Common Future. Oxford University Press. (full text available at http://en.wikisource.org/wiki/Brundtland_Report)

Wright J, Giovinazzo R (2000) Delphi-Uma ferramenta de apoio ao planejamento prospectivo. Cadernos de Pesquisas em Administração 1(12):54–65 (in Portuguese)

Zhong-Wu L, Guang-Ming Z, Hua Z, Bin Y, Sheng J (2007) The integrated eco-environment assessment of the red soil hilly region based on GIS–A case study in Changsha City. China Ecol Model 202(3–4):540–546

Transmission Network Expansion Planning Under Deliberate Outages

Natalia Alguacil, José M. Arroyo, and Miguel Carrión

Abstract This chapter sets forth a new approach for transmission network expansion planning that accounts for increasingly plausible deliberate outages. Malicious attacks expose the network planner, a centralized entity responsible for expansion decisions of the entire transmission network, to a new challenge: how to expand and reinforce the transmission network so that the vulnerability against intentional attacks is mitigated while meeting budgetary limits.

Two vulnerability-constrained transmission expansion models are presented in this chapter. The first model allows the network planner to analyze the tradeoff between economic- and vulnerability-related issues and its impact on the expansion plans. The uncertainty associated with intentional outages is modeled through scenarios. Vulnerability is measured in terms of the system load shed. In the second model, the risk associated with the nonrandom uncertainty of deliberate outages is incorporated through the minimax weighted regret criterion. The proposed models are formulated as mixed-integer linear programs for which efficient solvers are available. Illustrative examples show the performance of both models.

Keywords Deliberate outages · Minimax weighted regret · Mixed-integer linear programming · Risk · Scenario · Transmission network expansion planning · Vulnerability

1 Introduction

Nowadays, electric energy is an indispensable commodity in national economies worldwide. Consequently, security is a crucial aspect of power systems operation and planning. Recent blackouts worldwide reveal that contingencies in the transmission network may result in devastating effects for the society. This undesirable

---

N. Alguacil (✉)
Universidad de Castilla – La Mancha, ETSI Industriales, Campus Universitario s/n, 13071, Ciudad Real, Spain
e-mail: Natalia.Alguacil@uclm.es

S. Rebennack et al. (eds.), *Handbook of Power Systems I*, Energy Systems,
DOI 10.1007/978-3-642-02493-1_16, © Springer-Verlag Berlin Heidelberg 2010

level of vulnerability of the transmission network can be used by terrorist groups as an instrument to reach their goals.

This chapter describes the work carried out by the authors at the Universidad de Castilla – La Mancha, which is focused on the proposal of the reinforcement and expansion of the transmission network as a way of mitigating the impact of increasingly plausible deliberate outages. The traditional goal of the transmission network planner has been the minimization of investment and operation costs over the planning horizon. However, in the new context where destructive agents come into play, we argue that transmission expansion planning should also be driven by security and vulnerability issues.

Two transmission network expansion planning models are presented in this chapter. The uncertainty associated with intentional outages is characterized through scenarios in both models. The first model is useful to show that additional investments in the transmission network might help mitigate the severe consequences of potential intentional outages. The level of vulnerability is defined as the average system load shed weighted over the considered scenarios. Furthermore, this vulnerability-constrained model allows the network planner to analyze the tradeoff between economic- and vulnerability-related issues and its impact on the expansion plans.

This introductory model is risk neutral, that is, the risk associated with the nonrandom uncertainty of deliberate outages is not accounted for. Selecting the investment using weighted average system load shed works well for most scenarios but may perform poorly for some scenarios corresponding to low likelihood but potentially catastrophic intentional outages. To overcome this shortcoming, a second approach is presented in which risk aversion is explicitly incorporated through the minimax weighted regret criterion.

The remainder of the chapter is organized as follows. Section 2 describes the traditional transmission network expansion planning problem. Section 3 motivates the need for new models accounting for deliberate outages. Section 4 presents the salient features of the decision framework in which the proposed vulnerability-constrained transmission expansion planning is embedded. This section also provides the risk measure adopted in this work. Section 5 includes the mathematical formulation of the proposed models. Section 6 gives numerical results to illustrate the performance of the proposed approaches. In Sect. 7 relevant conclusions are drawn. In Appendix 1 the vulnerability analysis used to generate scenarios is formulated. Appendix 2 provides the equivalent linear expressions of nonlinear constraints appearing in the original formulation of the proposed models. Finally, the deterministic expansion problem associated with each scenario is formulated in Appendix 3.

2 Traditional Transmission Network Expansion Planning

Network planning plays a key role in power system planning. The traditional transmission expansion planning problem consists in determining the optimal timing, location, and sizing of transmission facilities to be installed in an existing network so that power is supplied to consumers in a reliable and economic fashion over the planning horizon. Other decisions, such as the modification of the network topology or the interconnection of isolated systems, may also be considered part of this problem. The optimal network configuration is determined accounting for the forecast load growth and the generation planning scheme for the planning horizon (Wang and McDonald 1994).

Network planning is divided into static and dynamic network planning. Static network planning determines the network configuration for a single future period, and thus it does not consider when new transmission assets are to be built. Dynamic network planning deals with a longer planning horizon, which is divided into several periods. Dynamic network planning addresses where and when to build new transmission assets.

The transmission expansion problem arises in both centralized and competitive frameworks. Without loss of generality, this chapter addresses the static network planning problem in which a central entity, referred to as the network planner, is responsible for expansion decisions of the entire transmission network. The network planner may be the transmission company, the independent system operator, or the regional transmission organization. A planning horizon of 1 year is considered. As is commonly assumed in static transmission expansion planning (Latorre et al. 2003), during this target year, generation sites are known and a single load scenario is modeled, typically corresponding to the highest load demand forecast for the considered planning horizon.

The traditional static transmission expansion problem can be formulated in the following compact way:

$$\text{Minimize}_{s_\ell, x^{\text{O}}, x^{\text{C}}} \quad f(s_\ell, x^{\text{O}}, x^{\text{C}}) \tag{1}$$

Subject to:

$$g(s_\ell, x^{\text{O}}, x^{\text{C}}) \leq 0 \tag{2}$$

$$s_\ell \in \{0, 1\}; \qquad \forall \ell \in L^{\text{C}}, \tag{3}$$

where s_ℓ is a binary variable that is equal to 1 if candidate line ℓ is built, being 0 otherwise; x^{O} and x^{C}, respectively, denote continuous variables associated with the original and the expanded networks such as power flows, nodal generations, etc.; and L^{C} is the set of indices of candidate lines. The traditional goal of the network planner is to minimize an objective function f typically comprising economic-related terms (1). This optimization problem is subject to a set of constrained functions g expressing the operation of the system (2). The discrete nature of expansion variables s_ℓ is imposed by (3).

This large-scale, mixed-integer programming problem has received extensive attention for the past 40 years. Technical references can be classified according to the methodology used to solve the problem. Linear programming techniques were applied in (Garver 1970; Kaltenbach et al. 1970; Villasana et al. 1985). References (Alguacil et al. 2003; Bahiense et al. 2001; Oliveira et al. 2007; Seifu et al. 1989; Sharifnia and Aashtiani 1985) dealt with a mixed-integer linear formulation of the problem. Dynamic programming (Dusonchet and El-Abiad 1973) and decomposition techniques (Binato et al. 2001; Romero and Monticelli 1994a,b) have also been applied. Heuristic approaches (Monticelli et al. 1982; Oliveira et al. 1995; Romero et al. 1996), genetic algorithms (Silva et al. 2005; da Silva et al. 2000), and game theory (Contreras and Wu 1999, 2000; Zolezzi and Rudnick 2002) have been proposed as alternative methods to optimization-based tools. In addition to the above deterministic approaches, uncertainty aspects of the problem have also been analyzed in (Choi et al. 2005), where uncertainties associated with the reliability of the network components were modeled.

In the competitive electricity market environment, the solution of the transmission improvement/expansion problem requires some important modifications, such as the introduction of a new objective, for example, social welfare maximization (Sauma and Oren 2007; de la Torre et al. 2008; Wu et al. 2006), and the consideration of market-related uncertainties (Buygi et al. 2004, 2006; Fang and Hill 2003; Thomas et al. 2005; de la Torre et al. 2008, 1999; Wu et al. 2006).

References (Latorre et al. 2003) and (Lee et al. 2006) present comprehensive reviews of the models on transmission expansion planning available in the technical literature.

3 Vulnerability-Constrained Transmission Expansion Planning

Over the last years, the consumption of electricity has increased above forecasts in many countries. However, the transmission network has not been expanded accordingly due to economic, environmental, and political reasons. As a consequence, the transmission network is being operated close to its static and dynamic limits, yielding a vulnerable operation. This higher level of vulnerability, joined to its crucial significance as a strategic infrastructure, makes the transmission network appealing for intentional attacks (MIPT 2009).

This issue has raised the concern of governments, and several initiatives have been launched worldwide to assess and mitigate the vulnerability of such a critical infrastructure (Commission of the European Communities 2005; CIPC 2009; Gheorghe et al. 2006; JIIRP 2009). Vulnerability of the transmission network against intentional attacks has also drawn the interest of researchers. Recently published works (Arroyo and Galiana 2005; Bier et al. 2007; Holmgren et al. 2007; Motto et al. 2005; Salmeron et al. 2004) propose vulnerability analysis techniques to identify critical network components that are potential targets for destructive agents.

Within this framework of increasingly plausible intentional attacks on the transmission network, network planners are exposed to a new challenge: how to expand and reinforce the transmission network so that the vulnerability against intentional attacks is mitigated while meeting budgetary limits.

Security aspects were pioneered in network planning in (Monticelli et al. 1982; Seifu et al. 1989; Sharifnia and Aashtiani 1985). In (Monticelli et al. 1982), a two-phase heuristic algorithm was proposed to solve the security-constrained transmission expansion planning problem. In (Sharifnia and Aashtiani 1985), the effect of single line contingencies on the transmission capability of the system was indirectly accounted for through additional constraints that were iteratively incorporated in the original model. In (Seifu et al. 1989), a contingency-constrained transmission expansion planning model based on mixed-integer linear programming was first proposed. Built on the formulation presented in (Seifu et al. 1989), an N-1 security-constrained transmission expansion planning model was solved with a genetic algorithm in (Silva et al. 2005). In these approaches (Monticelli et al. 1982; Seifu et al. 1989; Sharifnia and Aashtiani 1985; Silva et al. 2005) the uncertainty associated with outages was not modeled. In (Oliveira et al. 2007), the model of (Seifu et al. 1989) was extended to incorporate the uncertainty of contingencies and demand through scenarios.

Based on the contingency-constrained transmission expansion model described in (Seifu et al. 1989), this chapter presents a new approach referred to as transmission network expansion planning under deliberate outages (Alguacil et al. 2008; Carrión et al. 2007). The reduction of the vulnerability of the network against deliberate outages is one of the objectives of the network planner. To account for the uncertainty associated with deliberate outages, a set of scenarios is considered, where each scenario represents a credible attack plan resulting in a particular level of system load shed. Weights are assigned to scenarios to represent their perceived relative likelihood. These weights are based on the level of system load shed and on the number of network components down in the corresponding scenario.

This approach is subsequently extended to incorporate the risk associated with deliberate outages. Risk modeling is particularly relevant in decision-making problems under uncertainty wherein scenarios with very low likelihood and high impact such as intentional attacks have to be accounted for. Decision-analysis tools considering risk have been successfully applied in power systems planning (Buygi et al. 2004, 2006; Crousillat et al. 1993; Fang and Hill 2003; Gorenstin et al. 1993; Merrill and Wood 1991; Miranda and Proença 1998a,b; de la Torre et al. 1999). Here, risk aversion is characterized based on the notion of regret, which is a well-established risk measure (Bell 1982; Loomes and Sugden 1982).

4 Decision Framework, Uncertainty Characterization, and Risk Modeling

This section describes the main modeling features of the proposed transmission network expansion approaches.

4.1 Decision-Making Process

The vulnerability-constrained transmission expansion planning involves two types of decisions: (1) expansion decisions made by the network planner and (2) operation decisions made by the system operator. Expansion plans consist in building new lines from a candidate set. We assume that the investment happens at the beginning of the planning horizon and new circuits are immediately ready for use. These expansion decisions are made under the uncertainty associated with intentional attacks. In addition, if a deliberate outage occurs, the system operator reacts so that the damage is minimized. Thus, operation decisions such as power generation and load shedding, and the resulting power flows depend on the attack plan.

Intentional attacks to the transmission network can be classified as nonrandom uncertain events (Buygi et al. 2004). They are uncertain because they are unknown a priori. Furthermore, they are nonrandom because they cannot be modeled using a known probability distribution based on past observations. The uncertainty of attack plans is characterized through a set of scenarios Ω, where each scenario represents a credible attack plan resulting in a particular level of damage. The level of damage is measured in terms of the system load shed. Although any network component is a potential target for destructive agents, for the sake of clarity and conciseness, this chapter considers only intentional outages of transmission lines. The set of scenarios Ω is made up of vectors $v(\omega)$ of 0s and 1s as follows:

$$v(\omega) = \{v_1(\omega), \dots, v_{n_\mathrm{L}}(\omega)\}; \qquad \omega = 0, \dots, n_\Omega, \tag{4}$$

where n_L is the number of lines in the original transmission network, n_Ω is the number of attack plans considered as scenarios, and $v_\ell(\omega)$ is a constant equal to 0 if line ℓ is attacked in scenario ω, being 1 otherwise. A scenario $\omega = 0$ with no attacks is included in Ω when the estimated demand requires transmission expansion.

Each attack plan selected as a scenario is associated with a weight or degree of importance to represent its perceived relative likelihood.

Under this uncertainty characterization, the vulnerability of the transmission network against intentional attacks is defined as the weighted average system load shed caused by the considered attack plans. Figure 1 depicts the decision framework of the vulnerability-constrained transmission expansion planning.

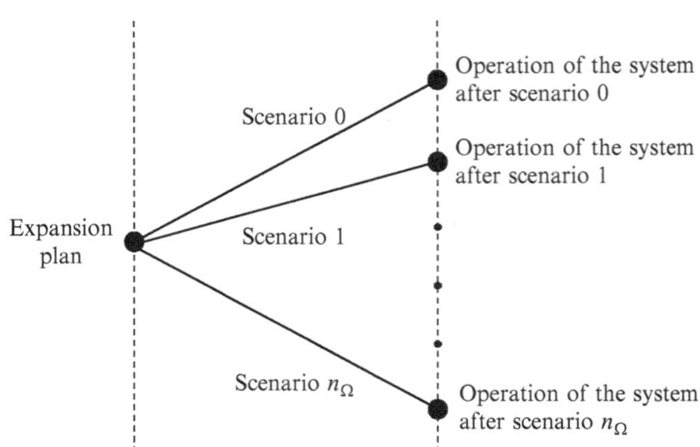

Fig. 1 Decision framework

As is common in security-constrained transmission expansion planning (Oliveira et al. 2007; Seifu et al. 1989; Silva et al. 2005), the above decision framework assumes outages only in existing lines, recognizing that the use of such a simplified model leads to results that may be optimistic and that a complete study should also consider candidate lines as potential targets. This generalization would, however, render the problem essentially intractable through optimization. Notwithstanding these modeling limitations, the solution of the proposed model considering only attacks to existing lines provides the network planner with valuable information to mitigate the vulnerability of the transmission network against deliberate outages.

4.2 Scenario Generation Procedure

The scenario generation procedure is based on the solution of the so-called terrorist threat problem (Arroyo and Galiana 2005; Motto et al. 2005; Salmeron et al. 2004). The terrorist threat problem is a static vulnerability analysis of the transmission network considering intentional outages. The objective of this problem is to determine the attack plan causing the largest disruption in the network, given limited destructive resources.

Scenarios are selected according to the level of damage caused by the corresponding attack plan. The scenario generation procedure iteratively finds the most damaging attack plans by solving an instance of the terrorist threat problem formulated in (Motto et al. 2005), which is hereinafter denoted by MTTP (Modified Terrorist Threat Problem). The formulation of MTTP can be found in Appendix 1.

The scenario generation procedure works as follows. Let n_{A} be the counter of destroyed lines that is initialized to 1. The scenario generation procedure starts by solving MTTP to find the most disruptive attack plan consisting in the destruction of n_{A} lines. If the optimal solution of MTTP results in a level of system load shed less than or equal to the maximum system load shed achieved by destroying $n_{\mathrm{A}}-1$ lines, then the attack plan is discarded as a scenario, and the counter of destroyed lines is increased. On the other hand, if the level of system load shed determined by MTTP exceeds that achieved with the destruction of $n_{\mathrm{A}}-1$ lines, the attack plan is selected as a scenario and MTTP is again solved for the current value of n_{A}. As explained in Appendix 1, the formulation of MTTP avoids finding attack plans already selected for the same value of n_{A}.

It should be noted that, for each number of lines down n_{A}, the proposed procedure selects only those attack plans yielding levels of system load shed exceeding the maximum system load shed achieved with fewer lines down. This scenario generation procedure is motivated by the fact that destructive agents have limited resources and thus, expensive but lowly disruptive attack plans are unlikely.

The above procedure is repeated until a pre-specified number of scenarios is reached. Figure 2 shows the flowchart of the scenario generation procedure.

Once the set of scenarios Ω is obtained, a weight or degree of importance $\pi(\omega)$ is assigned to each attack plan ω based on the following practical considerations:

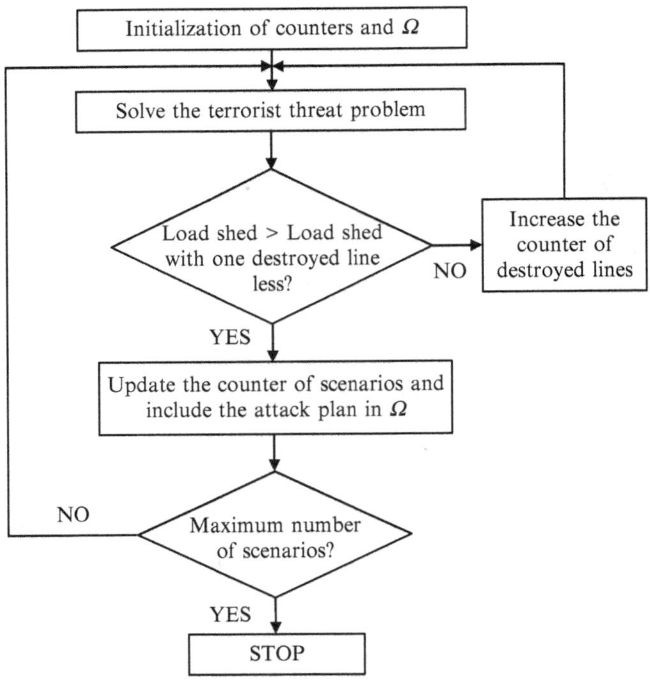

Fig. 2 Scenario generation procedure

1. The weight assigned to an attack plan ω is directly proportional to the damage caused $\Delta D(\omega)$, which is obtained in the scenario generation procedure.
2. The weight assigned to an attack plan ω is inversely proportional to the number of destroyed lines $I(\omega)$, which constitutes a measure of the disruptive resources required by the destructive agent.

Under these two assumptions, the weight of each attack plan is calculated as follows:

$$\pi(\omega) = \frac{\dfrac{\Delta D(\omega)}{I(\omega)}}{\displaystyle\sum_{\omega'=1}^{n_\Omega} \frac{\Delta D(\omega')}{I(\omega')}}; \qquad \omega = 1, \ldots, n_\Omega. \tag{5}$$

With (5) we model the tradeoff faced by the destructive agents between the level of damage achieved and the effort required to reach that level of destruction. The destructive effort might be a function of the disruptive resources, that is, number of agents, cost of explosives, etc., and here has been modeled by the number of destroyed lines. Note that the sum of the weights over all attack plans is equal to 1.

It should be emphasized that the model results may depend on the set of attack plans selected as scenarios and thus on the scenario generation procedure. This scenario dependency is shared by other security-constrained transmission expansion planning approaches (Oliveira et al. 2007; Seifu et al. 1989; Silva et al. 2005). Alternative ways of generating scenarios can be straightforwardly used, such as those including subjective aspects through Bayesian networks (Tranchita et al. 2006) or regression-based models (Simonoff et al. 2007). Since the scenario generation procedure is external to the proposed tool, the analysis of the sensitivity of the model with respect to the scenario set is beyond the scope of this chapter.

4.3 Risk Modeling

The risk associated with the uncertainty of low frequency phenomena such as deliberate outages is modeled by the regret felt by the network planner after verifying that the selected decision is not optimal, given the future that actually occurs. The notion of regret is a well-established approach to measure risk in decision making under uncertainty (Bell 1982; Loomes and Sugden 1982). Within the framework of vulnerability-constrained transmission network expansion planning, the regret or loss for an expansion plan in a given scenario is defined as the difference between the system load shed for this expansion plan under the considered scenario and the minimum system load shed in this scenario that would have been attained by the network planner if there were prior knowledge that this scenario would take place. Therefore, the network planner is influenced by the possibility that the level of damage will not be sufficiently close to the lowest possible (ideal) value, that is, what would have occurred if the actual scenario had been known at the time of decision making.

Mathematically, the regret of expansion plan j and scenario ω, $R_j(\omega)$, is formulated as

$$R_j(\omega) = \Delta D^j(\omega) - \Delta D^{\min}(\omega); \qquad \forall j \in J, \omega = 0, \ldots, n_\Omega, \tag{6}$$

where $\Delta D^j(\omega)$ is the system load shed associated with expansion plan j and scenario ω, J is the index set of expansion plans, and $\Delta D^{\min}(\omega)$ is the minimum system load shed attainable in scenario ω, that is,

$$\Delta D^{\min}(\omega) = \min_{j \in J} \left\{\Delta D^j(\omega)\right\}; \qquad \omega = 0, \ldots, n_\Omega. \tag{7}$$

The weighted regret models the impact of the perceived likelihood of attack plans. Hence, the weighted regret of expansion plan j and attack plan ω, $WR_j(\omega)$, is formulated as

$$WR_j(\omega) = \pi(\omega) R_j(\omega); \qquad \forall j \in J, \omega = 1, \ldots, n_\Omega. \tag{8}$$

It should be noted that, as formulated in Sect. 5, no load shedding is allowed in the no-attack scenario $\omega = 0$. Thus, the regrets and weighted regrets in this scenario are all 0, that is, $R_j(0) = 0, \forall j \in J$ and $WR_j(0) = 0, \forall j \in J$.

Moreover, the maximum weighted regret of expansion plan j, $WR_j^{\max}$, is expressed as

$$WR_j^{\max} = \max_{\omega=0,\ldots,n_\Omega} \{WR_j(\omega)\}; \qquad \forall j \in J. \tag{9}$$

Under this risk characterization, the optimal expansion plan is the one that minimizes the maximum weighted regret over all scenarios considered. The minimax weighted regret criterion focuses on avoiding regrets resulting from making a nonoptimal decision. Therefore, it is an appropriate way of quantifying risk when considering low frequency and potentially catastrophic phenomena such as deliberate outages (Miranda and Proença 1998a,b). The minimax weighted regret, WR^*, can be formulated as follows:

$$WR^* = \min_{j \in J} \{WR_j^{\max}\}. \tag{10}$$

Risk-neutral solutions perform well for most scenarios but very poorly for some, and consequently a high regret may be attained. As opposed to risk-neutral decisions, the minimax weighted regret criterion does not allow selecting an expansion plan that may have serious consequences in any of the plausible scenarios considered. Furthermore, this criterion also accounts for the relative importance of scenarios. Thus, the decisions derived from this criterion are robust in terms of their acceptability under all futures anticipated.

5 Formulation

This section presents the formulation of the two vulnerability-constrained transmission expansion approaches proposed in this chapter, namely the risk-neutral model and the risk-averse model.

5.1 *Risk-Neutral Approach*

The mathematical formulation of the risk-neutral transmission expansion problem under deliberate outages is stated below:

$$\text{Minimize}_{\Delta D_n(\omega), s_\ell, P_g^{\mathrm{G}}(\omega), P_\ell^{\mathrm{L}}(\omega), \delta_n(\omega)} \quad \sum_{\omega=1}^{n_\Omega} \pi(\omega) \left[\sum_{n \in N} \Delta D_n(\omega) \right] + \beta \sum_{\ell \in L^{\mathrm{C}}} C_\ell^{\mathrm{L}} s_\ell \tag{11}$$

Subject to:

$$\sum_{\ell \in L^{\mathrm{C}}} C_\ell^{\mathrm{L}} s_\ell \leq C_{\mathrm{T}}^{\mathrm{L}} \tag{12}$$

$$\sum_{g \in G_n} P_g^{\mathrm{G}}(\omega) - \sum_{\ell | O(\ell)=n} P_\ell^{\mathrm{L}}(\omega) + \sum_{\ell | R(\ell)=n} P_\ell^{\mathrm{L}}(\omega)$$
$$= D_n - \Delta D_n(\omega); \qquad \omega = 0, \ldots, n_\Omega, \forall n \in N \tag{13}$$

$$P_\ell^{\mathrm{L}}(\omega) = \frac{1}{x_\ell}\left[\delta_{O(\ell)}(\omega) - \delta_{R(\ell)}(\omega)\right] v_\ell(\omega); \qquad \omega = 0, \ldots, n_\Omega, \forall \ell \in L^{\mathrm{O}} \tag{14}$$

$$P_\ell^{\mathrm{L}}(\omega) = \frac{1}{x_\ell}\left[\delta_{O(\ell)}(\omega) - \delta_{R(\ell)}(\omega)\right] s_\ell; \qquad \omega = 0, \ldots, n_\Omega, \forall \ell \in L^{\mathrm{C}} \tag{15}$$

$$-\overline{P}_\ell^{\mathrm{L}} \le P_\ell^{\mathrm{L}}(\omega) \le \overline{P}_\ell^{\mathrm{L}}; \qquad \omega = 0, \ldots, n_\Omega, \forall \ell \in \left\{L^{\mathrm{O}} \cup L^{\mathrm{C}}\right\} \tag{16}$$

$$0 \le P_g^{\mathrm{G}}(\omega) \le \overline{P}_g^{\mathrm{G}}; \qquad \omega = 0, \ldots, n_\Omega, \forall g \in G \tag{17}$$

$$\underline{\delta} \le \delta_n(\omega) \le \overline{\delta}; \qquad \omega = 0, \ldots, n_\Omega, \forall n \in N \tag{18}$$

$$\Delta D_n(\omega) = 0; \qquad \omega = 0, \forall n \in N \tag{19}$$

$$0 \le \Delta D_n(\omega) \le D_n; \qquad \omega = 1, \ldots, n_\Omega, \forall n \in N \tag{20}$$

$$s_\ell \in \{0, 1\}; \qquad \forall \ell \in L^{\mathrm{C}}, \tag{21}$$

where $\Delta D_n(\omega)$ is the load shed in node n and scenario ω; $P_g^{\mathrm{G}}(\omega)$ is the power output of generator g in scenario ω; $P_\ell^{\mathrm{L}}(\omega)$ is the power flow in line ℓ and scenario ω; $\delta_n(\omega)$ is the phase angle in node n and scenario ω; N is the set of node indices; β is a weighting factor for the investment cost; C_ℓ^{L} is the investment cost of candidate line ℓ; $C_{\mathrm{T}}^{\mathrm{L}}$ is the expansion planning budget; G_n is the set of indices of generators connected to node n; $O(\ell)$ and $R(\ell)$ are the sending and receiving nodes of line ℓ, respectively; D_n is the demand in node n; x_ℓ is the reactance of line ℓ; L^{O} is the set of indices of lines in the original transmission network; $\overline{P}_\ell^{\mathrm{L}}$ is the power flow capacity of line ℓ; $\overline{P}_g^{\mathrm{G}}$ is the capacity of generator g; G is the set of generator indices; and $\underline{\delta}$ and $\overline{\delta}$ are the lower and upper bounds for the nodal phase angles, respectively.

The objective function (11) comprises two terms. The first term represents the vulnerability of the transmission network against intentional attacks, which is calculated as the sum over all attack plans of the system load shed associated with each attack plan multiplied by its degree of importance. The investment cost is expressed by the second term. The weighting parameter β models the tradeoff between vulnerability- and economic-related goals and thus depends on the preferences of the network planner. The multiobjective function (11) differs from that in classical transmission expansion approaches where the goal is to minimize the investment and operation costs. In contrast, the proposed objective function incorporates the fundamental concern of the network planner within the context of this chapter, that is, the reduction in the vulnerability level against deliberate outages. Investment costs have also been accounted for in (12), in which an upper economic bound is set on expansion plans.

Constraints (13) enforce the power balance at every node and in every scenario. Using a dc load flow model, constraints (14) represent the power flows in the original network for each scenario as a function of the nodal phase angles. Note that for lines destroyed in scenario ω, that is, $v_\ell(\omega) = 0$, the corresponding power flows are 0. Analogously, constraints (15) express the line flows in the candidate lines for each scenario in terms of the nodal phase angles and the expansion variables, s_ℓ. Note that if candidate line $\ell \in L^C$ is not built, that is, $s_\ell = 0$, constraints (15) set the associated line flow to 0.

Constraints (16) provide the bounds for the line flows of the original and prospective lines for each scenario. Constraints (17) and (18) set the limits of generation and nodal phase angles, respectively, in each scenario. Note that, for the expansion problem, it is practical to assume that the lower bound on the power output of each generating unit is zero, thereby neglecting the effects of unit commitment and decommitment. This assumption is consistent with previously published works and is also applied throughout this chapter. Likewise, constraints (19) and (20) bound the nodal load shed in each scenario. Note that the system load shed for the no-attack scenario $\omega = 0$ is set to 0 by constraints (19), that is, the network is expanded to supply at least the forecasted demand. Finally, the integrality of variables s_ℓ is expressed in (21).

Problem (11)–(21) is a mixed-integer nonlinear programming problem. Nonlinearities arise in (15) due to the product of the binary variables s_ℓ and the continuous variables $\delta_{O(\ell)}(\omega)$ and $\delta_{R(\ell)}(\omega)$. However, the product of a binary variable and a continuous variable can be equivalently transformed into linear expressions (Floudas 1995). Such a transformation, known as modeling with disjunctive constraints, has been widely used both in the integer optimization and in the power systems literature (Alguacil et al. 2003; Arroyo and Galiana 2005; Bahiense et al. 2001; Motto et al. 2005; Oliveira et al. 2007). The equivalent linear formulation of constraints (15) is provided in Appendix 2.

Considering the equivalent linear expressions of constraints (15), problem (11)–(21) becomes a mixed-integer linear programming problem that can be efficiently solved by commercial branch-and-cut software (Dash XPRESS 2009; ILOG CPLEX 2009). The solution of problem (11)–(21) helps the network planner to determine appropriate values for β.

5.2 Risk-Averse Approach

This section presents the mathematical formulation of the proposed risk-averse transmission expansion problem. Previous risk-based planning approaches (Buygi et al. 2004, 2006; Crousillat et al. 1993; Fang and Hill 2003; Miranda and Proença 1998a,b; de la Torre et al. 1999) computed regrets ex post based on the complete enumeration of a reduced number of candidate plans. In contrast, the minimax problem is formulated here as the following standard mathematical programming problem with a single optimization, which allows considering a larger number of candidate plans:

$$\text{Minimize}_{WR^{\max}, s_\ell, WR(\omega), \Delta D_n(\omega), P_g^{\mathrm{G}}(\omega), P_\ell^{\mathrm{L}}(\omega), \delta_n(\omega)} \quad WR^{\max} + \beta \sum_{\ell \in L^{\mathrm{C}}} C_\ell^{\mathrm{L}} s_\ell \tag{22}$$

Subject to:

$$\sum_{\ell \in L^{\mathrm{C}}} C_\ell^{\mathrm{L}} s_\ell \leq C_{\mathrm{T}}^{\mathrm{L}} \tag{23}$$

$$WR(\omega) = \pi(\omega) \left[\sum_{n \in N} \Delta D_n(\omega) - \Delta D^{\min}(\omega) \right]; \qquad \omega = 1, \ldots, n_\Omega \tag{24}$$

$$WR^{\max} \geq WR(\omega); \qquad \omega = 1, \ldots, n_\Omega \tag{25}$$

$$\begin{aligned} &\sum_{g \in G_n} P_g^{\mathrm{G}}(\omega) - \sum_{\ell | O(\ell) = n} P_\ell^{\mathrm{L}}(\omega) + \sum_{\ell | R(\ell) = n} P_\ell^{\mathrm{L}}(\omega) \\ &= D_n - \Delta D_n(\omega); \qquad \omega = 0, \ldots, n_\Omega, \forall n \in N \end{aligned} \tag{26}$$

$$P_\ell^{\mathrm{L}}(\omega) = \frac{1}{x_\ell} \left[\delta_{O(\ell)}(\omega) - \delta_{R(\ell)}(\omega) \right] v_\ell(\omega); \qquad \omega = 0, \ldots, n_\Omega, \forall \ell \in L^{\mathrm{O}} \tag{27}$$

$$P_\ell^{\mathrm{L}}(\omega) = \frac{1}{x_\ell} \left[\delta_{O(\ell)}(\omega) - \delta_{R(\ell)}(\omega) \right] s_\ell; \qquad \omega = 0, \ldots, n_\Omega, \forall \ell \in L^{\mathrm{C}} \tag{28}$$

$$-\overline{P}_\ell^{\mathrm{L}} \leq P_\ell^{\mathrm{L}}(\omega) \leq \overline{P}_\ell^{\mathrm{L}}; \qquad \omega = 0, \ldots, n_\Omega, \forall \ell \in \left\{ L^{\mathrm{O}} \cup L^{\mathrm{C}} \right\} \tag{29}$$

$$0 \leq P_g^{\mathrm{G}}(\omega) \leq \overline{P}_g^{\mathrm{G}}; \qquad \omega = 0, \ldots, n_\Omega, \forall g \in G \tag{30}$$

$$\underline{\delta} \leq \delta_n(\omega) \leq \overline{\delta}; \qquad \omega = 0, \ldots, n_\Omega, \forall n \in N \tag{31}$$

$$\Delta D_n(\omega) = 0; \qquad \omega = 0, \forall n \in N \tag{32}$$

$$0 \leq \Delta D_n(\omega) \leq D_n; \qquad \omega = 1, \ldots, n_\Omega, \forall n \in N \tag{33}$$

$$s_\ell \in \{0, 1\}; \qquad \forall \ell \in L^{\mathrm{C}}, \tag{34}$$

where $WR^{\max}$ is the maximum weighted regret and $WR(\omega)$ is the weighted regret in scenario ω.

The objective function (22) comprises two terms. The first term represents the risk of vulnerability of the transmission network against intentional attacks, that is, the maximum weighted regret. The investment cost is expressed by the second term. The weighting parameter β balances the concern of the network planner on vulnerability- and economic-related issues. As in the risk-neutral model, the investment cost is bounded by a budget (23).

Constraints (24) represent the weighted regrets associated with each attack plan ω. Similar to the scenario generation procedure that requires the solution of multiple instances of MTTP, parameters $\Delta D^{\min}(\omega)$ in (24) result from solving a vulnerability-constrained transmission expansion planning problem for each attack plan ω. In other words, n_Ω instances of this single-scenario transmission expansion problem are solved to obtain $\Delta D^{\min}(\omega)$. The formulation of this deterministic transmission expansion problem can be found in Appendix 3.

By (25), $WR^{\max}$ is greater than or equal to the weighted regret in each attack plan. Hence, $WR^{\max}$ is greater than or equal to the maximum weighted regret. Since $WR^{\max}$ is minimized in the objective function (22), the maximum weighted regret is also minimized. Constraints (26)–(34) are, respectively, identical to (13)–(21), which were described in Sect. 5.1.

Note that index j of expansion plans is embedded in the expansion variables s_ℓ and consequently it is implicitly included in the problem formulation. Thus, the proposed formulation simultaneously handles 2^{n_P} candidate plans, n_P being the number of prospective lines.

Problem (22)–(34) is a mixed-integer nonlinear programming problem. As explained in Sect. 5.1 and Appendix 2, by transforming the nonlinear expressions (28) into equivalent linear expressions, the resulting risk-based transmission expansion problem under deliberate outages becomes a mixed-integer linear programming problem suitable for off-the-shell branch-and-cut software (Dash XPRESS 2009; ILOG CPLEX 2009). The solution of problem (22)–(34) is useful for the network planner to determine appropriate values for β.

5.3 Computational Issues

Table 1 provides the size of the resulting mixed-integer linear programming problems expressed as the number of constraints, real variables, and binary variables. Parameters n_G and n_N represent the number of generators and buses, respectively. As can be seen, the computational dimension is highly dependent on the number of scenarios. Although computational issues are not a primary concern in this kind of planning problems, a large number of scenarios may result in intractable optimization models. Tractability can be regained in two ways: (1) by reducing the number of scenarios so that the resulting tractable set of scenarios yields an optimal solution close in value to the solution of the original problem, or (2) by limiting the number of candidate lines, that is, the number of binary variables, so that only those potentially effective lines are included.

Table 1 Computational size of the proposed models

	Risk-neutral model	Risk-averse model
Number of constraints	$(n_\Omega+1)(2n_\mathrm{G}+3n_\mathrm{N}+3n_\mathrm{L}+13n_\mathrm{P}) + (2n_\Omega+1)\,n_\mathrm{N}+1$	$(n_\Omega+1)(2n_\mathrm{G}+3n_\mathrm{N}+3n_\mathrm{L}+13n_\mathrm{P}) + (2n_\Omega+1)(n_\mathrm{N}+1)$
Number of real variables	$(n_\Omega+1)(n_\mathrm{G}+2n_\mathrm{N}+n_\mathrm{L}+5n_\mathrm{P})$	$(n_\Omega+1)(n_\mathrm{G}+2n_\mathrm{N}+n_\mathrm{L}+5n_\mathrm{P}+1)$
Number of binary variables	n_P	n_P

6 Numerical Results

The proposed vulnerability-constrained formulations have been tested on Garver's six-node test system (Garver 1970). Garver's example is a network with six nodes and six installed lines. The topology of this system, the nodal demands, and the upper generation bounds are shown in Fig. 3. The data of every corridor are listed in Table 2 (obtained from (Alguacil et al. 2003)).

In all the simulations, $\overline{\delta}$ and $\underline{\delta}$ have been set to $\pi/2$ rad and $-\pi/2$ rad, respectively. The models have been implemented on a Sun Fire X4600 M2 with four processors at 2.60 GHz and 32 GB of RAM using CPLEX 10.2 (ILOG CPLEX 2009) under GAMS (GAMS Development Corporation 2009).

The maximum system load shed in this network is obtained after destroying at least five lines and is equal to 640 MW, representing 84.2% of the total demand (760 MW). After applying the procedure presented in Sect. 4.2, four attack plans have been considered as scenarios (Table 3). In the worst scenario, all existing lines are destroyed except for line 2–4. The system load shed associated with this scenario is 640 MW, which is the maximum system load shed attainable by the destructive agents. The last column of Table 3 presents the minimum system load shed associated with each attack plan ω, $\Delta D^{\min}(\omega)$, which corresponds to the optimal solution

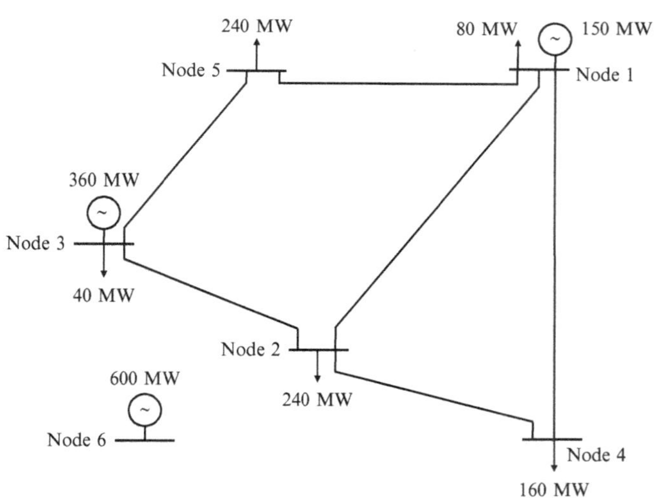

Fig. 3 Garver's system

Table 2 Corridor data

Corridor	1–2	1–3	1–4	1–5	1–6	2–3	2–4	2–5	2–6	3–4	3–5	3–6	4–5	4–6	5–6
x_ℓ (pu)	0.40	0.38	0.60	0.20	0.68	0.20	0.40	0.31	0.30	0.59	0.20	0.48	0.63	0.30	0.61
C_ℓ^{L} ($)	40	38	60	20	68	20	40	31	30	59	20	48	63	30	61
$\overline{P}_\ell^{\mathrm{L}}$ (MW)	100	100	80	100	70	100	100	100	100	82	100	100	75	100	78

Table 3 Attack plans

ω	Destroyed lines	$\Delta D(\omega)$ (MW)	$\pi(\omega)$	$\Delta D^{\min}(\omega)$ (MW)
1	2–3	470	0.3474	205.7
2	3–5	470	0.3474	226.1
3	2–3, 3–5	570	0.2106	270.0
4	1–2, 1–4, 1–5, 2–3, 3–5	640	0.0946	370.6

to the deterministic expansion planning problem described in Appendix 3. The scenario set Ω also includes a scenario $\omega = 0$ with no attacks because the original system with isolated Node 6 is unable to supply the demand in the remaining nodes.

6.1 Risk-Neutral Analysis

This section presents some results of the risk-neutral model. To analyze the trade-off faced by the network planner, different values of the weighting parameter β and a range of expansion budgets have been tested. The maximum number of lines (prospective plus installed) per corridor is three, that is, the number of candidate lines is 39.

If the network is expanded following the traditional investment cost minimization approach without considering deliberate outages and without allowing load shedding, the resulting expansion plan consists in building three lines in corridor 4–6 and reinforcing corridor 3–5 with one new line. The total investment cost of this economic-driven solution is \$110; however, the level of vulnerability (weighted average system load shed) under the set of considered attack plans is equal to 115.1 MW. Moreover, to require no load shed under any scenario, the network planner should build eight new lines: one line in corridors 1–5 and 2–6, and two lines in corridors 2–3, 3–5, and 4–6. The total cost incurred by this expansion plan is \$190.

Figure 4 represents the variation of the vulnerability with the expansion budget for different values of β. Note that for each β the reduction of vulnerability stops at a certain value, regardless of the expansion budget. This limit on the vulnerability reduction increases with the value of β, reaching a maximum for $\beta = 0.05$ with which the expansion plan is identical to that achieved by the traditional approach, and the vulnerability is 115.1 MW for all budgets. In other words, high values of β characterize a transmission planner mainly concerned with economic issues, and therefore, the investment costs are low at the expense of high levels of vulnerability. In contrast, low values of β imply lower levels of vulnerability with higher investment costs.

Table 4 lists the investment cost, the level of vulnerability, and the corresponding expansion plan for an expansion budget of \$170 and different values of β ranging between 0.00 and 0.05. The figures in brackets in the last column represent the number of parallel lines built in the corresponding corridor. As stated earlier, high values of β yield low investment costs, whereas low values of β imply lower levels

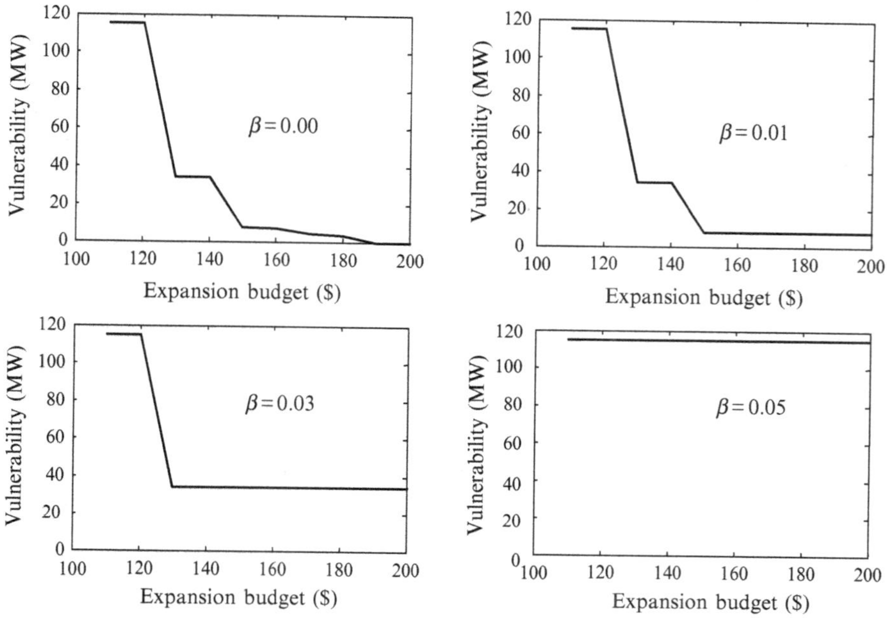

Fig. 4 Risk-neutral model. Vulnerability vs. expansion budget

Table 4 Risk-neutral model

β	Investment cost ($)	Vulnerability (MW)	Expansion plan
0.00	170	4.6	2–3 (2), 2–6, 3–5 (2), 4–6 (2)
0.01	150	7.6	2–3, 3–5 (2), 4–6 (3)
0.03	130	34.2	2–6 (3), 3–5 (2)
0.05	110	115.1	3–5, 4–6 (3)

Results for a $170 expansion budget

of vulnerability with higher investment costs. As an example, with $\beta = 0.00$ the level of vulnerability experiences a 96.0% reduction with respect to the economic-driven solution ($\beta = 0.05$) by building one line in corridor 2–6 and two lines in corridors 2–3, 3–5, and 4–6. The total investment cost incurred by this expansion plan is $170.

The average computing time required to attain the optimal solutions to all simulations was 0.7 s.

6.2 Risk-based Analysis

This section focuses on the application of the minimax weighted regret criterion. Hence, the proposed risk-based transmission expansion approach has been solved for $\beta = 0$. For expository purposes, only one candidate line is considered in corridors 1–3, 2–6, 3–4, and 4–6, that is, the number of prospective lines is 4. The reduced size of this problem (16 possible expansion plans and five scenarios) allows its solution by complete enumeration. Tables 5 and 6 provide the results for an expansion budget of \$150. Scenario 0 is not included in these tables since the corresponding system load shed is 0, as imposed by (32). It should be noted that expansion plan 16, consisting in the construction of all candidate lines, is infeasible for this expansion budget.

Table 5 lists the system load shed associated with each pair expansion plan-attack plan, $\Delta D^j(\omega)$. The figures in brackets in the first column represent the lines corresponding to each expansion plan j. As can be seen, expansion plan 1 consists in building no candidate line. The last column of Table 5 provides the weighted average system load shed for each expansion plan j over the considered set of attack plans, that is, $WD_j = \sum_{\omega=1,\ldots,n_\Omega} \pi(\omega)\Delta D^j(\omega); \forall j \in J$. The ideal expansion plan for scenarios 1, 2, and 3 is plan 14 (lines 1–3, 2–6, and 4–6) while expansion plan 8 (lines 2–6, 3–4, and 4–6) is the optimal decision for the worst-case scenario 4. Note that the optimal risk-neutral solution is expansion plan 14, which has the lowest weighted average system load shed (248.5 MW). The investment cost associated with the optimal risk-neutral solution is \$98.

Table 6 shows the weighted regret associated with each pair of expansion plan-attack plan, $WR_j(\omega)$. The last column of Table 6 lists the maximum weighted regret for each expansion plan j. Note that expansion plan 8 has the lowest maximum weighted regret (5.0 MW), and thus constitutes the optimal risk-based solution. In contrast, the optimal risk-neutral solution (expansion plan 14) has a maximum weighted regret equal to 6.6 MW. Although expansion plan 14 would be optimal should scenarios 1, 2, or 3 occur, its performance under scenario 4 is worse than the

Table 5 Risk-averse model

Expansion plan	ω				WD_j
	1	2	3	4	
1 (–)	470.0	470.0	570.0	640.0	507.2
2 (4–6)	370.0	370.0	470.0	540.0	407.2
3 (3–4)	388.0	392.7	488.0	558.0	426.8
4 (3–4, 4–6)	288.0	323.7	388.0	458.0	337.5
5 (2–6)	370.0	370.0	470.0	540.0	407.2
6 (2–6, 4–6)	270.0	270.0	370.0	440.0	307.2
7 (2–6, 3–4)	288.0	291.0	388.0	458.0	326.2
8 (2–6, 3–4, 4–6)	220.1	236.1	292.9	370.6	255.2
9 (1–3)	397.6	403.7	470.0	640.0	437.9
10 (1–3, 4–6)	303.1	316.8	370.0	540.0	344.4
11 (1–3, 3–4)	328.4	340.5	388.0	558.0	366.9
12 (1–3, 3–4, 4–6)	240.8	283.4	288.0	458.0	286.1
13 (1–3, 2–6)	297.6	300.3	370.0	540.0	336.7
14 (1–3, 2–6, 4–6)	205.7	226.1	270.0	440.0	248.5
15 (1–3, 2–6, 3–4)	228.4	240.5	288.0	458.0	266.9
16 (1–3, 2–6, 3–4, 4–6)	–	–	–	–	–

System load shed (MW) for a \$150 expansion budget

Table 6 Risk-averse model

Expansion plan	ω				$WR_j^{\max}$
	1	2	3	4	
1 (–)	91.8	84.7	63.2	25.5	91.8
2 (4–6)	57.1	50.0	42.1	16.0	57.1
3 (3–4)	63.3	57.9	46.0	17.7	63.3
4 (3–4, 4–6)	28.6	33.9	24.9	8.3	33.9
5 (2–6)	57.1	50.0	42.1	16.0	57.1
6 (2–6, 4–6)	22.4	15.2	21.1	6.6	22.4
7 (2–6, 3–4)	28.6	22.6	24.9	8.3	28.6
8 (2–6, 3–4, 4–6)	5.0	3.5	4.8	0.0	5.0
9 (1–3)	66.7	61.7	42.1	25.5	66.7
10 (1–3, 4–6)	33.9	31.5	21.1	16.0	33.9
11 (1–3, 3–4)	42.6	39.7	24.9	17.7	42.6
12 (1–3, 3–4, 4–6)	12.2	19.9	3.8	8.3	19.9
13 (1–3, 2–6)	31.9	25.8	21.1	16.0	31.9
14 (1–3, 2–6, 4–6)	0.0	0.0	0.0	6.6	6.6
15 (1–3, 2–6, 3–4)	7.9	5.0	3.8	8.3	8.3
16 (1–3, 2–6, 3–4, 4–6)	–	–	–	–	–

Weighted regret (MW) for a \$150 expansion budget

worst performance of expansion plan 8 under any plausible scenario. The investment cost associated with the optimal risk-averse solution is \$119, and its weighted average system load shed is 255.2 MW (Table 5). Thus, a 24.2% reduction in risk is achieved by slightly increasing the weighted average system load shed by 2.7%.

The proposed risk-averse optimization model (22)–(34) was applied to this illustrative example and the optimal expansion plan 8 was achieved in 0.1 s.

7 Conclusions

This chapter has presented a methodology to expand and reinforce the transmission network accounting for both economic issues and the impact of increasingly plausible deliberate outages. To characterize the uncertainty associated with intentional outages, a set of scenarios has been generated based on a recently reported method for vulnerability analysis.

Two models have been described, namely a risk-neutral model and a risk-averse model. In the former, the network planner is provided with relevant information on expansion plans while taking into account the tradeoff between vulnerability mitigation and investment cost reduction. The risk-based approach allows the network planner to quantify and hedge risk and to identify robust expansion plans. The risk aversion of the network planner is modeled by the minimax weighted regret criterion. This risk paradigm is an appropriate framework to model the impact of low likelihood but potentially catastrophic events such as intentional outages.

The proposed transmission expansion problems are formulated as mixed-integer nonlinear programs, which are equivalently transformed into mixed-integer linear programming problems. Thus, the resulting problems can be efficiently solved using available commercial branch-and-cut solvers. Garver's six-node system has been used to illustrate the performance of both models. First, the tradeoff between vulnerability mitigation and investment cost reduction has been analyzed. In addition, results from the risk-based model show the interest of this methodology since risk is reduced with small increments in weighted average system load shed.

Research is currently underway to account for other sources of uncertainty such as demand fluctuation, reliability of network components, and those associated with competition in power markets. Another interesting avenue of research is to study the sensitivity of the expansion plans with respect to different scenario generation procedures and weight assignments. Finally, further research will also be devoted to modeling candidate lines as potential targets.

Acknowledgements This work was supported in part by the Ministry of Education and Science of Spain under CICYT Projects DPI2006-01501 and DPI2006-08001, and by the Junta de Comunidades de Castilla – La Mancha, under Projects PAI08-0077-6243 and PCI08-0102.

Appendix 1

The scenario generation procedure presented in Sect. 4.2 requires the resolution of the modified terrorist threat problem (MTTP). The formulation of MTTP, which is based on the terrorist threat problem reported in (Motto et al. 2005), is provided below:

(MTTP)

(Upper-Level Problem)

$$\text{Maximize}_{v_\ell} \quad \Delta D \tag{35}$$

Subject to:

$$\sum_{\ell \in L^{\mathrm{O}}} (1 - v_\ell) = n_{\mathrm{A}} \tag{36}$$

$$\sum_{\ell \in L^{\mathrm{O}}} [1 - v_\ell(\omega)](1 - v_\ell) \le n_{\mathrm{A}} - 1; \qquad \forall \omega \in \Omega_{n_{\mathrm{A}}} \tag{37}$$

$$v_\ell \in \{0, 1\}; \qquad \forall \ell \in L^{\mathrm{O}} \tag{38}$$

(Lower-Level Problem)

$$\Delta D = \text{Minimize}_{\Delta D_n, P_g^{\mathrm{G}}, P_\ell^{\mathrm{L}}, \delta_n} \quad \sum_{n \in N} \Delta D_n \tag{39}$$

Subject to:

$$\sum_{g \in G_n} P_g^{\mathrm{G}} - \sum_{\ell \in L^{\mathrm{O}} | O(\ell)=n} P_\ell^{\mathrm{L}} + \sum_{\ell \in L^{\mathrm{O}} | R(\ell)=n} P_\ell^{\mathrm{L}} = D_n - \Delta D_n; \qquad \forall n \in N \quad (40)$$

$$P_\ell^{\mathrm{L}} = \frac{1}{x_\ell}\left[\delta_{O(\ell)} - \delta_{R(\ell)}\right] v_\ell; \qquad \forall \ell \in L^{\mathrm{O}} \quad (41)$$

$$-\overline{P}_\ell^{\mathrm{L}} \le P_\ell^{\mathrm{L}} \le \overline{P}_\ell^{\mathrm{L}}; \qquad \forall \ell \in L^{\mathrm{O}} \quad (42)$$

$$\underline{\delta} \le \delta_n \le \overline{\delta}; \qquad \forall n \in N \quad (43)$$

$$0 \le P_g^{\mathrm{G}} \le \overline{P}_g^{\mathrm{G}}; \qquad \forall g \in G \quad (44)$$

$$0 \le \Delta D_n \le D_n; \qquad \forall n \in N, \quad (45)$$

where $\Omega_{n_{\mathrm{A}}}$ is the index set of scenarios with n_{A} lines down. Note that the notation used in this appendix is consistent with that used in Sects. 4 and 5.

The mathematical programming problem (35)–(45) models a decision problem involving two agents who try to optimize their respective objective functions over a jointly dependent set. The above bilevel problem consists of an upper-level optimization (35)–(38) associated with the destructive agent and a lower-level optimization (39)–(45) associated with the system operator.

Binary variables v_ℓ are the upper-level decision variables controlled by the destructive agent, where v_ℓ is equal to 0 if line $\ell \in L^{\mathrm{O}}$ is attacked, being 1 otherwise. The upper-level objective (35) is to maximize the system load shed so as to meet a specific number of destroyed lines defined by (36). Constraints (37) avoid finding attack plans already constituting scenarios in $\Omega_{n_{\mathrm{A}}}$. Finally, constraints (38) express the binary nature of the upper-level decisions. Constraints (36) and (37) are the main changes of MTTP with respect to the terrorist threat problem formulated in (Motto et al. 2005).

In contrast, the system operator controls variables ΔD_n, P_g^{G}, P_ℓ^{L}, and δ_n. The lower-level objective (39) is to minimize the system load shed under the combination of destroyed lines chosen by the upper-level agent. The constraint set of the lower-level problem comprises network constraints (40)–(43) and bounds on generator power outputs (44) and load shedding (45).

As explained in (Motto et al. 2005), MTTP can be equivalently transformed into a single-level mixed-integer linear program that can be efficiently solved by available commercial software. The optimal solutions of MTTP, v^* and ΔD^*, are used to decide whether the attack plan constitutes a scenario, as described in Sect. 4.2.

Appendix 2

In (15) and (28) there are two products of binary and continuous variables per line and scenario: (1) s_ℓ and the phase angle of the sending node of line ℓ in scenario ω, denoted as $\delta_{O(\ell)}(\omega)$, and (2) s_ℓ and the phase angle of the receiving node of line ℓ

in scenario ω, denoted as $\delta_{R(\ell)}(\omega)$. As explained in (Floudas 1995), by introducing four new sets of continuous variables $\delta^{\mathrm{Q}}_{O(\ell)}(\omega)$, $\delta^{\mathrm{Q}}_{R(\ell)}(\omega)$ (representing the products $s_\ell\delta_{O(\ell)}(\omega)$ and $s_\ell\delta_{R(\ell)}(\omega)$, respectively), $\delta^{\mathrm{A}}_{O(\ell)}(\omega)$, and $\delta^{\mathrm{A}}_{R(\ell)}(\omega)$, the nonlinear constraints (15) and (28) are equivalently replaced by

$$P^{\mathrm{L}}_\ell(\omega) = \frac{1}{x_\ell}\left[\delta^{\mathrm{Q}}_{O(\ell)}(\omega) - \delta^{\mathrm{Q}}_{R(\ell)}(\omega)\right]; \qquad \omega = 0, \ldots, n_\Omega, \forall \ell \in L^{\mathrm{C}} \tag{46}$$

$$\delta^{\mathrm{Q}}_{O(\ell)}(\omega) = \delta_{O(\ell)}(\omega) - \delta^{\mathrm{A}}_{O(\ell)}(\omega); \qquad \omega = 0, \ldots, n_\Omega, \forall \ell \in L^{\mathrm{C}} \tag{47}$$

$$\delta^{\mathrm{Q}}_{R(\ell)}(\omega) = \delta_{R(\ell)}(\omega) - \delta^{\mathrm{A}}_{R(\ell)}(\omega); \qquad \omega = 0, \ldots, n_\Omega, \forall \ell \in L^{\mathrm{C}} \tag{48}$$

$$\underline{\delta} s_\ell \le \delta^{\mathrm{Q}}_{O(\ell)}(\omega) \le \overline{\delta} s_\ell; \qquad \omega = 0, \ldots, n_\Omega, \forall \ell \in L^{\mathrm{C}} \tag{49}$$

$$\underline{\delta} s_\ell \le \delta^{\mathrm{Q}}_{R(\ell)}(\omega) \le \overline{\delta} s_\ell; \qquad \omega = 0, \ldots, n_\Omega, \forall \ell \in L^{\mathrm{C}} \tag{50}$$

$$\underline{\delta}(1 - s_\ell) \le \delta^{\mathrm{A}}_{O(\ell)}(\omega) \le \overline{\delta}(1 - s_\ell); \qquad \omega = 0, \ldots, n_\Omega, \forall \ell \in L^{\mathrm{C}} \tag{51}$$

$$\underline{\delta}(1 - s_\ell) \le \delta^{\mathrm{A}}_{R(\ell)}(\omega) \le \overline{\delta}(1 - s_\ell); \qquad \omega = 0, \ldots, n_\Omega, \forall \ell \in L^{\mathrm{C}}. \tag{52}$$

Constraints (46) are the new linear expressions of the line power flows. Expressions (47) and (48) relate the nodal phase angles with the new variables $\delta^{\mathrm{Q}}_{O(\ell)}(\omega)$, $\delta^{\mathrm{A}}_{O(\ell)}(\omega)$, and $\delta^{\mathrm{Q}}_{R(\ell)}(\omega)$, $\delta^{\mathrm{A}}_{R(\ell)}(\omega)$, respectively. Finally, lower and upper bounds on variables $\delta^{\mathrm{Q}}_{O(\ell)}(\omega)$, $\delta^{\mathrm{Q}}_{R(\ell)}(\omega)$, $\delta^{\mathrm{A}}_{O(\ell)}(\omega)$, and $\delta^{\mathrm{A}}_{R(\ell)}(\omega)$ are imposed in (49)–(52), respectively.

If candidate line ℓ is not built ($s_\ell = 0$), variables $\delta^{\mathrm{Q}}_{O(\ell)}(\omega)$ and $\delta^{\mathrm{Q}}_{R(\ell)}(\omega)$ are set to 0 by (49) and (50), and consequently, the power flow through line ℓ is equal to 0 by (46). In addition, variables $\delta^{\mathrm{A}}_{O(\ell)}(\omega)$ and $\delta^{\mathrm{A}}_{R(\ell)}(\omega)$ are, respectively, equal to the phase angles at the sending and receiving nodes by (47) and (48).

Similarly, if candidate line ℓ is built ($s_\ell = 1$), variables $\delta^{\mathrm{A}}_{O(\ell)}(\omega)$ and $\delta^{\mathrm{A}}_{R(\ell)}(\omega)$ are both equal to 0 by (51) and (52), and variables $\delta^{\mathrm{Q}}_{O(\ell)}(\omega)$ and $\delta^{\mathrm{Q}}_{R(\ell)}(\omega)$ are, respectively, equal to $\delta_{O(\ell)}(\omega)$ and $\delta_{R(\ell)}(\omega)$ by (47) and (48). Hence, the power flow is determined by the difference of phase angles at the sending and receiving nodes (46).

Appendix 3

The minimum system load shed associated with each scenario $\omega = 1, \ldots, n_\Omega$, $\Delta D^{\min}(\omega)$, is obtained from the solution to the following deterministic transmission expansion problem:

$$\Delta D^{\min}(\omega) = \text{Minimize}_{\Delta D_n(\omega), s_\ell, P^{\mathrm{G}}_g(\omega), P^{\mathrm{L}}_\ell(\omega), \delta_n(\omega)} \quad \sum_{n \in N} \Delta D_n(\omega) \tag{53}$$

Subject to:

$$\sum_{\ell \in L^{C}} C_{\ell}^{L} s_{\ell} \leq C_{T}^{L} \tag{54}$$

$$\sum_{g \in G_n} P_g^{G}(\omega) - \sum_{\ell | O(\ell)=n} P_{\ell}^{L}(\omega) + \sum_{\ell | R(\ell)=n} P_{\ell}^{L}(\omega) = D_n - \Delta D_n(\omega); \quad \forall n \in N \tag{55}$$

$$P_{\ell}^{L}(\omega) = \frac{1}{x_{\ell}}\left[\delta_{O(\ell)}(\omega) - \delta_{R(\ell)}(\omega)\right] v_{\ell}(\omega); \quad \forall \ell \in L^{O} \tag{56}$$

$$P_{\ell}^{L}(\omega) = \frac{1}{x_{\ell}}\left[\delta_{O(\ell)}(\omega) - \delta_{R(\ell)}(\omega)\right] s_{\ell}; \quad \forall \ell \in L^{C} \tag{57}$$

$$-\overline{P}_{\ell}^{L} \leq P_{\ell}^{L}(\omega) \leq \overline{P}_{\ell}^{L}; \quad \forall \ell \in \left\{L^{O} \cup L^{C}\right\} \tag{58}$$

$$0 \leq P_g^{G}(\omega) \leq \overline{P}_g^{G}; \quad \forall g \in G \tag{59}$$

$$\underline{\delta} \leq \delta_n(\omega) \leq \overline{\delta}; \quad \forall n \in N \tag{60}$$

$$0 \leq \Delta D_n(\omega) \leq D_n; \quad \forall n \in N \tag{61}$$

$$s_{\ell} \in \{0, 1\}; \quad \forall \ell \in L^{C}. \tag{62}$$

Problem (53)–(62) is a vulnerability-constrained transmission expansion problem for each scenario $\omega = 1, \ldots, n_{\Omega}$. Note that for scenario $\omega = 0$, the system load shed is set to 0 by (32) and therefore, $\Delta D^{\min}(0) = 0$.

The objective function (53) represents the system load shed associated with scenario ω. Constraints (54) are identical to (23), constraints (55)–(61), respectively, correspond to (26)–(31) and (33) for the considered scenario ω, and constraints (62) are identical to constraints (34).

References

Alguacil N, Motto AL, Conejo AJ (2003) Transmission expansion planning: a mixed-integer LP approach. IEEE Trans Power Syst 18:1070–1077

Alguacil N, Carrión M, Arroyo JM (2008) Transmission network expansion planning under deliberate outages. In: Proceedings of the 16th Power Syst Comput Conf, PSCC'08, Paper no. 23, Glasgow

Arroyo JM, Galiana FD (2005) On the solution of the bilevel programming formulation of the terrorist threat problem. IEEE Trans Power Syst 20:789–797

Bahiense L, Oliveira GC, Pereira M, Granville S (2001) A mixed integer disjunctive model for transmission network expansion. IEEE Trans Power Syst 16:560–565

Bell DE (1982) Regret in decision making under uncertainty. Oper Res 30:961–981

Bier VM, Gratz ER, Haphuriwat NJ, Magua W, Wierzbicki KR (2007) Methodology for identifying near-optimal interdiction strategies for a power transmission system. Reliab Eng Syst Saf 92:1155–1161

Binato S, Pereira MVF, Granville S (2001) A new Benders decomposition approach to solve power transmission network design problems. IEEE Trans Power Syst 16:235–240

Buygi MO, Balzer G, Shanechi HM, Shahidehpour M (2004) Market-based transmission expansion planning. IEEE Trans Power Syst 19:2060–2067

Buygi MO, Shanechi HM, Balzer G, Shahidehpour M, Pariz N (2006) Network planning in unbundled power systems. IEEE Trans Power Syst 21:1379–1387

Carrión M, Arroyo JM, Alguacil N (2007) Vulnerability-constrained transmission expansion planning: a stochastic programming approach. IEEE Trans Power Syst 22:1436–1445

Choi J, Tran T, El-Keib AA, Thomas R, Oh H, Billinton R (2005) A method for transmission system expansion planning considering probabilistic reliability criteria. IEEE Trans Power Syst 20:1606–1615

Commission of the European Communities (2005) Green paper on a European programme for critical infrastructure protection. http://eur-lex.europa.eu/LexUriServ/site/en/com/2005/com2005_0576en01.pdf

Contreras J, Wu FF (1999) Coalition formation in transmission expansion planning. IEEE Trans Power Syst 14:1144–1152

Contreras J, Wu FF (2000) A kernel-oriented algorithm for transmission expansion planning. IEEE Trans Power Syst 15:1434–1440

Critical Infrastructure Protection Committee, CIPC (2009) North American Electric Reliability Council, NERC. http://www.nerc.com/page.php?cid=1|117|139

Crousillat EO, Dorfner P, Alvarado P, Merrill HM (1993) Conflicting objectives and risk in power system planning. IEEE Trans Power Syst 8:887–893

da Silva EL, Gil HA, Areiza JM (2000) Transmission network expansion planning under an improved genetic algorithm. IEEE Trans Power Syst 15:1168–1175

Dash XPRESS (2009). http://www.dashoptimization.com/home/products/products_optimizer.html

de la Torre S, Conejo AJ, Contreras J (2008) Transmission expansion planning in electricity markets. IEEE Trans Power Syst 23:238–248

de la Torre T, Feltes JW, Roman TGS, Merrill HM (1999) Deregulation, privatization, and competition: Transmission planning under uncertainty. IEEE Trans Power Syst 14:460–465

Dusonchet YP, El-Abiad A (1973) Transmission planning using discrete dynamic optimizing. IEEE Trans Power App Syst PAS-92:1358–1371

Fang R, Hill DJ (2003) A new strategy for transmission expansion in competitive electricity markets. IEEE Trans Power Syst 18:374–380

Floudas CA (1995) Nonlinear and mixed-integer optimization: fundamentals and applications. Oxford University Press, New York

GAMS Development Corporation (2009). http://www.gams.com/

Garver LL (1970) Transmission network estimation using linear programming. IEEE Trans Power App Syst PAS-89:1688–1697

Gheorghe AV, Masera M, Weijnen M, de Vries L (2006) Critical infrastructures at risk: securing the European electric power system. Springer, Dordrecht

Gorenstin BG, Campodonico NM, Costa JP, Pereira MVF (1993) Power system expansion planning under uncertainty. IEEE Trans Power Syst 8:129–136

Holmgren ÅJ, Jenelius E, Westin J (2007) Evaluating strategies for defending electric power networks against antagonistic attacks. IEEE Trans Power Syst 22:76–84

ILOG CPLEX (2009). http://www.ilog.com/products/cplex/

Joint Infrastructure Interdependencies Research Program, JIIRP (2009) Department of Public Safety and Emergency Preparedness Canada, PSEPC, and Natural Sciences and Engineering Research Council, NSERC. http://www.publicsafety.gc.ca/prg/em/jiirp/index-eng.aspx

Kaltenbach JC, Peschon J, Gehrig EH (1970) A mathematical optimization technique for the expansion of electric power transmission systems. IEEE Trans Power App Syst PAS-89: 113–119

Latorre G, Cruz RD, Areiza JM, Villegas A (2003) Classification of publications and models on transmission expansion planning. IEEE Trans Power Syst 18:938–946

Lee CW, Ng SKK, Zhong J, Wu FF (2006) Transmission expansion planning from past to future. In: Proceedings of the 2006 IEEE PES Power Syst Conf Expo, PSCE'06:257–265, Atlanta

Loomes G, Sugden R (1982) Regret theory: an alternative theory of rational choice under uncertainty. Econ J 92:805–824

Memorial Institute for the Prevention of Terrorism, MIPT (2009). http://www.mipt.org

Merrill HM, Wood AJ (1991) Risk and uncertainty in power system planning. Int J Electr Power Energy Syst 13:81–90

Miranda V, Proença LM (1998a) Why risk analysis outperforms probabilistic choice as the effective decision support paradigm for power system planning. IEEE Trans Power Syst 13:643–648

Miranda V, Proença LM (1998b) Probabilistic choice vs. risk analysis – Conflicts and synthesis in power system planning. IEEE Trans Power Syst 13:1038–1043

Monticelli A, Santos A Jr, Pereira MVF, Cunha SH, Parker BJ, Praça JCG (1982) Interactive transmission network planning using a least-effort criterion. IEEE Trans Power App Syst PAS-101:3919–3925

Motto AL, Arroyo JM, Galiana FD (2005) A mixed-integer LP procedure for the analysis of electric grid security under disruptive threat. IEEE Trans Power Syst 20:1357–1365

Oliveira GC, Costa APC, Binato S (1995) Large scale transmission network planning using optimization and heuristic techniques. IEEE Trans Power Syst 10:1828–1834

Oliveira GC, Binato S, Pereira MVF (2007) Value-based transmission expansion planning of hydrothermal systems under uncertainty. IEEE Trans Power Syst 22:1429–1435

Romero R, Monticelli A (1994a) A hierarchical decomposition approach for transmission network expansion planning. IEEE Trans Power Syst 9:373–380

Romero R, Monticelli A (1994b) A zero-one implicit enumeration method for optimizing investments in transmission expansion planning. IEEE Trans Power Syst 9:1385–1391

Romero R, Gallego RA, Monticelli A (1996) Transmission system expansion planning by simulated annealing. IEEE Trans Power Syst 11:364–369

Salmeron J, Wood K, Baldick R (2004) Analysis of electric grid security under terrorist threat. IEEE Trans Power Syst 19:905–912

Sauma EE, Oren SS (2007) Economic criteria for planning transmission investment in restructured electricity markets. IEEE Trans Power Syst 22:1394–1405

Seifu A, Salon S, List G (1989) Optimization of transmission line planning including security constraints. IEEE Trans Power Syst 4:1507–1513

Sharifnia A, Aashtiani HZ (1985) Transmission network planning: a method for synthesis of minimum-cost secure networks. IEEE Trans Power App Syst PAS-104:2026–2034

Silva ID, Rider MJ, Romero R, Garcia AV, Murari CA (2005) Transmission network expansion planning with security constraints. IEE Proc Gener Transm Distrib 6:828–836

Simonoff JS, Restrepo CE, Zimmerman R (2007) Risk-management and risk-analysis-based decision tools for attacks on electric power. Risk Anal 27:547–570

Thomas RJ, Whitehead JT, Outhred H, Mount TD (2005) Transmission system planning – The old world meets the new. Proc IEEE 93:2026–2035

Tranchita C, Hadjsaid N, Torres A (2006) Ranking contingency resulting from terrorism by utilization of Bayesian networks. In: Proceedings of the 13th IEEE Mediterr Electrotech Conf, MELECON 2006:964–967, Málaga

Villasana R, Garver LL, Salon SJ (1985) Transmission network planning using linear programming. IEEE Trans Power App Syst PAS-104:349–356

Wang X, McDonald JR (1994) Modern power system planning. McGraw-Hill, London

Wu FF, Zheng FL, Wen FS (2006) Transmission investment and expansion planning in a restructured electricity market. Energy 31:954–966

Zolezzi M, Rudnick H (2002) Transmission cost allocation by cooperative games and coalition formation. IEEE Trans Power Syst 17:1008–1015

Long-term and Expansion Planning for Electrical Networks Considering Uncertainties

T. Paulun and H.-J. Haubrich

Abstract The impending regulation of European electricity markets has led to an increasing cost pressure for network system operators. This applies in equal measure to transmission and distribution network operators. At the same time, boundary conditions of network planning are becoming more and more uncertain as a consequence of unbundling formerly integrated generation, transmission and distribution companies. To reduce network costs without worsening security and quality of supply, planning of transmission and distribution networks needs to be improved. For this, several computer-based optimization methods have been developed over the last years. In this chapter, boundary conditions and degrees of freedom that need to be taken into account during network optimization are discussed at first. Following that, the most commonly used optimization algorithms are presented, and the practical application of those methods is summarized.

Keywords Ant colony optimization · Electrical networks · Genetic algorithms · Long-term and expansion planning · Network Optimization · Network planning · Transmission and distribution · Uncertainties

1 Introduction

In recent years, the situation of transmission and distribution network system operators in European countries has changed significantly. The liberalization of European electricity markets has led to an increase in public interest in quality of supply and network costs and thus to an increase in cost pressure on network system operators (Commission of the European Communities 2009). At the same time, boundary conditions of network planning are becoming more and more uncertain as a result

T. Paulun (✉)
Institute of Power Systems and Power Economics (IAEW), RWTH Aachen University, Schinkelstraße 6, 52056 Aachen, Germany
e-mail: tp@iaew.rwth-aachen.de

S. Rebennack et al. (eds.), *Handbook of Power Systems I*, Energy Systems,
DOI 10.1007/978-3-642-02493-1_17, © Springer-Verlag Berlin Heidelberg 2010

of unbundling formerly integrated generation and transmission companies. Unlike before, investments in generation sites are evaluated and planned without taking into account the possible impacts on the network, and may therefore increase the need for additional investments by the network operator (Neimane 2001).

The situation of markets in Europe is different from that in other countries or areas, for example, the electricity market in the US or Brazil. In some countries, the network operator has only limited power over the network and investment or maintenance decisions. However, a technical and economical optimal development of the network is of interest in those countries as well. Since it is equal, whether the network operator or a national authority uses methods to determine optimal investment and maintenance measures, the methods described in this chapter are applicable to electricity networks outside Europe as well. However, the methods presented strictly follow a cost minimization paradigm according to the situation in Europe. Other aspects of network planning – for example the available transmission capacity and quality of supply – are treated as boundary conditions. Thus it may be necessary to adapt the objective function of the methods presented in the following in order to apply those methods to transmission and distribution networks outside Europe.

Electricity transmission and distribution networks have certain characteristics of natural monopolies, as high investments are required for installing new lines. Moreover, two or more electrical networks supplying the same area in parallel are neither economically efficient nor acceptable from an environmental point of view. Thus liberalization of electricity markets has not led to effective competition between network operators.

As a result, European countries have implemented regulatory authorities that compare costs of different network operators and limit revenues to the level of costs of an efficient network operator. Thus evaluating and measuring the efficiency of network system operators in an objective manner is one of the most important tasks in liberalized and regulated electricity markets (Cooper and Tone 1997).

To increase the efficiency of their networks and to earn adequate revenues in the future, network operators need to improve the process of network planning. For this, one of the most promising approaches is the use of computer-based network optimization methods (Oliveira et al. 2002; Binato et al. 2001; Da Vilva et al. 2002; Sauma and Oren 2005, 2007; Latorre et al. 2003; Romero and Monticelli 1994; Maurer 2004; Maurer et al. 2006; Paulun 2006, 2007). In recent years, several optimization algorithms have been developed and successfully applied to practical planning problems. Additionally, some regulatory authorities such as the German *Federal Network Agency* have adopted those methods for calculating the efficiency of electrical networks in an appropriate way (Bundesnetzagentur 2006).

Most of the methods developed during the last 10 years model the problem of network planning by analytical approaches (Oliveira et al. 2002; Binato et al. 2001; Da Vilva et al. 2002; Sauma and Oren 2005, 2007; Latorre et al. 2003; Romero and Monticelli 1994). The equations resulting from this may be solved using exact methods (Oliveira et al. 2002; Binato et al. 2001) or heuristic algorithms (Sauma and Oren 2007). However, analytical formulations are usually not capable of taking into account all boundary conditions of network planning that need to be considered by network planners in reality. This applies, in particular, if the boundary conditions are subjected to significant uncertainties like the future development of load and generation or specific investment costs. Thus, this chapter implies an alternative modelling approach, which is based on an object-oriented computer model in which network elements such as substations, transformers and lines are modelled as individual objects. Hence, the goal of network planning is to find the optimal combination of those elements that fulfill all planning criteria such as technical and geographical restrictions, and optimize the value of an objective function, which is usually given by a minimization of the overall network costs.

In the following, planning of electricity transmission and distribution networks is discussed and the available methods for solving the optimization problem based on an object-oriented model are presented. For this, boundary conditions and degrees of freedom during network planning are summarized at first. Following that, optimization algorithms for different aspects of network planning are presented and explained in detail. At the end of the chapter, practical application of those methods as well as unanswered questions and unsolved problems regarding practical planning problems are discussed.

2 Planning of Transmission and Distribution Networks

Planning of electrical networks usually consists of several steps that focus on different aspects of network planning (Maurer 2004; Maurer et al. 2006). On the one hand, the optimal development of a given network in the long run needs to be optimized. For this, usually a period of 20–30 years or even more decades is considered. This comprises basic planning decisions, such as the voltage level or the general structure of the network as well as general decisions about the types of equipment to be used. On the other hand, the task of short- and mid-term network planning is to optimize the development of an existing network with regard to the specifications of long-term planning. Because of this, and in order to reduce the computational effort of optimization, network planning is separated into *long-term* and *expansion planning* (see Fig. 1).

In the following sections, uncertainties that need to be considered during network planning are presented at first. Afterwards, technical restrictions and boundary conditions are presented. Finally, long-term and expansion planning are explained in detail.

2.1 *Uncertainties of Network Planning*

Uncertain boundary conditions of network planning can be classified according to the origin of their uncertainty, as there are *technical*, *economical* and *regulatory* uncertainties (Paulun 2006, 2007).

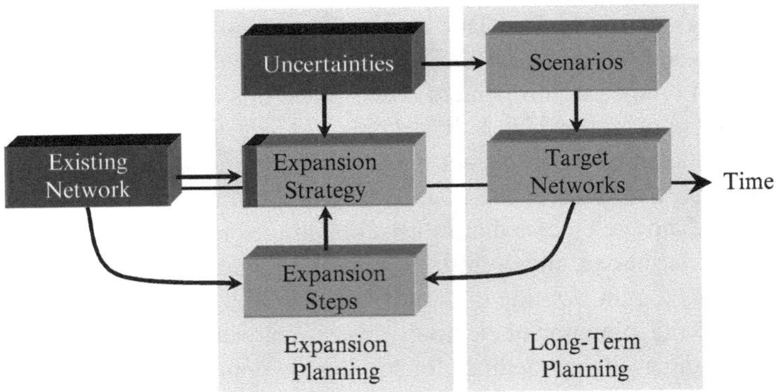

Fig. 1 Long-term and expansion planning of electrical networks

2.1.1 Technical Uncertainties

Technical uncertainties are mainly given by the uncertain development of load and generation in the supply area. This development can be described by a customer-specific load increase rate per year and by the additional data about the connection or disconnection of individual customers. The uncertain load increase rate as well as the uncertain individual development may be modelled by different scenarios, which are then taken into account during network planning.

In electricity distribution networks, the future development of distributed generation is especially relevant, as this influences the development of load and generation significantly. Basically, this development may be modelled by an individual load increase rate for every customer, but this leads to high efforts when the future development is forecasted. Because of this, customers are usually classified by different groups. The development of distributed generation is then assumed to be identical for all customers assigned to the same group.

Another technical uncertainty is given by the uncertain useful lifetime of equipment. While the expected lifetime of every asset is equal to the average lifetime of all assets of the same type (e.g. all identical transformers), the actual deviation from this value is usually uncertain. Thus the useful lifetime needs to be modelled by a distribution function. Usually the useful lifetime is assumed to be normally distributed, but other distribution functions may be used as well. The expected value and the standard deviation of the function used can be obtained from statistics (Paulun 2006, 2007).

2.1.2 Economical Uncertainties

Economical uncertainties emerge from the uncertain interest rate during the period under consideration, the uncertain development of investment and maintenance costs for the equipment used and the specific costs of losses.

Although a volatile development of the interest rate may be very unlikely, it may still have a significant impact on optimal planning decisions and needs therefore to be considered during network optimization. Besides, the replacement of old equipment may become economically feasible due to increasing maintenance costs of aging utilities. Furthermore, investing in new lines to increase the transmission capacity of the network may become reasonable if the specific costs of losses exceed a boundary value.

Except for the interest rate during the period under consideration, all economical uncertainties can be described by a standard distribution with good approximation. For example, the specific costs of losses can be defined by the expected value and the standard deviation, respectively, for every year of the period under consideration. To model the increasing uncertainty of the forecast from the beginning to the end of this period, an increasing standard deviation may be assumed.

2.1.3 Regulatory Uncertainties

The optimal future network development may be considerably influenced by regulatory decisions. However, most of the uncertainties resulting from this can be modelled as technical uncertainties as their impact on the network is quite similar. For example, regulatory support for distributed generation affects the future development of distributed generation, which has already been discussed above. Similarly, the regulatory decision in Germany to phase out nuclear energy affects the future development of generation in the whole country. This leads to additional technical uncertainties regarding the future development of load and generation for every network operator.

Other regulatory uncertainties that may also be modelled as technical uncertainties affect the useful lifetime of the types of equipment used (e.g. a prohibition of oil-isolated cables). Such uncertainties increase the standard deviation of the distribution function describing the useful lifetime of the assets.

Additional uncertainties are given by the required time for realizing network expansion projects. For example, planning and constructing new overhead lines in high or extra high voltage networks may take several years due to the time-consuming application for permission. This is especially relevant as new lines are required urgently in the European transmission network to meet the demands of the liberalized market. In worst case scenarios, even stability of the network may be endangered if required lines are not approved and authorized in time.

2.2 *Technical Boundary Conditions of Network Planning*

Technical restrictions that have to be taken into account during network planning are usually the minimum requirements that need to be fulfilled. Nevertheless, a network operator may choose to define additional criteria to improve security and quality of supply in his supply area. However, minimum requirements that must not be violated are given by the types of equipment used and the demands of network customers and consist of the following:

- Maximum thermal currents
- Maximum and minimum voltages
- Maximum and minimum short circuit currents
- Minimum requirements regarding quality of supply

Maximum thermal current of lines as well as maximum power of transformers must not be exceeded during normal operation. Only during short-term disturbances, violations of boundary values are acceptable to a level depending on the network operator's individual planning criteria. Usually, a maximum loading of the overhead lines and transformers up to 120% of the rated values is tolerated for several minutes.

To reduce losses on lines, the operational voltage is chosen preferably high. Maximum values depend on the size of the insulators and the installation arrangement of equipment. Minimum values are defined by the demands of network customers. Even more, the effective value of the voltage does need to be not only high enough on average but also constant over time because sensitive apparatus may be switched off even in the case of voltage drops far below a duration of 1 s.

Short-circuit currents are an adequate measure for the network strength at a given connection point and for the voltage quality at that point. Guidelines for calculating short-circuits on the basis of detailed network models can be found in Berizzi et al. (1993) and in international standards. During network planning, usually three-phase short-circuits are considered as the only relevant faults, as those failure types lead to the highest short-circuit currents in most cases.

High short-circuit currents improve the voltage quality at a customers' connection point as disturbances caused by other customers are avoided. In contrast, low short-circuit currents reduce equipment stress and enable the use of cheaper equipment. Thus similar to boundary values for voltages, typically maximum and minimum values for short-circuit currents have to be considered during network planning.

Quality of supply is commonly defined by the number of outages per customer per year (*System Average Interruption Frequency Index, SAIFI*), the average interruption duration (*Customer Average Interruption Duration Index, CAIDI*) and the probability for an interruption (*System Average Interruption Duration Index, SAIDI*), which is also referred to as *non-availability* and is given in minutes per year. The non-availability can be calculated as the product of SAIFI and CAIDI (Ward 2001).

Calculating quality of supply indices is done by means of probabilistic quality of supply simulations that require high computational effort. During network planning, quality of supply is therefore taken into account in a simplified way by qualitative criteria like the commonly used (n − 1)-criterion. If this criterion is fulfilled, any single unplanned outage of equipment does not lead to an interruption of customers (Ward 2001).

2.3 Long-term Planning

As mentioned earlier, the aim of long-term network planning is to calculate networks that should be realized in the long run to minimize network costs. Thus network structures that fulfil a given supply task with minimum costs need to be determined (Maurer 2004; Maurer et al. 2006).

Since the existing network of a network system operator may not be optimal from a long-term point of view, most of the existing assets are neglected during long-term planning. As a consequence, calculated networks may differ significantly from the existing network structure and may be realized only after several decades. Nonetheless, those networks are *target networks* for future development of the existing network.

Besides the technical restrictions explained above, boundary conditions during long-term planning consist of geographical restrictions such as the position of load and generators and possible connections to neighbouring or overlaying networks. For all network customers, load or feed-in, respectively, need to be forecasted for the point in time when target networks are to be calculated for. Since the future development of load and generation or even the connection or disconnection of customers is usually subject to extensive uncertainties (see Sect. 2.1), several alternative scenarios should be considered during long-term planning. For example, alternative target networks may be calculated for 110% and for 120% of today's maximum load. This would correspond to a linear annual load increase rate of 0.5% or 1%, respectively, of today's maximum load, given a period under consideration of 20 years.

Degrees of freedom during long-term planning are which of the available routes should be used for overhead lines or cables and where substations, switchgears and transformers should be installed. Additionally, dimensioning of equipment and the choice of switchgear concepts such as block or branch-arrangements, single or double busbar concepts are also degrees of freedom for the network operator.

As target networks are calculated for distinct points in time, they are not developed dynamically over long periods of time as the existing networks are. In particular, target networks are assumed to be constructed entirely at the point in time under consideration without any delay. As a consequence, it is not possible to define the age of equipment used in target networks as its time of construction is not optimized during long-term planning.

Costs of target networks comprise equipment investment and maintenance costs as well as costs of losses. Since the construction time of equipment is not known, annuity investment costs are used instead. These costs are equal to the average annual costs that correspond to the net present value given by the equipment investment costs and depend on the interest rate used for calculations and the expected useful lifetime of equipment. For example, the annuity A of an investment I under the assumption of an interest rate r and an useful lifetime N can be calculated according to formula (1) (Haubrich 2001).

$$A = \frac{(1+r)^N \cdot r}{(1+r)^N - 1} \cdot I \tag{1}$$

Maintenance costs are usually given in percent per year related to the total investment costs. A rough estimation of the investment costs of overhead lines, transformers or switchgears is 0.5–2%/a of the investment costs, and maintenance costs of cables can often be neglected compared to their investment costs.

For calculating costs of losses, a load flow calculation needs to be carried out. This does not cause additional computational effort as this calculation has to be carried out anyway in order to verify that all technical boundary conditions such as equipment load or voltage limits are fulfilled. The network losses in kilowatt or megawatt are then multiplied with the specific costs of losses in EUR/MWh. If the specific costs of losses can only be forecasted with significant uncertainty, alternative target networks should be calculated for different scenarios analogously to load scenarios mentioned above. This is reasonable because investing in additional lines or transformers may reduce network losses and thereby costs of losses, and thus may lead to a reduction of costs in total.

The results of long-term planning are given by alternative network structures that are target networks for different scenarios of uncertainties. At the end of this planning step, it is not known whether those target networks can actually be reached during several years. Furthermore, investments needed to reach long-term target networks are not known, since this requires a comparison of target networks and the existing network, and the latter has not been taken into account during long-term planning. Thus network expansion planning needs to be applied in an additional separate planning step.

2.4 Expansion Planning

Unlike long-term planning, network expansion planning focuses not only on distinct points in time but also on a complete period of usually up to 20 or 30 years. The objective of optimization is to minimize the net present value of the total costs during the period under consideration by optimizing the future development of the existing network. Analogously to long-term planning, investment and maintenance costs of equipment as well as costs of losses are taken into account (Paulun 2006, 2007).

In this planning step, long-term target networks that have been calculated beforehand are input parameters of optimization. Additionally, the existing network structure is also taken into account in this planning step as the optimal future development of this network is to be determined. In particular, the age of the existing assets needs to be known to take necessary reinvestments as well as costs for deconstruction of equipment into account during optimization. Resulting input parameters of network expansion planning are shown in Fig. 2.

Degrees of freedom during network expansion planning are defined by the existing network and long-term target networks. Only expansion steps that develop the existing network towards one or more target networks may be carried out during the period under consideration. For example, the construction of a line or transformer

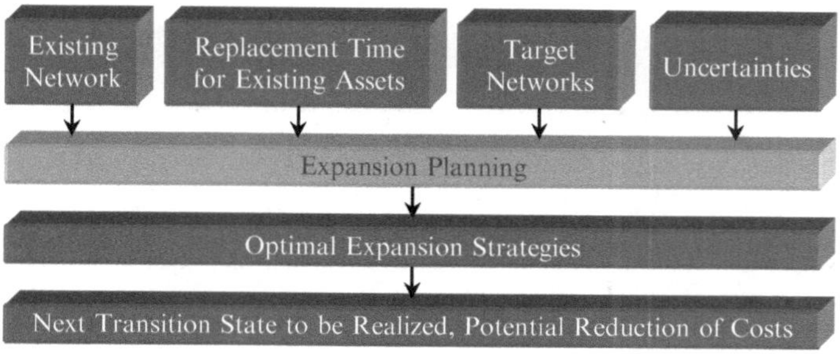

Fig. 2 Input parameters and results of network expansion planning

that exists in one or more target networks but does not exist today is one possible expansion step. The same applies to the deconstruction of existing assets that are not existent in one or more target networks.

Hence, the realization of each expansion step that develops the existing network towards one or more target networks is a degree of freedom during long-term planning, as the network operator needs to decide which expansion steps should be realized during the period under consideration. Moreover, the optimal points in time for realizing expansion steps need to be determined.

To estimate future consequences of expansion steps, the development of the network needs to be evaluated for the complete period under consideration. Obviously, it is not possible to define which expansion steps should be realized in 10 or 20 years from now on, as those decisions may need to be withdrawn if uncertain boundary conditions of network planning like load and generation develop in an unexpected way. Thus future planning decisions need to be roughly estimated, but the focus of network expansion planning is to optimize the expansion steps that should be carried out immediately, that is within the next one or two years.

Up to now, the future development of an existing network has been described by *expansion plans*. An expansion plan defines in detail which expansion steps and which reinvestments are carried out during the period under consideration and when those projects need to be realized. As a consequence, the total costs during the period under consideration as well as the budget that needs to be spent in each year of this period are known as a result of network expansion planning.

However, practical experience has shown that this approach is not sufficient to meet the demands resulting from increasing uncertainties in liberalized electricity markets. As those markets develop very fast and especially future development of load and generation is even more uncertain than before, expansion plans have to be adapted to changing boundary conditions almost every year. In that case, long-dated expansion plans provide no additional benefit compared to short-term optimization without taking long-term target networks into account.

This problem can be avoided by taking uncertainties of network planning into account not only during optimization, but also when defining network expansion plans. Expansion plans with respect to uncertain boundary conditions are referred to as *expansion strategies* and do not define the year in which certain expansion steps

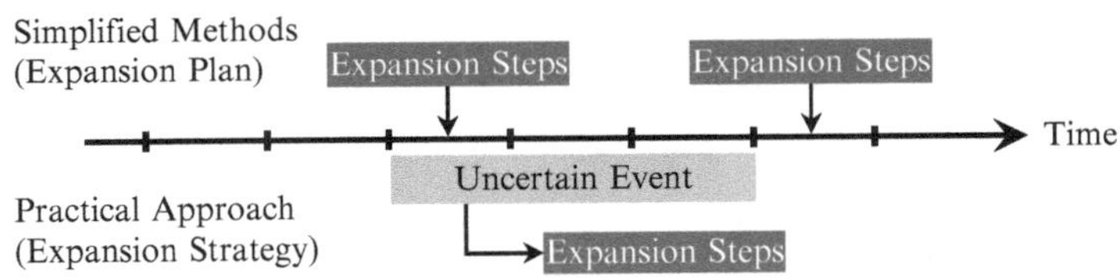

Fig. 3 Dependency between uncertainties of network planning and expansion steps

should be carried out. Instead, they link expansion steps to uncertain events that initiate the realization of this step. For example, the construction of new lines may be reasonable if the load increases above a certain threshold, and a new substation may be required if new customers are to be connected to the network. If the year in which the uncertain event occurs is uncertain, the year in which the corresponding expansion steps are realized is uncertain as well (see Fig. 3).

As a result, the net present value of the total costs during the period under consideration is also uncertain after expansion strategies have been optimized. Uncertainties arise from the discount factor that depends on the year the uncertain event – and thus investments for realizing corresponding expansion steps – occurs and on the fact that some events may not occur at all. Nevertheless, expansion strategies describe the optimal future development of the existing network under uncertainty. Thus the probability distribution of the net present value may be used to quantify the risk related to a given expansion strategy using methods from decision theory such as value at risk or conditional value at risk.

Besides the economical evaluation of network expansion strategies, fulfilment of technical boundary conditions needs to be checked during optimization. Technical constraints are identical to those considered during long-term planning and comprise maximum equipment load, voltage, and short-circuit current limits as well as qualitative criteria regarding quality of supply. Again, in most cases, consideration of the $(n-1)$-criterion is sufficient to assure an adequate level of quality of supply.

Technical boundary conditions are tested by carrying out load flow and short-circuit current analyses for every network state that may be reached during the period under consideration. Since network development depends strongly on the future development of uncertainties, a huge number of different network structures may be reached and thus needs to be evaluated. Computational effort for network expansion planning is therefore significantly higher compared to long-term planning.

3 Algorithms for Long-term and Expansion Planning

In recent years, several algorithms for computer-based optimization of network planning problems have been developed. Those algorithms can be classified into *exact* and *heuristic* optimization methods (Neumann and Morlock 1993).

- Exact methods determine the optimal solution of a given maximization or minimization problem in a finite period of time. However, in many cases, computational effort increases exponentially with the number of variables or boundary conditions.
- Heuristic methods limit the size of the solution space that is searched for the optimal solution depending on the available computing time. Thus the best solution found by the algorithm is not necessarily the optimal solution of the complete optimization problem. Nevertheless, in most cases adequate solutions close to the global optimum are found by heuristic methods within a short period of time.

The first algorithms that have been applied for network planning problems were based on exact optimization methods, but in recent years more and more heuristic approaches have been used. The main advantage of the latter is that many alternative solutions may be calculated under different boundary conditions with small computational effort. This enables the use of extensive sensitivity analyses that quantify the coherence between restrictions of network planning and network costs. For example, in long-term planning, several target networks can be calculated for different load scenarios, providing minimum network costs for each scenario.

The main disadvantage of heuristic algorithms compared to the exact ones is the lack of optimality. However, many studies and research projects have proved that the difference between the best solution found by heuristic algorithms and the global optimum of the optimization problem is usually less than 1% of the objective function value (Maurer et al. 2006; Paulun 2006; Michalewicz 1994; Stützle and Dorigo 1999). As the impact of the prediction error that arises from forecasting the development of uncertain boundary conditions is usually significantly higher, differences below this scale can be neglected during network planning.

Because of this, exact optimization algorithms are only rarely used since heuristic methods are available for long-term and for expansion planning of all different voltage levels and network structures. The most frequently used methods are based on *Genetic Algorithms* and *Ant Colony Optimization*, which are explained in the following sections.

3.1 Genetic Algorithms

Genetic algorithms are based on the evolution of creatures in nature and try to approach the optimal solution of the optimization problem in an iterative manner (see Fig. 4) (Michalewicz 1994).

In each iteration, different alternative solutions are created and evaluated. According to evolution in nature, each solution is represented by an *individual*, and the number of individuals created in one iteration is called *population*.

Variables of the optimization problem are represented by *genes*. Every individual consists of a *string of genes* that represents all variables of the optimization problem. A possible solution is therefore unambiguously defined by an individual.

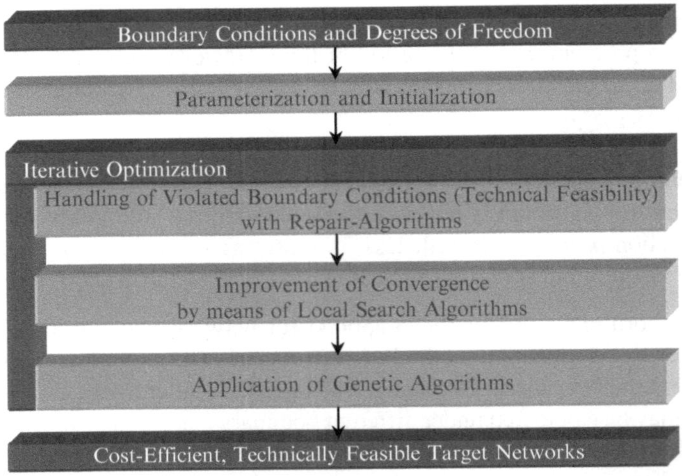

Fig. 4 Iterative optimization by means of genetic algorithms

At the beginning of optimization, the population needs to be initialized by the algorithm. One possibility is to assign a random value to every gene and thus every variable of the optimization problem. This approach is reasonable if nothing is known about good solutions for the given problem. As an alternative, the population may be initialized with possible solutions that have been calculated beforehand. For example, in long-term planning, one target network may be calculated for the expected scenario of uncertain boundary conditions and then be used as the initial value for optimizing target networks for the remaining scenarios. This would lead to alternative network structures that are similar to the target network calculated at first and can thus be realized with small additional effort if the development of uncertain boundary conditions differs from the expected scenario.

In every iteration, all possible solutions represented by the population are evaluated technically and economically. In this step, fulfilment of technical boundary conditions needs to be verified by carrying out load flow and short-circuit current analyses. If violated boundary conditions are identified, *repair algorithms* are used to transfer the individual into the valid solution space. For example, overloading of equipment may be avoided by adding additional lines or transformers to the network structure or by increasing the dimensioning of the affected equipment.

Economical evaluation of individuals yields the annuity network costs of the network represented by the corresponding individual. As a result, individuals can be sorted according to their costs in ascending order. New individuals for the next iteration are then created out of the best individuals known so far.

For creating new individuals, methods like *selection, crossover*, and *mutation* are used analogously to evolution in nature (see Fig. 5). First, individuals who should be considered for reproduction are selected from the population. Second, gene strings of the selected individuals are crossed and combined to new strings for new individuals. Finally, single genes of newly created individuals are mutated in a stochastic

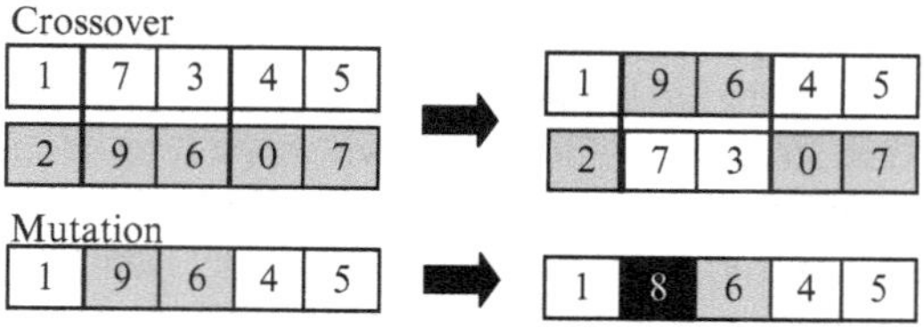

Fig. 5 Crossover and mutation operators in genetic algorithms

manner. This last step assures that new parts of the solution space are taken into account in every iteration and thus avoids convergence in local optima.

Optimization aborts if the best individual with minimum costs could not be improved over several iterations. A boundary value can usually be chosen by the user and depends on the available computing time. On the one hand, many iterations as well as many individuals per population assure that good solutions close to the global optimum are found and local optima are avoided. On the other hand, solutions found after a small number of iterations may be sufficient to estimate the impact of boundary conditions on network planning.

3.2 Ant Colony Optimization

Besides genetic algorithms, *Ant Colony Optimization* has been successfully applied to network planning problems recently. This algorithm has been adapted from the strategy ants use when searching for food (Stützle and Dorigo 1999; Dorigo et al. 1996). It is quite fascinating that a colony of ants – that are cooperating agents from a technical point of view – is capable of finding the shortest path between their nest and a food source in their environment within a short period of time.

To achieve this, every ant marks its way used for searching with *pheromone* that can be detected by other ants. At the same time, following ants choose their individual way for searching according to the pheromone intensity on different paths. Paths with high pheromone intensity are chosen with a higher probability while paths with low pheromone intensity are chosen only rarely. If an ant detects foods, it immediately returns to the nest, marking its way back again.

Since every ant moves approximately with the same speed, short paths between the nest and a possible food source are used more frequently compared to longer paths. As a consequence, the pheromone intensity on short paths increases faster than on the longer paths. Thus short paths are selected by more ants, which in turn increase pheromone intensity and so on. The environment that is used for searching does therefore store information about good solutions that is coded by the pheromone intensity on the available paths. Because of this, the environment is also called *knowledge base* of the algorithm.

This principle of optimization can be used for solving technical optimization problems as well. A solution of the optimization problem is represented by a path between the origin and the destination of searching. In network planning, the possible value of every optimization variable would be part of such a path. For example, a network expansion strategy would define which path to use for developing the existing network in the future. Variables of expansion planning are, as mentioned earlier, the point in time for realizing expansion projects and the type of new lines, transformers, or substations that should be built.

The knowledge base of optimization needs to store information about the potential benefit that is achieved by carrying out given expansion steps at a given point in time when uncertain events occur. Thus the knowledge base consists of n times m entries, if n is the number of uncertain events and m the number of possible expansion steps. The potential benefit is calculated by evaluating the solutions that are analyzed as optimization proceeds and increases with the number of good solutions that link the given expansion step to the given uncertain event. For example, if all solutions that lead to a low net present value of the total costs during the period under consideration would define a certain line to be built if a given uncertain event – for example, the end of the useful lifetime of another line – occurs, the potential goodness and thus the entry in the knowledge base corresponding to this combination would be very high. In Fig. 6, an overview of the iterative optimization procedure is shown.

At the beginning of optimization, the algorithms need to be initialized. Unlike genetic algorithms, new solutions are not created out of the best solutions that have been calculated so far but on the basis of the knowledge base. Thus it is sufficient to initialize the knowledge base for starting the optimization.

Similar to genetic algorithms, ant colony optimization tends to converge against local optima if the first solutions calculated are very similar. It is therefore advantageous to ensure a high variability of the solutions at the beginning of optimization. This is achieved by initializing all entries in the knowledge base with the same value, as this does not prefer certain combinations of expansion steps and uncertain events.

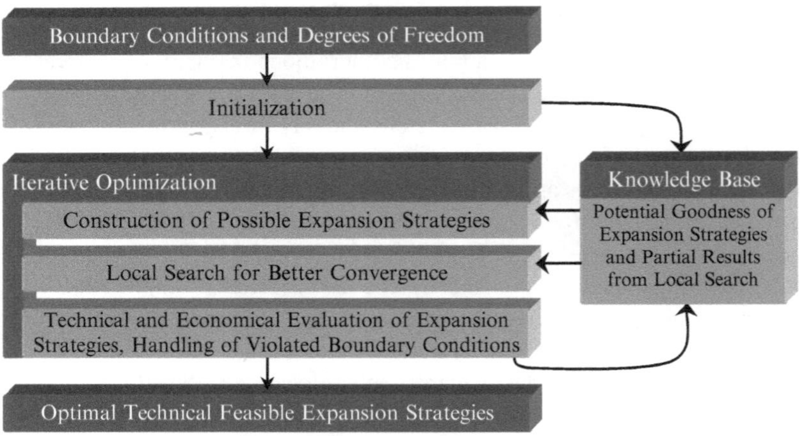

Fig. 6 Iterative optimization by means of ant colony optimization

Additionally, the number of solutions created at the beginning of optimization on the basis of this neutral knowledge base should not be too low. Studies regarding the optimal number of solutions depending on the number of variables are presented in Paulun (2007).

All solutions that have been calculated are evaluated technically and economically. If violations of boundary conditions are detected, the corresponding solution is corrected by means of repair algorithms analogously to genetic algorithms. Finally, the net present value of the total costs during the period under consideration is compared for all technically valid solutions, and the knowledge base is updated according to the information that can be obtained by this comparison. The algorithm then proceeds to the next iteration and calculates new solutions on the basis of the new knowledge base. As a consequence, new solutions are similar to the best solutions calculated beforehand and the algorithm converges against the optimal solution with an increasing number of iterations.

3.3 Comparison of Algorithms

Genetic algorithms and ant colony optimization have different advantages and disadvantages that limit their field of application during network planning. One of the most important differences is that genetic algorithms use the population of one iteration to store information about the goodness of possible solutions, that is, the costs of the network represented by one individual. As a consequence, solutions that have been evaluated in several previous iterations are no longer accessible and thus cannot be used to improve the efficiency of repair algorithms. This leads to the typical situation that repair algorithms need to repair violations of boundary conditions that have already been repaired several times for other individuals. This is particularly critical if the computing time for selecting the optimal measure that fixes the violation with minimum costs is very high.

During long-term planning of electrical networks, violations of boundary conditions can be repaired quite easily. Given the example mentioned earlier, overloading of equipment can be avoided by installing additional lines or transformers or by increasing the dimensioning of equipment. Different alternatives are economically compared with regard to the costs of equipment, and this does not require high computational effort.

In network expansion planning, violations of boundary conditions are more complex. Generally, overloading of equipment may be avoided by the same measures as in long-term planning, but now the additional problem of *when* an additional line or transformers should be built needs to be solved. For example, if the technical evaluation of an expansion strategy reveals that a certain line will be overloaded in 20 years from now on, additional lines may be installed at any point in time in advance. However, this changes the network expansion strategy and does not only influence the overloaded network state under consideration but also any other network state that may be reached after the additional line has been constructed. Thus, correcting violated boundary conditions during network expansion planning requires more computational effort compared to long-term planning. As a consequence, ant colony optimization is used for network expansion planning while genetic algorithms are the most efficient approach for long-term planning of electrical networks.

4 Practical Application of Network Optimization Algorithms

In recent years, several optimization methods mainly based on genetic algorithms or on ant colony optimization have been developed. As the implementation of expert knowledge is a key element of heuristic optimization approaches to achieve effective local search algorithms, those methods need to be adapted to individual characteristics of the optimization task. In particular, the network structure and the voltage level that shall be optimized need to be considered when optimization algorithms are developed and implemented.

High and extra high voltage networks usually consist of a meshed network structure as security of supply is a crucial factor for layout and dimensioning of networks in these voltage levels. In medium voltage networks, usually ring or interconnected network structures are used and low voltage networks are most often built up as radial networks. This topology evolves from the fact that medium and low voltage networks cause a large fraction of the overall network costs, and thus reducing network costs in those voltage levels is very important during network planning. What is more, disturbances in medium and low voltage networks do only affect a local area and do not endanger security of supply in the transmission network.

During network optimization, technical evaluation of different planning alternatives can be simplified if the network is known to be a ring, interconnected, or radial network. This is due to the fact that load flow and short-circuit current analyses does not require a complete system of equations to be solved to calculate branch currents or nodal voltages if the network actually consists of a tree structure. This reduces the computational effort during optimization significantly and leads to very low computing times. For example, medium voltage networks with approximately 300 substations that shall be connected by a ring or an interconnected network structure can be calculated within several minutes on a standard desktop computer. Computing times for optimization of meshed high voltage networks with up to 100 substations are approximately 1–2 h. For extra high voltages, computational effort is even higher and optimization of near-to-practice networks can take up to 24 h. This is also due to the fact that in European extra high voltage networks 380 kV as well as 220 kV can be applied, which increases the degrees of freedom during network optimization significantly.

With the methods presented in this chapter, it is possible to solve the complete network planning process in an objective manner. Unlike before, results of network planning do not depend on subjective influences like operating experience of the network planner. This enables the use of network optimization algorithms for calculating optimal cost-efficient networks and optimal network expansion strategies that may also be used as a benchmark by regulatory authorities. For example, the German regulator uses tools for long-term planning of electrical networks to calculate minimal costs that are required for fulfilling the supply task of a given network operator. However, this application is usually limited due to the high effort that is required for defining all necessary input data and for carrying out network expansion planning.

5 Conclusion

In this chapter, computer-based methods for planning of electricity transmission and distribution networks are presented. The chapter focuses on heuristic optimization algorithms, since the computational effort of those methods is much lower compared to mathematically exact optimization algorithms. Thus for planning problems of practical size, heuristic methods are much more common.

The goal of network optimization algorithms presented in this chapter is to minimize costs for investments, maintenance, and operation of an electricity network in a given area, while fulfilling all boundary conditions of network planning at the same time. Boundary conditions are given by technical, geographical and operational restrictions, for example, maximum thermal currents, voltage and short-circuit current limits, useable routes in the network area, and minimum quality of supply requirements.

Usually, several boundary conditions of network planning are subject to remarkable uncertainties. For example, the development of load and generation or the development of specific investment and maintenance costs is unknown for the period under consideration during network planning. This increases the effort of network planning significantly. To respond to those uncertainties, the process of network planning may be divided into long-term and expansion planning, the first focusing on fundamental planning decisions and the second focusing on short- and medium-term measures.

For long-term planning as well as for expansion planning of electrical networks, computer-based methods are available and have proven excellent performance for practical planning problems during the last decade. Methods for long-term planning are also used by several regulatory authorities in Europe as a benchmark for the efficiency of network system operators. Individual methods are available for all voltage levels and network structures of relevance in reality.

References

Berizzi A, Silvestri A, Zaninelli D, Massucco S (1993) Short-circuit current calculation: A comparison between methods of IEC and ANSI standards using dynamic simulation as reference. Industry Applications Society Annual Meeting, IEEE Conference Record 2:1420–1427

Binato S, Pereira MVF, Grnaville S (2001) A new Benders decomposition approach to solve power transmission network design problems. IEEE Trans Power Syst 16:235–240

Bundesnetzagentur (2006). Bericht der Bundesnetzagentur nach §112a EnWG zur Einführung der Anreizregulierung nach §21a EnWG, http://www.bundesnetzagentur.de/media/archive/6715.pdf

Commission of the European Communities (2009) A European Strategy for Sustainable, Competitive and Secure Energy. http://ec.europa.eu/energy/green-paper-energy/index_en.htm

Cooper W, Tone K (1997) Measures of inefficiency in data envelopment analysis and stochastic frontier estimation. Eur J Oper Res 99(1):72–88

Da Silva EL, Ortiz JMA, De Oliveira GC, Binato S (2002) Transmission network expansion planning under a Tabu Search approach. IEEE Trans Power Syst 16(1):62–68

Dorigo M, Maniezzo V, Colorni A (1996) Ant system: Optimization by a colony of cooperating agents. IEEE Transactions on Systems, Man and Cybernetics 26:29–41

Haubrich H-J (2001) Elektrische Energieversorgungssysteme – Technische und wirtschaftliche Zusammenhänge. Scriptum zur Vorlesung *Elektrische Anlagen*. Institute of Power Systems and Power Economics (IAEW), RWTH Aachen University

Latorre G, Cruz RD, Areiza JM, Villegas A (2003) Classification of publications and models on transmission expansion planning. IEEE Trans Power Syst 18(2):938–946

Maurer C (2004) Integrierte Grundsatz- und Ausbauplanung für Hochspannungsnetze. Doctoral Ph.D. thesis, RWTH Aachen University

Maurer C, Paulun T, Haubrich H-J (2006) Planning of High Voltage Networks under Special Consideration of Uncertainties of Load and Generation. Proceedings of CIGRE Session 41

Michalewicz Z (1994) Genetic algorithms + data structures = evolution programs, 2nd edn. Springer Verlag, Berlin Heidelberg New York

Neimane V (2001) On development planning of electricity distribution networks. Doctoral Ph.D. thesis, Royal Institute of Technology Stockholm

Neumann K, Morlock M (1993) Operations research, Carl Hanser Verlag, Wien, München

Oliveira GC, Costa APC, Binato S (2002) Large scale transmission network planning using optimization and heuristic techniques. IEEE Trans Power Syst 10(4):1828–1834

Paulun T (2006) Strategic expansion planning for electrical networks considering uncertainties. Eur Trans on Electrical Power 16(6):661–671

Paulun T (2007) Strategische Ausbauplanung für elektrische Netze unter Unsicherheit. Doctoral Ph.D. thesis, RWTH Aachen University

Romero R, Monticelli A (1994) A hierarchical decomposition approach for transmission network expansion planning. IEEE Trans Power Syst 9(1):373–380

Sauma EE, Oren SS (2005) Conflicting investment incentives in electricity transmission. IEEE Power Engineering Society General Meeting 3:2789–2790

Sauma EE, Oren SS (2007) Economic criteria for planning transmission investment in restructured electricity markets. IEEE Trans Power Syst 22(4):1394–1405

Stützle T, Dorigo M (1999) ACO algorithms for the quadratic assignment problem. In: Corne D, Dorigo M, Glover F (ed) New ideas in optimization. McGraw-Hill, NY, USA

Ward DJ (2001) Power quality and the security of electricity supply, IEEE Proceedings. http://ieeexplore.ieee.org

Differential Evolution Solution to Transmission Expansion Planning Problem

Pavlos S. Georgilakis

Abstract Restructuring and deregulation have exposed the transmission planner to new objectives and uncertainties. As a result, new criteria and approaches are needed for transmission expansion planning (TEP) in deregulated electricity markets. This chapter proposes a new market-based approach for TEP. An improved differential evolution (IDE) model is proposed for the solution of this new market-based TEP problem. The modifications of IDE in comparison to the simple differential evolution method are the following: (1) the scaling factor F is varied randomly within some range, (2) an auxiliary set is employed to enhance the diversity of the population, (3) the newly generated trial vector is compared with the nearest parent, and (4) the simple feasibility rule is used to treat the constraints. Results from the application of the proposed method on the IEEE 30-bus, 57-bus, and 118-bus test systems demonstrate the feasibility and practicality of the proposed IDE for the solution of TEP problem.

Keywords Differential evolution · Electricity markets · Power systems · Reference network · Transmission expansion planning

1 Introduction

In regulated electricity markets, the transmission expansion planning (TEP) problem consists in minimizing the investment costs in new transmission lines, subject to operational constraints, to meet the power system requirements for a future demand and for a future generation configuration. The TEP problem in regulated electricity markets has been addressed by mathematical optimization as well as by heuristic models (Alguacil et al. 2003; Dechamps and Jamoulle 1980; Latorre et al. 2003; Latorre-Bayona and Pérez-Arriaga 1994; Monticelli et al. 1982; Oliveira

P.S. Georgilakis
National Technical University of Athens (NTUA), GR-15780, Athens, Greece
e-mail: pgeorg@power.ece.ntua.gr

S. Rebennack et al. (eds.), *Handbook of Power Systems I*, Energy Systems,
DOI 10.1007/978-3-642-02493-1_18,

et al. 1995; Padiyar and Shanbhag 1988; Pereira and Pinto 1985; Romero et al. 2002). Mathematical optimization models for TEP problem include linear programming (Garver 1970; Villasana et al. 1985), dynamic programming (Dusonchet and El-Abiad 1973), nonlinear programming (Youssef and Hackam 1989), mixed integer programming (Alguacil et al. 2003; Bahiense et al. 2001), branch and bound (Haffner et al. 2001), Bender's decomposition (Binato et al. 2001), and hierarchical decomposition (Romero and Monticelli 1994). Heuristic models for the solution of TEP problem include sensitivity analysis (Bennon et al. 1982), simulated annealing (Gallego et al. 1997; Romero et al. 1996), expert systems (Teive et al. 1998), greedy randomized adaptive search procedure (Binato et al. 2000), tabu search (da Silva et al. 2001; Gallego et al. 2000; Wen and Chang 1997), genetic algorithms (GAs) (da Silva et al. 2000; Gallego et al. 1998b), and hybrid heuristic models (Gallego et al. 1998a).

There are two main differences between planning in regulated and deregulated electricity markets from the point of view of the transmission planner: (1) the objectives of TEP in deregulated power systems differ from those of the regulated ones, and (2) the uncertainties in deregulated power systems are much more than in regulated ones.

The main objective of TEP in deregulated power systems is to provide a nondiscriminatory and competitive environment for all stakeholders while maintaining power system reliability. TEP affects the interests of market participants unequally and this should be considered in transmission planning. The TEP problem in deregulated electricity markets has been addressed by probabilistic and stochastic methods (Buygi et al. 2003). Probabilistic methods for the solution of TEP problem include probabilistic reliability criteria method (Li et al. 1995), market simulation (Chao et al. 1999), and risk assessment (Buygi et al. 2003, 2004). Stochastic methods for the solution of TEP problem include game theory (Contreras and Wu 2000), fuzzy set theory (Sun and Yu 2000), GA (Georgilakis et al. 2008), and differential evolution (DE) (Georgilakis 2008b).

Nowadays, the TEP problem has become even more challenging because the integration of wind power into power systems often requires new transmission lines to be built (Georgilakis 2008a).

This chapter proposes a general formulation of the transmission expansion problem in deregulated market environment. The main purpose of this formulation is to support decisions regarding regulation, investments, and pricing (Farmer et al. 1995; Kirschen and Strbac 2004; Mutale and Strbac 2000). This chapter proposes an improved differential evolution (IDE) model for the solution of the market-based TEP problem. In particular, the DE algorithm is used to solve the overall TEP problem, whereas in an inner level, that is, for each individual of this evolution-inspired approach, an iterative solution algorithm is required to solve a reference network subproblem.

Evolutionary optimization algorithms have been successfully applied for the solution of difficult power system problems (Georgilakis 2009; Lee and El-Sharkawi 2008). DE is a relatively new evolutionary optimization algorithm (Price et al. 2005; Storn and Price 1997). Many studies demonstrated that DE converges fast and is robust, simple in implementation and use, and requires only a few control parameters. In spite of the prominent merits, sometimes DE shows the premature convergence and slowing down of convergence as the region of global optimum is approached. In this chapter, to remedy these defects, some modifications are made to the simple DE. An auxiliary set is employed to increase the diversity of population and to prevent the premature convergence. In the simple DE, the trial vector, or offspring, is compared with the target vector having the same running index, while in this chapter, the trial vector is compared with the nearest parent in the sense of Euclidean distance. Moreover, the comparison scheme is changed according to the convergence characteristics. The scaling factor F, which is constant in the original DE, is varied randomly within some specified range. The above modifications form an IDE algorithm, which is applied for the solution of TEP problem. The proposed IDE algorithm is extensively tested on the IEEE 30-bus, 57-bus, and 118-bus test systems, and the results of the proposed IDE are compared with the results of the simple DE (Georgilakis 2008b) as well as with the results obtained by the GA method (Georgilakis et al. 2008).

2 Problem Formulation

This section presents a general formulation of market-based TEP problem. The main purpose of this formulation is to support decisions regarding regulation, investments, and pricing (Farmer et al. 1995; Kirschen and Strbac 2004; Mutale and Strbac 2000), and so the main users of this model are regulatory authorities. This formulation is based on the concept of a reference network (Farmer et al. 1995). The determination of such a reference network requires the solution of a type of security-constrained optimal power flow (OPF) problem (Kirschen and Strbac 2004). A market-based TEP problem that optimizes the line capacities of an existing network has been formulated in Kirschen and Strbac (2004) and Mutale and Strbac (2000). This section extends the work presented in Kirschen and Strbac (2004) and Mutale and Strbac (2000) by formulating a more complex market-based TEP problem that optimizes the topology and the line capacities of a transmission network.

2.1 Overall TEP Problem

The objective of the overall TEP problem is to select the new transmission lines that should be added to an existing transmission network (intact system) so as to minimize the overall generation and transmission cost (1), subject to constraints defined by (2)–(9). Alternatively, a different objective also could be considered, such as maximizing the social welfare (de la Torre et al. 2008; Sauma and Oren 2007; Wu et al. 2006).

The objective function of the overall TEP problem is expressed as follows:

$$\min \text{AGTIC} = \min_{w_b,\, P_{pg},\, P_{pg}^c,\, T_b,\, T_b^c,\, F_p^0,\, F_p^c} \left[\sum_{p=1}^{np} \tau_p \cdot \sum_{g=1}^{ng} C_g \cdot P_{pg} + \sum_{b=1}^{nl} w_b \cdot k_b \cdot l_b \cdot T_b \right], \tag{1}$$

where AGTIC ($) is the annual generation and transmission investment cost, P_{pg} (MW) is the output of generator g during demand period p, T_b (MW) is the capacity of transmission line b, np is the number of demand periods, τ_p is the duration of demand period p, ng is the number of generators, C_g is the operating cost of generator g, nl is the number of prospective transmission lines, k_b is the annuitized investment cost for transmission line b in $/(MW km year), l_b is the length of transmission line b in km, and w_b is a binary variable ($w_b = 1$ if line b is built; $w_b = 0$ if line b is not built).

This optimization is constrained by Kirchhoff's current law, which requires that the total power flowing into a node must be equal to the total power flowing out of the node:

$$A^0 \cdot F_p^0 - P_p + D_p = 0 \ , \quad \forall\, p = 1, \ldots, np, \tag{2}$$

where A^0 is the node-branch incidence matrix for the intact system, F_p^0 is the vector of transmission line flows for the intact system during demand period p, P_p is the vector of nodal generations for demand period p, and D_p is the nodal demand vector for period p.

The Kirchhoff's voltage law implies the constraint (3) that relates flows and injections:

$$F_p^0 = H^0 \cdot (P_p - D_p) \ , \quad \forall\, p = 1, \ldots, np, \tag{3}$$

where H^0 is the sensitivity matrix for the intact system.

The thermal constraints on the transmission line flows also have to be satisfied:

$$-T \le F_p^0 \le T \ , \quad \forall\, p = 1, \ldots, np, \tag{4}$$

where T is the vector of transmission line capacities.

It should be noted that the constraints (2)–(4) have been derived using a dc power flow formulation neglecting losses.

The constraints (2)–(4) must also be satisfied for contingencies, that is, for credible outages of transmission and generation facilities. As a result, the constraints (5)–(7) also have to be satisfied:

$$A^c \cdot F_p^c - P_p^c + D_p = 0 \ , \quad \forall\, p = 1, \ldots, np \ ; \ c = 1, \ldots, nc, \tag{5}$$

$$F_p^c = H^c \cdot (P_p^c - D_p) \ , \quad \forall\, p = 1, \ldots, np \ ; \ c = 1, \ldots, nc, \tag{6}$$

$$-T^c \le F_p^c \le T^c \ , \quad \forall\, p = 1, \ldots, np \ ; \ c = 1, \ldots, nc, \tag{7}$$

where A^c is the node–branch incidence matrix for contingency c, F_p^c is the vector of transmission line flows for contingency c during demand period p, P_p^c is the vector of nodal generations for demand period p and contingency c, H^c is the sensitivity matrix for contingency c, T^c is the vector of transmission line capacities for contingency c, and nc is the number of contingencies.

The optimization must respect the limits on the output of the generators:

$$P^{\min} \le P_p \le P^{\max} \quad , \quad \forall\, p = 1, \ldots, np, \tag{8a}$$

$$P^{\min} \le P_p^c \le P^{\max} \quad , \quad \forall\, p = 1, \ldots, np \;\; ; \;\; c = 1, \ldots, nc, \tag{8b}$$

where P^{min} is the vector of minimum nodal generations and P^{max} is the vector of maximum nodal generations.

Since the objective of the optimization is to find the optimal thermal capacity of the lines, these variables can take any positive value:

$$T \ge 0, \tag{9a}$$

$$T^c \ge 0 \quad , \quad \forall\, c = 1, \ldots, nc. \tag{9b}$$

Network security constraints include generator output constraints and line thermal limits (Kirschen and Strbac 2004; Mutale and Strbac 2000), that is, constraints (4), (7), (8a), and (8b). The solution of the optimization problem of (1)–(3), (5), (6), and (9) provides the capacity for pure transport of each line, T^{pt}. On the other hand, the solution of the optimization problem of (1)–(9) provides the optimal capacity of each line, T^{tot}. The capacity for security of each line, T^s, is defined as $T^s = T^{tot} - T^{pt}$.

2.2 Reference Network Subproblem

For a practical power system and for a given number of *nl* prospective transmission lines, the solution of the overall TEP problem by complete enumeration of prospective transmission network topologies is not realistic, that is why it is proposed to solve the overall TEP problem by DE method, whereas in an inner level, that is, for each individual of this evolution-inspired approach, the *reference network subproblem* is formulated and solved.

The *reference network* is topologically identical to an existing (or expanding) transmission network, and the generators and loads are unchanged. The reference network subproblem determines the optimal capacities of transmission lines by minimizing the sum of the annual generation cost and the annuitized investment cost of new transmission lines (10), subject to constraints defined by (2)–(9).

The objective function of the reference network subproblem is expressed as follows (Kirschen and Strbac 2004):

$$\min \mathrm{AGTIC^r} = \min_{P_{pg}, P_{pg}^c, T_b, T_b^c, F_p^0, F_p^c} \left[\sum_{p=1}^{np} \tau_p \cdot \sum_{g=1}^{ng} C_g \cdot P_{pg} + \sum_{b=1}^{nl^r} k_b \cdot l_b \cdot T_b \right], \tag{10}$$

where AGTIC^r ($) is the annual generation and transmission investment cost of the reference network and nl^r is the number of prospective transmission lines of the reference network. It should be noted that, for the reference network, it is supposed that all nl^r lines are built as well as $0 \leq nl^r \leq nl$. It should be also mentioned that for each demand period the reference network subproblem is in fact a type of security-constrained OPF problem (Kirschen and Strbac 2004).

3 Solution of Reference Network Subproblem

Because of its size, the reference network subproblem is solved using the iterative algorithm shown in Fig. 1 (Kirschen and Strbac 2004). At the start of each iteration, a generation dispatch is established and the capacity of each line is calculated in such a way that the demand is met during each period and that the transmission constraints are satisfied. Note that at the beginning of the process there are no transmission constraints. The feasibility of this dispatch is then evaluated by performing a power flow analysis for all contingent networks in each demand period (Kirschen and Strbac 2004). If any of the line flows is greater than the proposed capacity of the

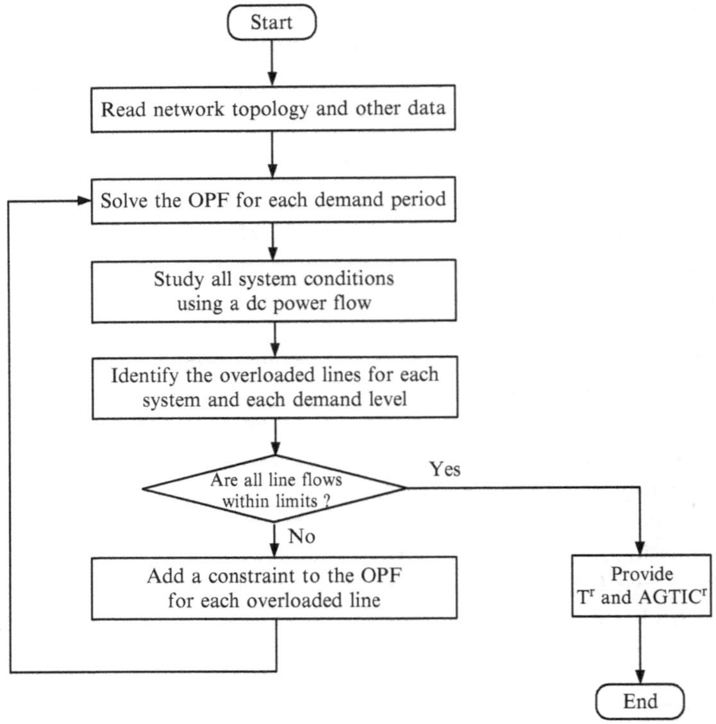

Fig. 1 Flowchart of the algorithm used to solve the reference network subproblem

line, a constraint is created and inserted in the OPF at the next iteration. This process is repeated until all line overloads are eliminated. At the end, the algorithm provides the optimal capacities T^r of the transmission lines and the minimum $AGTIC^r$ for the reference network.

4 Simple Differential Evolution

The procedure of DE is almost the same as that of the GA, whose main process has selection, crossover, and mutation. The main difference between DE and GA lies in the mutation process. In GA, mutation is caused by the small changes in the genes, whereas in DE, the arithmetic combinations of the selected individuals carry out mutation. An additional difference between DE and GA is the order in which operators are used. It should be noted that DE maintains a population of constant size that consists of *NP* real-valued vectors $\mathbf{x}_i^G$, $i = 1, 2, \ldots, NP$, where i indicates the index of the individual and G is the generation index. The evolution process of the DE algorithm is as follows.

4.1 Initialization

To construct a starting point for the optimization process, the population with *NP* individuals should be initialized. Usually, the population is initialized by randomly generated individuals within the boundary constraints

$$x_{j,i}^{0} = rand_{j,i}[0,\ 1] \cdot \left(x_j^{(U)} - x_j^{(L)}\right) + x_j^{(L)}, \tag{11}$$

where $i = 1, 2, \ldots, NP$, $j = 1, 2, \ldots, D$, D is the variable dimension, $x_j^{(L)}$ and $x_j^{(U)}$ are the lower and upper boundary of the j component, respectively, and $rand_{j,i}[0,\ 1]$ denotes a uniformly distributed random value in the range [0, 1].

4.2 Mutation

For each target vector, or parent vector $\mathbf{x}_i^G$, a mutant vector is generated according to

$$\mathbf{v}_i^{G+1} = \mathbf{x}_{n1}^{G} + F \cdot \left(\mathbf{x}_{n2}^{G} - \mathbf{x}_{n3}^{G}\right), \tag{12}$$

where random indexes $n1$, $n2$, and $n3$ are integers, mutually different and also chosen to be different from the running index i. In the initial DE scheme (Storn and Price 1997), the parameter F is a real and constant factor during the entire optimization process, whose range is $F \in (0,\ 2]$.

4.3 Crossover

The trial vector $\mathbf{u}_i^{G+1}$ is generated using the parent and mutated vectors as follows:

$$\mathbf{u}_{j,i}^{G+1} = \begin{cases} \mathbf{v}_{j,i}^{G+1}, & \text{if } \text{rand}_{j,i}[0,\ 1) \leq CR \text{ or } j = k \\ \mathbf{x}_{j,i}^{G}, & \text{otherwise} \end{cases}, \tag{13}$$

where $k \in \{1,\ 2, \ldots,\ D\}$ is the randomly selected index chosen once for each i, and CR is the parameter that is a real-valued crossover factor in the range [0, 1] and controls the probability that a trial vector component comes from the randomly chosen, mutated vector $\mathbf{v}_{j,i}^{G+1}$, instead of the current vector $\mathbf{x}_{j,i}^{G}$. If CR is 1, then the trial vector $\mathbf{u}_i^{G+1}$ is the replica of the mutated vector $\mathbf{v}_i^{G+1}$.

4.4 Selection

To select the population for the next generation, the trial vector $\mathbf{u}_i^{G+1}$ and the target vector $\mathbf{x}_i^{G}$ are compared, and the individual of the next generation $\mathbf{x}_i^{G+1}$ is obtained according to the following rule for minimization problems:

$$\mathbf{x}_i^{G+1} = \begin{cases} \mathbf{u}_i^{G+1}, & \text{if } f\left(\mathbf{u}_i^{G+1}\right) \leq f\left(\mathbf{x}_i^{G}\right) \\ \mathbf{x}_i^{G}, & \text{otherwise} \end{cases}. \tag{14}$$

The feature of DE selection scheme is that a trial vector is compared with only one individual, not all the individuals in the current population. Because of the greedy selection scheme, all the individuals of the next generation are as good as or better than their counterparts in the current generation.

5 Improved Differential Evolution

This section presents the modifications to the simple DE method that lead to an IDE algorithm.

5.1 Scaling Factor F

In the initial DE, the scaling factor F in (12) is constant during the optimization process and F takes values in the range (0, 2]. However, no optimal choice of F has been proposed in the bibliography for DE. All the studies used an empirically derived value, and in most cases F varies from 0.4 to 1. This means F is strongly problem-dependent and the user should choose F carefully after some trial and error tests. In this chapter, F is varied randomly within some specified range, as follows:

$$F = a + b \cdot rand_i[0,\ 1], \tag{15}$$

where a and b are positive and real-valued constants, the sum of a and b is less than 1, and $rand_i[0,\ 1]$ denotes a uniformly distributed random value in the range [0, 1].

Consequently, F is different for each generation, and the computation of F by (15) is effective when the optimal value of F is difficult to be determined for complicated problems like TEP.

5.2 Selection Scheme

In the original DE, the trial vector or offspring $\mathbf{u}_i^{G+1}$ is compared with the target vector $\mathbf{x}_i^G$, whose index is the same as the running index i, using (14). In the modified DE, the trial vector is compared with the nearest target vector in the sense of Euclidean distance. This comparison scheme is employed in the crowding DE algorithm for multimodal function optimization (Thomsen 2004). By this scheme, as the optimization proceeds, the individuals are scattered and gathered around the local optimal points. However, in this chapter, only global optimization is considered, and if there is no improvement of the optimal value during a predefined number of generations, then the comparison scheme is changed to that of the original DE. Therefore, in the initial period of optimization, the DE algorithm explores to find not only global but also local optima, and in the later stage, it searches only for the global optima with greedy selection scheme.

5.3 Auxiliary Set

In the selection of the next generation individual, if the trial vector is worse than the target vector, then the trial vector is discarded. To enhance the explorative search and the diversity of the population, an auxiliary set is employed. The auxiliary set P_a has the same population size *NP*, and the initialization process is the same as that of the main set, using (11). At each generation, if the trial vector $\mathbf{u}_i^{G+1}$ when compared with the corresponding target vector in the main set is found to be worse than its target vector, then the rejected trial vector is compared with the point $\mathbf{w}_i^G$ with the same running index i in the auxiliary set P_a. If $f\left(\mathbf{u}_i^{G+1}\right) < f\left(\mathbf{w}_i^G\right)$, then $\mathbf{u}_i^{G+1}$ replaces $\mathbf{w}_i^G$.

To use the solutions in P_a, after a predefined number of generations, several of the worst solutions in the main set are periodically replaced with the best ones in the auxiliary set by comparing the objective function value.

5.4 Treatment of Constraints

Most optimization problems in the real world have constraints to be satisfied. One common approach to deal with constraints is to penalize constraint violations using an appropriate penalty function (Runarsson and Yao 2000). In this approach, considerable effort is required to tune the penalty coefficients. In this chapter, three selection criteria are used to handle the constraints of the TEP problem:

1. If two solutions are in the feasible region, then the one with the better fitness value is selected.
2. If one solution is feasible and the other is infeasible, then the feasible one is selected.
3. If both solutions are infeasible, then the one with the lowest amount of constraint violation is selected.

It should be noted that the final (best) solution provided by IDE is accepted only if it is feasible; otherwise, the execution of IDE algorithm is repeated.

5.5 Handling of Integer Variables

DE in its initial form is a continuous variables optimization algorithm, and was extended to mixed variables problems (Lampinen and Zelinka 1999). During the evolution process, the integer variable is treated as a real variable, and in evaluating the objective function, the real value is transformed to the nearest integer value as follows:

$$f = f(\mathrm{Y}) \ : \ \mathrm{Y} = y_j, \tag{16}$$

where

$$y_j = \begin{cases} x_j, & \text{if } x_j \text{ is integer} \\ \mathrm{INT}(x_j), & \text{if } x_j \text{ is continuous} \end{cases}, \tag{17}$$

where $\mathrm{INT}(x_j)$ function gives the nearest integer to x_j, and the solution vector is $\mathbf{x} = [x_1, x_2, \ldots, x_D]$.

6 Overview of the IDE Solution to TEP Problem

The IDE algorithm is used to solve the overall TEP problem, whereas in an inner level, that is, for each individual of this evolution-inspired approach, the iterative solution algorithm of Sect. 3 is required to solve the reference network subproblem. In particular, the proposed IDE solution for the market-based TEP problem is composed of the following steps:

1. Given the initial transmission network topology and the planned new generators, create an exhaustive list of candidate new transmission lines.

2. Create an initial population of candidate solutions. The initial population is randomly created from the exhaustive list of candidate new transmission lines using (11).
3. While the termination criterion is not met, the DE algorithm iterates over the following three phases:

   (a) Evaluation of the candidate solutions by solving the reference network sub-problem (Sect. 3)
   (b) Mutation (with randomly varied scaling factor F) and crossover
   (c) Selection by using the auxiliary set concept

4. As soon as the termination criterion is met (maximum number of generations), the solution proposed by the IDE is the one with minimum operating and investment cost, which simultaneously satisfies all the constraints.

Figure 2 presents the flowchart of the proposed IDE solution to TEP problem.

7 Results and Discussion

The proposed IDE algorithm has been extensively tested on the IEEE 30-bus, 57-bus, and 118-bus test systems (PSTCA 1999) that are named as case 30, case 57, and case 118, respectively. In particular, case 30 is a modified version of IEEE 30-bus system (Alomoush 2000; Buygi et al. 2004). Actual cost data of the Hellenic transmission system have been used in the computations. The results of the proposed IDE have been compared with the results of the simple DE (Georgilakis 2008b) as well as with the results obtained by the GA method (Georgilakis et al. 2008). Pentium 4, 3.20 GHz processor was used in the simulations.

7.1 Parameter Values for IDE

The population size and the maximum number of generations are set to 30 and 200, respectively. The best parameter values for IDE were selected after 100 trials of IDE method with varied values of IDE parameters. The average AGTIC of the final solutions for different values of IDE parameters are shown in Table 1. The best settings are $a = 0.4$, $b = 0.4$, and $CR = 0.9$, since they provide the minimum AGTIC for case 30 test system, as shown in Table 1. These settings were also confirmed for case 57 and case 118 test systems.

7.2 Comparison of TEP Methods

7.2.1 Case 30

As can be seen in Fig. 3, the initial transmission network is composed of 32 transmission lines and 28 buses. Bus 11 is a new power plant to be connected to the

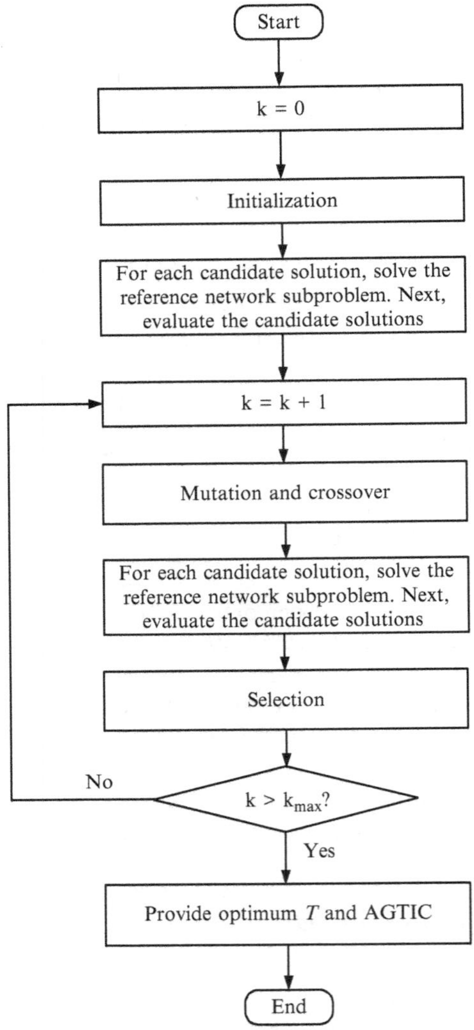

Fig. 2 Flowchart of the proposed improved differential evolution (IDE) solution to transmission expansion planning (TEP) problem

Table 1 Impact of improved differential evolution (IDE) parameters on the computed final solution for case 30 test system

IDE parameters			AGTIC (M$)
a	*b*	*CR*	
0.2	0.3	0.8	7,203
0.3	0.3	0.9	7,153
0.3	0.4	0.8	7,046
0.4	0.4	0.9	7,043
0.4	0.5	0.8	7,114

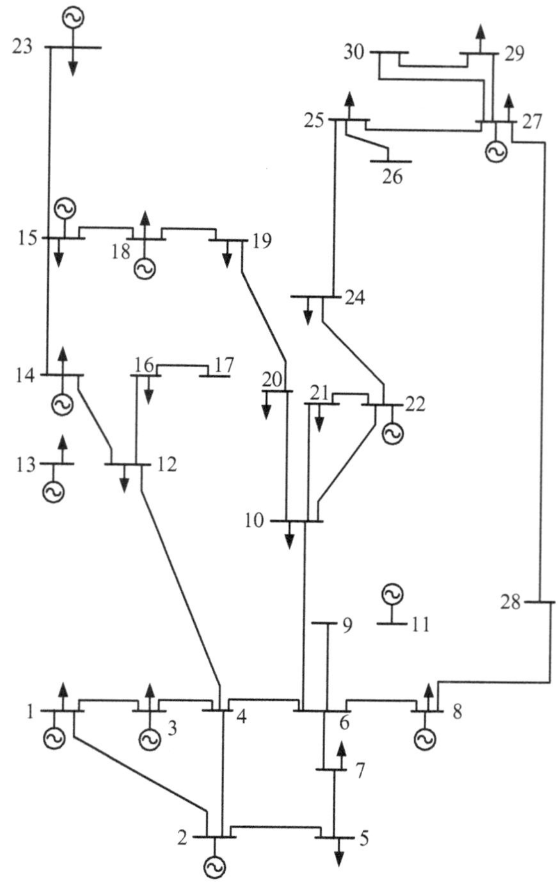

Fig. 3 Single line diagram of the initial transmission network for the modified IEEE 30-bus system

network, and so initially there is no existing transmission line between bus 11 and any bus in the initial network. Bus 13 also corresponds to a new power plant. Table 2 presents the codes of the 32 transmission lines of the initial network of Fig. 3, together with the list of 24 candidate new transmission lines that have been considered for the solution of the transmission expansion problem for the power system in Fig. 3.

The statistic results of the proposed IDE, the simple DE (Georgilakis 2008b), and the GA (Georgilakis et al. 2008) over 100 trials are shown in Table 3. It can be seen in Table 3 that only the proposed IDE technique converges to the best solution, that is, $7,043 million minimum AGTIC. The success rate of IDE is 85%, that is, for 85 times out of the 100 trial runs, the same best solution is obtained. It can be seen from Table 3 that the minimum AGTIC provided by the IDE is 1.2% lower than that obtained by the GA. The application of IDE leads to significant AGTIC savings of $86 million in comparison with GA and $61 million savings in comparison with

Table 2 Transmission lines of the initial network (Type = I), contingencies of transmission lines of the initial network (Type = O), and candidate new transmission lines (Type = C)

Code	Line	Type	Reactance (per unit)	Capacity (MW)	Code	Line	Type	Reactance (per unit)	Capacity (MW)
1	1–2	O	0.0575	250	29	25–27	I	0.2087	15
2	1–3	I	0.1652	100	30	27–28	O	0.3960	50
3	2–4	I	0.1737	60	31	27–29	I	0.4153	15
4	2–5	I	0.1983	100	32	27–30	I	0.6027	15
5	3–4	I	0.0379	90	33	2–6	C	0.1763	70
6	4–6	I	0.0414	80	34	6–28	C	0.0599	25
7	4–12	O	0.2560	50	35	9–10	C	0.1100	30
8	5–7	I	0.1160	40	36	9–11	C	0.2080	20
9	6–7	I	0.0820	40	37	10–17	C	0.0845	15
10	6–8	I	0.0420	40	38	12–13	C	0.1400	15
11	6–9	I	0.2080	40	39	12–15	C	0.1304	25
12	6–10	I	0.5560	25	40	23–24	C	0.2700	15
13	8–28	I	0.2000	10	41	5–6	C	0.1525	50
14	10–20	I	0.2090	20	42	6–11	C	0.1982	40
15	10–21	I	0.0749	25	43	10–11	C	0.1400	30
16	10–22	I	0.1499	15	44	10–12	C	0.0930	20
17	12–14	I	0.2559	15	45	10–16	C	0.0940	30
18	12–16	I	0.1987	15	46	10–28	C	0.0650	50
19	14–15	I	0.1997	15	47	11–28	C	0.2230	25
20	15–18	I	0.2185	15	48	12–18	C	0.1400	30
21	15–23	I	0.2020	15	49	13–14	C	0.2700	20
22	16–17	I	0.1923	15	50	13–16	C	0.2900	20
23	18–19	I	0.1292	15	51	15–16	C	0.1800	25
24	19–20	I	0.0680	15	52	16–18	C	0.1750	30
25	21–22	I	0.0236	15	53	17–20	C	0.2150	20
26	22–24	I	0.1790	15	54	19–24	C	0.1560	20
27	24–25	I	0.3292	15	55	20–24	C	0.1450	30
28	25–26	I	0.3800	15	56	23–25	C	0.1750	30

simple DE. Moreover, both DE methods, the simple DE and the IDE, are faster than the GA method, as Table 3 shows. Consequently, the proposed IDE is very suitable for the solution of the TEP problem.

By applying the proposed IDE method, it has been found that the best-expanded transmission network has selected 7 out of the 24 candidate new transmission lines of Table 2. These 7 transmission lines are shown in Table 4. Figure 4 presents the best-expanded transmission network for the modified IEEE 30-bus system. As can be seen in Fig. 4, the best-expanded transmission network is composed of 39 transmission lines and 30 buses.

Figure 5 presents the capacity for pure transport in each one of the 39 transmission lines of the best-expanded transmission network (Fig. 4) as a percentage of the optimal capacity of the respective transmission line, where the optimal capacity is the sum of two components: (1) the capacity for pure transport and (2) the capacity

Table 3 Comparison of optimization results for the solution of TEP problem

Parameter	Method		
	GA	DE	IDE
Minimum AGTIC (M$)	7, 129	7, 104	7, 043[a]
Minimum AGTIC (% of minimum AGTIC by GA)	100.0	99.6	98.8
Success rate (%)	0	0	85
CPU time (min)	6.3	5.3	5.4
CPU time (% of GA)	100.0	84.1	85.7

[a]$7,043 million is considered as the best solution

Table 4 New transmission lines selected by the proposed IDE

Code	Line
33	2–6
34	6–28
35	9–10
36	9–11
37	10–17
38	12–13
39	12–15

for security. For example, Fig. 5 shows that the transmission line with code 7, that is, the transmission line between buses 4 and 12 (Table 2), has 38% capacity for pure transport, while the rest 62% is its capacity for security. It can be concluded from Fig. 5 that, except for a small number of transmission lines, capacities for pure transport are well below 50% of the optimal capacities even during the period of maximum demand. This observation confirms the importance of taking security into consideration when solving the transmission expansion problem.

7.2.2 Case 57 and Case 118

Figure 6 shows the results obtained by GA, DE, and IDE methods for case 30, case 57, and case 118 test systems. The computing times were 5.4, 21.8, and 87.8 min for case 30, case 57, and case 118 test systems, respectively. As can be seen in Fig. 6, for the three test systems examined, the IDE method is the best as it provides a TEP solution with minimum AGTIC, which is 0.7–1.2% lower than the AGTIC of GA and 0.5–0.8% lower than the AGTIC of DE.

8 Conclusions

A general formulation of the transmission expansion problem in deregulated market environment is proposed in this chapter. The main purpose of this formulation is to support decisions regarding regulation, investments, and pricing. This chapter

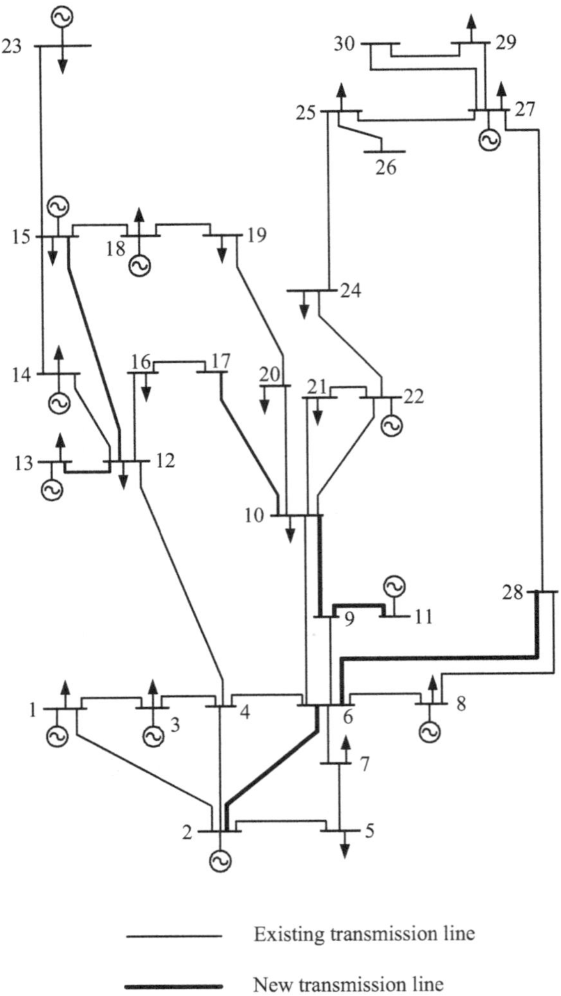

Fig. 4 Single line diagram of the best-expanded transmission network for the modified IEEE 30-bus system

proposes an IDE model for the solution of the market-based TEP problem. The proposed IDE has the following four modifications in comparison to the simple DE: (1) the scaling factor F is varied randomly within some range, (2) an auxiliary set is employed to enhance the diversity of the population, (3) the newly generated trial vector is compared with the nearest parent, and (4) the simple feasibility rule is used to treat the constraints. In particular, the IDE algorithm is used to solve the overall TEP problem, whereas in an inner level, that is, for each individual of this evolution-inspired approach, an iterative solution algorithm is required to solve a reference network subproblem. The proposed method is applied on the IEEE 30-bus, 57-bus, and 118-bus test systems, and the results show that the proposed IDE attains better

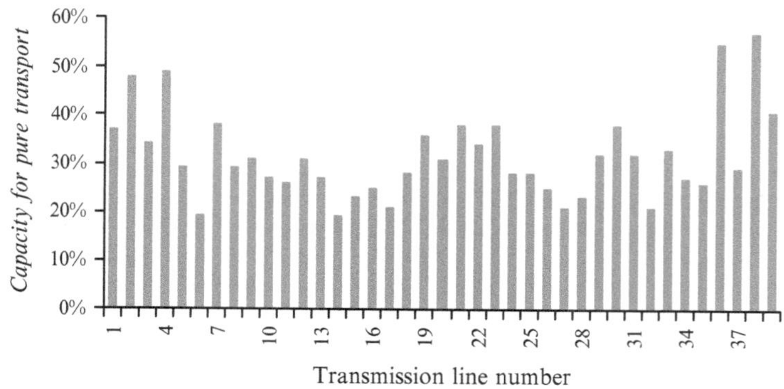

Fig. 5 Capacity needed for pure transport as a percentage of the optimal capacity of each transmission line

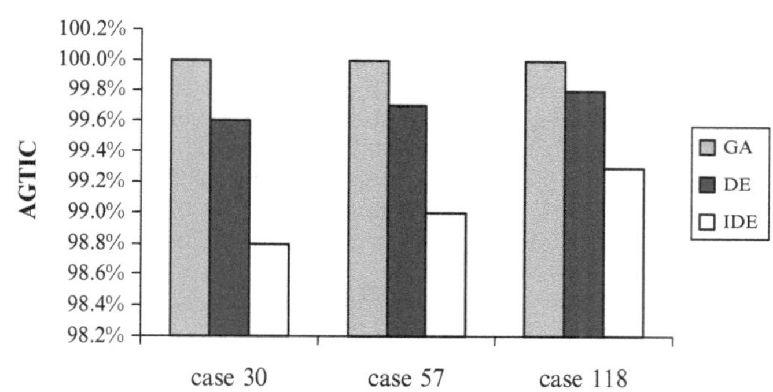

Fig. 6 Annual generation and transmission investment cost (AGTIC) by genetic algorithm (GA), differential evolution (DE), and IDE as a percentage of AGTIC obtained by GA for case 30, case 57, and case 118 test systems

solutions than those found by simple DE and GA. The above four modifications are the possible reasons why IDE outperforms simple DE. Because of its advanced features, IDE also outperforms simple GA. The IDE results show that, except for a small number of transmission lines, capacities for pure transport are well below 50% of the optimal capacities and this observation confirms the importance of taking security into consideration when solving the transmission expansion problem. A proposal for future work includes comparison of the results obtained by IDE with the results of mixed-integer linear programming formulation of TEP problem (Alguacil et al. 2003; Bahiense et al. 2001; de la Torre et al. 2008).

References

Alguacil N, Motto AL, Conejo AJ (2003) Transmission expansion planning: a mixed-integer LP approach. IEEE Trans Power Syst 18(3):1070–1077

Alomoush M (2000) Auctionable fixed transmission rights for congestion management. Ph.D. dissertation, Illinois Institute of Technology

Bahiense L, Oliveira GC, Pereira M, Granville S (2001) A mixed integer disjunctive model for transmission network expansion. IEEE Trans Power Syst 16(3):560–565

Bennon RJ, Juves JA, Meliopoulos AP (1982) Use of sensitivity analysis in automated transmission planning. IEEE Trans Power Apparatus Syst 101(1):53–59

Binato S, de Oliveira GC, de Araújo JL (2000) A greedy randomized adaptive search procedure for transmission expansion planning. IEEE Trans Power Syst 16(2):247–253

Binato S, Pereira MVF, Granville S (2001) A new Benders decomposition approach to solve power transmission network design problems. IEEE Trans Power Syst 16(2):235–240

Buygi MO, Shanechi HM, Balzer G, Shahidehpour M (2003) Transmission planning approaches in restructured power systems. Proc IEEE Bologna Power Tech Conf, Bologna, Italy, June 23–26

Buygi MO, Balzer G, Shanechi HM, Shahidehpour M (2004) Market-based transmission expansion planning. IEEE Trans Power Syst 19(4):2060–2067

Chao XY, Feng XM, Slump DJ (1999) Impact of deregulation on power delivery planning. Proc IEEE Transm Distrib Conf, pp. 340–344, New Orleans, LA, USA, April 11–16

Contreras J, Wu FF (2000) A kernel-oriented algorithm for transmission expansion planning. IEEE Trans Power Syst 15(4):1434–1440

da Silva EL, Gil HA, Areiza JM (2000) Transmission network expansion planning under an improved genetic algorithm. IEEE Trans Power Syst 15(3):1168–1175

da Silva EL, Ortiz JMA, de Oliveira GC, Binato S (2001) Transmission network expansion planning under a tabu search approach. IEEE Trans Power Syst 16(1):62–68

de la Torre S, Conejo AJ, Contreras J (2008) Transmission expansion planning in electricity markets. IEEE Trans Power Syst 23(1):238–248

Dechamps C, Jamoulle E (1980) Interactive computer program for planning the expansion of meshed transmission networks. Electr Power Energy Syst 2(2):103–108

Dusonchet YP, El-Abiad A (1973) Transmission planning using discrete dynamic optimizing. IEEE Trans Power Apparatus Syst 92(4):1358–1371

Farmer ED, Cory BJ, Perera BLPP (1995) Optimal pricing of transmission and distribution services in electricity supply. IEE Proc Generat Transm Distrib 142(1):1–8

Gallego RA, Alves AB, Monticelli A, Romero R (1997) Parallel simulated annealing applied to long term transmission network expansion planning. IEEE Trans Power Syst 12(1):181–188

Gallego RA, Monticelli A, Romero R (1998a) Comparative studies of non-convex optimization methods for transmission network expansion planning. IEEE Trans Power Syst 13(3):822–828

Gallego RA, Monticelli A, Romero R (1998b) Transmission system expansion planning by extended genetic algorithm. IEE Proc Generat Transm Distrib 145(3):329–335

Gallego RA, Romero R, Monticelli AJ (2000) Tabu search algorithm for network synthesis. IEEE Trans Power Syst 15(2):490–495

Garver LL (1970) Transmission network estimation using linear programming. IEEE Trans Power Apparat Syst 89(7):1688–1697

Georgilakis PS (2008a) Technical challenges associated with the integration of wind power into power systems. Renew Sustain Energy Rev 12(3):852–863

Georgilakis PS (2008b) Differential evolution solution to the market-based transmission expansion planning problem. Proc Mediterranean Conf Power Gener Transm Distrib Energy Conversion (MedPower 2008), Thessaloniki, Greece, November 2–5

Georgilakis PS (2009) Spotlight on modern transformer design. Springer, London, UK

Georgilakis PS, Karytsas C, Vernados PG (2008) Genetic algorithm solution to the market-based transmission expansion planning problem. J Optoelectronics Adv Mater 10(5):1120–1125

Haffner S, Monticelli A, Garcia A, Romero R (2001) Specialised branch-and-bound algorithm for transmission network expansion planning. IEE Proc Generat Transm Distrib 148(5):482–488

Kirschen DS, Strbac G (2004) Fundamentals of power system economics. Wiley, Chichester

Lampinen J, Zelinka I (1999) Mixed integer-discrete-continuous optimization by differential evolution, Part 1: the optimization method. Proc 5th Int Conf Soft Computing, pp. 77–81, Brno, Czech Republic, June 9–12

Latorre G, Cruz RD, Areiza JM, Villegas A (2003) Classification of publications and models on transmission expansion planning. IEEE Trans Power Syst 18(2):938–946

Latorre-Bayona G, Pérez-Arriaga IJ (1994) CHOPIN, a heuristic model for long term transmission expansion planning. IEEE Trans Power Syst 9(4):1886–1894

Lee KY, El-Sharkawi MA (2008) Modern heuristic optimization techniques: theory and applications to power systems. Wiley, Hoboken

Li W, Mansour Y, Korczynski JK, Mills BJ (1995) Application of transmission reliability assessment in probabilistic planning of BC Hydro Vancouver South Metro system. IEEE Trans Power Syst 10(2):964–970

Monticelli A, Santos A Jr, Pereira MVF, Cunha SH, Parker BJ, Praça JCG (1982) Interactive transmission network planning using a least-effort criterion. IEEE Trans Power Apparat Syst 101(10):3909–3925

Mutale J, Strbac G (2000) Transmission network reinforcement versus FACTS: an economic assessment. IEEE Trans Power Syst 15(3):961–967

Oliveira GC, Costa APC, Binato S (1995) Large scale transmission network planning using optimization and heuristic techniques. IEEE Trans Power Syst 10(4):1828–1834

Padiyar KR, Shanbhag RS (1988) Comparison of methods for transmission system expansion using network flow and DC load flow models. Electr Power Energy Syst 10(1):17–24

Pereira MVF, Pinto LMVG (1985) Application of sensitivity analysis of load supplying capability to interactive transmission expansion planning. IEEE Trans Power Apparat Syst 104(2): 381–389

Price KV, Storn RM, Lampinen JA (2005) Differential evolution: a practical approach to global optimization. Springer, Berlin

PSTCA (1999) Power systems test case archive. University of Washington. Available: http://www.ee.washington.edu/research/pstca/

Romero R, Monticelli A (1994) A hierarchical decomposition approach for transmission network expansion planning. IEEE Trans Power Syst 9(1):373–380

Romero R, Gallego RA, Monticelli A (1996) Transmission system expansion planning by simulated annealing. IEEE Trans Power Syst 11(1):364–369

Romero R, Monticelli A, Garcia A, Haffner S (2002) Test systems and mathematical models for transmission network expansion planning. IEE Proc Generat Transm Distrib 149(1):27–36

Runarsson TP, Yao X (2000) Stochastic ranking for constrained evolutionary optimization. IEEE Trans Evol Comput 4(3):284–294

Sauma EE, Oren SS (2007) Economic criteria for planning transmission investment in restructured electricity markets. IEEE Trans Power Syst 22(4):1394–1405

Storn R, Price K (1997) Differential evolution – a simple and efficient heuristic for global optimization over continuous spaces. J Global Optim 11(4):341–359

Sun H, Yu DC (2000) A multiple-objective optimization model of transmission enhancement planning for independent transmission company (ITC). Proc IEEE Power Engineering Society Summer Meeting, pp. 2033–2038, Seattle, WA, USA, July 16–20

Teive RCG, Silva EL, Fonseca LGS (1998) A cooperative expert system for transmission expansion planning of electrical power systems. IEEE Trans Power Syst 13(2):636–642

Thomsen R (2004) Multimodal optimization using crowding-based differential evolution. Proc Evol Comput Conf, pp. 1382–1389, Portland, Oregon, USA, June 20–23

Villasana R, Garver LL, Salon SL (1985) Transmission network planning using linear programming. IEEE Trans Power Apparatus Syst 104(2):349–356

Wen F, Chang CS (1997) Transmission network optimal planning using the tabu search method. Elec Power Syst Res 42(2):153–163

Wu FF, Zheng FL, Wen FS (2006) Transmission investment and expansion planning in a restructured electricity market. Energy 31(6–7):954–966

Youssef HK, Hackam R (1989) New transmission planning model. IEEE Trans Power Syst 4(1): 9–18

Agent-based Global Energy Management Systems for the Process Industry

Y. Gao, Z. Shang, F. Cecelja, A. Yang, and A.C. Kokossis

Abstract Energy utility systems are typically responsible for satisfying internal customers (e.g., the various process plants in the industrial complex). The increasing independence of business units in the complex matches an emerging trend in the utility systems to operate for own economic viability and for the encouragement to trade with both internal and external customers. The paper presents a dynamic management system supporting autonomy and the optimal operation of the utility system. The management system comprises three functional components, which support negotiation, short-term (tactical) and long-term (strategic) optimisation. The negotiation component involves an agent-based system exploiting the knowledge base established with real-time and historical data, whereas the optimisation provides a primal front (operational changes) and background front (structural changes) to account for the tactical and strategic decisions.

Keywords Off-line decision support · Multilevel optimization · Multiagent system · Negotiation · Online decision support · Utility system

1 Introduction

Most process plants operate in the context of a Total Site where production processes consume heat and power supplied by a central utility system. The processes consume or generate steam at various levels. Generated steam can be supplied to the steam mains and consumed by other processes. The potential for indirect interaction between processes may lead to significant savings. The problem to define solutions becomes quite complex though. Conventional problem descriptions would assume a set of tasks, steam, and power demands. The central utility system would then

---

A. Yang (✉)
Department of Chemical and Process Engineering, University of Surrey,
Guildford, GU2 7XH, UK
e-mail: a.yang@surrey.ac.uk

S. Rebennack et al. (eds.), *Handbook of Power Systems I*, Energy Systems,
DOI 10.1007/978-3-642-02493-1_19,

be designed to determine optimal pressure levels and optimal configuration of the units involved in the system. In a conventional description of the problem, the utility system addresses the strict needs of the site. In more competitive environments, utility systems, like other parts of the site, operate for their own economic viability. Consequently, they become increasingly independent and are expected to scope for additional profits by trading services outside the premises of their corporate environment. To accomplish such an objective, a utility system needs to monitor both internal and external opportunities and to integrate online trading with decisions on the changes to the system.

The above type of integration, referred to as dynamic management in the sequel, requires proper enabling technology to support decisions and the dynamic trading of supplies. This includes in the first place an efficient negotiation mechanism to support the formation of agreements on service provision between a utility system and its internal and external customers under particular circumstances of demand. Such negotiations require support from a set of modeling and optimisation tools. These tools are expected to compute the economic potential of the utility system under particular deals with its customers, as well as the paths to realizing the potential in terms of the changes to make on the design and/or operation of the utility system. Utility systems modeling and optimisation is a well-researched area, in which a number of useful models and optimisation methods developed, including, for example, those of Papoulias and Grossmann (1983a), Hui and Natori (1996), Mavromatis and Kokossis (1998a,b), Iyer and Grossmann (1998), Varbanov et al. (2004) and Aguilar et al. (2007a,b). In particular, this work will make use of the tools developed by the author's group. Shang and Kokossis (2004) developed systematic approaches to the optimisation of stream levels based on a transhipment model. In Shang and Kokossis (2005), a method is developed for the synthesis and design of utility systems, which is based on an effective combination mathematical optimisation and thermodynamic analysis. In relevance to this present work, the tool that implements the above method supports the optimisation of both the retrofit design as well as the operation of an existing utility system.

The paper presents the conceptual stages of the decision support structure to offer online and off-line support, highlighting service functions that are possible to integrate and install. An agent-based realisation of the decision support structure is reported. Two illustrative examples are finally given to demonstrate applications of the system.

A number of efforts related to the optimisation of utility systems have been reported more recently, most of which focus on specific applications including small-scale CHP by Savola and Fogelholm (2007), multi-fuel boiler by Dunn and Du (2009), emission and biomass utilisation by Martinez and Eliceche (2009) and Mohan and El-Halwagi (2007), and oil refineries by Micheletto et al. (2007) and Zhang and Hua (2007).

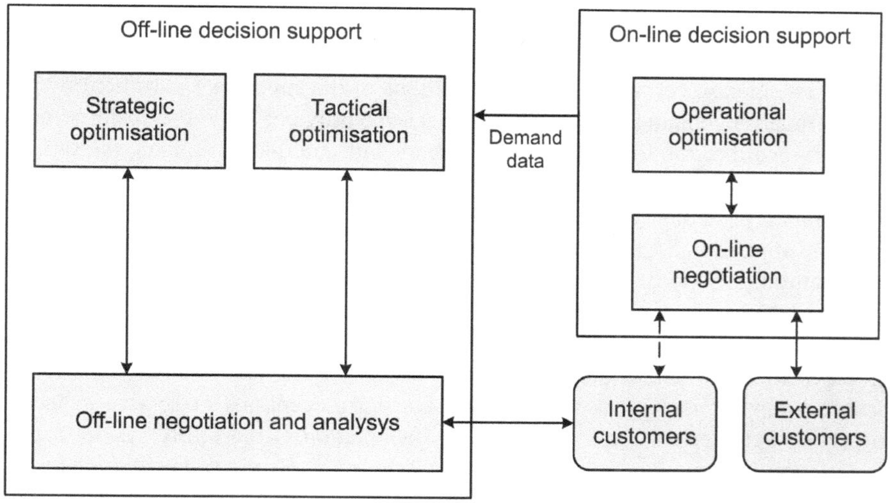

Fig. 1 Conceptual representation of dynamic management system

2 Conceptual Representation of a Dynamic Management System

The paper presumes that a utility service management system should feature capabilities to seize short-term and long-term opportunities in external markets. In reference to its internal customers the system secures quality of service, building capabilities to negotiate on solid evidence as available from data in the utility markets. The management system should interact with internal and external customers providing decision support at both the online and the off-line levels illustrated in Fig. 1. Each level involves utility services negotiation with the customers and optimisation; the latter provides rational input to the former to ensure optimal negotiation outcomes. These two aspects are explained in the following two subsections, respectively.

2.1 Utility Services Negotiation

Negotiation occurs with both internal customers, including the internal process plants (to which the utility system has the obligation to supply steams and electricity), and external customers who may include single users or clusters of users (local or public grids). In general, the utility system has very different relationships with internal and external customers. Accordingly, different negotiation strategies may be applied by the utility system for dealing with these two types of customers.

With external customers the negotiations are speculative, driven by the occasional profitability and power availability, and can be commissioned through online decision support embedded to the day-to-day operation of the utility provision system. Additionally, online negotiation may also take place with some internal customers, in case certain short-term variations of the supply are permitted by the long-term arrangement between the utility service provider and these customers.

With regard to the long term arrangements with internal customers, the typical commitment would be to provide 100% demand at standard prices. Any negotiations that may establish terms different from this typical commitment are dictated by strategic objectives. Instead of being part of the day-to-day operation, such negotiations naturally follow the cycle of internal production planning carried out to provide off-line decision support. In principle, when economics favor the trade with external markets, the utility system may offer discounted prices to internal customers prepared to accept lesser demands from the system. The key task for the off-line decision support is to generate off-line, long term scenarios of service capacity allocation between different customers and adaptation of the utility system itself, and use the information embedded in these scenarios to support negotiations with internal customers.

From a utility system's perspective, the overall objective of the negotiation, either online or off-line, is to always secure the most profitable transactions. While the agreements with customers largely determine the revenue of the utility system, its profit is still affected by production costs. Therefore, the coordination of the negotiation stages with operational (adjusting routine operations), tactical (scoping for profitable assignments), and strategic (building a capacity matching to the need) optimisation becomes essential.

2.2 Utility System Optimisation

As shown in Fig. 1, optimisation supports the two negotiation stages at three different levels. The task of the optimisation at each level is to determine the economic potential of the utility system and the corresponding changes in its operations or structural configuration, in response to the profile for the demand negotiated with the customers.

At the *operational level*, optimisation is performed in response to real-time events that occur during the negotiation with the external (and possible the internal) customers. Based on the existing configuration of the utility system in terms of the steam levels and the number, size, and locations of the units (i.e., boilers, turbines, etc.), the optimisation determines the most profitable operating strategy, usually in terms of the specification of fuel consumption (types and amounts) and the distribution of steams (or the loads of individual turbines). Varbanov et al. (2004) and Shang and Kokossis (2004, 2005) report mathematical models suitable for this type of optimisation. In principle, against the occasional profiles of customer demands, such optimisation produces instant results in terms of the amount of power priced and traded (online function of optimisation).

In contrast, optimisation at the *tactical* and the *strategic* levels is more involved. At the tactical level, the pressures of the steam mains are readjusted from the existing operation specification, triggering a series of rather significant changes to some components of the utility system as well as the served process systems. At the strategic level, more significant changes will be suggested particularly with respect to the structure of the utility system (e.g., number of steam levels, the number, size, and locations of the units). Mathematical models suitable for these two types of optimisation can be found in, for example, Shang and Kokossis (2004, 2005). In comparison with operational optimisation, optimisation at the tactical and strategic levels is usually triggered to address long-term perspectives. The optimisation will take as input the future demand and the pricing scenarios over a sufficiently long period projected on the basis of the historical data or the specification of the deal being negotiated. The output of the optimisation will offer scenarios corresponding to different positions of capital investment and utility service allocation between internal and external customers. These mathematical model-backed scenarios will form a solid basis to support off-line negotiation and decision.

3 Mathematical Formulations, Optimisation Models, and Integration

The optimisation of the daily utility operation could be managed with mathematical programming formulations used widely in industry in the form of LP or SLP models. In general, these models are formulated as follows:

Given

1. A set equipment units (boilers, turbines of different types)
2. The layout of steam headers at different pressures
3. Utility demands for steam and power

Optimally determine

- The steam flows across the headers
- The loadings of each power unit
- The minimum energy required to satisfy the demands (units are assumed of fixed thermodynamic efficiencies)

Relevant models have been discussed extensively with formulations spanning two decades of continuous improvements by Papoulias and Grossmann (1983b), Shang and Kokossis (2004), and a large number of in-house industrial and in-use models available. However, the link with the tactical and the strategic decisions are by no means established. Tactical decisions put the question on item 2 that address the steam pressures, whereas strategic decisions on item 3 that address the number, types, and sizes of the selected units. The remaining of this section is accordingly devoted to explain the optimisation approaches and the models required to support tactical and strategic decisions.

3.1 Level I: Tactical Level

3.1.1 Problem Statement

The problem assumes a given structure for the chemical processes. The objective is to find the optimum locations for the steam levels considering the total site. The proposed method follows the approach presented in Shang and Kokossis (2004).

Different operation scenarios are given for the chemical processes along with forecasts for prices of utilities over a finite number of time periods. The different operation scenarios of the total site can be described by the sets of total site profiles (TSP) proposed by Dhole and Linnhoff (1993). The site heat source rejects heat by raising steam at different levels; the site heat sink absorbs heat also at different levels. The total amount of steam raised by the site heat source does not usually match the amount required by the site heat sink. Because of thermodynamic constraints and heat transfer constraints, auxiliary cooling and heating are required. These are available by the cold utility and the very high pressure (VHP) steam raised by the boiler. The timing of demands changes the profiles of heat source and sink over time and the duration of each time period is usually different.

Single operation scenarios will be discussed first and subsequently be generalised for multiple operation scenarios. As discussed by Shang and Kokossis (2004), if auxiliary fuel-boilers are required, they should operate at the highest-pressure level. The optimisation problem then needs to determine the temperature of the VHP steam, the temperature of each steam level, the auxiliary boiler duty, the cooling utility demand, and the shaft-work produced by the steam turbine network for each expansion zone. By using the turbine hardware model (THM) (Mavromatis and Kokossis 1998b) and boiler hardware model (BHM) developed in this paper (Shang and Kokossis 2004), one is able to target the overall fuel requirement, the cooling utility demand, and the co-generation potential. To obtain the optimal solution for minimum utility cost, we need to identify the correct compromise between heat recovery and co-generation. This can be found through the following optimisation methodology.

3.1.2 Problem Representation

The transportation model determines the optimum transfer of commodities from sources to destinations (Shang and Kokossis 2004). The transhipment model has been widely used in the operation research that deals with the optimum allocation of resources and represents a variation of the transportation problem. Papoulias and Grossmann (1983a) proposed a transhipment model for the synthesis of heat exchanger networks.

In this work, a transhipment network representation is developed for the total site to get the optimal steam levels. The total site heat flows can be represented by total site profiles. As demonstrated in Fig. 2, heat is regarded as a commodity to ship from process heat sources to steam levels and from steam levels to process heat

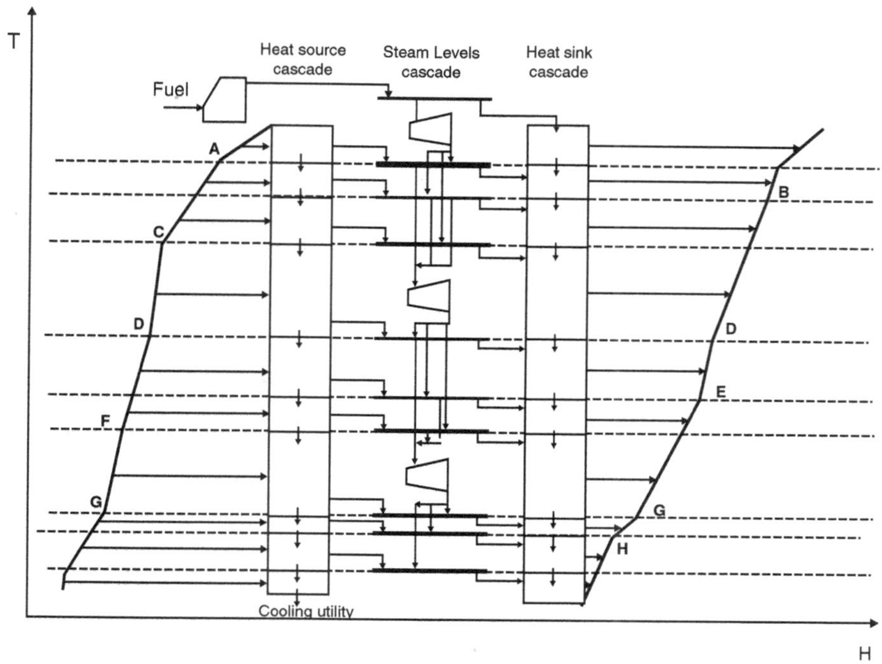

Fig. 2 Transhipment network representation of the total site heat flow

sinks through temperature intervals. These intervals account for thermodynamic constraints in the transfer of heat. In particular, the second law of thermodynamics requires that heat flows only from higher to lower temperatures, and therefore these thermodynamic constraints have to be accounted for in the network model. This is accomplished by partitioning the entire temperature range into temperature intervals. For the total site profiles, the interval temperatures are the temperatures of turning points (critical points) of each heat source and heat sink. These are all candidate locations of the optimum steam levels. The temperatures are listed in descending order. The optimal steam levels will be selected from all potential steam levels denoted by their temperatures.

As shown in Fig. 2, the points A, B, C, D, E, F, . . . are the turning points. This partitioning method guarantees the feasible transfer of heat in each interval, given the minimum temperature approach ΔT_{min}. In this way the total site heat flows are represented by the transhipment network, which comprises three cascades of temperature intervals: (1) heat source cascade, (2) steam level cascade, and (3) heat sink cascade.

The heat source cascade represents that heat flows from process heat sources to the corresponding temperature interval and then to the steam level in the same temperature interval, with residual going to the next lower temperature interval. For the heat sink cascade, it can be considered that heat flows from steam level to the corresponding temperature interval, and then to the process heat sinks in the same temperature interval, with residual going to the next lower temperature interval. The steam level cascade represents that heat flows from process heat sources to the corresponding steam level, and then to the process heat sinks in the same temperature interval, with residual passing a steam turbine to the next steam level.

It is assumed that the total number of steam levels for the site is I. The levels are labeled from the highest level ($i = 1$) down to the lowest level ($i = I$). The temperature range for each level is partitioned into J temperature intervals, which are labeled from the highest interval ($j = 1$) down to the lowest interval ($j = J$). In this way, the entire temperature range of the total site is partitioned into $I \times J$ temperature intervals. The intervals are labeled from the highest interval ($i = 1$, $j = 1$) down to the lowest interval ($i = I$, $j = J$).

The heat flow pattern of the temperature intervals for stream level i can be illustrated as shown in Fig. 3. It is represented by the three heat cascades:

(a) Heat source cascade

1. Heat flows into a particular interval from the process heat sources contributing to the temperature interval.
2. Heat flows out of a particular interval to raise steam with a temperature at the lower bound of the interval.
3. Heat flows out of a particular interval to the next lower temperature interval or the cooling utility. The heat is the residual heat that cannot be utilised in the present interval, and consequently has to flow to a lower temperature interval or the cooling utility.
4. Heat flows into a particular temperature interval from the previous interval that is at higher temperature. This heat is the residual heat that cannot be utilised in the higher temperature interval.

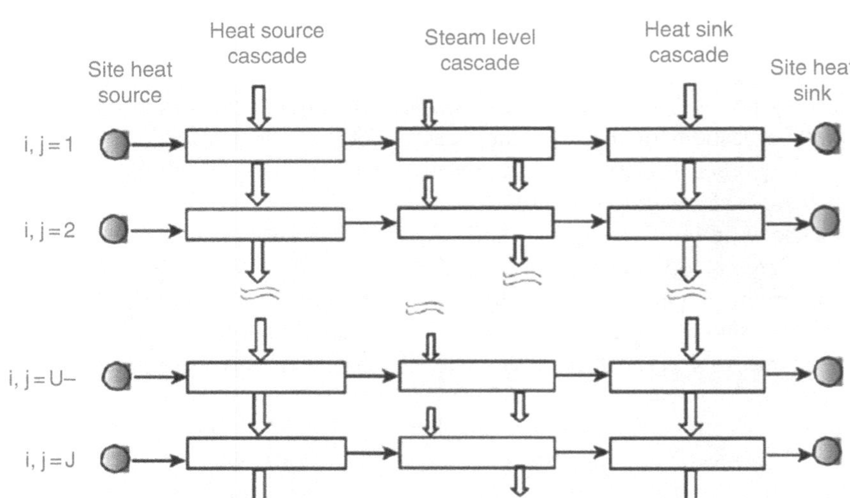

Fig. 3 Heat flow pattern of the temperature intervals for steam level i

(b) Steam level cascade

1. Heat flows into a particular level from the heat source cascade in the same temperature interval and VHP steam.
2. Heat flows out of a particular level to the heat sink cascade in the same temperature interval.
3. Heat flows out of a particular level passing a steam turbine to the next lower temperature steam level.
4. Heat flows into a particular level from the higher temperature steam level. This heat is the residual heat out of steam turbines.

(c) Heat sink cascade

1. Heat flows into a particular interval from the steam level in the same temperature interval or VHP steam.
2. Heat flows out of a particular interval to the process heat sinks within the temperature interval.
3. Heat flows out of a particular interval to the next lower temperature interval. This heat is the residual heat that cannot be utilised in the present interval, and consequently has to flow to a lower temperature interval.
4. Heat flows into a particular temperature interval from the previous interval that is at higher temperature. This heat is the residual heat that cannot be utilised in the higher temperature interval.

Different operations are favored by different sets of steam levels. Since it is impractical to vary the conditions of steam levels between different operation scenarios, the optimisation is searching for the conditions that minimise the total utility cost over the entire set of scenarios. For multiple scenarios the temperature intervals are extracted from each individual scenario following the previous analysis that is based on a single scenario; a general model is constructed next whereby intervals are listed in descending order. The selection of steam levels is made out of all possible cases.

3.1.3 Mathematical Formulation

In this section, the postulated transhipment representation is modeled as a multiperiod MILP model. The model minimises the utility cost for the total site utility system under multiple operation scenarios and incorporates the boiler hardware model (BHM) and turbine hardware model (THM) that predict the reliable equipment performance against a wide range of operating conditions. To develop the multiperiod MILP model, continuous and binary variables are associated with the transhipment network presented in Fig. 2. The binary variables assigned to steam levels represent the existence or nonexistence of the corresponding steam level at a given condition. The binary variables associated to units define the operating status of boilers and steam turbines for each scenario. The continuous variables represent the heat flows across temperature intervals, the boiler duty, the fuel requirement, the cooling utility demand, the power output of each steam turbine, etc. For these sets of parameters and variables, the mathematical model includes (1) heat balance for each temperature interval in the heat source cascade process, (2) heat balance for each temperature interval in the steam level cascade, (3) heat balance for each temperature interval in the heat sink cascade process, (4) heat balance of the heat sink cascade process above the temperature interval (1, 1), and (5) set of operating conditions for each steam level, as specified in Shang and Kokossis (2004). In addition, the model includes the annual cost of the fuel required for the boiler C^f as

$$C^f = \sum_{k \in K} U_k^f Q_k^f T_k^S H, \tag{1}$$

the annual cost of cooling utility C^c as

$$C^c = \sum_{k \in K} U_k^c (R_{I,J,k} + HL_k) T_k^S H, \tag{2}$$

and the annual cost of electricity $C^{p,tot}$ as

$$C^{p,tot} = \sum_{k \in K} U_k^p W_k^d T_k^S H - \sum_{i \in IS} C_i^p - C^{VHP}. \tag{3}$$

Here U_k^f is the unit cost of fuel under certain scenario, Q_k^f is fuel required by the boiler under selected scenario, T_k^S is time fraction of selected scenario, H is operating hours per year, U_k^c is unit cost of cooling utility under selected scenario, HL_k is total heat provided by all process heat sources below temperature interval (i, j), U_k^p is unit cost of electricity under selected scenario, W_k^d is electricity demand of the total site under selected scenario, and C_i^p and C^{VHP} are the savings from power cogeneration.

The objective function MUC used minimises the annual utility cost that includes the cost of fuel and cooling utilities, as well as the cost of electricity, and it is given by

$$min\ MUC = C^c + C^f + C^{p,tot}. \tag{4}$$

It should be noted that used MILP model can also be used to find the optimal steam levels for total site utility systems to minimise the fuel requirement. This can be accomplished by replacing the objective function (4) by

$$min\ MFR = C^f. \tag{5}$$

Normally, the two objectives (4) and (5) define different steam levels and this will be illustrated by Case Study one, which is introduced later in this paper.

The above formulation consists of linear constraints of continuous and integer variables. It comprises a multiperiod MILP model. The problem of synthesising a total site utility system corresponds to a model whose development requires the following information:

- Data on the total site profiles for each operation scenario.
- Specific heat load of VHP steam for each working condition; it is assumed that the specific heat load of steam expanded through a turbine remains approximately constant for all exhaust pressure values. The assumption is based on the observation by Mavromatis and Kokossis (1998b).
- Cost correlations for the available utilities.

3.2 *Level II: Strategic Level*

The problem at strategic level is formulated as follows. Given are

1. A set of chemical processes whose requirements for steam and power are addressed by a system of fixed steam levels
2. A presumed horizon of operation
3. Power and steam demands at each level (as from Level I)
4. A set of units that could generate steam and power.

The design problem is to determine the structure of the site utility system that minimises the total cost so that to satisfy utility demands across the selected operation horizon. The proposed method follows the approach presented in Shang and Kokossis (2005).

3.2.1 Superstructure Development

For each operation scenario the thermodynamic efficiency curve (TEC) is constructed and the curves are applied to identify candidate structures and potential capacities of the utility units. The steps to generate the superstructure are presented as follows.

Superset of Back-pressure (BP) Steam Turbines

As discussed by Mavromatis and Kokossis (1998b), both complex turbines and multistage turbines are equivalent to a cascade of simple turbines, each taking up potential from a single expansion zone, as shown in Fig. 4. On the grounds of the equivalence, all possible combinations of turbine layouts can be reduced to a single superset of component cylinders. This superset of design components is adequate to achieve the targets expected by the BP steam turbine network. The assumption is apparently correct from the thermodynamic point of view, as the total capital cost of a series of simple turbines is more expensive than a complex/multistage turbine

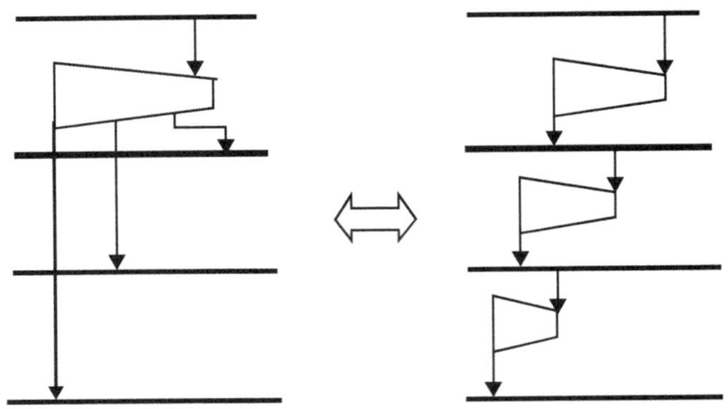

Fig. 4 Complex turbines are considered as a cascade of simple turbines

with the same capacity. Economic benefits to further integrate simple turbines into pass-out and complex turbines are addressed in the postoptimisation stage.

The sizes of component turbines for each scenario are determined at the thermodynamic analysis stage. For multiple operation scenarios the number and sizes of simple component cylinders are identified for each expansion zone by using the discretisation method proposed by Shang and Kokossis (2005).

Superset of Gas Turbines

The capacity of the gas turbine for each scenario is determined by using the TEC. For multiple operation scenarios the number, sizes, and types of candidate gas turbines of the superset depend on the specific problem.

The types of the gas turbine cycles are concerned with simple and regenerative gas turbine cycles. The major difference between the simple and regenerative gas turbine cycles is the addition of a recuperator for heat exchange between the turbine outlet and the compressor outlet, as shown in Shang and Kokossis (2005).

Superset of boilers

The boiler hardware model (BHM) is used to describe the performance of each fired boiler and waste heat boiler (Shang and Kokossis 2005). The design model is given by

$$Q_{fuel} = (C_p \Delta T_{sat} + q)((1 + b)M + aM_{max}), \tag{6}$$

where Q_{fuel} is the fuel requirement, C_p is the specific heat of boiler water, ΔT_{sat} is the temperature difference between the saturation temperature of the steam generated in the boiler and the temperature of the boiler inlet water, q is the specific heat load of the steam generated in the boiler, M is the steam load, M_{max} is the maximum steam load, and a and b are the regression parameters. The expression (6) relates to the fuel flow rate with the steam load and the boiler size. The waste heat is that from gas turbine cycles. The number, sizes, and fuel requirements of the boilers are determined by the optimisation.

Very High Pressure (VHP) Condensing Steam Turbines, Surplus Steam Condensing Turbines, and Reheat Cycles

For each scenario, the power output of the VHP condensing turbine is determined by the thermodynamic efficiency curves (TEC). For a single scenario, the optimum size matches exactly the power demand. For multiple scenarios, the number and sizes of candidate condensing turbines of the superset for multiple operation scenarios are determined by the discretisation scheme followed for the gas turbines.

For each steam level, the surplus heat of the processes is obtained by total site analysis. The number and sizes of the surplus condensing turbines are determined in a similar way as VHP condensing steam turbines. For multiple operation scenarios, the surplus steam condensing turbines are sized to match the demands of each individual scenario as well as all their different combinations.

3.2.2 Optimisation Model

The model incorporates the boiler hardware model (BHM), turbine hardware model (THM), condensing turbine hardware model (CTHM), and gas turbine hardware model (GTHM). The optimisation is a screening tool for the selected alternative design options by using the thermodynamic analysis, rather than for the exhaustive structures. The binary variables account for the selection of units and their operation status at each scenario. The continuous variables relate to the stream flow rates (steam, fuel), the power outputs, and the operating and capital costs. Given these parameters, sets, and variables, the design model includes consideration for the following models: (1) boilers: the BHM yields, (2) BP steam turbines: the THM applied for the power output of a BP steam turbine under selected scenario yields, (3) Condensing steam turbines: the CTHM applied for the power output of a condensing turbine under selected scenario yields, (4) Gas turbines: The GTHM applied for the power output of a gas turbine under selected scenario yields, (5) Steam mass balances: the mass balance across each expansion zone for selected scenario, which involves the steam through the turbines and the steam throttled through the let valves, and (6) Power balance: the electricity balance under selected scenario, as specified in Shang and Kokossis (2005). In addition, the model includes annual costs, that is, the total annual fuel cost of the boilers:

$$C^{B,f} = \sum_{k \in K} U_k^f \, Q_{ib,k}^{B,f} T_k^S H, \tag{7}$$

the total annual fuel cost of the gas turbines:

$$C^{GT,f} = \sum_{k \in K,\, ig \in IG} U_k^{GT,f} F_{ig,k}^f T_k^S H, \tag{8}$$

and the total capital cost incurred for the installation of the equipment:

$$C^{c,tot} = \sum_{i \in l} C_i^c \tag{9}$$

Here, U_k^f is unit cost of fuel for boilers under selected scenarios, $Q_{ib,k}^{B,f}$ is fired fuel load of boiler under selected scenario, T_k^S is time fraction of selected scenario, H is operating hours per year, $U_k^{GT,f}$ is unit cost of fuel for gas turbines under selected scenario, $F_{ig,k}^f$ is fuel load of gas turbine under selected scenario, and C_i^c is the capital cost of each unit.

The objective function that minimises the total annual cost is given by

$$min\ C^{tot} = C^{B,f} + C^{GT,f} + C^{c,tot}. \tag{10}$$

The total annual cost consists of the capital cost and the fuel cost. The optimisation model consists of linear constraints and integer variables and comprises a multiperiod MILP model. The structure and the operation strategy are optimised to minimise the total cost consisting of capital cost and operating cost. The development of the MILP model requires the following information:

- Steam level specifications
- Data on total site profiles for each scenario
- Power demand for each scenario
- Cost correlations for the utilities
- Capital cost correlations for the units.

Even though the primitive approach to the problem yields an MINLP formulation with a very large number of variables, the use of total site analysis and the TCE reduces it into a reasonably sized MILP. The optimisation yields a layout of simple turbines that are post-processed to synthesise complex or multistage turbines. For two cylinders to merge into a complex unit, they both have to be loaded during the same scenario. Depending on whether the steam flow through the upper cylinders is larger or smaller than that in the lower sections, the turbines can be of an extraction or induced type.

4 An Agent-enabled Realisation

The section presents the agent-enabled realisation of the utility service management system envisaged above. In general, a multiagent software system comprises a number of interacting software agents, each of which typically possesses some kind of information or knowledge and realises certain functions as a contribution to the behavior of the entire software system (Wooldridge 2002; Wooldridge and Jennings 1995). In the past, a number of applications of multiagent systems have been reported to support particular tasks in process engineering; a comprehensive review on the research in this area has been given by Batres and Chatterjee et al. (2002). In this present work, software agents are employed to emulate different departments of the total site, individual production processes, the utility system, and trading departments. The agents are embedded into the three-layer architecture illustrated by Fig. 5.

At the foundation layer, the knowledge base comprises process models, heuristic rules, process-related data, and contextual information. The models include mathematical models for utility systems optimisation at different levels discussed in Sect. 2.2. Such models are MILP formulations developed in previous research (Shang and Kokossis 2004, 2005). Data include historical and real-time operation data. Heuristic rules resolve trading decisions customisable to user preferences.

This foundation layer supports a multiagent system, which itself is composed of individual software agents located at two different layers. The top layer comprises of negotiating agents, including a number of customer agents, each representing an internal or external customer, and a broker agent, which implements utility services negotiation as outlined in Sect. 2.1, by means of interacting with the customer agents, coordinating the performance of the task agents, and assisting human decision-making. These agents negotiate according to available trade-offs between satisfying internal customers and making extra profit from external customers. Generally, the higher the external demand (where profit margins are always higher), the lower the potential to discount the internal prices (so that to exploit the potential of the external market). The generation of intermediate negotiation positions is based on a heuristic scheme that creates incremental discounts in prices. Prices are negotiated with internal and external users and could make use of limits to accept discounts or to satisfy contractual obligations to the internal users. The heuristics are simple and do not generate conflicts in the current version of the work.

Task agents, located below the negotiating agents, are designed to perform decision support tasks that accounts for optimisation at different levels and for business scenarios generations; the functioning of these task agents will be illustrated by means of case studies in Sect. 4.

The multiagent system has been developed using JADE (Bellifemine et al. 2005). The communication protocol depends on the type of agents and the information exchanged. Agents to database communication is based on ODBC/JDBC using SQL requests and commands. The cooperating agents communicate through FIPA ACL (2002), supplemented by a simple ontology as the common vocabulary of the agent community (Gruber 1993). The above agents could run on the same or different machines.

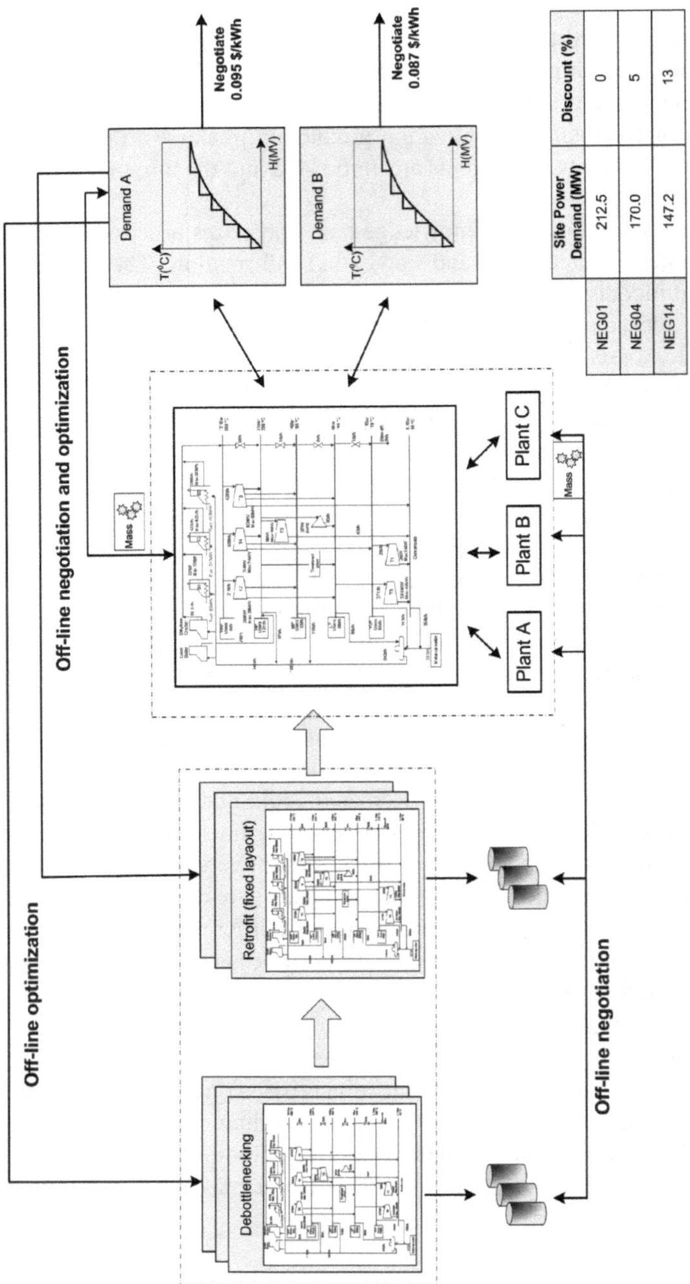

Fig. 5 An agent-enabled realisation of the dynamic management system

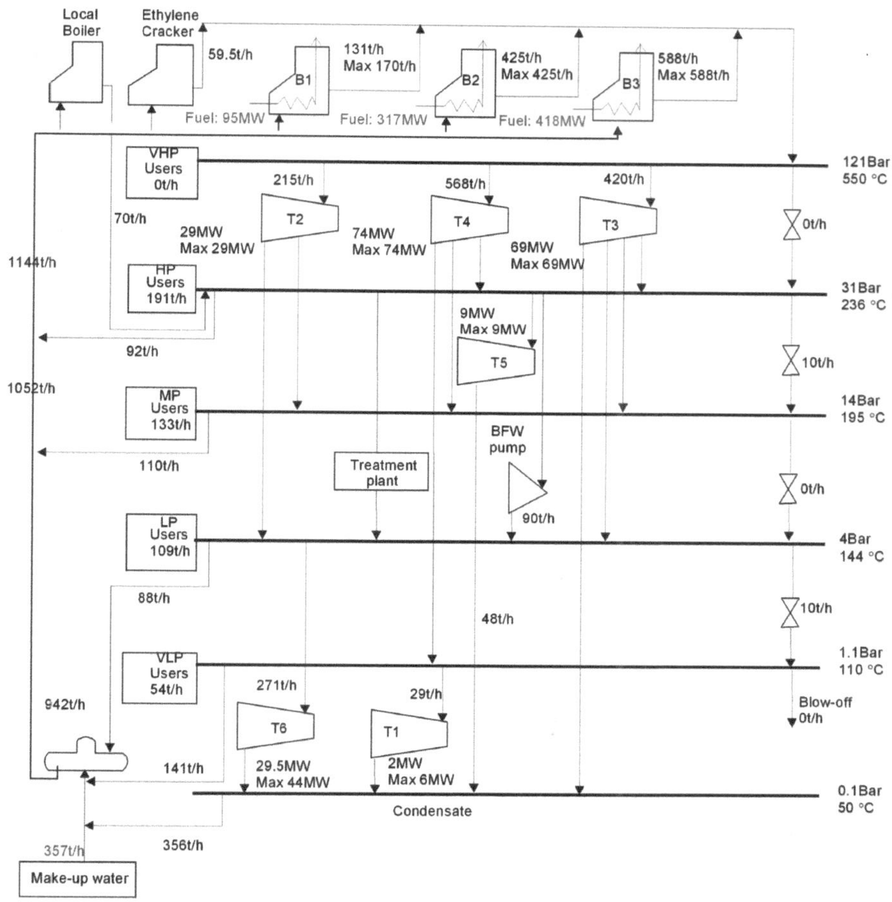

Fig. 6 Utility system utilised in the illustrated examples

5 Illustrative Examples

The application of the utility service management system is illustrated by two simplified case studies on a specific utility system (Fig. 6). Serving four plants as internal customers, this utility system features five steam levels and is composed of five boilers, six steam turbines, and one de-aerator. The first case study focuses on off-line decision support, considering both operational adjustment and retrofit design as possible options in response to external and internal demands. The second case focuses on online decision support and considers only the reconciliation of customer demands for maximum profit. The utility system is interconnected with the public grid. Occasional shortage of power is addressed by importing power. Cost data for the utilities are given in Table 1. The current operating cost is 82.2M$/year. The current site power demand is 212.5 MW. The demands of VLP, LP, MP, and HP

Table 1 Utility cost data (US$/kWh)

Fuel (B1)	Fuel (B1)	Fuel (B1)	Electricity	HP	MP	LP	VLP
0.0095	0.0097	0.0125	0.1	0.038	0.036	0.034	0.032

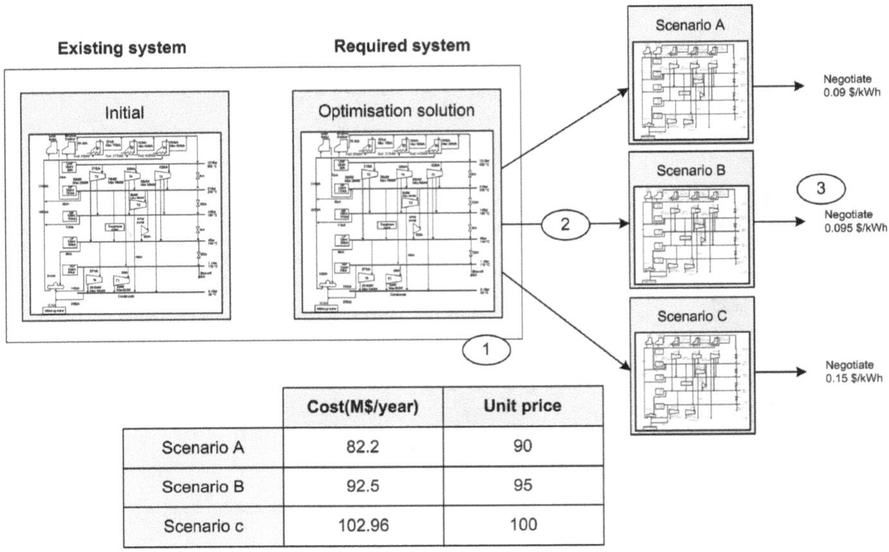

	Cost(M$/year)	Unit price
Scenario A	82.2	90
Scenario B	92.5	95
Scenario c	102.96	100

Fig. 7 Scenarios of case study one

are 40, 80, 100, and 150 MW, respectively. The maximum power to be allowed to export is 65 MW. The maximum capacity of the gas turbines used is 40 MW.

5.1 Case Study One

The objective of this case study is to identify the best investment scheme for the site utility system by negotiating the electricity demand and the price with its internal consumers. The site utility system is optimised for different negotiating scenarios, offering different positions for the decision maker to consider. The scenarios reviewed in the case are shown in the Fig. 7.

Scenario A (no new unit): For this scenario, no units are required. Out of negotiation, a new agreement can be achieved where the utility supplies 80% of the internal electricity demand; internal customers pay 90% of the standard price (0.1 US$/kWh). With this arrangement, the utility system will benefit from exporting 42.5 MW electricity to external users at 0.13 US$/kWh (or higher). Table 2 gives the economics of the scenario.

Table 2 Economics of Scenario A

Fuel (cost) (M$/year)	Revenue of selling electricity (ex.) (M$/year)	Revenue of selling electricity (in.) (M$/year)	Revenue of selling steam (M$/year)	Profit (M$/year)
82.2	48.4	134	116	217

Table 3 Economics of Scenario B

Fuel (cost) (M$/year)	Revenue of selling electricity (ex.) (M$/year)	Revenue of selling electricity (in.) (M$/year)	Revenue of selling steam (M$/year)	Profit (M$/year)
92.5	74	170	116	268

Table 4 Economics of Scenario C

Fuel (cost) (M$/year)	Revenue of selling electricity (ex.) (M$/year)	Revenue of selling electricity (in.) (M$/year)	Revenue of selling steam (M$/year)	Profit (M$/year)
103	74	186	116	274

- *Scenario B (one new gas turbine + one steam turbine)*: One new gas turbine and one steam turbine are required. By negotiating with both internal and external customers, an agreement has been achieved where the utility system supplies 90% of the internal electricity demand; in return, internal users pay only 95% of the standard price. The utility system could export 65 MW electricity to the external users at a price of 0.13 US$/kWh. Table 3 gives the economics of the scenario.
- *Scenario C (one new gas turbine + three steam turbines)*: One new gas turbine and three new steam turbines are required. This is the optimal design obtained by optimizing the utility system satisfying both internal and external demands. The utility system can supply all the internal electricity demand at the standard price and export 65 MW electricity to the external users at a price of 0.13 US$/kWh. Table 4 gives the economics of the scenario.

5.2 *Case Study Two*

Illustrating the function of online decision support, the objective of this case study is to identify the optimal distribution of the electricity produced out of the current operation of the utility system by negotiating the electricity demand and price with each internal consumer and external consumer. The site utility system is optimised for different negotiating scenarios. The scenarios reviewed in the case are shown in Fig. 8.

- *Scenario A (base case)*: For the base case, the utility system supplies only the internal electricity demand at the standard price (0.1 US$/kWh). Thus the

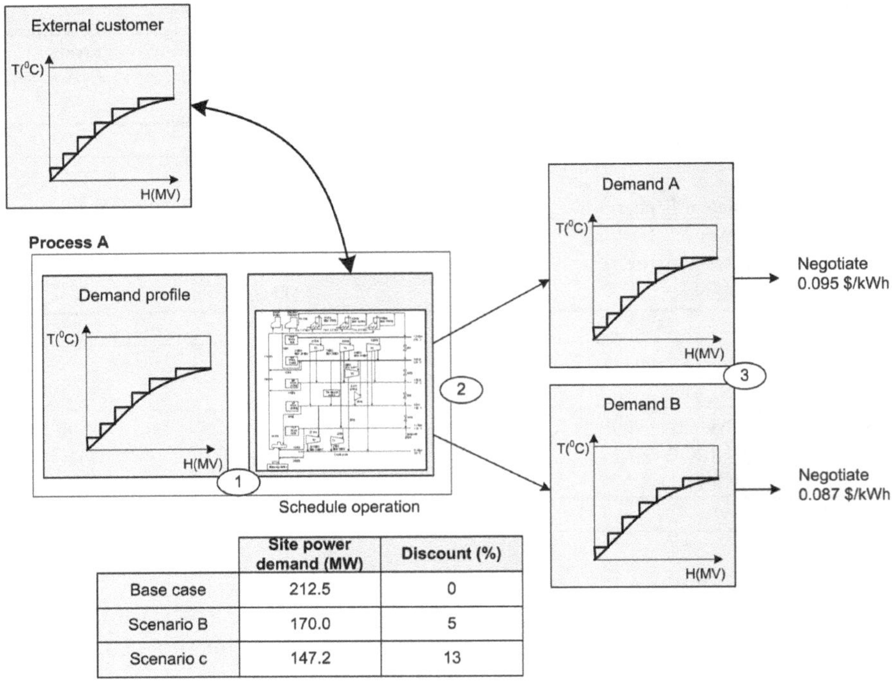

	Site power demand (MW)	Discount (%)
Base case	212.5	0
Scenario B	170.0	5
Scenario c	147.2	13

Fig. 8 Scenarios of case study two

Table 5 Economics of Scenario A

Fuel (cost) (M$/year)	Revenue of selling electricity (ex.) (M$/year)	Revenue of selling electricity (in.) (M$/year)	Revenue of selling steam (M$/year)	Profit (M$/year)
82.2	0	186	116	220

utility system does not have any additional power for export. Table 5 gives the economics of the scenario.

- *Scenario B:* By negotiating with both internal and external customers, the new agreement for Scenario B achieved is as follows: the utility system supplies only 90% of the internal electricity demand; in return, the internal users pay only 95% of the standard price (0.1 US$/kWh). Thus the utility system could export 21.25 MW electricity to the external users at a price of 0.13 US$/kWh. Table 6 gives the economics of the scenario.
- *Scenario C*: By negotiating with both internal and external customers, the new agreement for Scenario B achieved is as follows: the utility system supplies only 70% of the internal electricity demand; in return, the internal users pay only 88% of the standard price (0.1 US$/kWh). Thus the utility system could export 63.75 MW electricity to the external users at a price of 0.13 US$/kWh. Table 7 gives the economics of the scenario.

Table 6 Economics of Scenario B

Fuel (cost) (M$/year)	Revenue of selling electricity (ex.) (M$/year)	Revenue of selling electricity (in.) (M$/year)	Revenue of selling steam (M$/year)	Profit (M$/year)
82.2	24.2	159	116	218

Table 7 Economics of Scenario C

Fuel (cost) (M$/year)	Revenue of selling electricity (ex.) (M$/year)	Revenue of selling electricity (in.) (M$/year)	Revenue of selling steam (M$/year)	Profit (M$/year)
82.2	73	113	116	222

6 Conclusions

Utility networks are challenged to participate in open markets and competitive environments. Conventional formulations typically assume constant demands and search for the most economical ways to address the needs of internal customers. The paper presents the concept of a utility service management system, which offers both online and off-line decision support to enable these networks to uphold more aggressive policies, negotiate and settle prices, manage and coordinate demands from different customers, and settle for the most profitable prices at each time. The work combines optimisation capabilities with the ones in knowledge management, proposing an agent-enabled environment equipped with an authoring system to negotiate and trade. Indicative results of two case studies are provided on the basis of an actual network, demonstrating the underlying principles and potential benefits of such a system.

References

Aguilar O, Perry SJ, Kim J-K, Smith R (2007a) Design and optimization of flexible utility systems subject to variable conditions. Part 1: Modelling Framework. Chem Eng Res Design 85(A8):1136–1148

Aguilar O, Perry SJ, Kim J-K, Smith R (2007b) Design and optimization of flexible utility systems subject to variable conditions. Part 2: Methodology and applications. Chem Eng Res Design 85(A8):1149–1168

Batres R, Chatterjee R, Garcia-Flores R, Krobb C, Wang XZ, Yang A, Braunschweig B (2002) Software agents, In: Braunschweig B, Gani R (eds) Software architectures and tools for computer aided process engineering. Elsevier, Amsterdam, pp. 455–484

Bellifemin e F, Bergenti F, Caire G, Poggi A (2005) Jade – a java agent development framework. In: Bordini RH, Dastani M, Dix J, Seghrouchni AEF (eds) Multi-agent programming languages, platforms and applications. Springer US, pp. 125–147

Dhole VR, Linnhoff B (1993) Total site targets for fuel, co-generation, emissions and cooling. Comput Chem Eng 17:S101

Dunn AC, Du YY (2009) Optimal load allocation of multiple fuel boilers. ISA Trans 48(2): 190–195
FIPA (2002) FIPA ACL message structure specification. Published by Foundation of Intelligent Physical Agents (FIPA). http://www.fipa.org/specs/fipa00061/SC00061G.pDF. Accessed 6 Oct 2006
Gruber TR (1993) A translation approach to portable ontology specifications. Knowl Acquis 5:199–220
Hui CW, Natori Y (1996) An industrial application using mixed integer-programming technique: a multi-period utility system model. Comput Chem Eng 20:s1577–s1582
Iyer RR, Grossmann IE (1998) Synthesis and operational planning of utility systems for multi-period operation. Comput Chem Eng 22:979–993
Martinez P, Eliceche A (2009) Minimization of life cycle CO2 emissions in steam and power plants. Clean Technologies and Environmental 11(1):49–57
Mavromatis SP, Kokossis AC (1998a) Conceptual optimisation of utility networks for operational variations-1: targets and level optimisation. Chem Eng Sci 53:1585–1608
Mavromatis SP, Kokossis AC (1998b) Conceptual optimisation of utility networks for operational variations-2: network development and optimisation. Chem Eng Sci 53:1609–1630
Micheletto SR, Carvalho MCA, Pinto JM (2007). Operational optimization of the utility system of an oil refinery. Comput Chem Eng 32(1–2):170–185
Mohan T, El-Halwagi MM (2007) An algebraic targeting approach for effective utilization of biomass in combined heat and power systems through process integration. Clean Technologies and Environmental Policy 9(1):13–25
Papoulias SA, Grossmann IE (1983a) A structural optimization approach in process synthesis-I utility systems. Comput Chem Eng 7:695–706
Papoulias SA, Grossmann IE (1983b) A structural optimization approach in process synthesis. III. Total processing systems. Comput Chem Eng 7:723
Savola T, Fogelholm CJ (2007) MINLP optimisation model for increased power production in small-scale CHP plants. Appl Therm Eng 27(1):89–99
Shang Z, Kokossis AC (2004) A transhipment model for the optimization of steam levels of total site utilitysy stem for multiperiod operation. Comput Chem Eng 28:1673–1688
Shang Z, Kokossis A (2005) A systematic approach to the synthesis and design of flexible site utility systems. Chem Eng Sci 60:4431–4451
Varbanov PS, Doyle S, Smith R (2004) Modelling and optimization of utility systems, Institution of chemical engineers, Trans IchemE, Part A, May 2004, Chem Eng Res Design 82(A5):561–578
Wooldridge M, Jennings NR (1995) Intelligent agents: theory and practices. Knowl Eng Rev 10(2):115–152
Wooldridge MJ (2002) An introduction to multi-agent systems. Wiley, NY
Zhang BJ, Hua B (2007) Effective MILP model for oil refinery-wide production planning and better energy utilization. J Cleaner Prod 15(5):439–448

Optimal Planning of Distributed Generation via Nonlinear Optimization and Genetic Algorithms

Ioana Pisică, Petru Postolache, and Marcus M. Edvall

Abstract The paper proposes a comparison between a nonlinear optimization tool and genetic algorithms (GAs) for optimal location and sizing of distributed generation (DG) in a distribution network. The objective function comprises of both power losses and investment costs, and the methods are tested on the IEEE 69-bus system. The study covers a comparison between the proposed approaches, the influence of GAs parameters on their performance in the DG allocation problem and the importance of installing the right amount of DG in the best suited location.

Keywords Distributed generation allocation · Genetic algorithms · Nonlinear optimization

1 Introduction

The traditional model of power systems, based on conventional energy sources (mainly coal, oil, natural gas, and nuclear energy) was designed in a centralized manner: electricity was supplied exclusively from generating units to end users through extensive transmission and distribution networks (T&D). Over the years, demand started to grow, and load centers started to spread geographically, forcing utilities to increase electricity production, and also to build new facilities, both in generation and T&D networks to satisfy the new customer requirements.

According to the most recent study of the International Energy Agency (IEA), World Energy Outlook 2008 (World Energy Outlook 2008), the global electricity demand will increase by year 2030 with approximately 45% of present consumption.

Utilities have to make great economical efforts in building and conditioning of facilities so as to accommodate the 1.6% demand increase each year. IEA forecasts

I. Pisică (✉)
University Politehnica of Bucharest, Department of Electrical Power Engineering,
e-mail: ioanapisica@gmail.com

S. Rebennack et al. (eds.), *Handbook of Power Systems I*, Energy Systems,
DOI 10.1007/978-3-642-02493-1_20, © Springer-Verlag Berlin Heidelberg 2010

a generation capacity of 900 GW by 2030, with renewable energy sources (RES) of 500 GW at peak.

The investments in T&D due to ageing assets, expansion, and integration of RES and distributed generation (RES + DG) are estimated at about 600 billion euro (World Energy Outlook 2008).

An attractive solution in satisfying increasing demand and accommodating new generation is to invest locally and progressively in distributed generation. DG installed tactically, delivering electricity where it is needed, avoids or at least postpones and diminishes investments needed for building new facilities.

Furthermore, as will be shown in Sect. 2, DG also has technical benefits regarding power system operation that should also be considered.

As stated in the Electrical Power Research Institute's study forecasts, 25% of new generation will be distributed by 2010. The National Gas Foundation foresees 30% of new generation to be distributed (CIGRE 2000).

The European Renewable Energy Study (TERES) concluded that about 60% of the renewable energy potential that can be exploited by 2010 will be used in decentralized facilities (Grubb 1995). The study was conducted under the commission of the European Union (EU) to examine the feasibility of EU's goals in CO_2 reductions and renewable energy targets. Also, taking into consideration the Kyoto Protocol (United Nations 1997), renewable sources are very likely to overtake an important share of DG, given the incentives towards this goal. For example, as shown by IEA, the EU target shares of renewable energy by year 2020 will reach a maximum of 49%, for Sweden. All countries are expected to reach a greater amount of RES integrated into their power systems (Fig. 1 (World Energy Outlook 2008)).

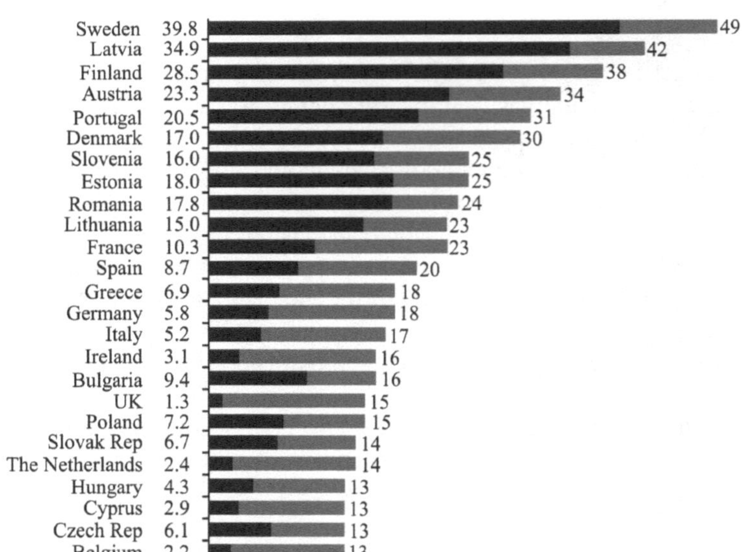

Fig. 1 EU's 2005 and 2020 target shares of renewable energy used

This paper addresses the problem of DG location and sizing. The object of this study is to assess the performances of two computational methods for solving the DG allocation problem, nonlinear optimization and genetic algorithms. The remaining of the paper is structured as follows: Sect. 2 gives an overview of technologies, advantages, and difficulties of DG; Sect. 3 focuses on issues emerging from DG location and sizing, giving a literature review on the subject; Sect. 4 formulates the mathematical model of the optimization problem, describing the objective function and its constraints; Sect. 5 presents the nonlinear optimization algorithm used in this study; Sect. 6 contains the Genetic Algorithms (GAs) approach, giving an extensive problem modeling for GAs; the case study is presented in Sect. 7, in which the computational results from both methods, with a comparison between them, detail the problems that may occur when applying each method. An analysis regarding the optimal location and sizing of DG is also made; conclusions are drawn in Sect. 8.

2 Distributed Generation

Distributed power generation refers to small generating units installed near load centers, avoiding the need to expand the network in order for it to cover new load areas or to uphold the increased energy transfers that would be necessary for satisfying the consumers' demand.

There is no explicit definition adopted at a global level; each organization is giving a different definition of DG, which, in the essence, express the same concept. IEA defines DG as a generating plant serving a customer on-site or providing support to a distribution network, connected to the grid at distribution-level voltage (Hammond and Kendrew 1997). CIGRE defines DG as the generation that has the following characteristics (Lasseter 1998): it is not centrally planned, it is not centrally dispatched at present, it is usually connected to the distribution network, and it is smaller than 50–100 MW. EPRI regards generation units from a few kilowatt up to 50 MW as distributed generation (International Energy Agency 1997).

The concept of DG is quite controversial when it comes to defining it, and a lot of literature has been written on this subject. For example, Ackerman et al. (2000) reviews different issues related to DG and aims at providing a general, consistent definition: distributed generation is an electric power source connected directly to the distribution network or on the customer side of the meter.

All in all, DG is small-scale generation.

DG can be implemented by using several technologies, powered by both conventional and RES. At the beginnings, DG used mainly fossil fuel (such as gasoline, natural gas, propane, methane, etc.), but photovoltaic (PV), wind turbines, and fuel cells are starting to catch-up and seem very promising.

DG can be implemented both in an isolated and an integrated way. In the first case, it only serves the local demand of a consumer, whereas in an integrated manner, it serves the entire electric power network. DG provides benefits for consumers and utilities when connected to a distribution network.

From the utilities' point of view, DG brings two types of benefits: economical (reduce delivery costs by delivering loads locally, reduce the penalty fares to customers resulting from power quality disturbances and interruptions, etc.) and technical (reduce power losses, grid reinforcement, improve system stability and efficiency, etc.).

From the customers' side, DG reduces the electricity bills, improves power quality, reduces emissions, etc. (Borges and Falcao 2006)

At first glance, DG might be considered the answer to many of the problems in today's distribution grids. Looking into more detail, however, there are several issues to be settled: the operational assimilation into the grid and into electricity market mechanisms, network, and protection scheme adaptation. Moreover, the type of DG technology adopted imposes new constraints and limitations.

Of great importance is also the problem of location and sizing of DG, which will be addressed in the following and makes the object of this study.

3 DG Location and Sizing Issues

The operating conditions of a power system after connecting DG sources can change drastically as compared to the base case. The planning of DG installations should, therefore, consider several factors: what would be the best technology to be used, how many units of DG and what capacities, where should they be installed, what connection type should be used, etc.

The problem of DG allocation and sizing should be approached carefully. If DG units are connected at nonoptimal locations, the system losses can increase, resulting in increased costs.

The challenge of identifying the optimal locations and sizes of DG has generated research interests all over the world and many efforts have been made in this direction.

Studies have indicated that inappropriate locations and sizes of DG may lead to higher system losses than the ones in the existing network (Griffin et al. 2000).

Numerous papers have been written on this subject, referring to either "optimal capacity allocation" (Vovos et al. 2005; Keane and O'Malley 2005), "DG placement" (Siano et al. 2007), or even "capacity evaluation" (Harrison and Wallace 2005).

The literature suggests a wide variety of objectives and constraints, but two main approaches emerge: finding optimal locations for defined DG capacity and finding optimal capacity at defined locations.

Of all benefits and objectives of DG implementation, the idea of implementing DG for loss reduction needs special attention. This is why many studies have been performed on this matter, only a few of them being briefly presented in the following.

A very detailed study on the influence of DG location and size upon system losses is given in Acharya et al. (2006). It is shown that, as the size of DG in increased at a particular bus, the losses are reduced, eventually reaching a minimum. If, however, the size is further increased, the losses start to increase as well and may become larger than the ones in the initial network. A conclusion that rises from this study is that DG size should only reach a capacity that can be consumed within the distribution substation boundary. This can be explained by the fact that the distribution system was initially designed for predicted power flows, and the new ones cannot be supported by the small-sized conductors. The need to prepare a methodology that is able to optimally designate DG allocation and sizing within a distribution network arises from the above-stated considerations.

Adaptations of genetic algorithms have been studied in papers like Shukla et al. (2008), Singh et al. (2007), Haensen et al. (2005), and Celli and Pillo (2001), with the objective of minimizing system losses and maintaining acceptable voltage levels.

An analytical approach is used in Shukla et al. (2008) to decide the appropriate DG location, based on losses and sensitivity analysis. Afterwards, a GA is used to compute the optimal DG size to be installed at that location. The objective is to minimize the active power losses and the methodology is tested on the IEEE 69-bus network. Studies are performed for one and two distinct connection points for DG units, showing that smaller capacities lead to less power losses.

The approach presented in Singh et al. (2007) also aims at minimizing the active power losses, but uses GAs to simultaneously search for both location and size of DG. The algorithm is run for different loading conditions (peak, medium, and low), for a 10-bus, 33-bus, and 75-bus system, concluding that losses vary with system loads.

Another study of DG location and sizing for loss minimization is presented in Haensen et al. (2005), looking into CHP and PV generation units in different load conditions, including season changing (summer, winter), based on measurements performed at a residential location. It is also pointed out that any nonoptimal solution, for any given scenario, increases power losses.

A hybrid method, using GA and optimal power flow (OPF) to allocate a predefined number of DG units is presented in Siano et al. (2007). GAs search a large number of combinations of locations, employing OPF to define available capacity for each combination. The novelty of the approach lies in studying the placement of multiple DG units. The algorithm quantifies economical benefits per year at each generation, using normalized geometric ranking selection. Simulations were performed on the IEEE 69-bus test system, considering 3, 5, 7, and 9 DG units to be placed, showing that 9 units would lead to larger benefits.

A multiobjective formulation for the DG allocation and sizing problem is proposed in Celli et al. (2005), using an evolutionary algorithm as solution method. The multiobjective function comprises of network upgrading costs, energy losses costs, cost of energy not supplied, and cost of purchased energy. The ε-constrained method used allows obtaining noninferior solutions. The solution coding is different than the general trend in the reviewed literature, each individual having the length equal to the number of buses in the system. Each gene value can be either 0, if there are no DG units connected to the corresponding bus, or a value from 1 to the maximum size index taken into consideration. Therefore, a predefined fixed number of generator sizes is assumed before the optimization process starts.

Other papers also address the economical concerns in the DG allocation problem, using several solution methods, such as Particle Swarm Optimization (PSO) (Hajizadeh and Hajizadeh 2008), Evolutionary Programming (de Souza and de Albuquerque 2006), or Sequential Quadratic Programming (Le et al. 2007).

Another type of approach to the optimal planning of DG is given in Rosehart and Nowicki (2002), where two optimization formulations are studied: one to determine generator locations based on minimizing operating/clearing costs and one based on enhancing system stability. Lagrangian multipliers associated with the active and reactive power flow equations are used to indicate buses where DG should be installed, using voltage stability constrained OPF formulations. Furthermore, the effect of DG size and location upon spot prices and power system stability is evaluated.

A recent study (Raj et al. 2008) proposes a new approach, based on PSO, considering an objective function consisting of voltage profile improvement index and line loss reduction index. Thus, the solution given by the PSO algorithm increases the maximum loadability of the system. The network used for testing was a 30-bus IEEE system.

Taking into consideration the general trends and framework that can be extracted from the literature (the presentation given here was by no means exhaustive), this paper proposes the methodology based on nonlinear optimization and one based on GAs to optimize the allocation and sizing of DG, so as to minimize the active power losses and investment cost, while certifying acceptable loading conditions and voltage profiles. A comparison is made between the outcomes of the two solvers for the same test system. For comparison accuracy, both implementations share the same power flow routine. Several studies were made concerning the performances of the two methods, as well as regarding the importance of placing DG units of optimal sizes at optimal locations.

4 Problem Formulation

Consider a distribution network, given by its impedance (depending on the characteristics of the conductor material and lengths), topology, and the connected loads.

The objective of this study is to reduce active power losses and investment costs by connecting a DG unit of optimal size, in an optimal location, keeping the voltage and branches loading within acceptable limits. This can be formulated as an optimization problem with the objective function depending on two integer variables: location and size of DG unit.

The general mathematical formulation of an optimization problem can be put as: let *St* be the vector of state variables and let *Sol* be the set of control variables. The problem lays in determining *St* and *Sol* so as to minimize or maximize a certain objective function $f(St, Sol)$ while verifying the following two types of constraints:

$$g(St, Sol) = 0 \text{ (equality)}, \tag{1}$$
$$h(St, Sol) \leq 0 \text{ (inequality)}. \tag{2}$$

4.1 *The Objective Function*

The function that has to be minimized consists of two objectives, one technical and one economical, as follows:

- Minimize the active power losses:

$$\begin{aligned} O_1 &= \sum_{i,j\in k} (P_i - P_j) = \\ &= \sum_{i=1}^{n}\sum_{j=1}^{n} (\frac{R_{ij}\cos(\delta_i - \delta_j)}{V_i V_j}(P_i P_j + Q_i Q_j) \\ &+ \frac{R_{ij}\sin(\delta_i - \delta_j)}{V_i V_j}(Q_i P_j - P_i Q_j), \end{aligned} \tag{3}$$

  where n is the number of buses, R_{ij} is the resistance of line between buses "i" and "j", P_i, Q_i are net real and reactive power injections in bus "i", and V_i, δ_i are the voltage magnitude and angle at bus "i".
- Minimize the investment costs.

The DG investment costs depend on the installed capacity and the cost per installed kW:

$$O_2 = \Omega \cdot P_{DG}, \tag{4}$$

where P_{DG} is the DG installed active power in kW and Ω represents the investment cost. The tests in this study were performed with a value for Ω of 950 UDS/kW (Madarshahian et al. 2009). The DG cost computing is more complex and it has been discussed in detail in Madarshahian et al. (2009), Zeljković et al. (2009), Berg et al. (2008), and Gil and Joos 2006. This, however, is not the object of this study and for simplicity reasons only an average cost is taken into consideration.

If the objective function comprised only of power losses, then the optimization mechanism would have always pointed to solutions near the upper limit of the DG size, as losses decrease when more power is connected. This, however, would not have the meaning of a truly optimal solution, as optimality implies minimum investment costs.

Optimality is achieved when the solution represents a compromise between network benefits and capital investments, and therefore the objective function is constituted by two contradictory objectives. To be able to mathematically aggregate the two objectives of different natures, the first one is also transformed into an economical factor:

$$f(bus, size) = O_1 \cdot ppkWh \cdot \tau + O_2, \tag{5}$$

where τ is the time length taken into consideration and *ppkWh* is the cost per kWh.

Trying to find a minimum value for the objective function, the algorithm will eventually find an optimal solution, which satisfies both objectives.

The time frame considered here was 1 year, in order to make the two terms comparable, as the cost of DG is usually higher than the savings generated by loss reduction. The power losses are computed by a distribution power flow routine.

4.2 Operational Constraints

- *Power flow balance equations*. The balance of active and reactive powers must be satisfied in each node:

$$\begin{aligned} P_i &= P_{DG,i} - P_{Di} \\ &= U_i \sum_{k=1}^{n} [U_k[G_{ik} \cos(\theta_i - \theta_k) + B_{ik} \sin(\theta_i - \theta_k)]] \\ Q_i &= -Q_{Di} \\ &= U_i \sum_{k=1}^{n} [U_k[G_{ik} \sin(\theta_i - \theta_k) + B_{ik} \cos(\theta_i - \theta_k)]], \end{aligned} \tag{6}$$

  where the $P_{DG,i}$ and Q_{Di} represent the active and reactive power at bus "i".
- *Power flow limits*. The apparent power that is transmitted through a branch l must not exceed a limit value, $S_{i\,\max}$, which represents the thermal limit of the line or transformer in steady-state operation:

$$S_i \leq S_{i\,\max}. \tag{7}$$

- *Bus voltages*. For several reasons (stability, power quality, etc.), the bus voltages must be maintained around the nominal value:

$$U_{i\,\min} \leq U_{inom} \leq U_{i\,\max}. \tag{8}$$

  In practice, the accepted deviations can reach up to 10% of the nominal values.
- DG size

$$P_{DG} \leq P_{DG\max}. \tag{9}$$

The optimization problem described by the objective function and constraints detailed in this section represents the mathematical model for the optimal location and sizing of a DG unit in a distribution network, minimizing power losses and investment costs.

5 Nonlinear Optimization Approach

Nonlinear optimization seeks to compute and characterize "near" optimum solutions of nonlinear programs in the presence of multiple local optima, mostly of NP-hard problems, including nonconvex continuous, mixed-integer, differential-algebraic, and nonfactorable. While in nonlinear programming, the quality of the obtained solution is mostly unknown, global optimization also provides a lower (upper) bound for minimization (maximization) problems. Given an objective function f to be minimized (or maximized) and a set of equality and inequality constraints, global optimization has the task of finding a set of parameters (also called variables) that globally minimize (or maximize) f, with theoretical guarantees (Floudas et al. 1999; Floudas and Pardalos 2003).

Global optimization has found an increased number of applications, in advanced engineering design, computational physics, chemistry, biology, data classification and visualization, economic and financial forecasting, environmental risk assessment and management, and numerous other areas. The global optimization strategies have been developing in two directions: exact and heuristic methods; heuristic methods cannot guarantee the quality of the solution, that is, they do not provide lower (upper) bounds for minimization (maximization) problems. Some of the exact methods include naive approaches (Horst and Pardalos 1995; Horst et al. 1995), exhaustive search strategies (Horst and Tuy 1996), homotopy, trajectory methods (Forster 1995), successive relaxation methods (Horst and Tuy 1996), branch and bound algorithms (Hansen 1992; Neumaier 1990), decomposition methods (Rebennack et al. 2009), Bayesian and adaptive stochastic search algorithms (Pinté 1996; Zabinsky 2003; Zhigljavsky 1991). For a review on recent advances in global optimization, see Pardalos and Chinchuluun 2005 and Floudas and Gounaris (2009).

Even though global optimization guarantees the optimality of the computed solution, it often cannot be applied to real world problems due to its limitations in problem size (e.g., number of variables, nonconvex terms, etc.). Furthermore, the whole problem has to be formulated as a mathematical programming model. Theoretical background on nonlinear optimization can be found in Bazaraa and Shetty (1979), Bertsekas (1999), and Himelblan (1972).

A nonlinear optimization technique was chosen over a global optimization approach for solving the optimal DG allocation problem in an electrical distribution network due to the structure of the problem, presented in Sect. 4, which does not allow a straightforward closed form modeling. The studies were conducted by using an advanced and extended implementation of the algorithm DIRECT, developed by Jones, Pettunen, and Stuckman (Jones et al. 1993). This is a modification of the standard Lipschitzian approach that eliminates the need to specify a Lipschitz constant, which is viewed as a weighting parameter that indicates how much emphasis should be put on global search in relation to local search. The search is carried out simultaneously for all possible constants.

The modified DIRECT algorithm was implemented by authors of Björkman and Holmström (1999) and is now part of the TOMLAB optimization toolbox (Holmström 2001; Holmström and Edvall 2004) as the *glcDirect* routine, which is briefly described in the following paragraphs.

The first step in the DIRECT algorithm is to transform the search space into the unit hypercube and sample the function at the center-point of this hypercube. This makes it easier to compute the function value for high dimensional problems, as it is not computed anymore at the vertices. The initial hypercube is divided into smaller hyperrectangles with sampled center-points. Instead of using a Lipschitz constant when determining the next rectangles to sample, the algorithm identifies at each iteration a set of potentially optimal rectangles. All potential rectangles are further divided into smaller ones and their centre-points are sampled. As no Lipschitz constant is used, the convergence cannot be assessed and so this procedure is performed for a predefined number of iterations or until a user-given goal has been achieved. The problem of determining all potentially optimal rectangles implies finding the extreme points on the lower convex hull of a set of points in the plane. This is done by a subroutine based on Rebennack et al. (2009) and it is detailed in Björkman and Holmström 1999.

The implemented algorithm is given in the following, where the tree structure proposed in Jones et al. (1993) for storing the information of each rectangle is replaced by a straightforward matrix/index-vector technique (Björkman and Holmström 1999).

Notations:

C – Matrix with all rectangle center-points
D – Vector with distances from center-point to vertices
F – Function values vector
I – Set of directions with maximum side length for the current rectangle
L – Matrix with all rectangle side lengths in each dimension
S – Index set of potentially optimal rectangles
T – Number of iterations to be performed
f_{min} – Current minimum function value
i_{min} – Rectangle index
ξ – Global/local search weight
δ – New side length in the current divided dimension

Algorithm

Set the global/local search weight parameter ξ
Set $C_{i1} = \frac{1}{2}$ and $L_{i1} = \frac{1}{2}, i = 1, 2, 3, \ldots, n$
Set $F_1 = f(x)$, where $x_i = x_{Li} + C_{i1}(x_{U_i} - x_{L_i}), i = 1, 2, 3, \ldots, n$
Set $D_1 = \sqrt{\sum_{k=1}^{n} L_k^2}$

Set $f_{\min} = F_1$ and $i_{min} = 1$
For $t = 1, 2, 3, \ldots, T$ do
Set $\hat{S} = \left\{ j : D_j \geq D_{imin} \wedge F_j = \min_i \{F_i : D_i = D_j\} \right\}$
Define α and β by letting the line $y = \alpha x' + \beta$ pass through the points (D_{imin}, F_{imin}) and $\left(\max_j (D_j), \min_i \left\{ F_i : D_i = \max_j (D_j) \right\} \right)$
Let $\tilde{S}$ be the set of all rectangles $j \in \hat{S}$ fulfilling $F_j \leq \alpha D_j + \beta + 10^{-12}$
Let S be the set of all rectangles in $\tilde{S}$being extreme points in the convex hull of the set $\{(D_j, F_j) : j \in \tilde{S}\}$
While $S \neq \emptyset$ do
Select j as the first element in S, set $S = S/\{j\}$
Let I be the set of dimensions with maximum side length, i.e. $I = \left\{ i : D_{ij} = \max_j (D_{kj}) \right\}$
Let δ be equal to two thirds of the maximum side length, i.e. $\delta = \frac{2}{3} \max_k (D_{kj})$
For all $i \in I$ do
Set $c_k = C_{kj}, k = 1, 2, 3, \ldots, n$
Set $\hat{c} = c + \delta e_i$ and $\tilde{c} = c - \delta e_i$ where e_i is the ith unit vector
Compute $\hat{f} = f(\hat{x})$ and $\tilde{f} = f(\tilde{x})$ where $\hat{x}_k = x_{L_k} + \hat{c}_k (x_{U_k} - x_{L_k})$ and $\tilde{x}_k = x_{L_k} + \tilde{x}_k (x_{U_k} - x_{L_k})$
Set $\omega_i = \min(\hat{f}, \tilde{f})$
Set $C = (C\ \hat{c}\ \tilde{c})$ and $F = (F\ \hat{f}\ \tilde{f})$
End for
While $I \neq \emptyset$ do
Select the dimension $i \in I$ with the lowest value of ω_i and set $I = I \backslash \{j\}$
Set $L_{ij} = \frac{1}{2}\delta$
Let $\hat{j}$ and $\tilde{j}$ be the indices corresponding to the points $\hat{c}$ and $\tilde{c}$, i.e. $F_{\hat{j}} = \hat{f}$ and $F_{\tilde{f}} = \tilde{f}$
Set $L_{k\hat{j}} = L_{kj}$ and $L_{k\tilde{j}} = L_{kj}, k = 1, 2, 3, \ldots, n$
Set $D_j = \sqrt{\sum_{k=1}^{n} L_{kj}^2}$
Set $D_{\hat{j}} = D_j$ and $D_{\tilde{j}} = D_j$
End while
End while
Set $f_{\min} = \min_j (F_j)$
Set $i_{\min} = \arg\min \left(\frac{F_j - f_{\min} + E}{D_j} \right)$, where $E = \max(\xi\, |f_{\min}|\, , 10^{-8})$
End for

In recent work the *glcDirect* algorithm has been extended to handle binary and integer variables. It is also possible to supply a set of linear constraints and enforce feasibility for each function evaluation.

A more detailed implementation of *glcDirect* and a comparison with the DIRECT algorithm is given in Björkman and Holmström (1999) and Holmström (2001). The computational results for the DG allocation problem using this algorithm are given in Sect. 7.

6 Genetic Algorithms Approach

GAs are a way of solving problems by emulating the mechanism of evolution as found in natural processes. They use the same principles of selection, recombination, and mutation to evolve a set of solutions toward a "best" one.

Genetic algorithms have their starting point in John Holland's studies on cellular automata, at the University of Michigan. He was also the first to explicitly propose crossover and other recombination operators. However, the expansion of work in the GAs field started in 1975, with the publication of the book *Adaptation in Natural and Artificial Systems* (Holland 1975). Research in GAs remained largely theoretical until the 1980s, when the First International Conference on Genetic Algorithms was held at the University of Illinois. Since then, genetic algorithms have been applied to a broad range of subjects, from abstract mathematical problems like knapsack and graph coloring to tangible engineering issues (Mitchell 1996; Goldberg 1989; Baker 1987; Whitely 1989). In power systems, GAs' applications include economic dispatch, unit commitment, reliability studies, resource allocation problems, system planning, maintenance scheduling, state estimation, FACTS devices control, stability studies, OPF, and many others.

Because of the intrinsic parallelism of GAs, they can explore the search space in multiple directions at the same time. This is why they are suited for nonlinear problems of high complexity and dimensionality, for which the objective function (transposed into a fitness measure) has a complex look (discontinuous, noisy, time-dependent, with many local optima). GAs avoid getting trapped into local optima because they use populations of candidate solutions and therefore there are always multiple comparison values, unlike, for example, hill-climbing or gradient-based methods (Michalewicz 1996).

Before using any of the GA models, the problem must be represented in a suitable format that allows the application of genetic operators. GAs work by optimizing a single entity, the fitness function. Hence, the objective function and the constraints of the problem at hand must be transformed into some measure of fitness.

Encodings. The first feature that should be defined is the type of representation to be used, so that an individual represents one and only one of the candidate solutions. A candidate solution (or chromosome) designed in this paper for the problem of finding the optimal location and size of one DG unit is a two-component vector (Fig. 2a). The first component represents the location, the node in which the DG

a

Position(node number)	Size (max. 2 MW)

b

Unit 1 Position(node number)	Unit 1 Size (max. 2 MW)	Unit 2 Position(node number)	Unit 2 Size (max. 2 MW)

Fig. 2 Chromosome encoding for one DG unit (**a**) and two DG units (**b**) to be allocated

should be connected, and can take values from 1 to the number of buses in the network. The second component represents the DG size and can take values from 0 to 2,000 kW. For two DG units to be allocated, a new pair of genes is added to the chromosome, like in Fig. 2b. A population of possible solutions will be evolved from one generation to another to obtain an optimum setup, that is, a very well fitted individual.

Fitness Function. This function is responsible for measuring the quality of chromosomes and it is closely related to the objective function. The objective function for this paper is computed using (5). The constraints of this particular problem do not explicitly contain the variables (the genes in this case) and therefore the effect of the constraints must be included in the value of the fitness function. The constraints are checked separately and the violations are handled using a penalty function approach. The overall fitness function designed during this study is

$$f(x) = O_1 \cdot ppkWh \cdot \tau + O_2 + \xi \cdot \sum_{i=1}^{nr} bal_i + \zeta \cdot \sum_{k=1}^{n} thermal_k + \vartheta \cdot \sum_{k=1}^{n} voltage_k, \quad (10)$$

where the first two terms are the ones in the objective function and the following are penalty functions. The element bal_i is a factor equal to 0 if the power balance constraint at bus i is not violated and 1 otherwise. The sum of these violations represents the total number of buses in the network that do not follow constraint (6) and it is multiplied by a penalty factor meant to increase the fitness function up to an unacceptable figure, therefore making the solution unfeasible. The second and third sums in the fitness function represent the total number of violations of constraints (7) and (8), respectively, and they are also multiplied by penalty factors. The last three sums in this fitness function are a measure of unfeasibility for each candidate solution x. The penalty factors used in this study were set to 10,000.

The constraint expressed in (9) is satisfied for each run, as the limits for each individual are set within the main GA routine: the first component (location) varies between 0 and the number of buses and the second component (size) varies according to (9).

The genetic algorithm proposed for solving the optimal DG placement under the above-described problem formulation can be written in the following simplified form (Pisica et al. 2009):

Begin

Read network data
Run power flow and store results for base case
Encode network data
Set genetic parameters
Create initial population
While <stopping condition not met> execute
For each individual in current generation
Run power flow and evaluate fitness
EndFor
Select(current_generation, population_size)
Crossover(selected_parents, crossover_rate)
Mutation(current_generation, mutation_rate)
current_generation ++
EndWhile
Show solution
End.

Selection Methods. The selection methods specify how the genetic algorithm chooses parents for the next generation. In this study, two selection methods were tested. The first method was Roulette Wheel Selection, which chooses parents by simulating a roulette wheel with different-sized slots, proportional to the individuals' fitness. The second method tested was tournament selection. Each parent is chosen as the best individual from a random selection of k individuals, where k is a preset number – here 2 proved to be a suited tournament size.

Crossover Mechanism. The one-point and scattered crossover mechanisms were tested in this study. The one-point crossover exchanges the genetic information found after a random position in the two selected parents. The scattered crossover mechanism works as follows: for each pair of selected parents, the algorithm generates a set of binary components. The number of components is equal to the number of genes in an individual. This is a mask that will guide the crossover: if the mask value for the ith gene is 0, then this gene of the offspring will inherit the ith gene from the first parent, otherwise, the corresponding gene from the second parent. This mechanism is applied for each gene. For example, if the number of genes is set to 4, then a possible mask would be 0110. Let ABCD and XYZW be the two selected parents. The scattered crossover would lead in this case to the following two offspring: AYZD and XBCW.

The scattered crossover proved to work better for the problem at hand.

The crossover is applied in each successive generation with a certain probability, known as the crossover fraction or rate. A large crossover rate decreases the population diversity, but in this problem a higher exchange of genetic material is needed.

Mutation Mechanisms. This mechanism is very important from the genetic diversity point of view, and it prevents landing a local, sub-optimal solution. The mutation rate is highly connected with the crossover fraction. The mutation mechanism used in this study implies generating a random gene number and flipping the bit found at that position.

Initial Population. GAs are theoretically able to find global optimum solutions, but the initial population must contain individuals with good genetic material for the problem at hand. This paper uses an initial population randomly generated, with individuals within the bounds set for each independent variable of the problem. The unfeasible solutions are discarded by penalizing the fitness function. Although GAs are theoretically able to find global optimum solutions in optimization problems, the initial population plays an important role and therefore has to contain individuals with good genetic material. An alternative to the randomly generated initial population is to run the GA several times and using each time as initial population the final population of the previous run. This approach avoids results with unfeasible genes and landing local minima and increases the probability of finding optimal solutions.

Stopping Criteria. Other important decision variables are the stopping criteria. Some of the most widely used stopping criteria are the following:

- The *maximum number of generations* that the GA will compute: after computing this preset number of generations, the GA stops and the best result until then is considered to be optimal.
- *Time limit*: specifying the maximum number of seconds the algorithm will run, this criterion stops the GA after a predefined computational time.
- *Fitness limit*: the algorithm stops when encounters a fitness value smaller than a preset target value.
- *Stall generations*: the GA terminates when no improvements in the best fitness values take place for a predefined number of generations. This can be regarded as a stagnation in the evolution process.
- *Stall time*: acts the same as the stall generation's criterion, but the predefined parameter is the computational time.

For example, if the computational time for each generation is high (due to a large number of individuals or the nature of the problem – like in the case of DG placement, where power flows are computed for each individual in each generation) and the stall time limit is set to a low value, the algorithm will not get the chance to explore the whole space, as the GA will terminate even before few generations will be computed. If the maximum number of generations is set to a small number and the population size is also small, then the algorithm will not be able to compute all the generations needed to find the optimal solution, as it will stop after completing the specified number of generations. The same argument is also applicable for the time limit criterion. The most accurate way to stop the GA is after finding a fitness value lower than the targeted one, but there are some problems for which the solution is not known a priori, and so a fitness target is impossible to be set. On the other hand, the algorithm may never land a solution with the fitness lower than the targeted one, making the criterion unfeasible. The GA stops when any of the stopping criteria is met, on a first come first served basis.

The implementation presented in this paper uses a maximum number of generations of 100 and a stall limit of 15 s and the computational results are given in Sect. 7.

7 Case Study

To assess the performances of the proposed algorithms in solving the DG allocation and sizing problem, the IEEE 69-bus distribution test system has been considered. The system has 68 sections with a total load of 3,800 kW and 2,690 kVAr (Fig. 3). The network data can be found in Baran and Wu (1989). As the power flow routine is run for every individual in every generation, a fast-decoupled power flow routine for distribution networks was nested both in the GA and in the nonlinear optimization. This routine was initially run on the base case (without DG) and resulted in total active power losses of 225 kW and total reactive power losses of 102.2 kVAr.

The testing methodology adopted to compare the two solution methods proposed in Sects. 5 and 6 was to start from one DG unit to be allocated and to increase the number of DG units until one of the methods failed, as the increased problem dimensionality overpowers the solution method.

7.1 Nonlinear Optimization Algorithm

Table 1 presents the solutions obtained with the nonlinear optimization algorithm for one and two DG units. For three DG units, the algorithm fails to provide a solution. As it results, bus number 61 is the most suited for DG installation, with a size

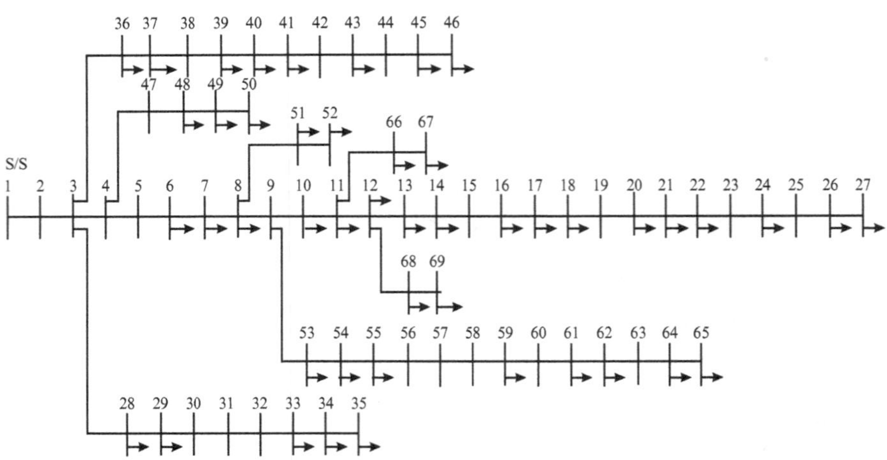

Fig. 3 69-Bus radial distribution network

Table 1 Computational results with the nonlinear optimization algorithm

No. of DG units	Losses (kW)	Comp. time (s)	Solution (bus, size (kW))			
1	83.4252	260.3594	61	1794		
2	84.233	797.1875	1	6	62	1,794
3						

of 1,794 kW in each case. The losses are higher in the case of two DG units, leading to the idea that the algorithm performs poorly once the number of variables increase. A more detailed solution analysis is made in Sect. 7.3.

7.2 Genetic Algorithm

Before presenting the GAs results, it is necessary to make some considerations. As it was shown in Sect. 2, the process of solving the DG allocation problem with GAs implies a number of parameters that have to be specified. The population size is a discrete parameter that sets the number of individuals that the GA evolves in each generation. It comes naturally that a small number of individuals in a population may result in a premature convergence, may not provide enough covering of the search space, and so the algorithm would become unreliable. On the other hand, using a very large number of individuals means that a very large number of possible solutions have to be assessed, and so the computational time increases drastically. The crossover rate and mutation rate are continuous variables, defined over the interval [0, 1]. If the mutation rate becomes too high, then the search becomes a random one; if the crossover rate becomes too high, the search can get trapped within local minima. A balance between these two values has to be found to improve the algorithm's performances. The selection method is a discrete parameter, referring to different methods, like tournament or roulette wheel, mentioned above. This variable is also accompanied by the parameters concerning the selection method. As an example, the tournament selection also implies the tournament size. The crossover mechanism can be viewed in a similar way.

According to the above remarks, a GA could be fully specified by a set of bounded parameters, which influence its performances.

Finding the optimal values for each of the parameters in the above-described set becomes a *problem within a problem*. The parameters are dependent on one another. If an algorithm gives good results with a set-up, for example, roulette wheel selection, crossover rate of 0.85, single-point crossover method and a population size of 40, changing just one of the parameters (e.g., a value of 20 for population size) can make the algorithm to perform poorly.

It must be specified that the number of tuning methods are virtually infinite and the following study is an empirical approach, with the sole intention of showing the importance of choosing the proper values for these parameters, highlighting their influence on the performances of GAs. For simplicity reasons, only the case of one DG unit is addressed.

7.2.1 Selection Mechanism

Because GAs are based on random numbers, one cannot be sure that a first run would be sufficient to obtain the optimal solution. Therefore, to overcome this problem, the algorithm was run 50 times for tournament selection and 50 times for roulette wheel selection. Figure 4 shows the voltage profiles for each of the 50 solutions ((a) tournament, (b) roulette wheel).

As it can be seen, the base case (dashed line) has very poor voltage levels, the voltage value at bus 65 reaching as low as 0.91 p.u. After DG installation, in all 50 cases, the voltage level improves, and in some cases it reaches more than 0.96 p.u., all voltages being in the admissible strip (0.95, 1.05 p.u).

At first sight, the two selection methods seem to produce similar results. More detailed analyses are needed to decide which is better for DG location and sizing.

Two data bases have been created, one for each selection method, containing the resulting location, size, power losses, and bus voltages for all 50 runs. To make a comparison, we extracted the best and worst solutions generated with each selection mechanism (Table 2).

We can conclude that the roulette wheel selection leads to smaller amounts of power losses in both min and max situations.

A comparison between the locations resulted with the two methods is given in Table 3, giving the number of occurrences and overall ratio.

As it can be seen, the roulette wheel method proves to be more accurate, 98% of the results pointing to bus 61 as the optimal location. With the tournament selection, on the other hand, bus number 61 resulted in only 82% of the cases. We can conclude, however, that bus 61 is the most suitable location for DG.

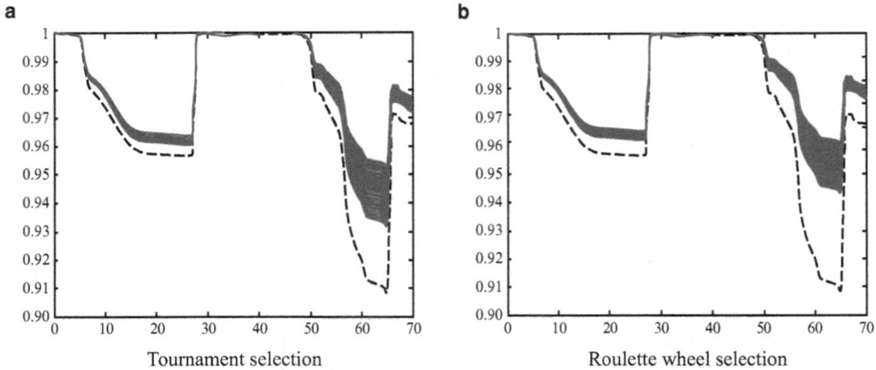

Fig. 4 Voltage profiles for 50 runs

Table 2 Maximum and minimum power losses for the two selection methods

		Bus	DG size	Losses	Time
Tournament	Min	61	1,132	103.45	84.266
	Max	62	583	147.21	83.266
Roulette	Min	61	1,500	88.206	95.906
	Max	61	878	120.41	88.297

Table 3 Optimal locations occurrences

Bus	Roulette wheel		Tournament	
	Occurrences	Ratio (%)	Occurrences	Ratio (%)
61	49	0.98	41	0.82
62	1	0.02	8	0.16
63	0	0	1	0.02

7.2.2 Population Size

To set the best value for the population size, the algorithm was run 10 times for each population size between 20 and 80, with an increment of 10. The empirical cumulative distribution functions for all cases are plotted in Fig. 5a. As it can be seen, the minimum losses values are obtained for 80 individuals in each generation. However, looking at the computational time (Fig. 5b), it increases with the population size, as more fitness functions have to be computed for each generation. A balance has to be found between these two aspects. Figure 5b shows that a population size of 80 would lead to unacceptable computational time. As Fig. 5a shows similar results for population sizes of 50, 60, and 70, taking into consideration the corresponding computational time, a value of 50 for this parameter can be considered as suitable.

7.2.3 Crossover Fraction

Having set the population size set to 50, one can analyze the crossover rate and its importance on the performance of the algorithm. This parameter specifies the percentage of individuals that enter the mating pool to exchange genetic material. They will produce crossover children. The algorithm was run 20 times for each crossover fraction between 0 and 1, with an increment of 0.1. Figure 6 plots the minimum fitness value obtained during the 20 runs for each crossover rate against the respective crossover rate value.

The plot shows that the crossover rate for the DG allocation problem for the 69-bus network should be around 0.7. A higher crossover fraction implies better genetic information exchange between parents, guiding the search, but a lower one increases diversity within the generations and provides the algorithm a better chance of finding the optimal solution by better covering the search space. A fraction of 1 means that all children, other than elite individuals, are crossover children, while a crossover fraction of 0 means that all children are obtained from mutation. The tests

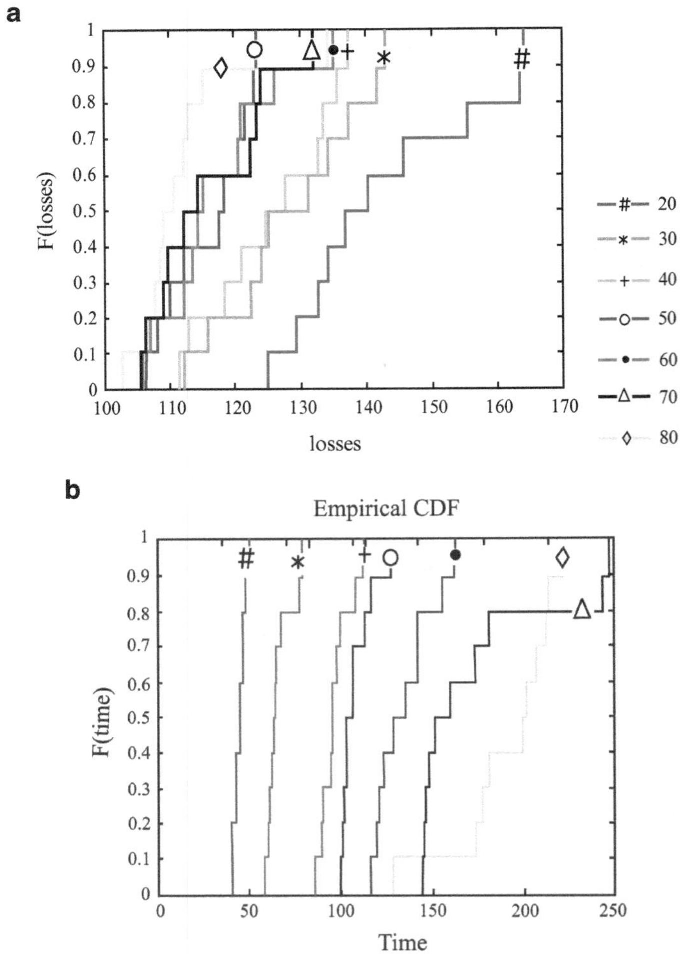

Fig. 5 CDF for losses (**a**) and computational time (**b**)

show that neither of these extremes is an efficient strategy for optimizing a function. The results of a run for a crossover rate of 1 are presented in Fig. 7, showing both the evolution of the mean and best fitness values at each generation and the average distance between individuals. The average distance shows how the search space is explored by the individuals. A small distance means less exploring. If the crossover rate is set to 1, then all the individuals in the next generations are obtained by crossover, except the elite ones, meaning that no mutation takes place whatsoever.

The algorithm gets trapped in the same best solution, as no diversity mechanisms occur. The search is over-guided and the initial solution guides the algorithm throughout the run without allowing it to explore the search space.

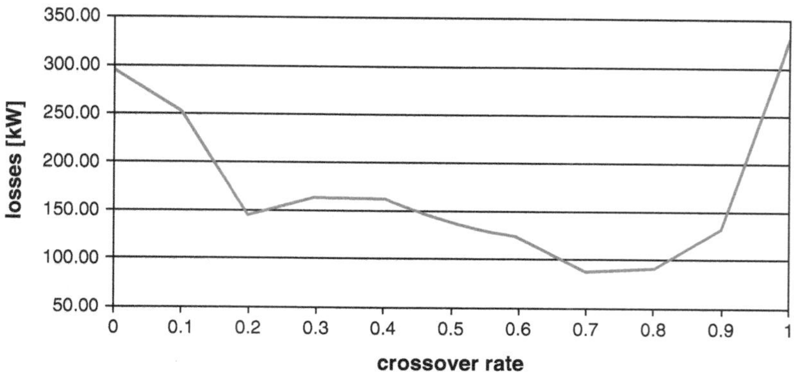

Fig. 6 Crossover fraction importance upon the performance of the GA

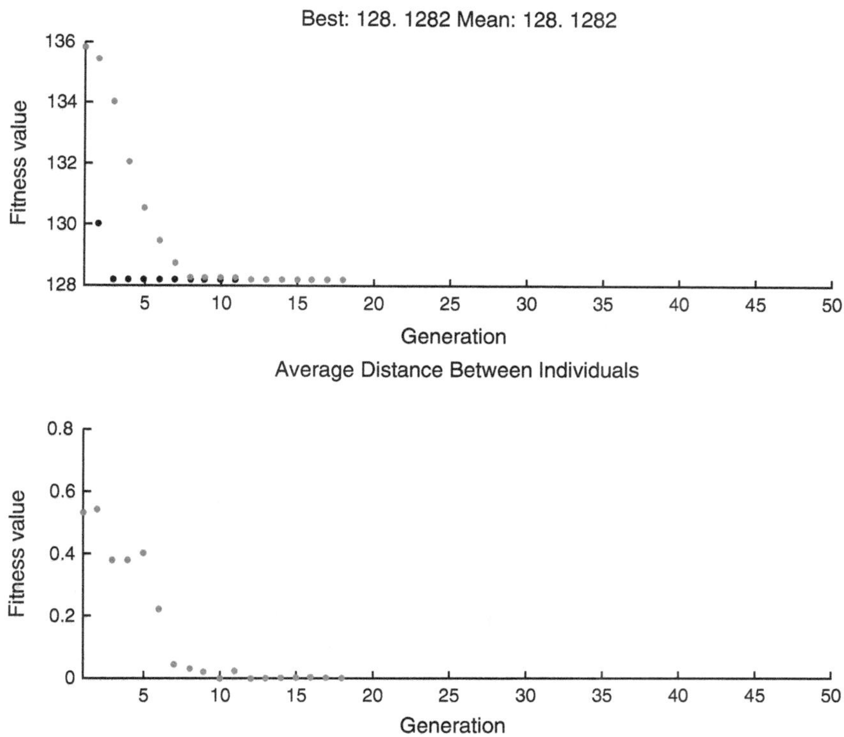

Fig. 7 GA results for a crossover rate of 1

The only genetic material is the one of the individuals from the first generation, randomly generated. The algorithm recombines this material and no new genes are created, because no mutation takes place. The average distance between the individuals becomes zero, as they are all identical. The algorithm runs until the stall generations parameter value is reached.

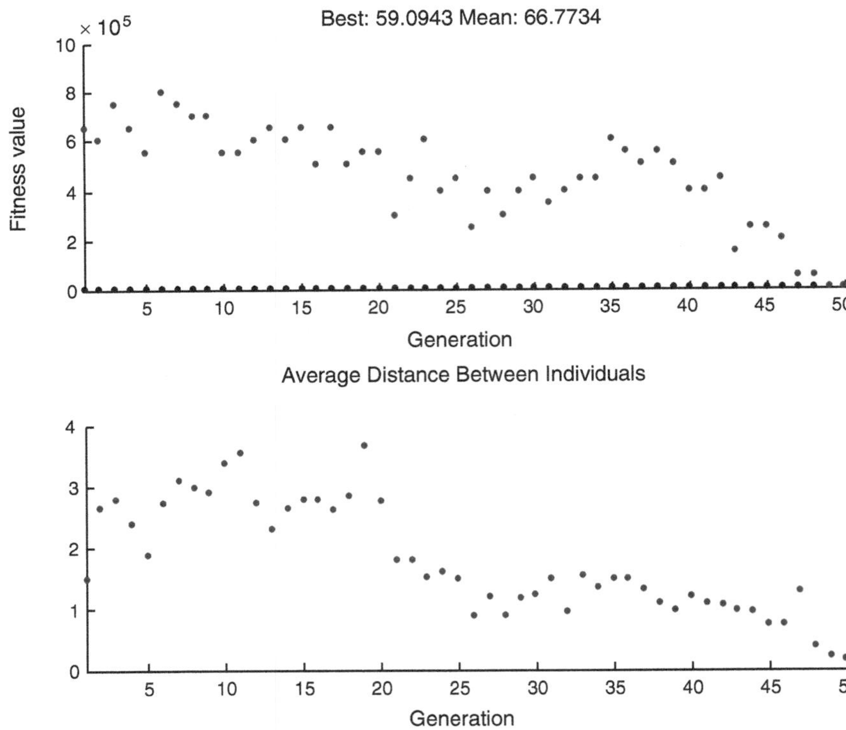

Fig. 8 GA results for a crossover rate of 0

Figure 8 shows the plot for a run with a crossover rate of 0, meaning that the individuals in each generation are exclusively created by mutation.

In this case, the random changes that the algorithm applies only slightly improve the fitness value of the best individual from the first generation. The upper plot shows some improvement in the mean fitness value in some of the generations, but no crossover takes place and so the method more likely becomes a random search. The best fitness plot from Fig. 8 demonstrates that the algorithm does not converge.

The above interpretation concerning GA parameters shows the strong link between the parameters and the performances of GAs. The operations described above can, however, take place in any other random order, each case resulting in different outputs. The tuning of parameters in GAs is still an open topic, especially because dynamic methods have to be applied due to the strong interconnections between these parameters.

The results presented in Table 4 were obtained for roulette wheel selection, a crossover rate of 0.7 and the population size set to 50. The best results from 50 runs were kept.

Table 4 Computational results with the genetic algorithm approach

No. of DG units	Losses (kW)	Comp. time (s)	Solution (bus, size (kW))					
1	88.206	262.1215	61	1,500				
2	83.9092	864.8327	62	861	61	886		
3	73.764	1543.3623	62	736	18	519	61	809

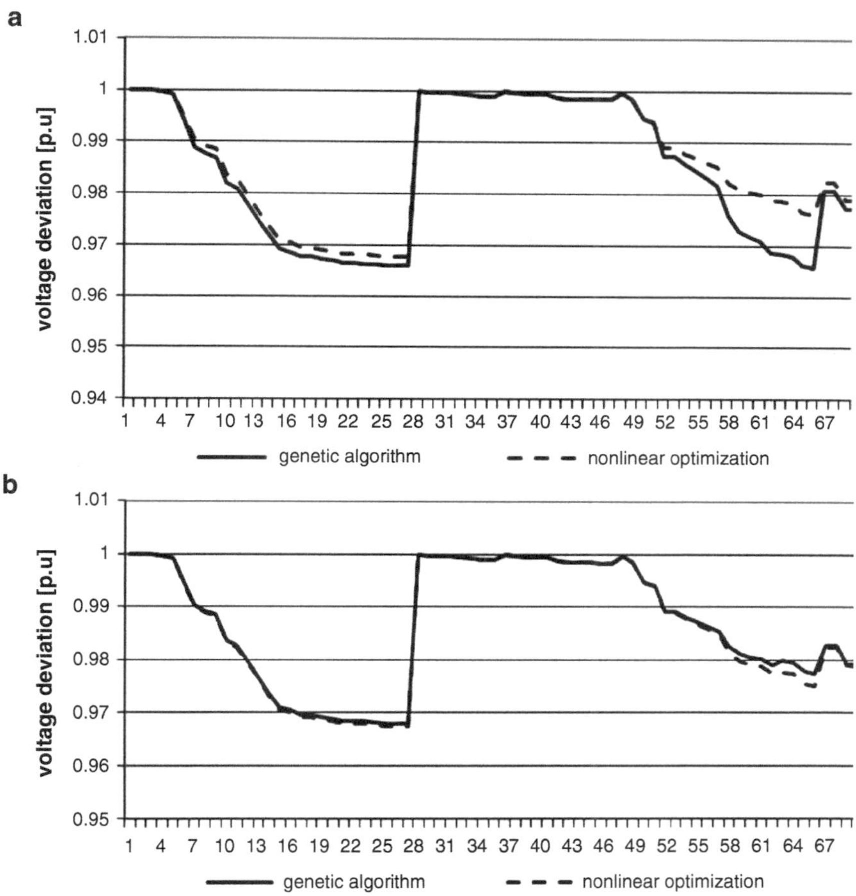

Fig. 9 Voltage levels for solutions given by nonlinear optimization and genetic algorithms for one DG unit (**a**) and two DG units (**b**)

7.3 Solution Analysis

As it results from Tables 1 and 4, the nonlinear optimization algorithm cannot face the high complexity problem of allocating more than 2 DG units, in comparison with genetic algorithms. Even though the losses in the case of GAs for one DG unit are slightly higher than the ones resulted with the global optimization algorithm, the superiority of GAs is proven when the problem complexity increases and the global optimization algorithm fails to provide a solution. For a better insight of the solutions supplied by the two methods, a voltage level analysis is made in Fig. 9, showing the voltage levels in the network for one (Fig. 9a) and two (Fig. 9b) DG units – the two cases that were solved by both methods.

The solution provided by the nonlinear optimization method for placing and sizing one DG unit results in a better voltage profile than the one in the case of genetic algorithms, both having all buses with voltages within the admissible strip (Fig. 9a). For two DG units, the genetic algorithm leads to slightly better voltage levels (Fig. 9b), proving that genetic algorithms perform better than nonlinear optimization when the number of variables increases. From the computational time point of view, the nonlinear optimization algorithm is faster than the genetic algorithm. However, the results for two and even three DG units indicate genetic algorithms to be more suited as solution method for the DG location and sizing problem. Even though genetic algorithms require thorough analyses for optimally tuning the parameters involved in the process, presented in Sect. 7.2, their superiority in relation to the proposed nonlinear optimization algorithm is obvious when the dimensionality of the problem increases. Therefore, the GAs will be considered in the following as the appropriate solution method for the DG placing and sizing problem. Their results are analyzed from the distributed generation allocation point of view, looking into outcomes for one DG unit. The optimal solution is thus considered to be bus 61 and a DG size of 1,500.

Taking into consideration only solutions indicating bus 61 as optimal, the dependence of losses on DG size can be represented graphically, plotting the resulted power losses against the size of DG (Fig. 10). The active power losses increase when the DG size decreases. The values are taken from the 49 runs that result in bus 61 as optimal, for roulette wheel selection (Table 3).

Figures 11 and 12 present the voltage levels and voltage deviations for the following three cases: the network without DG, 1,500 kW installed in bus 61, and 878 kW installed in bus 61, representing the base case, the best case, and the worst case, respectively.

The base case is far the worst as regarding the voltage profile, the minimum voltage in the system being reached at bus 65, rating 0.9092 p.u. The worst case scenario regarding DG improves the voltage level through the network, the most significant increase being registered at bus 65, reaching a voltage value of 0.9433. This value is far better than the base case scenario, but it is not sufficient, as it still remains out of the admissible limits. The best case scenario, which will be adopted in the following as optimum and therefore the DG allocation problem solution, proves an admissible value for the voltage at bus 65, of 0.9659 p.u. This case, as it can be seen in Fig. 12, allows all voltages to be within the admissible limits (voltage deviations less than 0.05 p.u.).

It has been stated in Sect. 3 that the problem of allocating DG units involves two variables: location and size. These are not independent. If DG is placed at the

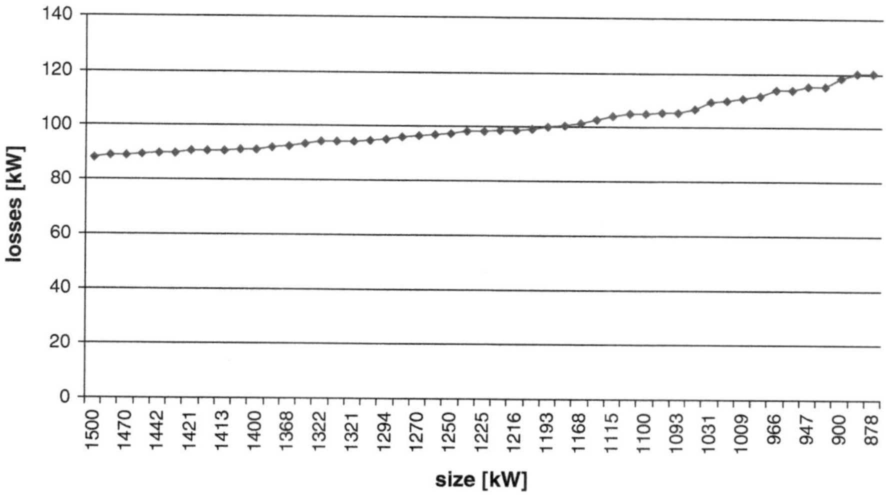

Fig. 10 Dependency of power losses on DG size

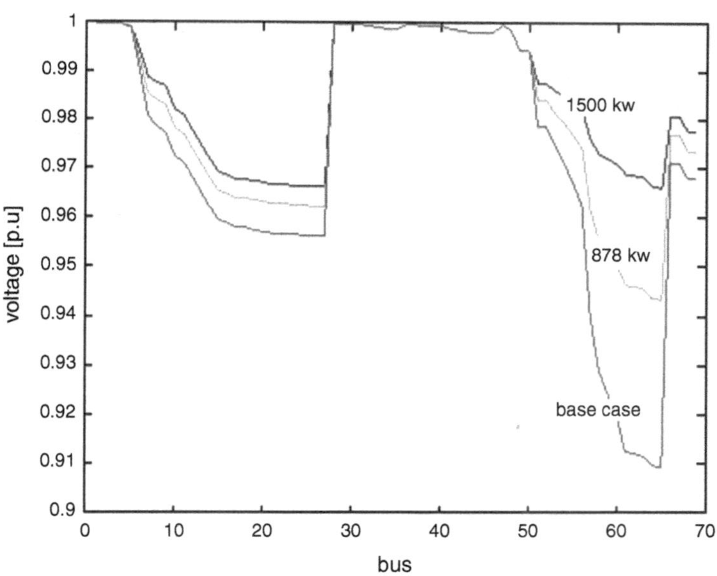

Fig. 11 Voltage profiles for the three cases

optimal location, but with a different capacity, system losses will increase. Moreover, if the size of DG is optimal, but it is connected to a bus different from the optimal one, losses also increase. This is why the solution method for DG allocation is very important and it has to provide simultaneously both values of the variables.

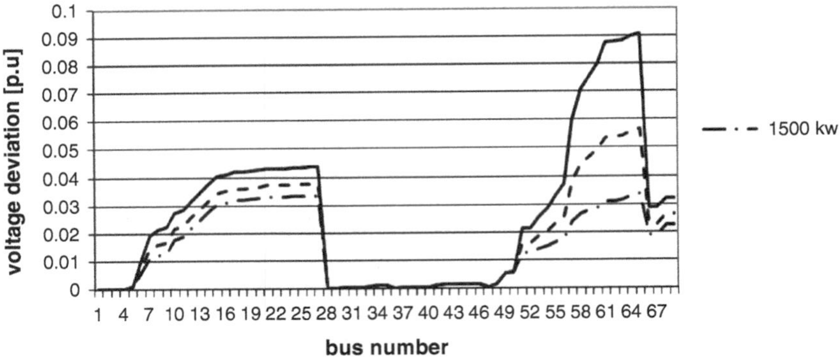

Fig. 12 Voltage deviations for the three cases (p.u)

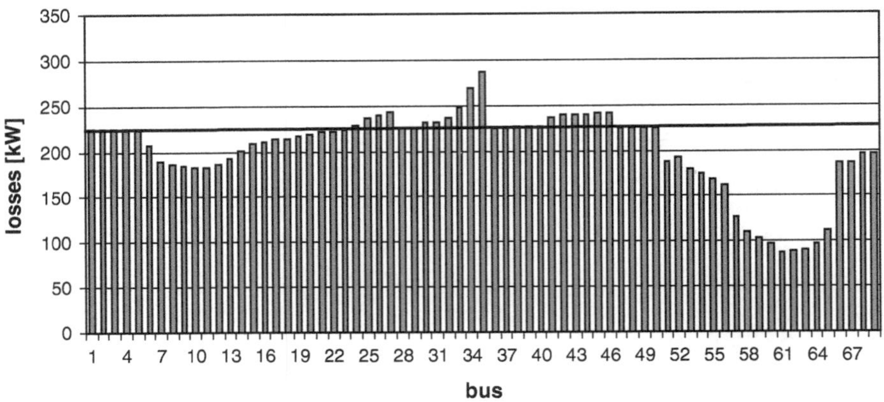

Fig. 13 Losses variation with the bus where a DG unit of optimal size is connected

To validate the GA as solution method, the following studies are carried out:

– Assume the optimal size, as resulting from the GA, of 1,500 kW. This capacity is allocated successively at each of the 69 buses in the system and a distribution power flow routine will compute the active power losses. The graphical representation in Fig. 13 is obtained by putting together the 69 power losses values. The horizontal line represents the power losses in the original system, without DG.

As it can be observed, the minimum power losses are obtained when DG is placed near bus 61, this being the global optimum. There is another optimal location, near bus 10, but this is clearly a suboptimal solution, representing only a local optimum. Figure 13 not only proves the performances of the proposed solution method and its implementation, but also highlights a very important issue in DG planning: there are buses (the most representative being here is the bus number 35) where DG connection results in power losses increasing to values that are higher than the power losses

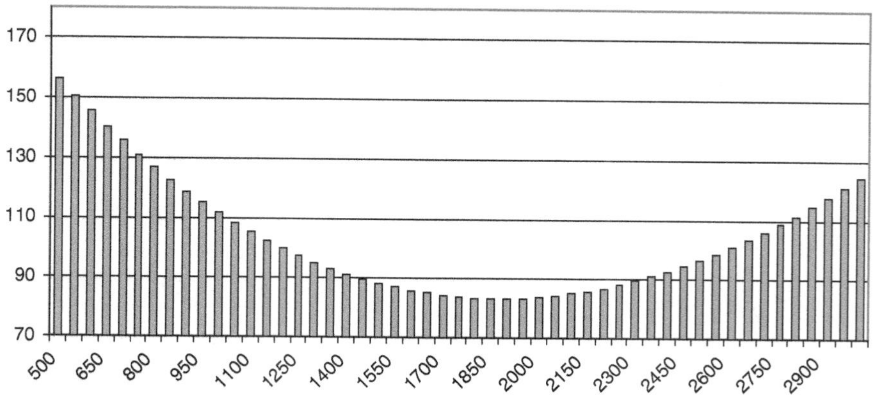

Fig. 14 Losses variation with DG size for a DG unit connected to bus 61

before installing DG. This is an effect exactly opposite to the one sought after in the DG optimal allocation. A similar proof is constructed starting from the optimal bus.

- Assume the optimal bus that resulted from the GA, number 61. To test whether the resulted capacity is also optimal, successive power flows and system losses are computed for connected DG units of different sizes. For example, losses are computed with DG installed at bus 61 with sizes starting from 500 to 3,000 kW, with an increment of 50 kW. The results are presented in Fig. 14.

The plot in Fig. 14 can be divided into three areas: the first one, for DG sizes of 500–1,300 kW presents a steep decreasing trend; the second one, from 1,350 to 2,250 kW, is relatively smooth, suggesting that increases or decreases of DG sizes within this interval do not lead to significant power losses reductions; the last area, from 2,300 to 3,000 kW, is more steep, power losses increasing progressively for DG sizes beyond 2,300–2,400 kW.

This proves that even though the location is optimal, the size of DG influences the power losses. The values in the central area of the plot suggest that minimum power losses (83.242 kW) are obtained when 1,850 kW are connected to bus 61. However, the power losses differences are very small when connecting DG units of any size from this central area of the graph. The optimization procedure proposed in this paper also takes into account the investment costs, and therefore the GA has chosen a smaller size, as the power losses improvements when upgrading to larger sizes are insignificant.

The voltage levels that resulted when connecting the DG unit of optimal size to the buses that lead to losses higher than the ones in the base case (buses with losses above the horizontal line in Fig. 13) are represented graphically in Fig. 15. The dashed line represents the voltage levels when the DG unit is connected at the optimal bus. All cases, except the optimal setup, result in inadmissible voltage levels, reaching as low as 0.91 p.u. This proves that not only power losses increase

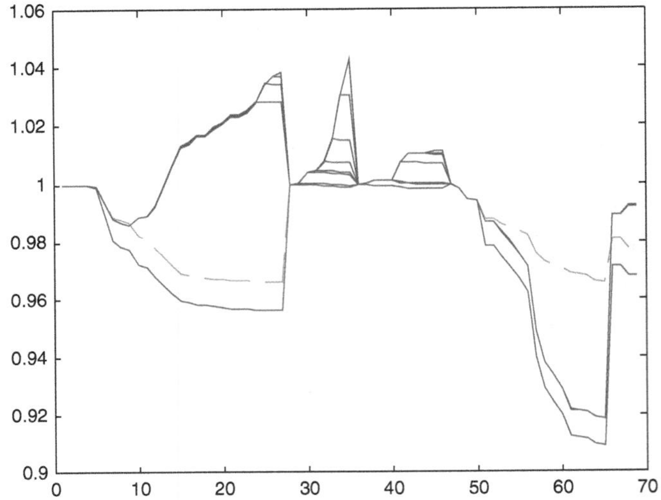

Fig. 15 Voltage deviations when DG unit is installed in buses that lead to losses higher than the base case

when the DG unit is not properly installed, but also the voltage levels are negatively affected.

As it results, the DG allocation implies simultaneously searching for both optimal location and size, and the GA approach is suited for solving this problem. Furthermore, if DG is placed in multiple locations, the voltage profile and power losses can be additionally improved. Figure 16 shows the voltage profiles that resulted after installing one, two, and three DG units of locations and sizes from Table 2, which also highlights that power losses decrease when the number of DG units increases.

8 Conclusions

The paper addresses the problem of optimal DG location and sizing in a distribution network. Two solution methods are proposed, one based on a nonlinear optimization algorithm and one based on genetic algorithms. The objective function comprises of both power losses and investment costs. The studies are performed on multiple levels: a comparison between the proposed approaches, the influence of GAs parameters on their performance in the DG allocation problem, and the importance of installing the right amount of DG in the best suited location.

To compare the solution methods, tests were performed successively for an increasing number of DG units to be allocated. The first run was made for one DG unit, and both methods provided similar results in similar periods of time, the nonlinear solver having a small advantage regarding the power losses and computational

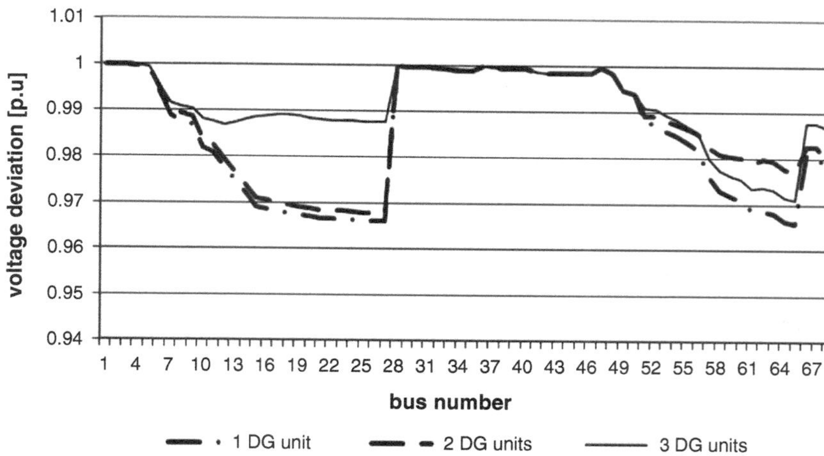

Fig. 16 Voltage levels for multiple DG units

time. Next, the number of DG units was increased to two, and the GA proved to provide better results than the nonlinear solver, in a slightly higher computational time. After increasing the number of units to three, the GA outperformed the nonlinear optimization algorithm, which failed to land a solution.

Studies showed that different values of GA parameters lead to different outputs, with examples on selection mechanisms, population size, and crossover fraction. The network used for tests is the IEEE 69-bus distribution system, for which the GA results in a DG size of 1,500 kW, installed at bus 61. Connecting this amount of DG at the optimal bus leads to a power loss reduction from approximately 225 kW to about 88 kW. The voltage profile is substantially improved as well.

The tight bound between the optimal location and size is proved by allocating the optimal size at different buses in the network and by allocating different DG capacities at the optimal bus that resulted from the GA. Both studies show that system losses increase drastically. In some cases they become even larger than the ones in the base case. Furthermore, the voltage profiles are also degraded if the optimal solution is not implemented.

Even though GAs are very much dependent on the parameters, they work when other methods fail. First, the algorithm is a multipath that searches many peaks in parallel, and hence reduces the possibility of local minimum trapping. Second, GA works with a coding of parameters instead of the parameters themselves. The coding of parameter will help the genetic operator to evolve the current state into the next state with minimum computations. Third, GA evaluates the fitness of each string to guide its search instead of the optimization function. The GA only needs to evaluate the objective function (fitness) to guide its search. Hence, there is no need for computation of derivatives or other auxiliary functions. Finally, GA explores the search space in regions where the probability of finding improved performance is high.

The solution method is therefore proved to be suited for the DG allocation problem. Several improvements should be taken into consideration in future studies: using load profiles and assessing the benefits accordingly to the DG type.

References

Acharya N, Mahat P, Mithualananthan N (2006) An analytical approach for DG allocation in primary distribution network. Electr Power Energy Syst 28

Ackerman T, Andersson G, Soder L (2000) Distributed generation: a definition. Electr Power Syst Res 57.

Baker J (1987) Reducing bias and inefficiency in the selection algorithm Proceedings of the second international conference on genetic algorithms and their Applications

Baran ME, Wu FF (1989) Optimum sizing of capacitor placed on radial distributions systems. IEEE Trans PWRD

Bazaraa MS, Shetty CM (1979) Nonlinear programming: theory and algorithms. Wiley, New York

Berg A Krahl S Paulun T (2008) Cost-efficient integration of distributed generation into medium voltage networks by optimized network planning. Smartgrids for distribution, IET-CIRED. CIRED Seminar, Frankfurt

Bertsekas D (1999) Nonlinear programming. 2nd edn. Athena Scientific, Belmont

Björkman M, Holmström K (1999) Global optimization using the DIRECT algorithm in matlab. Electron Int J Ad Model Optim 1(2)

Borges C, Falcao D (2006) Optimal distributed generation allocation for reliability, losses and voltage improvement. Electr Power and Energy Syst 28

Celli G, Ghiani E, Mocci S, Pilo F (2005) A multiobjective evolutionary algorithm for the sizing and siting of distributed generation. IEEE Trans Power Syst 20(2)

Celli G, Pillo F (2001) Optimal distributed generation allocation in MV distribution networks. IEEE.

CIGRE (2000) CIGRE technical brochure on modeling new forms of generation and storage, November 2000. Available from http://microgrids.power.ece.ntua.gr/documents/CIGRE-TF-380110.pdf

de Souza B, de Albuquerque J (2006) Optimal placement of distributed generators networks using evolutionary programming. IEEE PES transmission and distribution conference and exposition Latin America.

Floudas CA, Pardalos PM (eds) (2003) Frontiers in global optimization. Kluwer, Dordrecht

Floudas CA, Gounaris CE (2009) A review of recent advances in global optimization. J Global Optim 45(1):3–38

Floudas CA, Pardalos PM, Adjiman CS, Esposito WR, Gumus Z, Harding ST, Klepeis JL, Meyer CA, Schweiger CA (1999) Handbook of test problems for local and global optimization. Kluwer, Dordrecht

Forster W (1995) Homotopy methods. In: Horst R, Pardalos PM (eds) Handbook of global optimization, vol. 1. Kluwer, Dordrecht

Gil HA, Joos G (2006) On the quantification of the network capacity deferral value of distributed generation. IEEE Trans Power Syst 21(4) 1592–1599

Goldberg DE (1989) Genetic algorithms in search, optimization, and machine learning. Addison - Wesley, Reading, MA

Griffin T, Tomsovic K, Secrest D, Law A (2000) Placement of dispersed generations systems for reduced losses. Proceedings of the 33rd Hawaii international conference of sciences, Hawaii

Grubb M 1995 Renewable energy strategies for Europe – Volume I. Foundations and context. The Royal Institute of International Affairs, London, UK

Haensen E et al. (2005) Optimal placement and sizing of distributed generation units using genetic optimization algorithms. Electr Power Qual Util J XI(1)

Hajizadeh A, Hajizadeh E (2008) PSO-based planning of distribution systems with distributed generations. Int J Electr Comput Syst Eng 2(1)
Hammond D, Kendrew J (1997) The real value of avoided transmission costs. Conference proceedings of energy efficiency and conservation Authority/New Zeeland wind energy association, New Zeeland
Hansen ER (1992) Global optimization using interval analysis. Dekker, New York
Harrison GP, Wallace AR (2005) OPF evaluation of distribution network capacity for the connection of distributed generation. Proc Inst Elect Eng -Gen Transm Dist 152(1):115–122
Himelblan D (1972) Applied non-linear programming. Mc GrawHill, New York
Holland JH (1975) Adaptation in natural and artificial systems. University of Michigan Press, Michigan, NM
Holmström K (2001) Practical optimization with the TOMLAB environment in matlab. Scandinavian simulation society conference on simulation and modelling, SIMS
Holmström K, Edvall M (2004) Chapter 19: The TOMLAB optimization environment. In Kallrath J (ed) Modeling languages in mathematical optimization, Applied optimization, vol. 88. Kluwer, Boston. ISBN 1–4020–7547–2
Horst R, Pardalos PM (eds) (1995) Handbook of global optimization. 1. Kluwer, Dordrecht
Horst R, Pardalos PM, Thoai NV (1995) Introduction to global optimization. Kluwer, Dordrecht
Horst R, Tuy H (1996) Global optimization: deterministic approaches. 3rd edn. Springer, Berlin
International Energy Agency (1997) Energy technologies for the 21st century. OECD, Paris
Jones DR, Pettunen CD, Stuckman BE (1993) Lipschitzian optimization without the Lipschitz constant. J Optim Theor Appl 79 (1)
Keane A, O'Malley M (2005) Optimal allocation of embedded generation on distribution networks, IEEE Trans Power Syst, 20:1640–1646
Lasseter RH (1998) Control of distributed resources. In: Fink LH, Vournas CD, (eds) Proceedings: bulk power system dynamics and control IV, Greece
Le A, Kashem MA, Negnevitsky M, Ledwich G (2007) Optimal distributed generation parameters for reducing losses with economic consideration. IEEE
Madarshahian MM, Afsharnia S, Ghazizadeh MS (2009) Optimal investment of distributed generation in restructured power system. IEEE bucharest power tech conference, Romania .
Michalewicz Z 1996 Genetic Algorithms + Data Structures = Evolution Programs, third extended ed. Springer, Heidelberg
Mitchell M (1996) An introduction to genetic algorithms. MIT, Cambridge, MA
Neumaier A (1990) Interval methods for systems of equations. Cambridge University Press, Cambridge
Pardalos PM, Chinchuluun A (2005) Some recent developments in deterministic global optimization. Oper Res Int J 4:3–28
Pinté JD (1996) Global optimization in action. Kluwer, Dordrecht
Pisica I, Bulac C, Toma., Eremia M (2009) Optimal SVC placement in electric power systems using a genetic algorithms based method. IEEE bucharest power tech conference. Romania
Raj P et al (2008) Optimization of distributed generation capacity for line loss reduction and voltage profile improvement using PSO.
Rebennack S, Kallrath J, Pardalos PM (2009) Column enumeration based decomposition techniques for a class of non-convex MINLP problems. J Global Optim 43(2–3):277–297
Rosehart W, Nowicki E (2002) Optimal placement of distributed generation. 14th PSCC, Sevilla
Shukla TN, Singh SP, Naik KB (2008) Allocation of distributed generation using GA for minimum system losses. Fifteenth national power systems conference (NPSC), IIT Bombay.
Siano P, Harrison G, Piccolo A, Wallace AR (2007) Strategic placement of distributed generation capacity. 19th Conference on Electricity Distribution, Austria
Singh D, Verma KS (2007) GA based optimal sizing&placement of distributed generation for loss minimization. World Academy of Science. Eng Tech 35
Singh RK, Goswami SK (2009) A genetic algorithm based approach for optimal allocation of distributed generations in power systems for voltage sensitive loads. Asian Research publishing Network J Eng Appl Sci 4(2)

United Nations (1997) Kyoto Protocol to the United Nations Framework Convention on climate change. http://unfccc.int/resource/docs/convkp/kpeng.html
Vovos PN, Harrison GP, Wallace AR, Bialek JW (2005) Optimal power flow as tool for fault level constrained network capacity analysis. IEEE Trans Power Syst 20(2):734–741
Whitely D (1989) The GENITOR algorithm and selective pressure Proceedings of the 3rd international conference on genetic algorithms. Morgan- Kaufmann, pp. 116–121
World Energy Outlook (2008) International energy agency, 2 rue andre pascal. OECD Publications, Paris
Zabinsky ZB (2003) Stochastic adaptive search for global optimization. Kluwer, Dordrecht
Zeljković ČV, Rajaković N Lj Zubi SJ (2009) A method for cost minimization applicable to load centers containing distributed generation. Bucharest power tech conference, Romania
Zhigljavsky AA (1991) Theory of global random search. Kluwer, Dordrecht

Index

Abandon option, 332, 333, 344, 346,
Active set method, *16*
Agent-based computational economics (ACE), *243–251, 255, 256, 258, 260, 262–265, 273, 274, 279–281*
Agent-based modeling and simulation, *241–281*
Aggregation and distribution, 101, 103–107
Agriculture, 58, 64, 68, 353
Algorithm, 7, 34, 60, 81, 122, 155, 166, 177, 211, 238, 257, 298, 324,392, 410, 453, *16, 31, 56, 107, 131, 176, 214, 248, 297, 317, 352, 407*
Allocation problem, 73, 129–131, 137, 453, 456, 459, 462, 467, 469, 474, 478, 480, *342, 344, 345, 349*
Analytic hierarchy process (AHP), 343–362
Analytical simulation, 211, *458, 459, 461, 468–469, 472, 478*
Ancillary services, 97, 99, 108, 115, 117, 307, 326, *85, 162, 242, 316*
Ant colony optimization (ACO), 401, 403–406
ARIMA, *134, 137–139, 144, 151, 163, 165, 171, 173, 227*
Artificial intelligence (AI), *16, 86, 137, 139–140, 150, 152–153*
Asset management, *190, 449–478*
Attack plan, 369–373, 375, 379, 380, 382, 385
Augmentation, 11, 17, 19–21, 23, 24
Australia, 308, 309, 316, 318, *274, 278*
Auto-regressive model, 37, 50, *144*
Automated generation control (AGC), 311, 316, 317, 324
Autoregressive process, 27
Average social index (ASI), 359, 360
Average value-at-risk (AVaR), *391, 395, 398, 400, 406, 421–423, 428*

Backward reduction, *412*
Backward tree construction, *414–416*
Backwards recursion, 5, 8, 9, 18, 43, 44, 51, 53, 68, 173–175
Bathtub curve, *454*
Bellman equation, 42
Bellman function, 78–80, 82–86, 335
Benchmarking, *191*
Bender's partitioning method, *339–341*
Benders decomposition, 6, 190, 191, 193, 194, 199, 201, 205, 257, *294, 317, 325, 339, 340, 342, 343*
Benefit function, 61
Bilateral contracts, *130, 146, 162, 168*
Black-Scholes, *119*
Boiler, 259, 293, 295, 297–300, 317, 430, 432–434, 437, 438, 440–442, 445
Boolean expressions, 237–238, 243–246
Branch and cut, 122, 125
Branch-and-bound, 191, 193, 194, 197–199, 201, 205
Branch-and-fix, 206

Candidate line, 367, 371, 376, 378, 381, 382, 384, 386
Capacity functions, *33, 34, 38, 43, 49*
Capacity limits, 219, 241, 257, 259, 261, 263, 324, *348, 349, 353*
Capacity reserve, 101, 107, 116–118, 325, *197*
Capital cost, 262, 297, 300, 302, 303, 439, 441, 442, *33, 455, 468, 471, 476*
Cash flows, *195, 406–408, 424, 425, 468, 470*
CDDP, 6
Change-point detection, *106, 107*
Choice of network risk methodology, *450, 451, 453, 469–471*
City centre networks, *474*

Clustering, 37, 212, *131, 217, 317, 323, 360, 387, 388, 391, 414, 416*
CO_2 emission, 95, 210, 360, 362, *250, 271, 277, 278, 281, 324*
prices, *198, 370*
Co-optimization, 307–326
Coal, 6, 23, 49, 78, 79, 86, 107, 121, 128, 150, 152, 352, 353, 355, 356, 358–361, 451, *102, 190, 191, 197–199, 201, 205, 209, 211, 242, 259, 272, 273, 324, 462*
prices, *201, 209, 211*
Column generation, 201
Commodity, 323, 365, 434, *31, 33–43, 46, 49–51, 102–109, 114, 119, 120, 146, 148, 191, 193, 242, 317, 339, 341, 343, 347–349*
COMPAS, 80–82, 84–86
Complementarily, 101
Complicating variables, *340–342, 346, 347*
Conditional distributions, *27, 419*
Conditional value-at-risk (CVaR), *400, 290, 291, 293, 295, 297, 302, 306, 311–312, 321, 333, 391, 421, 433–437, 440–444*
CONDOR, 299
Confidence level, *177, 293–295, 297, 306, 311, 435*
Congestion, 108, 114, 116, 223, 227–230, *4, 168–170, 175, 183–184, 248, 249, 258, 267, 269, 349, 354, 367*
Conic quadratic model, *5–6*
Conjectural variation, *317, 320, 360*
Conjectured response, *351–377*
Cournot as specific case, 25, 141, *368*
estimation, *352, 354, 359*
bid-based, *359*
cost-based, *360*
implicit, *359*
symmetry, *366*
Constrained power flow, *13–15, 20, 23*
Constraints, 5, 40, 58, 81, 123 150, 154, 178, 211, 236, 257, 297, 310, 337, 366, 400, 409, 434, 453, *3, 34, 162, 197, 215, 242, 291, 319, 341, 352, 384, 407, 433*
Constructive dual dynamic programming (CDDP), 4, 6–13, 15–17, 19, 21–24, 27, 30, 60
Contingency reserve, 308, 309, 311, 315, 317, 322
Contingency response, 308–310, 317, 318, 324–326
Convexification, 193–200
Correlation, 5–7, 17, 20, 21, 25, 36, 39, 42, 50, 51, 79, 104–106, 109, 110, 300, 336, 439, 442, *109, 111, 114, 115, 117, 118, 120–122, 130, 134, 144, 154, 163, 165, 178, 183, 184, 194, 227, 228, 235, 331, 386*
Cost function, 39, 42–44, 46, 52, 131, 133, 166, 169–171, 203, 214, 217, 221, 224, 256, 264, 270, 273, 279, *19, 32–35, 38, 39, 42, 43, 45, 49, 222, 232, 233, 258–261, 266, 268, 269, 356–358, 361–363, 366, 370–372, 375*
Cost minimization, 392, *315, 322, 377*
Cost of carry, *193*
Coupling constraints, 81, 188, 191, 258, 264, *420*
Cournot, 13, 25, 141, *167, 215, 222–224, 232, 233, 235, 247, 248, 256, 257, 265, 267, 274, 277, 320, 353, 354, 361, 368, 370*
Cournot gaming, 13, 25, *222, 233, 265*
Critical regions, 217–225
Crop yield, 63, 64
Crossover, 402, 403, 415, 416, 419, 462, 464, 467, 469–473, 479, *98, 257, 272*
Curse of dimensionality, 5, 6, 22, 30, 62, 155, 258
Curtailment, 14, 98, 101, 108, 116, 118
Customers, 113, 138, 139, 293, 307, 346, 394, 396, 397, 400, 430–433, 443, 445–449, 451, 453, 454, *7, 84, 85, 245, 264, 347, 348, 350, 397, 398, 441, 449–452, 454, 455–457, 459, 460, 462–465, 467, 469, 470, 472, 474, 475*
external, 293, 430, 431, 433, 443, 448
internal, 431, 432, 443, 445, 446, 449
Cutting-plane, 190

Day-ahead energy market, *153, 179*
DC load flow, 375, *364, 368*
Decision making, 5, 68, 210, 326, 344–351, 353, 361, 362, 369–371, 373, 443, *85, 146, 155, 162, 169, 184, 215, 219, 266, 278, 280, 287, 290, 292, 294, 295, 300, 301, 303, 352, 365*
Decision variables, 37, 145, 156–158, 162, 203, 215, 250, 385, 465, *255, 320, 361–363, 367, 385, 435, 439*
Decomposition, 6, 60, 62, 63, 177, 178, 180, 188–201, 203, 205, 257, 410, 459, *64, 88–92, 94, 96, 97, 115, 134, 294,*

302, 310, 322, 325, 331, 333, 339, 340, 342, 343, 399, 409, 419, 420
Defer option, 332–334, 338
Degree of certainty (DOC), *177*
Deliberate outages, 365–387
Delphi, 343–362
Demand curve adding, 7, 11, 13, 20, 24, 26
Demand curve for release, 16
Demand curve for storage, 16
Demand function, *38, 196, 198, 257, 317, 319, 331, 332, 338, 357, 373*
Demand response (DR), 28, 112–114, 116, 118
Demand-side management (DSM), 101, 111–114, 116, 118
Demand-supply function, *33, 34, 38, 42, 49*
Deregulated market, 13, 99, 122, 410, 423
Destructive agent, 366, 368, 370–372, 385
Deterministic problem, 16, 200, 217, 218, *293, 308*
Differential evolution (DE), 131, 409–425
  crossover, 416, 472
  initialization, 415
  mutation, 415
  selection, 416, 417
DIRECT algorithm, 459, 460
Discrepancy distance, 205
Discrete variables, 190, 196, *339*
Disjunctive programming, 193, 195
Dispatchable power, 97, 110
Distributed generation, 101, 108, 117, 394, 395, 451–480, *245, 451*
Distributed generation allocation, 474
Distributed generation location, 453–456, 468, 474, 478
Distributed generation planning, 476
Distributed generation sizing, 453–458, 463, 468, 474, 475, 477, 479
Distribution networks, 391–400, 407, 451, 453, 455, 456, 459, 466, *449, 450, 463, 465, 478*
Dual, 3–30, 34, 44, 60, 61, 66, 67, 126, 134, 179, 190, 193, 195, 197, 198, 200, 201, 203–205, 217–221, 224–226, 229, 258, 262, 264–267, 274–277, 319, 321, *18, 338–349, 358, 361, 363, 364, 374, 375, 420, 467*
Dual dynamic programming (DDP), 3–30, 34, 35, 39, 42–44, 60, 79, 80, 136, 151, 171–175
Duality theory, *339, 340, 421, 423, 436*
DUBLIN, 7, 16, 25–30
Dynamic, 46, 59, 72, 73, 82, 110, 117, 236, 268, 332, 340, 367, 368, 430–433, 444, 472, *31–52, 103, 104, 119, 131, 134, 135, 140, 142, 143, 145, 148, 151, 152, 154, 163, 165, 171, 192, 194, 209, 215, 226, 244, 245, 267, 268, 321, 326, 332, 355, 405–410, 419, 434, 437–441, 444, 447, 457*
Dynamic constraints, 82, 434, *407, 438*
Dynamic network, 367, *31–52*
Dynamic programming (DP), 4, 34, 35, 39, 42–44, 79, 80, 136, 151, 171, 204, 257, 258, 267, 270, 273, 280, 292, 368, 410, *377, 437, 439–443, 447*

ECON BID, 7, 9, 17, 19, 24–30
ECON SPOT, 7
Econometric, *163, 165, 214, 216, 220, 223, 386–388, 399*
Economic dispatching, 204, 258, *12–13, 16, 20, 141, 154, 166, 316, 373*
Economic evaluation of alternatives, *338*
Efficiency, 5, 9, 11, 12, 16, 22, 58, 64, 65, 114, 131, 135, 136, 159, 199, 210, 216, 262, 308, 310, 311, 345, 346, 362, 392, 405, 407, 454, *24, 85, 119, 182, 197, 220, 246–248, 250, 259, 260, 263, 266, 267, 279, 316, 324, 406*
Efficient frontier, *292*
Electricity, 7, 58, 59, 78, 86, 96, 98, 99, 101, 113, 115, 122, 141–145, 149–165, 201, 202, 210, 232, 295, 300, 307, 308, 332, 344, 391, 409, 451, *33, 85, 102, 129, 162, 189, 214, 242, 287, 316, 351, 383, 405, 434, 451*
  derivatives, 194, 428
  generation options, 350, 351, 354, 357, 358
  markets, 7, 59, 116, 117, 151, 312, 326, 333, 368, 391, 392, 409, 410, *4, 85, 89, 130, 141, 142, 146, 148, 150, 151, 153, 154, 162, 168, 174, 179, 189–191, 193, 194, 199, 201, 204–207, 214, 215, 218, 221, 222, 224–235, 241–281, 287, 296, 305, 315–334, 351–377, 385–391*
  networks, 104, 142, 392, *267*
  planning, 343, 344, 346, 348, 361
  portfolio, *383–403, 405–429, 444*
  portfolio management, *428, 433–447*
  power market, 28, 117, 320, 384, *169, 185, 213–236, 242, 243, 246, 255, 258, 259, 261, 264–268, 270, 271, 278–280, 351–355, 366, 370*
  price forecasting, *130, 150, 152–155, 161–185, 216–221*

prices, 164, 165, 332, 336, 338, 339, *104–107, 120–122, 129, 130, 146–154, 163, 167, 173, 176, 182, 189–191, 197, 198, 201, 204, 214–222, 224, 225, 236, 249, 271, 278, 319, 320, 388*
trading, *242, 264, 265, 287, 428*
Energy planning, 100, 344–348, 350, 353, 359
Energy storage, 101, 114–115, 117, 118
Equilibrium, 13, 70, 139–141, 158, 237, *108, 162, 165, 167–170, 185, 191, 194, 195, 198, 218, 233, 243, 245, 249, 255, 257, 264, 316–322, 325, 338, 339, 341–343, 350, 352–374, 377*
Equipment dimensioning, 397, 402, 405
Estimation, 250, 336, 337, 354, 398, 462, *15, 57, 60–62, 65, 66, 68–73, 81, 107, 120, 123, 145, 162, 191, 192, 199–211, 216–221, 223–224, 227–230, 259, 352, 354, 359, 399–400, 403*
Euphrates River, 58, 59, 68–70, 73
European Energy Exchange (EEX), *385, 424*
Evolutionary computation, *254–257*
Exclusive right, 332, 339
Expansion budget, 380–383
Expansion planning, 365–387, 391–407, 409–425
Expected future cost function, 42–44
Expected profit, *290, 292, 297–299, 305, 308, 309*
Expected utility, *422, 433, 436, 442*
Expert system, 259, 274, 279–280, 285, 287, 291, 292, 410, *137, 139*
Expert-based scenarios, 336
Experts, 24, 279, 347, 349–351, 353–357, 359, 361, 362, 406, 410, *139, 176, 177, 244*
Exponential smoothing models, *137, 138*

FACTS, 462, *4, 9–12, 14, 15, 23*
Feasibility, 37, 88, 151, 154, 201, 217, 219, 220, 222, 259, 325, 402, 414, 424, 452, 462, *16, 19, 291, 300, 301, 326*
Feasible dynamic flow, 201, 276–277, 317, 402, 404, *34, 35, 39*
Feasible dynamic multicommodity flow, *43, 44*
Financial transmission rights (FTRs), *102, 103, 108, 279*
Fitness function, 463
Flow storage at nodes, *33, 38–40*
Forecasting, 25, 27, 96, 100–102, 116, 401, 459, *86, 87, 89, 94–97, 129–155, 161–185, 192, 203, 208, 214, 216–221, 224, 229, 231–232, 247, 264, 288, 326, 329*
Forward contracts, 47, *190–193, 195, 200, 201, 214, 233, 268, 288–290, 294–301, 303, 304, 306–309, 352, 355, 358, 365*
Forward curve, *104, 209*
Forward positions, *355, 360, 361, 365*
CfD (contract for differences), *260*
forward contracts, *355*
influence on market equilibrium, *317, 319*
Forward prices, *189–211*
Forward selection, 188, *93, 95, 411*
Forward tree construction, *412–414*
Forwards, 335, *108, 114, 191, 193–195, 198, 200*
Fourtet-Mourier distance, 183, 186
Frequency control, 115, 309
Frequency control ancillary services (FCAS), 309
Fuel price, 58, 210, 217, 294, 295, *168, 211, 215, 216, 218, 221, 222, 233, 235, 317, 320, 333, 370*
Fundamental price drivers, *129, 195, 220, 224, 352*
Futures, 4, 59, 79, 117, 150, 295, 332, 344, 367, 392, 409, 433, *4, 85, 104, 130, 165, 190, 214, 287, 317, 360, 384, 405, 441*
Futures market, *130, 287–312, 374*
Fuzzy inference system (FIS), *162, 166, 171, 174–184*
Fuzzy logic, *137, 140, 145, 152, 153, 162, 165, 166, 170, 176*
Fuzzy optimal power flow, 210, 214–225
Fuzzy sets, 210, 212–214, 410

Gas, 58, 79, 121, 150, 410, 451, *190, 319, 323*
market, 114, 122, 123, 125, 131, 136–145, 146, 150–154, 158, 165
production, 114, 122–128, 131, 151, 152, 154, 158
recovery, 122, 123, 126–128
transmission, 131, 136, 139
transportation, 122
turbine unit, 259, 260, 273, 279, 280, 285, 286
wells, 128
Generalized Auto Regressive Conditional Heterokedastic (GARCH), *148, 150, 165, 170–174, 176, 179–182, 217, 227*

Generating, 59, 70, 78, 129, 237, 256–259, 295, 310, 316, 337, 352, 354, 357, 373, 453, 465, *32, 33, 85, 92, 189, 194, 215, 219, 223, 224, 233, 251, 258, 260, 261, 263–265, 267, 271, 274, 278, 316, 317, 339, 342, 370, 375, 409, 452, 462*
  units, 237, 256–261, 263, 264, 266, 280, 285, 376, 451, 453, *13, 32, 85, 221, 251, 253, 257, 260, 316, 319, 320, 370, 375*
Generation, 3, 34, 59, 77, 96, 122, 150, 210, 236, 257, 293, 331, 343, 367, 392, 409, 443, 451, *3, 31, 84, 103, 162, 207, 214, 243, 296, 315, 339, 352, 385, 409, 433, 451*
  cost uncertainties, 209–232
  operation, *315–334, 370*
  planning, 212, 367, *315–334, 434, 441*
  rescheduling, 251
Generator capability, *6, 13, 24, 26*
Generator shedding, 244, 251
Genetic algorithm (GA), 136, 257, 368, 369, 401–406, 410, 411, 415, 419, 421–423, 425, 451–470, 472–479, *86, 93, 94, 251, 253–257, 266, 272*
Geographic decomposition, 191, 203
Global optimizer, 79, 80, 93, 305
Gröbner basis, 191
Graph, 86, 90, 462, 477, *33, 43, 48, 122, 327, 425*
Grid environment, *161–185*
Grid integration, 100

Hasse diagram, 246
Head variation, 35, 36, 45, 46, 48
Hedging demand, *191*
Here-and-now, *289*
Hessian, *4, 19*
Heuristic, 4, 7, 24, 26, 30, 34, 36, 39, 44, 46, 47, 54, 78, 80, 86–92, 136, 146, 188, 189, 191, 201, 204, 205, 237, 257, 368, 392, 400, 401, 406, 407, 409, 410, 443, 459, *16, 176, 178, 179, 185, 254, 260, 355, 390, 411, 414, 415, 459, 478*
Heuristic rules, 443
High voltage risk, *478*
Horizon of simulation, 81
HVDC, 309, 322, 324, *225*
Hybrid, 42, 117, 212, 280, 410, *55, 84–98, 145, 150, 161, 162, 178, 179–182, 185, 213–236*
Hydraulic reservoir, 78–80, 82, *441*
Hydro
  power, 4, 33–36, 45, 49, 53, 57–73, 99, 107, 108, 110, 114, 115, 117, 118, 151, 152, 154, 155, 352, *147, 192, 206, 322*
  reservoir management, 78, 79
  scheduling, 33, 49, 53, 149–165, 166–169, *316, 321, 322, 326*
  units, 82, 255, 259, 261, 264, 274, 278, 280, 286, 287, 292, 310, *316, 324, 328, 330, 361, 362*
Hydro scheduling, 33, 49, 53, 149–165, 166–169, 316, 321, 322, 326
Hydrothermal, 3, 33, 34, 59, 62, 150, 151, 154, 155, 157, 159, 162, 163, 166–168, 174, 259, 274, 277–285, 292, 352, *316, 321–326, 332, 356, 361, 370*
Hydrothermal power system, 3, 33, *365*
Hyperplane, 35, 42–44, 46, 47, 52, 60, 190, 193, 195, 198, *339, 342, 343*

Impact on birds and wildlife, 353
Improved differential evolution (IDE), 410, 411, 416–425
  auxiliary set, 411, 417
  handling of integer variables, 418
  scaling factor F, 416, 417
  selection scheme, 417
  treatment of constraints, 418
Individual capacity function, *43*
Indivisibilities in production, *338*
Inelasticity, 63, *387*
Inflow, 170, *327*
Information structure, *408–410, 417–419*
Installed capacity, 71, 97, 105, 106, 150, 151, 225, 457, *141, 145, 197, 264, 319*
Integrated risk management, *319*
Intelligent systems, *166, 175, 185*
Interior-point method, *4, 10, 14, 16*
Intermittence, 95–97, 108, 116–118
Interpretability, *140, 153, 175, 176, 183, 185*
Interpretable prices, *337–350*
Investment
  costs, 128, 133, 223, 300, 331–339, 375, 377, 380–384, 393, 397, 398, 409, 412–414, 419, 456, 457, 459, 477
  flexibility, 332, 333
  uncertainty, 332, 333, 336, 337
  value, 333, 335, 336
IP-prices, *339, 342–346, 349, 350*
Irrigation, 57–73

Jump diffusion, *148, 150, 165, 216*
Jump diffusion models, *150, 165*

Lagrangian decomposition, 205
Lagrangian relaxation, 203, 257, 258, 264–274, 277, 291, *294, 322, 420*
Large scale integration, 95–118
Least-squares estimation (LSE), *162, 175, 178–184, 185*
Lift-and-project, 190, 196–199
Linear commodity prices, *339, 343*
Linear interpolation, 20, 44, 267–277
Linear programming (LP), 4, 35, 44, 52, 63, 123–125, 132, 133, 137, 139, 141, 143, 145, 298, 309, 367, 369, 376, 378, 425, *4, 16, 19, 294, 302, 310, 321, 340, 341, 343, 391*
  solver, 52, 68, 78, 86, 146, 338, 384, 478, 479, *19, 24, 355, 364, 385, 399, 400, 477*
Linear transfer function models (LTF), *139, 151, 172, 182*
Load, 4, 9, 150, 235, 257, 293, 366, 413, *56, 101, 129, 162*
  shedding, 235–252, 370, 373, 385, 452
  uncertainties, 212, 214, 216–224, 226–229, 231
Load and generation, 224, 229, 241, 323
Load-following, 98–100
Local optimizer, 79, 305
Locational marginal pricing (LMP), *269, 274, 277, 280*
Long-term, 34, 157, 393, 431, *162, 190, 217, 316, 352*
  load, *441, 451, 452*
  planning, 96, 157, 397–398, 400–402, 405–407, *389*
  risk, *451–452*
  scheduling, 34, 35
  supply, 155, 157, *465*
Loss aversion, *436–437, 442, 443*
Losses, 23, 62, 64, 65, 98, 108, 152, 222–224, 226–231, 267, 312, 320, 335, 394–398, 412, 454–458, 466–470, 473–479, *10, 272, 320, 321, 451*
Low-discrepancy approximations, *114*
Lower response, 315

Maintenance issues, *455–457, 478*
Marginal supplier, 320, 321
Marginal water value (MWV), 6, 10, 34, 63, 71, 155, *222, 223, 226*
Marginalistic DP, 9
Market, 7, 34, 79, 122, 150, 210, 294, 307, 331, 368, 391, 392, 409, *102, 129, 189, 214, 241, 287, 315, 338, 351, 405, 433*
Market clearing engine (MCE), 312
Market equilibrium conditions, *355*
  effective cost, 320, 348, *85, 98, 358, 363, 366, 372, 450, 468, 476*
  equivalent optimization problem, *358, 366, 420*
  in a power network, 240, 241, 453, *3, 9, 15, 270, 354–356, 364–369, 462*
    anticipating market clearing conditions, *367*
    as a function of the network status, 7
  under exogenous stochasticity, 34, 313, 325, *133, 134, 139, 216, 219–222, 257, 258, 364*
Market power, 28, 117, 320, 384, *169, 185, 214, 242, 246, 255, 258, 259, 261, 264, 265, 266, 268, 270, 271, 352, 354*
Market simulation, 25, 410, *219, 265*
Markets with non-convexities, *339, 341*
Markov, 5, 20, 21, 40, 80, 336, 337, 338, *107, 152, 153, 165, 272, 360, 454, 457, 459, 467, 471, 478*
Markov chain estimation, 80, 336, 337, *107*
Markov model, 5, 20, 40, 338, *153, 165, 360*
Markov modelling, *454, 457, 459, 467, 471, 478*
Master problem, 193–196, 198, 199, 205, *341, 342*
Mathematical model, 34, 47, 109, 138, 432, 433, 438, 443, 453, *38, 39, 316*
MATLAB, 68, 299, *20, 62, 97, 171, 173, 180, 182, 399*
Maximum likelihood, 337, *217, 218*
Mean average percentage error (MAPE), *140, 141, 173, 174, 180–182*
Mean reversion, 337, *148, 150, 165, 191, 214, 217, 225, 229, 234*
Medium-term, 33–54, 101, 407, *146, 162, 185, 292, 301, 316, 321–330, 332, 351–377, 389, 407*
  coordination with short term, *373*
  scheduling, 34, 45–46
Minimax weighted regret, 366, 374, 381, 383
Minimum cost dynamic flow problem, *33–35*
Minimum cost dynamic multicommodity flow problem, *33–47*
Minimum down time, 91, 260, 263, 269
Minimum up and down times, 202, 273, *169*

Minimum up time, 257, 260, 263, 269
Mixed integer linear programming (MILP), 123–125, 143, 145, 369, 376, 378, 425, *302, 310, 321, 340*
Mixed integer program, 125, 257, 258, 367, *331, 339, 340, 342*
Modeling, 4, 78, 162, 177, 338, *4, 87, 163, 214, 242, 351*
Modified IP-prices, *339, 342–344, 349, 350*
Monte Carlo, 174, 188, 211, 212, 339, 340, *103, 113–116, 167, 409, 461, 474*
Monte-Carlo simulation, *114, 116, 167, 224, 231, 317, 370, 458, 460, 461, 469, 473, 478*
Multi-agent system, 443
Multi-period risk functional, 204, *406, 408, 417, 418*
Multi-stage decision making, 68
Multi-stage stochastic optimization, 155, *383, 385, 400*
Multi-stage stochastic program, *290, 390, 395, 403*
Multicommodity modeling, *103*
Multicriteria, 346, 347
Multiparametric problem, 214, 217, 219
Municipal power utility, *423*
Mutation, 402, 403, 415, 419, 420, 462, 464, 467, 469–472, *254, 257, 272*
Mutual capacity function, *43, 49*
MWV, 6, *228*

National electricity market (NEM), *232, 256, 274, 278*
National regulation, *455*
Natural gas, 121–146, 150–154, 158, 159, 164, 165, 300, 301, 352, 353, 356, 360, 451, 453, *190, 197, 198, 242, 274*
  prices, 139, *191*
Negotiation, 430–433, 443, 446, *33, 245*
Net present value, 332, 336, 338, 397, 398, 400, 404, 405, *119*
Network, 151, 274, 365, 391, 410, *87, 129, 219, 354, 449*
  modeling, 141–143
  planner, 366–369, 371, 373, 375–378, 380, 383, 393, 406
  planning, 366–368, 391–401, 403, 405–407, *455*
Neural networks (NN), 61, *87, 92, 95–98, 140, 145, 153, 165, 166, 170, 171, 173, 185, 219, 251*
New Zealand, 4, 7, 20, 23, 25, 26, 60, 155, 171, 308–310, 312–314, 316, 317, 319, 322, 324, *152, 216, 222–224*
New Zealand electricity market (NZEM), *216, 224*
Newton's method, *16*
Newton-Raphson, 280, *13, 14, 458*
Nodal marginal prices, 223–231
Noise impact, 353–355, 358, 360, 361
Non-anticipativity, 4, 5, *384, 405, 407*
Non-linear price structure, *339*
Nonconvex, 61, 123, 129, 133, 146, 180, 181, 191, 193, 196, 197, 199, 205, 459, *322, 324, 364*
Nonlinear, 4, 123, 157, 237, 267, 453, *5, 32, 166, 257, 342, 408*
Nonlinear optimization, 59, 125, 131, 237, 451–480
Nonlinear programming (NLP), 125, 126, 274, 376, 378, 410, 459, *14, 17, 321*
Nonrandom, 366, 370, *104*
Nonsmooth, 123, 129, 193, 200
Norway, 4, 7, 26, 28, 33, 40, 50, 60, 155, 171, 338, *199, 201*
NP-hardness, 246–248

Objective function, 37–39, 60, 61, 63–65, 84, 91, 124–128, 131, 133, 134, 138, 143, 156, 159, 181, 182, 189, 197, 200, 236, 258, 262–264, 266, 267, 281, 295–300, 302, 303, 312, 321–323, 337, 367, 375, 377, 378, 385, 387, 392, 393, 412, 413, 417, 418, 438, 442, 453, 456–459, 462, 463, 478, 479, *3, 4, 12, 15, 19, 20, 35, 44, 171, 180, 181, 221, 260, 261, 291–294, 297, 301, 312, 319, 322, 324, 325, 332, 341, 342, 349, 361, 363, 364, 375, 376, 384, 385, 395, 408, 419, 421*
Off-line decision support, 432, 445, 449
Offers, 13, 312, 314–318, 320, 321, 325, 430, 432, 433, *138, 174, 208, 211, 221, 222, 243, 245, 331, 391*
Oil
  prices, *200*
On-line decision support, 432, 445, 447
Operating cost, 44, 59, 125, 156, 166–169, 174, 258, 262, 277, 282, 287, 291, 295–297, 300, 302, 412, 442, 445, *316, 332*
Operating guidelines, 14, 15, 23
Operating security limit, *4, 12, 15, 20*

Operational incentives, *464*
Operational optimization, 294, 297–299, 305
Operational reserve, 101, 102, 108
Operations planning, 33–54, 154
Opportunity cost, 58, 71, 73, 210, 308, 311, 312, 321, 325, *168*
Optimal, 4–11, 14, 33, 52, 53, 61, 62, 80, 122, 150, 210, 257, 294, 367, 430, 453, *3, 31, 167, 322, 338, 383, 436*
Optimal power flow (OPF), 214, 231, *3–27*
Optimal scenario reduction, *409, 411*
Optimal timing, 366
Optimality, 6, 9, 42, 60, 86, 143, 194–196, 198, 199, 205, 217, 221, 257, 280, 325, 335, 401, 457, 459, *16, 318, 321, 334, 338, 357, 359, 361, 364, 366, 369*
Optimality cut, 194, 196–199
Optimization, 294, 297
  operational, 294, 297–299, 305
  strategic, 432, 433, 439, *167, 169, 190, 245, 259–261, 264–267, 269, 270, 317, 351–356, 365, 367, 453*
  tactical, 432, 433
  under uncertainty, 331–340
Optimization methods comparison, 401, 406, *15, 16*
Optimized certainty equivalent, *435*
Ordered binary decision diagram (OBDD), 235–252
Over-frequency, 315
Over-the-counter trading, *103, 190*

Panhandle equation, 129
Parameter tuning, 472, 474
Parametric problem, 214, 216–219
Partial order, 246
Peak day option, 82
Penalty factors, 323, 463
Perfect information, *289*
Phase-shifting transformer, *8*
Philippines, 309
Policy, 5, 6, 8, 14, 16, 17, 22, 23, 25, 30, 58, 159, 164, 210, 223, 297, 298, 346–348, 359, *141, 214, 215, 219, 231, 242, 262, 263, 266, 269, 273, 274, 456, 462, 464, 465*
Polyhedral risk functionals, *407, 419–424, 427, 428*
Pool, 469, *33, 256, 257, 263, 265, 268, 270, 287–292, 294–304, 309, 310, 353*
Pool price, *207, 287, 288, 290, 295, 297, 298, 302, 306, 307, 309, 310*
Portugal, 97, 109–112, 344, 351–353, 356, 359, 361
Power, 3, 33, 57, 77, 96, 141, 150, 177, 209, 236, 257, 293, 307, 332, 344, 365, 392, 409, 429, 451, *3, 31, 56, 101, 129, 162, 199, 214, 241, 295, 316, 351, 391, 406, 449*
Power balance, 34, 37–39, 41, 49, 95, 236, 241, 243–248, 256–259, 263, 264, 266, 275–277, 280, 375, 441, 463, *357, 358, 362–364, 366*
Power flow, 214–225, 231, 241, 243, *3–27*
Power generation, 77–93, 111, 116, 122, 154, 164, 201, 241, 246, 250, 278, 346, 349, 370, 453, *13, 14, 23, 32, 33, 52, 84, 130, 147, 265, 277, 296, 361–363, 373*
Power injection model, *9, 11*
Power system economics, *12*
Power tariffs, 295
Pre-computation, 7, 12
Prediction, 102, 103, 352, 401, *84–98, 122, 134, 136, 139, 141, 144–147, 150, 152–155, 162, 165, 167, 168, 229, 262, 264*
Price, 11, 34, 58, 79, 113, 125, 164, 193, 210, 276, 294, 311, 331, 432, 456, *85, 102, 129, 162, 189–211, 214, 241, 287, 316, 338, 352, 384, 405, 438, 450*
Price forecasting, *130, 150–155, 162–167, 170–185, 216–221*
Price functions, 193, *218, 339, 344, 349*
Price model, 34, 35, 39–41, 44, 53, 336, *109, 198–199, 215, 217, 218, 226–229, 235*
Price prediction, *150, 153, 155*
Price spike, *101–123, 165, 166, 194, 222, 234, 386*
Price volatility, 223, *162, 215, 218, 225, 232, 236, 268*
Pricing of supply contracts, *391, 403*
Pricing test, *338, 339*
Principal components analysis (PCA), 37
PRISM, 6
Probability distribution, 22, 27, 37, 40, 41, 154, 160, 178, 181–185, 188, 203, 205, 212, 213, 370, 400, *109, 165, 194, 262, 263, 289, 311, 406, 408, 410, 428, 460, 461, 472, 477*
Producer, 27, 113, 117, 131, 136, 138–141, 332, 336, 339, 352, *129, 141, 194, 197, 198, 207, 242, 256, 257, 261, 267, 269, 270, 287–312, 349,*

352–354, 360, 365, 367, 368, 392, 403
Productivity, 58, 59, 70, 71, *451*
Profile of risk against time, *471*
Profitability, 319, 331–333, 431
Pumped-hydro, 114, *361*

Quasi-Monte Carlo, *409*

RAGE, 7
Raise response, 315, 318
Ramp rate, 96, 97, 115, 256, 259–261, 263, 267, 270–273, 276, 312, 318, *332*
Ramping constraints, 317, 318
Random variable, 192, 296, 333, *106, 115, 117, 317, 408, 409, 417, 434, 435*
Real options, 332–336, 338, 339, *101–123*
Recourse, 178, 179, 181, 184, 190–193, 197, 205
Recourse function, 181, 190–193, 205
Recursive scenario reduction, *412*
Redistribution rule, *411, 412*
Reduced time-expanded network, *48, 49*
Refinement, *274*
Regime-switching models, *148, 151, 152*
Regret, 366, 369, 373, 374, 376–378, 381–383
Regulated market, 138, 139, *146*
Regulating transformer, *8, 9, 13, 14*
Regulation, 98, 115, 117, 136, 138, 307–309, 311, 313, 316–318, 324, 346, 410, 411, 423, *231, 233, 242, 258, 274, 316, 455*
Regulatory Framework, *464*
Relative storage level, 28, *226–229*
Reliability, 96–99, 102, 113, 117, 141, 142, 151, 212, 223, 384, 410, 462, *86, 130, 163, 278, 326, 450, 452–457, 467, 476*
Renewable energy, 78, 108, 114, 118, 210, 331, 332, 344, 345, 347, 352, 452, *278*
Renewable energy sources (RES), 114, 210, 344, 452
Renewable energy technologies, 331
Reserve, 80, 123, 152, 257, 344
  reserve price, 319, 321
Reservoir balance, 36, 44, 52, 145
Reservoir management, 3, 5, 79, 86, 87, *222, 225, 324, 325, 327, 329–330*
Reservoir optimization, 4, 6, 21, *232*
Reservoirs, 3, 5, 7, 22–24, 27–30, 34, 35, 39, 45, 47–49, 52–54, 59, 60, 62, 65, 66, 68, 69, 73, 78–80, 82, 83, 85, 86, 88, 114, 143–145, 151, 162, 166, 167, 169, 170, *206, 226, 316, 317, 322, 324–330, 373*
RESOP, 6, 7, 12, 13, 15, 18–23, 26
Retailer, *129, 155, 162, 214, 287–312, 423, 428*
Retrofit design, 294–297, 299, 300, 302, 305, 430, 445
Risk
  adjustment, *194, 195, 199*
  assessment, *162, 166, 449–478*
  aversion, 7, 26, 30, 46, 366, 369, 383, *194, 225, 251, 281, 295, 303, 308, 309, 377, 395, 400, 433*
  control, 46, 47
  functional, 177, 204, *384, 406–408, 416–428*
  management, 47, 54, 204, *85, 114, 150, 316, 317, 383–403, 405–429*
  measure, 366, 369, *291, 293, 295, 297, 301, 306, 321, 385, 391, 434–436, 439, 440, 442–444, 447*
  neutral, 332, 366, 374–377, 380–383, *104, 110, 115, 116, 119, 291, 292, 296, 297, 299, 300, 307, 398*
  premium, *191, 194, 195*
Risk analysis, 212, *465, 472, 478*
Risk averse gaming engine (RAGE), 7, 16, 26–29
Risk-free capital, *418*
Root mean squared error (RMSE), *140, 173, 174, 180–182*

Sampling, 37, 51, 53, 60, 174, 203, 243, 297, 298, 302, *61, 62, 64, 67, 79, 88, 177, 409*
Scaling, 333, 411, 416, 419, 424, 486, *4, 17–19, 57, 72, 74*
Scenario, 4, 36, 78, 99, 145, 150, 177, 299, 323, 336, 345, 366, 394, 432, 455, *114, 170, 288, 295, 297, 306, 385, 400, 405, 461*
  generation, 371–373, 377, 384, 443, *353, 390, 391*
  tree, 199, *288, 294, 302, 306, 317, 320, 323, 331, 354, 370, 371, 385, 391, 395, 400, 406, 408–417, 424*
  tree generation, *389–391, 399, 409*
Scenario reduction, 177, 178, 184–188, 203, 205, *294, 297, 302, 306, 409–413, 415*

Scheduling, 34, 35, 39, 45–48, 51, 53, 58, 98, 102, 116, 122–126, 144, 150, 151, 154, 155, 157, 158, 165–167, 173, 174, 202, 257, 261, 274, 280, 462, *31, 32, 85, 129, 130, 168, 281, 316, 320, 321, 322*
Seasonal, 29, 34, 139, *107, 120, 130, 131, 133, 134, 138, 148, 152, 153, 165, 195, 217, 218, 226, 229, 323*
Seasonal variations, 36, 110, 114
Security, 57, 95, 128, 129, 151, 211, 241, 312, 325, 360, 365, 366, 368, *4, 12, 15, 16, 20, 130, 141, 154, 168, 242, 462*
Security-constrained, 368, 369, 371, 372, 414
Selection methods, 464, 467–469
Self-commitment, 316
Sensitivity analysis, 237, 358, 410, 455
Shadow price, 319, 323, *183, 184, 338, 347, 349*
Short-circuit currents, 396, 400, 402, 406
Short-term, 34, 78, 101, 157, 311, 431, *84, 130, 225, 478*
Short-term risk, *450, 451*
Simulation, 4, 34, 59, 410, *103*
Simulator, 78–85, 92, 93, *255, 266, 274, 278*
Singapore, 308, 309, 311–313, 318, 319
Single risk studies, *453, 454*
Social acceptance, 353, 354, 356–361
Social impact, 346, 348, 349, 352, 354, 356–361
Social sustainability, 344, 347
Solar energy, 110
SPECTRA, 5–7, 12–14, 19, 22, 24, 26, 30
Spinning reserve, 256, 257, 259, 261, 263, 264, 266, 268, 274, 308, 310, 311, *169*
Splitting surface searching, 251
Spot price, 38, 39, 44, 59, 139, 145, 456, *104, 130, 146, 147, 155, 193, 194–200, 204–206, 211, 215, 218, 234, 235, 319–321, 332, 360, 384–390, 399, 400*
Spot price modeling, *218*
Spread option, *101–103, 108–111, 113, 114, 116, 119, 123*
Stability, 178, 182–184, 205, 212, 237, 249, 362, 456, 458, 462, *6, 8, 257, 316, 389, 406–410, 421*
Stability limits, *8, 169*
Stakeholders, 347, 351, 356, 410
Standard market design, 309
State space, 5–8, 11, 20, 22, 24, 30, 36, 43, 60, 169, 171, 334, 336, 337, *140, 269–271, 281*
State-of-the-art, *161–185, 216, 243*
Static network planning, 367, *39*
Statistical, 50, 70, 104, 105, 294, 296, 304, 337, 350, 351, 356, 357, 362, *104, 107, 137–140, 142, 144, 150–152, 162–175, 178, 180, 182, 214, 215, 229, 230, 259, 352, 385, 387, 407, 409, 417*
Statistical distribution, 294, 296, 304, *104*
Statistical learning, *163, 166*
Steam demand, 294–296, 301, 302, 439
Steam generation, 47, 59
Steam levels, 293, 294, 432–439, 441, 442, 445
Steam turbine unit, 259, 267–270, 279, 286, 287, 289
Stochastic, 5, 33, 59, 96, 149, 177, 295, 332, *104, 148, 165, 196, 214, 287, 316, 369, 383, 405, 434*
Stochastic dual dynamic programming (SDDP), 5, 33, 41, 59, 73, 150, *317, 322*
Stochastic dynamic programming (SDP), 7, 34, 36, 40, 60, 330–340, *322, 370, 439*
Stochastic optimization, 155, 165, 326, *330, 373, 383–385, 400, 405–429*
Stochastic process, 40, 53, 167, 203, 332, 336, 340, *104–106, 148, 216–218, 220–227, 230, 232, 235, 288–290, 293, 295, 303, 384, 390, 405–407, 438*
Stochastic programming
  mixed-integer two-stage, 177
  multistage, 199, *317, 384, 405, 407, 408, 410, 417, 434*
Stochastic solution, *292, 293, 300, 301, 308*
Strategic unit commitment, *334*
Structural optimization, 297–299
Sub-problem, 5, 11, 43, 44, 52, *339–343*
Subsidies for renewable energy, 331–334, 336–339, *324*
Subtransmission at 132 kV, *453, 474–478*
Suburban and rural networks, *472, 473*
Superstructure, 439
Supply function equilibrium, 13
Sustainable development, 344–348
SVM models, *137, 140, 145, 152, 166, 180, 182*
Swing option pricing, *384, 393, 397*
System loadability, *15*
System operator, 59, 96, 97, 102, 150, 279, 307, 369, 370, 385, 392, 397, 407,

102, 120, 129, 130, 141, 179, 200, 331, 355, 357, 364–366, 368
System planning, 98, 99, 366, 369, *129, 154*

Tap-changing, *8, 9, 13, 14, 24, 27*
Thermal limits, 458, *7, 8*
Thermal power unit, 81, 82, 84
Thermal units, 82, 84, 168, 201–205, 256, 258, 259, 264, 265, 267, 274, 277, 278, 281, 285, 287–290, 292, 317, *294, 297, 316, 317, 320, 322, 324, 325, 328, 330, 332, 441*
Threshold auto-regressive (TAR) models, *151*
Time series, 36, 39, 50, 105, 109, 203, 336, 337, *87–92, 94–97, 107, 121, 130, 131–134, 137–140, 142–146, 148–152, 154, 158, 163–166, 170, 171, 173, 176, 182, 202, 203, 214, 216–218, 222, 227, 229, 329, 360, 389, 409, 424*
Time series estimation, *145*
Time-expanded network, *32, 35–51*
Total cost of dynamic flow, *33, 35*
Total cost of dynamic multicommodity flow, *42–52*
Total site, 429, 434, 435, 437–439, 441–443
Transfer capability, *4, 15, 23, 24*
Transhipment network representation, 434, 435
Transient stability, 249, *8*
Transit time function, *33, 40–43, 49–52*
Transit time function dependent on flow and time, *33, 40–43, 49–52*
Transmission congestion, 223, *102, 168, 183–184, 258, 267*
Transmission expansion planning (TEP), 366–371, 377, 409, 420
  in deregulated electricity markets, *85, 148*
  in regulated electricity markets, 59, 392, 409
  problem formulation, 219, 296, 297, 378, 411, 456, 463, *32, 33–35, 43–44, 390*
  reference network, 410, 411, 413–415, 418–420
  results obtained by improved differential evolution (IDE), 418
  security constraints, 368, 369, 371, 372
  solution methods, *16, 341, 419*
Transmission loss, 108, 224, 226–231, 256, 259, 264, 265, 267, 274, 282, 291
Transmission network expansion planning, 365–387
Transmission networks, 141–143, 365–387, 395, 406, 486, *33, 242*
Transmission valuation, *101–123*
Transparency, *162, 175, 177, 189, 191*
Trust region, 299, *16*
Turbine, 70, 96, 157, 259, 293, 334, 351, 432, 453, *141–143, 259, 272, 324, 328, 330, 462*
Two-stage, 177–205, 305, *288, 291, 355, 364, 365, 367*

UK industrial context, *458, 461–465*
Uncertainty, 11, 16–21, 79, 102, 116, 154, 166, 169, 174, 201, 209, 232, 293, 305, 330–340, 351, 366, 368–370, 373, 383, 384, 393–395, 398, 400, *57, 103, 104, 145, 151, 166, 191, 196, 197, 220, 256, 288, 293, 302, 311, 320, 322, 331, 351, 354, 356, 369–374, 384–391, 408, 417, 418, 424, 461*
  generator, 79
  modelling, *385–391*
Uncertainty adjustment, 11, 19–21
Under-frequency, 236, 237
Unified power flow controller (UPFC), *9–11, 13, 14, 20, 21, 23, 24*
Unit commitment, 27, 28, 78–82, 98, 99, 102, 178, 192, 201–205, 255, 257, 259, 261, 263, 265, 267, 269, 271, 273, 275, 277, 279, 281, 283, 285, 287, 289, 291, 292, 316, 325, 376, 462, *12, 141, 154, 166, 330–332, 373*
Uplift fee, *339, 341, 343*
Utility functions, 46, *254, 278, 359, 363, 372, 422, 433–437, 439–444*
Utility system, 285, 293–301, 304, 305, 429–433, 438, 439, 443, 445–448

Valid inequality, *339, 341–349*
Value function, 7–9, 17, 18, 20, 53, 54, 190–194, 196, 198, 199, 205, 335, *223, 440, 447*
Value of the stochastic solution (VSS), *292, 293, 300, 301, 308*
Value-at-risk, *391, 406, 421, 436–437, 442–443*
Variable cost, 64, 214, *316, 320, 321, 325, 362*
Variable metric, 266, 272, 274–276
Verbal rule, *177*
Visual impact, 353

Volatility, 104, 105, 210, 217, 223, 232, *104, 106, 111, 115, 130, 141–143, 147, 148, 151, 154, 162, 163, 165–167, 195, 214–218, 221, 222, 225, 227, 232, 234, 257, 267, 268, 287, 289, 295, 303, 309, 317*
Voltage limits, 398, *7, 24*
Vulnerability, 366, 368–371, 373–375, 377, 379–381, 383, 387, *453*

Wait-and-see, *289*
Water allocation, 72, 73
Water transfers, 58, 63, 70, 72, 73
Wholesale electricity markets, *146, 241–281*
Wind, 12, 95, 332, 344, 410, 453, *130, 196, 407, 453*
Wind power, 29, 95, 97, 99, 101, 103, 105, 107, 109, 111, 113, 115, 117, 119, 330–340, 352, 353, 359–361, 410, *130, 141–145, 407*
Wind power forecasting, 101–103, *141–146*

Zonal reserve requirement, 322

FSC
www.fsc.org
MIX
Papier | Fördert gute Waldnutzung
FSC® C083411

Kolibri 360 GmbH
Ferdinand-Jühlke-Straße 7
99095 Erfurt, Deutschland
produktsicherheit@kolibri360.de